P9-DCA-740

Fundamentals of Air Pollution

THIRD EDITION

RICHARD W. BOUBEL

Department of Mechanical Engineering
Oregon State University
Corvallis, Oregon

DONALD L. FOX

Department of Environmental Science
School of Public Health
University of North Carolina
Chapel Hill, North Carolina

D. BRUCE TURNER

Trinity Consultants, Inc.
Chapel Hill, North Carolina

ARTHUR C. STERN

(14 March 1909–17 April 1992)

Fundamentals of Air Pollution

THIRD EDITION

ACADEMIC PRESS
San Diego New York Boston London Sydney Tokyo Toronto

Cover photograph: Testing an air pollution source for particulate emissions. Photo courtesy of BWR Associates, Medford, Oregon.

This book is printed on acid-free paper. ∞

Academic Press, Inc.
A Division of Harcourt Brace & Company
525 B Street, Suite 1900, San Diego, California 92101-4495

United Kingdom Edition published by
Academic Press Limited
24-28 Oval Road, London NW1 7DX

Library of Congress Cataloging-in-Publication Data

Fundamentals of air pollution / Richard W. Boubel ... [et al.]. -- 3rd ed.

p. cm.
Includes bibliographical references and index.
ISBN 0-12-118930-9
1. Air--Pollution. 2. Air--Pollution--Law and legislation--United States. I. Boubel, Richard W. (Richard William), DATE.
TD883.F86 1994
363.73'92--dc20 94-2121
CIP

PRINTED IN THE UNITED STATES OF AMERICA
95 96 97 98 99 MM 9 8 7 6 5 4 3 2

Dedication

Dedicated to Arthur C. Stern, the pioneer in air pollution science and engineering, by three humble followers. He was our mentor, taskmaster, and friend.

Contents

Part II

The Effects of Air Pollution

Part VI

The Engineering Control of Air Pollution

Preface

The authors of this book include a chemist (Donald L. Fox), a meteorologist (D. Bruce Turner), and a mechanical engineer (Richard W. Boubel). This 1:1:1 ratio has some relevance in that it approximates the ratio of those professionally involved in the field of air pollution. In the environmental protection and management field, the experience of the recent past has been that physicists and electrical engineers have been most attracted to the radiation, nuclear, and noise areas; biologists and civil engineers to the aquatic and solid waste areas; chemists, meteorologists, and chemical and mechanical engineers to the area of air pollution and its control. These remarks are not intended to exclude all others from the party (or from this course). The control of air pollution requires the combined efforts of all the professions mentioned, in addition to the input of physicians, lawyers, and social scientists. However, the professional mix of the authors, and their expectation of a not-too-dissimilar mix of students using this book, forewarns the tenor of its contents and presentation.

Although this book consists of six parts and three authors, it is not to be considered six short books put together back-to-back to make one large one. By and large, the several parts are the work of more than one author. Obviously, the meteorologist member of the author team is principally responsible for the part of the book concerned with the meteorology of air pollution, the chemist author for the chapters on chemistry, and the engineer author for those on engineering. However, as you will see, no chapters are signed, and all authors accept responsibility for the strengths and weaknesses of the chapters and for the book as a whole.

In the twenty years since publication of the first edition of *Fundamentals of Air Pollution* (1973), and the nine years since the second edition (1984), the fundamentals have not changed. The basic physics, chemistry, and engineering are still the same, but there is now a greater in-depth understanding of their application to air pollution. This edition has been edited, revised, and updated to include the new technology available to air pollution practitioners. Its contents are also influenced to a great extent by the passage of the U.S. Clean Air Act Amendments of 1990 (CAAA 90). These amendments have changed the health and risk-based regulations of the U.S. Clean Air Act to technology-driven regulations with extensive penalty provisions for noncompliance.

We have added more detailed discussion of areas that have been under intensive study during the past decade. There has been a similar need to add discussion of CAAA 90 and its regulatory concepts, such as control of air toxics, indoor air pollution, pollution prevention, and trading and banking of emission rights. Ten more years of new data on air quality have required the updating of the tables and figures presenting these data.

We have expanded some subject areas, which previously were of concern to only a few scientists, but which have been popularized by the media to the point where they are common discussion subjects. These include "Global Warming," "The Ozone Hole," "Energy Conservation," "Renewable Resources," and "Quality of Life."

With each passing decade, more and more pollution sources of earlier decades become obsolete and are replaced by processes and equipment that produce less pollution. At the same time, population and the demand for products and services increase. Students must keep these concepts in mind as they study from this text, knowing that the world in which they will practice their profession will be different from the world today.

By virtue of its division into six sections, this text may be used in several ways. Part I, by itself, provides the material for a short course to introduce a diverse group of students to the subject—with the other five parts serving as a built-in reference book. Parts I, II, and II, which define the problem, can provide the basis for a semester's work, while Parts IV, V, and VI, which resolve the problem, provide the material for a second semester's work. Part IV may well be used separately as the basis for a course on the meteorology of air pollution, and the book as a whole may be used for an intensive one-semester course.

The viewpoint of this book is first that most of the students who elect to receive some training in air pollution will have previously taken courses in chemistry at the high school or university level, and that those few who have not would be well advised to defer the study of air pollution until they catch up on their chemistry.

The second point of view is that the engineering design of control systems for stationary and mobile sources requires a command of the principles of chemical and mechanical engineering beyond that which can be included in a one-volume textbook on air pollution. Before venturing into the field of engineering control of air pollution, a student should, as a minimum, master courses in internal combustion engines, power plant engineering, the unit processes of chemical engineering, engineering thermodynamics, and kinetics. However, this does not have to be accomplished before taking a course based on this book but can well be done simultaneously with or after doing so.

The third point of view is that *no one,* regardless of their professional background, should be in the field of air pollution control unless they sufficiently understand the behavior of the atmosphere, which is the feature that differentiates *air* pollution from the other aspects of environmental protection and management. This requires a knowledge of some basic atmospheric chemistry in addition to some rather specialized air pollution meteorology. The viewpoint presented in the textbook is that very few students using it will have previously studied basic meteorology. It is hoped that exposure to air pollution meteorology at this stage will excite a handful of students to delve deeper into the subject. Therefore, a relatively large proportion of this book has been devoted to meteorology because of its projected importance to the student.

The authors have tried to maintain a universal point of view so that the material presented would be equally applicable in all the countries of the world. Although a deliberate attempt has been made to keep American provincialism out of the book, it has inevitably crept in through the exclusive use of English language references and suggested reading lists, and the preponderant use of American data for the examples, tables, and figures. The saving grace in this respect is that the principles of chemistry, meteorology, and engineering are universal.

As persons who have dedicated all or significant parts of their professional careers to the field of air pollution, the authors believe in its importance and relevance. We believe that as the world's population increases, it will become increasingly important to have an adequate number of well-trained professions engaged in air pollution control. If we did not believe this, it would have been pointless for us to have written this textbook.

We recognize that, in terms of short-term urgency, many nations and communities may rightly assign a lower priority to air pollution control than to problems of population, poverty, nutrition, housing, education, water supply, communicable disease control, civil rights, mental health, aging, or crime. Air pollution control is more likely to have a higher priority for a person or a community already reaping the benefits of society in the form of adequate income, food, housing, education, and health care than for persons who have not and may never reap these benefits.

However, in terms of long-term needs, nations and communities can ignore air pollution control only at their peril. A population can subsist, albeit poorly, with inadequate housing, schools, police, and care of the ill, insane, and aged; it can also subsist with a primitive water supply. The ultimate determinants for survival are its food and air supplies. Conversely, even were society to succeed in providing in a completely adequate manner all of its other needs, it would be of no avail if the result were an atmosphere so befouled as not to sustain life. The long-term objective of air pollution control is to allow the world's population to meet all its needs for energy, goods, and services without sullying its air supply.

Part I

The Elements of Air Pollution

1

The History of Air Pollution

I. BEFORE THE INDUSTRIAL REVOLUTION

One of the reasons the tribes of early history were nomadic was to move periodically away from the stench of the animal, vegetable, and human wastes they generated. When the tribesmen learned to use fire, they used it for millennia in a way that filled the air inside their living quarters with the products of incomplete combustion. Examples of this can still be seen today in some of the more primitive parts of the world. After its invention, the chimney removed the combustion products and cooking smells from the living quarters, but for centuries the open fire in the fireplace caused its emission to be smoky. In AD 61 the Roman philosopher Seneca reported thus on conditions in Rome:

> As soon as I had gotten out of the heavy air of Rome and from the stink of the smoky chimneys thereof, which, being stirred, poured forth whatever pestilential vapors and soot they had enclosed in them, I felt an alteration of my disposition.

Air pollution, associated with burning wood in Tutbury Castle in Nottingham, was considered "unendurable" by Eleanor of Aquitaine, the wife of King Henry II of England, and caused her to move in the year 1157. One hundred-sixteen years later, coal burning was prohibited in London; and in 1306, Edward I issued a royal proclamation enjoining the use of *sea-coal* in furnaces. Elizabeth I barred the burning of coal in London when Parliament was in session. The repeated necessity for such royal action would seem to indicate that coal continued to be burned despite these edicts. By 1661 the pollution of London had become bad enough to prompt John Evelyn to submit a brochure "Fumifugium, or the Inconvenience of the Aer, and Smoake of London Dissipated (together with some remedies humbly proposed)" to King Charles II and Parliament. This brochure has been reprinted and is recommended to students of air pollution (1). It proposes means of air pollution control that are still viable in the twentieth century.

The principal industries associated with the production of air pollution in the centuries preceding the Industrial Revolution were metallurgy, ceramics, and preservation of animal products. In the bronze and iron ages, villages were exposed to dust and fumes from many sources. Native copper and gold were forged, and clay was baked and glazed to form pottery and bricks before 4000 BC Iron was in common use and leather was tanned before 1000 BC. Most of the methods of modern metallurgy were known before AD 1. They relied on charcoal rather than coal or coke. However, coal was mined and used for fuel before AD 1000, although it was not made into coke until about 1600; and coke did not enter metallurgical practice significantly until about 1700. These industries and their effluents as they existed before 1556 are best described in the book "De Re Metallica" published in that year by Georg Bauer, known as Georgius Agricola (Fig. 1-1). This book was translated into English and published in 1912 by Herbert Clark Hoover and his wife (2).

Examples of the air pollution associated with the ceramic and animal product preservation industries are shown in Figs. 1-2 and 1-3, respectively.

II. THE INDUSTRIAL REVOLUTION

The Industrial Revolution was the consequence of the harnessing of steam to provide power to pump water and move machinery. This began in the early years of the eighteenth century, when Savery, Papin, and Newcomen designed their pumping engines, and culminated in 1784 in Watt's reciprocating engine. The reciprocating steam engine reigned supreme until it was displaced by the steam turbine in the twentieth century.

Steam engines and steam turbines require steam boilers, which, until the advent of the nuclear reactor, were fired by vegetable or fossil fuels. During most of the nineteenth century, coal was the principal fuel, although some oil was used for steam generation late in the century.

Fig. 1-1. Lead smelting furnace. Source: G. Agricola, "De Re Metallica," Book X, p. 481, Basel, Switzerland, 1556. Translated by H. C. Hoover and L. H. Hoover, *Mining Magazine*, London, 1912. Reprinted by Dover Publications, New York, 1950.

The predominant air pollution problem of the nineteenth century was smoke and ash from the burning of coal or oil in the boiler furnaces of stationary power plants, locomotives, and marine vessels, and in home heating fireplaces and furnaces. Great Britain took the lead in addressing this problem, and, in the words of Sir Hugh Beaver (3):

> By 1819, there was sufficient pressure for Parliament to appoint the first of a whole dynasty of committees "to consider how far persons using steam engines and furnaces could work them in a manner less prejudicial to public health and comfort." This committee confirmed the practicability of smoke prevention, as so many succeeding committees were to do, but as was often again to be experienced, nothing was done.
>
> In 1843, there was another Parliamentary Select Committee, and in 1845, a third. In that same year, during the height of the great railway boom, an act of Parliament disposed of trouble from locomotives once and for all (!) by laying down the dictum that they must consume their own smoke. The Town Improvement Clauses Act

Fig. 1-2. A pottery kiln. Source: Cipriano Piccolpasso, "The Three Books of the Potters's Art," fol. 35C, 1550. Translated by B. Rackham and A. Van de Put, Victoria and Albert Museum, London, 1934.

two years later applied the same panacea to factory furnaces. Then 1853 and 1856 witnessed two acts of Parliament dealing specifically with London and empowering the police to enforce provisions against smoke from furnaces, public baths, and washhouses and furnaces used in the working of steam vessels on the Thames.

Smoke and ash abatement in Great Britain was considered to be a health agency responsibility and was so confirmed by the first Public Health Act of 1848 and the later ones of 1866 and 1875. Air pollution from the emerging chemical industry was considered a separate matter and was made the responsibility of the Alkali Inspectorate created by the Alkali Act of 1863.

Fig. 1-3. A kiln for smoking red herring. Source: H. L. Duhamel due Monceau, "Traité général des pêches," Vol. 2, Sect. III, Plate XV, Fig. 1, Paris, 1772.

In the United States, smoke abatement (as air pollution control was then known) was considered a municipal responsibility. There were no federal or state smoke abatement laws or regulations. The first municipal ordinances and regulations limiting the emission of black smoke and ash appeared in the 1880s and were directed toward industrial, locomotive, and marine rather than domestic sources. As the nineteenth century drew to a close, the pollution of the air of mill towns the world over had risen to a peak (Fig. 1-4); damage to vegetation from the smelting of sulfide ores was recognized as a problem everywhere it was practiced.

The principal technological developments in the control of air pollution by engineering during the nineteenth century were the stoker for mechanical firing of coal, the scrubber for removing acid gases from effluent gas streams, cyclone and bag house dust collectors, and the introduction of physical and chemical principles into process design.

Fig. 1-4. Engraving (1876) of a metal foundry refining department in the industrial Saar region of West Germany. Source: The Bettmann Archive, Inc.

III. THE TWENTIETH CENTURY

A. 1900–1925

During the period 1900–1925 there were great changes in the technology of both the production of air pollution and its engineering control, but no significant changes in legislation, regulations, understanding of the problem, or public attitudes toward the problem. As cities and factories grew in size, the severity of the pollution problem increased.

One of the principal technological changes in the production of pollution was the replacement of the steam engine by the electric motor as the means of operating machinery and pumping water. This change transferred the smoke and ash emission from the boiler house of the factory to the boiler house of the electric generating station. At the start of this period, coal was hand-fired in the boiler house; by the middle of the period, it was mechanically fired by stokers; by the end of the period, pulverized coal, oil, and gas firing had begun to take over. Each form of firing produced its own characteristic emissions to the atmosphere.

At the start of this period, steam locomotives came into the heart of the larger cities. By the end of the period, the urban terminals of many railroads had been electrified, thereby transferring much air pollution from the railroad right-of-way to the electric generating station. The replacement of coal by oil in many applications decreased ash emissions from those sources. There was rapid technological change in industry. However, the most significant change was the rapid increase in the number of automobiles from almost none at the turn of the century to millions by 1925 (Table 1-1).

The principal technological changes in the engineering control of air pollution were the perfection of the motor-driven fan, which allowed large-scale gas-treating systems to be built; the invention of the electrostatic precipitator, which made particulate control in many processes feasible; and the development of a chemical engineering capability for the design of process equipment, which made the control of gas and vapor effluents feasible.

B. 1925–1950

In this period, present-day air pollution problems and solutions emerged. The Meuse Valley (Belgium) episode (4) occurred in 1930; the Donora, Pennsylvania, episode (5) occurred in 1948; and the Poza Rica, Mexico, episode (6) in 1950. Smog first appeared in Los Angeles in the 1940s (Fig. 1-5). The Trail, British Columbia, smelter arbitration (7) was completed in 1941. The first National Air Pollution Symposium in the United States was held in Pasadena, California, in 1949 (8), and the first United States Technical Conference on Air Pollution was held in Washington, D.C., in 1950 (9). The first large-scale surveys of air pollution were undertaken—Salt Lake City, Utah (1926) (10); New York, New York (1937) (11); and Leicester, England (1939) (12).

Air pollution research got a start in California. The Technical Foundation for Air Pollution Meteorology was established in the search for means of disseminating and protecting against chemical, biological, and nuclear warfare agents. Toxicology came of age. The stage was set for the air pollution scientific and technological explosion of the second half of the twentieth century.

A major technological change was the building of natural gas pipelines, and where this occurred, there was rapid displacement of coal and oil as home heating fuels with dramatic improvement in air quality; witness the much publicized decrease in black smoke in Pittsburgh (Fig. 1-6) and St. Louis. The diesel locomotive began to displace the steam locomotive, thereby slowing the pace of railroad electrification. The internal combustion engine bus started its displacement of the electrified streetcar. The automobile continued to proliferate (Table 1-1).

Fig. 1-5. Los Angeles smog. Source: Los Angeles County, California.

During this period, no significant national air pollution legislation or regulations were adopted anywhere in the world. However, the first state air pollution law in the United States was adopted by California in 1947.

C. 1950–1980

In Great Britain, a major air pollution disaster hit London in 1952 (13), resulting in the passage of the Clean Air Act in 1956 and an expansion of the authority of the Alkali Inspectorate. The principal changes that resulted were in the means of heating homes. Previously, most heating was done by burning soft coal on grates in separate fireplaces in each room. A successful effort was made to substitute smokeless fuels for the soft coal used in this manner, and central or electrical heating for fireplace heating. The outcome was a decrease in "smoke" concentration, as measured by the blackness of paper filters through which British air was passed from 175 $\mu g/m^3$ in 1958 to 75 $\mu g/m^3$ in 1968 (14).

During these two decades, almost every country in Europe, as well as Japan, Australia, and New Zealand, experienced serious air pollution in its larger cities. As a result, these countries were the first to enact national air pollution control legislation. By 1980, major national air pollution research centers had been set up at the Warren Springs Laboratory, Stevenage, England; the Institut National de la Santé et de las Recherche Medicale at

Le Visinet, France; the Rijksinstituut Voor de Volksgezondheid, Bilthoven and the Instituut voor Gezondheidstechniek-TNO, Delft, The Netherlands; the Statens Naturvardsverk, Solna, Sweden; the Institut für Wasser-Boden- und Luft-hygiene, Berlin and the Landensanstalt für Immissions und Bodennutzungsshutz, Essen, Germany. The important air pollution research centers in Japan are too numerous to mention.

In the United States, the smog problem continued to worsen in Los Angeles and appeared in large cities throughout the nation (Fig. 1-7). In 1955 the first federal air pollution legislation was enacted, providing federal support for air pollution research, training, and technical assistance. Responsibility for the administration of the federal program was given to the Public Health Service (PHS) of the United States Department of Health, Education, and Welfare, and remained there until 1970, when it was transferred to the new United States Environmental Protection Agency (EPA). The initial federal legislation was amended and extended several times between 1955 and 1980, greatly increasing federal authority, particularly in the area of control (15). The automobile continued to proliferate (Table 1-1).

As in Europe, air pollution research activity expanded tremendously in the United States during these three decades. The headquarters of federal research activity was at the Robert A. Taft Sanitary Engineering Center of the PHS in Cincinnati, Ohio, during the early years of the period and at the National Environmental Research Center in Triangle Park, North Carolina, at the end of the period.

An International Air Pollution Congress was held in New York City in 1955 (16). Three National Air Pollution Conferences were held in Washington, D.C., in 1958 (17), 1962 (18), and 1966 (19). In 1959, an International Clean Air Conference was held in London (20).

TABLE 1-1

Annual Motor Vehicle Sales in the United States[a]

Year	Total	Year	Total
1900	4,192	1945	725,215
1905	25,000	1950	8,003,056
1910	187,000	1955	9,169,292
1915	969,930	1960	7,869,221
1920	2,227,347	1965	11,057,366
1925	4,265,830	1970	8,239,257
1930	3,362,820	1975	8,985,012
1935	3,971,241	1980	8,067,309
1940	4,472,286	1985	11,045,784
		1990	9,295,732

[a] Data include foreign and domestic sales for trucks, buses, and automobiles.

Fig. 1-6. (Right) Pittsburgh before the decrease in black smoke. Source: Allegheny County, Pennsylvania. (Left) Pittsburgh after the decrease in black smoke. Source: Allegheny County, Pennsylvania

In 1964, the International Union of Air Pollution Prevention Associations (IUAPPA) was formed. IUAPPA has held International Clean Air Congresses in London in 1966 (21); Washington, D.C., in 1970 (22); Dusseldorf in 1973 (23); Tokyo in 1977 (24); Buenos Aires in 1980 (25); Paris in 1983 (26); Sydney in 1986 (27); The Hague in 1989 (28); and Montreal in 1992 (29).

Technological interest during these 30 years has focused on automotive air pollution and its control, on sulfur oxide pollution and its control by sulfur oxide removal from flue gases and fuel desulfurization, and on control of nitrogen oxides produced in combustion processes.

Air pollution meteorology came of age and, by 1980, mathematical models of the pollution of the atmosphere were being energetically developed. A start had been made in elucidating the photochemistry of air pollution. Air quality monitoring systems became operational throughout the world. A wide variety of measuring instruments became available.

Fig. 1-7. Smog in New York City. Source: Wide World Photos.

IV. THE 1980s

The highlight of the 1970s and 1980s was the emergence of the ecological, or total environmental, approach. Organizationally, this has taken the form of departments or ministries of the environment in governments at all levels throughout the world. In the United States there is a federal Environmental Protection Agency, and in most states and populous counties and cities, there are counterpart organizations charged with responsibility for air and water quality, sold waste sanitation, noise abatement, and control of the hazards associated with radiation and the use of pesticides. This is paralleled in industry, where formerly diffuse responsibility for these areas is increasingly the responsibility of an environmental protection coordinator. Similar changes are evident in research and education.

Pollution controls were being built into pollution sources—automobiles, power plants, factories—at the time of original construction rather than later on. Also, for the first time, serious attention was directed to the problems caused by the "greenhouse" effect of carbon dioxide and other gases building up in the atmosphere, possible depletion of the stratospheric ozone layer by fluorocarbons, long-range transport of pollution, prevention of significant deterioration (PSD), and acidic deposition.

V. THE 1990s

The most sweeping change, in the United States at least, in the decade of the 1990s was the passage of the Clean Air Act Amendments on November 15, 1990 (29). This was the only change in the Clean Air Act since 1977, even though the U.S. Congress had mandated that the Act be amended much earlier. Michigan Representative John Dingell referred to the amendments as "the most complex, comprehensive, and far-reaching environmental law any Congress has ever considered." John-Mark Stenvaag has stated in his book, "Clean Air Act 1990 Amendments, Law and Practice" (30), "The enormity of the 1990 amendments begs description. The prior Act, consisting of approximately 70,000 words, was widely recognized to be a remarkably complicated, unapproachable piece of legislation. If environmental attorneys, government officials, and regulated entities were awed by the prior Act, they will be astonished, even stupefied, by the 1990 amendments. In approximately 145,000 new words, Congress has essentially tripled the length of the prior Act and geometrically increased its complexity."

The 1990s saw the emergence, in the popular media, of two distinct but closely related global environmental crises, uncontrolled global climate changes and stratospheric ozone depletion. The climate changes of concern were both the warming trends caused by the buildup of greenhouse gases in the atmosphere and cooling trends caused by particulate matter and sulfates in the same atmosphere. Some researchers have suggested that these two trends will cancel each other. Other authors have written (31) that global warming may not be all bad. It is going to be an interesting decade as many theories are developed and tested during the 1990s. The "Earth Summit," really the U.N. Conference of Environment and Development, in Rio de Janeiro during June 1992 did little to resolve the problems, but it did indicate the magnitude of the concern and the differences expressed by the nations of the world.

The other global environmental problem, stratospheric ozone depletion, was less controversial and more imminent. The U.S. Senate Committee Report supporting the Clean Air Act Amendments of 1990 states, "Destruction of the ozone layer is caused primarily by the release into the atmosphere of chlorofluorocarbons (CFCs) and similar manufactured substances—persistent chemicals that rise into the stratosphere where they catalyze the destruction of stratospheric ozone. A decrease in stratospheric ozone will allow more ultraviolet (UV) radiation to reach Earth, resulting in increased rates of disease in humans, including increased incidence of skin cancer, cataracts, and, potentially, suppression of the immune system. Increased UV radiation has also been shown to damage crops and marine resources."

The Montreal Protocol of July 1987 resulted in an international treaty in which the industrialized nations agreed to halt the production of most ozone-destroying chlorofluorocarbons by the year 2000. This deadline was hastily changed to 1996, in February 1992, after a U.S. National Aeronautics and Space Administration (NASA) satellite and high-altitude sampling aircraft found levels of chlorine monoxide over North America that were 50% greater than that measured over Antarctica.

VI. THE FUTURE

The air pollution problems of the future are predicated on the use of more and more fossil and nuclear fuel as the population of the world increases. During the lifetime of the students using this book, partial respite may be offered by solar, photovoltaic, geothermal, wind, nonfossile fuel (hydrogen and biomass), and oceanic (thermal gradient, tidal, and wave) sources of energy. Still, many of the agonizing environmental decisions of the next decades will involve a choice between fossil fuel and nuclear power sources and the depletion of future fuel reserves for present needs. Serious questions will arise regarding whether to conserve or to use these reserves—whether to allow unlimited growth or to curb it.

Other problems concerning transportation systems, waste processing and recycling systems, national priorities, international economics, employment versus environmental quality, and personal freedoms will continue to surface. The choices will have to be made, ideally by educated citizens and charismatic leaders.

REFERENCES

1. The Smoake of London—Two Prophecies [Selected by James P. Lodge, Jr.]. Maxwell Reprint, Elmsford, NY, 1969.

2. Agricola, G., "De Re Metallica," Basel, 1556. [English translation and commentary by H. C. Hoover and L. H. Hoover, *Mining Magazine,* London 1912]. Dover, New York, 1950.
3. Beaver, Sir Hugh E. C., The growth of public opinion, *In* "Problems and Control of Air Pollution" (F. S. Mallette, ed.). Reinhold, New York, 1955.
4. Firket, J., *Bull. Acad. R. Med. Belg.* **11,** 683 (1931).
5. Schrenk, H. H., Heimann, H., Clayton, G. D., Gafefer, W. M., and Wexler, H., *U.S. Pub. Health Serv. Bull.* **306** (1949), 173 pp.
6. McCabe, L. C., and Clayton, G. D., *Arch. Ind. Hyg. Occup. Med.* **6,** 199 (1952).
7. Dean, R. S., and Swain, R. E., Report submitted to the Trail Smelter Arbitral Tribunal, *U.S. Bur. Mines Bull. 453.* U.S. Government Printing Office, Washington, DC, 1944.
8. Proceedings of the First National Air Pollution Symposium, Pasadena, CA. Sponsored by the Stanford Research Institute in cooperation with the California Institute of Technology, University of California, and University of Southern California, 1949.
9. McCabe, L. C. (ed.), "Air Pollution" (Proc. U.S. Tech. Conf. Air Pollution, 1950). McGraw-Hill, New York, 1952.
10. Monett, O., Perrott, G. St. J., and Clark, H. W., *U.S. Bur. Mines Bull. 254* (1926).
11. Stern, A. C., Buchbinder, L., and Siegel, J., *Heat. Piping Air Cond.* **17,** 7–10 (1945).
12. Atmospheric Pollution in Leicester—A Scientific Study, D.S.I.R. Atmospheric Research Technical Paper No. 1. H.M. Stationery Office, London, 1945.
13. Ministry of Health, Mortality and Morbidity during the London Fog of December 1952, Rep. of Public Health and Related Subject No. 95. H.M. Stationery Office, London, 1954.
14. Royal Commission on Environmental Pollution, First Report, H.M. Stationery Office, London, 1971.
15. Stern, A. C., *Air. Pollut. Control Assn.* **32,** 44 (1982).
16. Mallette, F. S. (ed.), "Problems and Control of Air Pollution," American Society of Mechanical Engineers, New York, 1955.
17. Proceedings of the National Conference on Air Pollution, Washington, DC, 1958. U.S. Public Health Serv. Pub. 654, 1959.
18. Proceedings of the National Conference on Air Pollution, Washington, DC, 1962. U.S. Public Health Serv. Pub. 1022, 1963.
19. Proceedings of the National Conference on Air Pollution, Washington, DC, 1966. U.S. Public Health Serv. Pub. 1699, 1967.
20. Proceedings of the Diamond Jubilee International Clean Air Conference 1959. National Society for Clean Air, London, 1960.
21. Proceedings of the International Clean Air Congress, London, Oct. 1966, Part I. National Society for Clean Air, London, 1966.
22. Englund, H., and Beery, W. T. (eds.), "Proceedings of the Second International Clean Air Congress," Washington, DC, 1970. Academic Press, New York, 1971.
23. "Proceedings of the Third International Clean Air Congress," Dusseldorf, Sept. 1973. Verein Deutcher Inginieure, Dusseldorf, 1974.
24. "Proceedings of the Fourth International Clean Air Congress," Tokyo, Japanese Union of Air Pollution Prevention Associations, Tokyo, 1977.
25. "Proceedings of the Fifth International Clean Air Congress," Buenos Aires, 1981, 2 vols. Associon Argentina Contra La Contaminacion del Aire, Buenos Aires, 1983.
26. "Proceedings of the Sixth International Clean Air Congress," Paris, Association Pour La Prevention de la Pollution Atmospherique, Paris, 1983.
27. Proceedings of the Seventh International Clean Air Congress, International Union of Air Pollution Prevention Associations, Sydney, 1986.

28. Proceedings of the Eighth International Clean Air Congress, International Union of Air Pollution Prevention Associations, The Hague, 1989.
29. Public Law 101-549, 101st Congress. November 15, 1990, an Act, to amend the Clean Air Act to provide for attainment and maintenance of health protective national ambient air quality standards, and for other purposes.
30. Stenvaag, J. M., "Clean Air Act 1990 Amendments, Law and Practice." Wiley, New York, 1991.
31. Ausubel, J. H., A second look at the impacts of climate change. *Am. Sci.* **79** (May-June, 1991).

SUGGESTED READING

Daumas, M. (ed.), "The History of Technology and Invention," 3 vols. Crown, New York, 1978.

Fishman, J., "Global Alert: The Ozone Pollution Crisis," Plenum, New York, 1990.

Singer, C., Holmyard, E. K., Hall, A. R., and Williams, T. I. (eds.), "A History of Technology," 5 vols. Oxford Univ. Press (Clarendon), New York, 1954–1958.

Williams, T. (ed.), "A History of Technology—Twentieth Century, 1900–1950," part I (Vol. 6), part II (Vol. 7). Oxford Univ. Press (Clarendon), New York, 1978.

Wolf, A., "A History of Sciency, Technology and Philosophy in the Sixteenth and Seventeenth Centuries." George Allen & Unwin, London, 1935.

Wolf, A., "A History of Science, Technology and Philosophy in the Eighteenth Century." George Allen & Unwin, London, 1961.

Gore, A., "Earth in Balance," Houghton Mifflin, New York, 1992.

QUESTIONS

1. Discuss the development of the use of enclosed space for human occupancy over the period of recorded history.
2. Discuss the development of the heating and cooling of enclosed space for human occupancy over the period of recorded history.
3. Discuss the development of the lighting of enclosed space for human occupancy over the period of recorded history.
4. Discuss the development of means to supplement human muscular power over the period of recorded history.
5. Discuss the development of transportation over the period of recorded history.
6. Discuss the development of agriculture over the period of recorded history.
7. Discuss the future alternative sources of energy for light, heat, and power.
8. Compare the so-called soft (i.e., widely distributed small sources) and hard (i.e., fewer very large sources) paths for the future provision of energy for light, heat, and power.
9. What have been the most important developments in the history of the air pollution problem since the publication of this edition of this book?

2

The Natural versus Polluted Atmosphere

Pollution, *n:* the action of polluting; the condition of being polluted. Pollute, *vt:* [L. *pollutus,* past part. or *polluere,* to make physically impure or unclean]; to defile, desecrate, profane-syn. see CONTAMINATE. Air, *n:* [fr. L *aer,* fr. Gr. *aer*] 1. the mixture of invisible tasteless gases which surrounds the earth . Atmosphere, *n:* [fr. Gr. (*atmo* and *sphaira*) 2. the whole mass of air surrounding the earth . . .[1]

I. THE ATMOSPHERE

On a macroscale (Fig. 2-1) as temperature varies with altitude, so does density (1). In general, the air grows progressively less dense as we move upward from the troposphere through the stratosphere and the chemosphere to the ionosphere. In the upper reaches of the ionosphere, the gaseous molecules are few and far between as compared with the troposphere.

The ionosphere and chemosphere are of interest to space scientists because they must be traversed by space vehicles en route to or from the

[1] Definitions are from "Webster's Tenth New Collegiate Dictionary," 10th ed. Merriam, Springfield, MA, 1993.

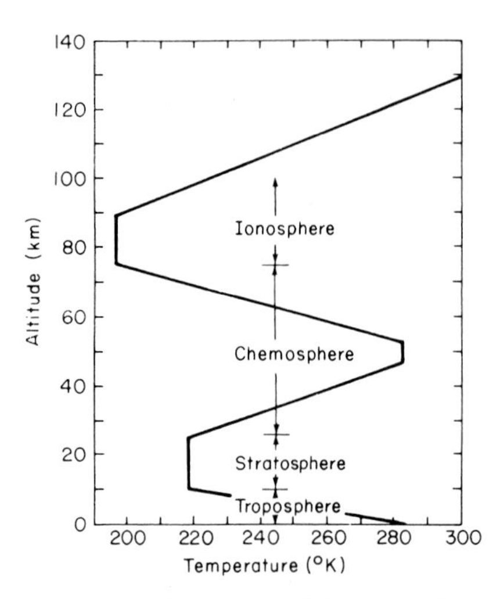

Fig. 2-1. The regions of the atmosphere.

moon or the planets, and they are regions in which satellites travel in the earth's orbit. These regions are of interest to communications scientists because of their influence on radio communications, and they are of interest to air pollution scientists primarily because of their absorption and scattering of solar energy, which influence the amount and spectral distribution of solar energy and cosmic rays reaching the stratosphere and troposphere.

The stratosphere is of interest to aeronautical scientists because it is traversed by airplanes; to communications scientists because of radio and television communications; and to air pollution scientists because global transport of pollution, particularly the debris of aboveground atomic bomb tests and volcanic eruptions, occurs in this region and because absorption and scattering of solar energy also occur there. The lower portion of this region contains the stratospheric ozone layer, which absorbs harmful ultraviolet (UV) solar radiation. Global change scientists are interested in modifications of this layer by long-term accumulation of chlorofluorocarbons (CFCs) and other gases released at the earth's surface or by high-altitude aircraft.

The troposphere is the region in which we live and is the primary focus of this book.

II. UNPOLLUTED AIR

The gaseous composition of unpolluted tropospheric air is given in Table 2-1. Unpolluted air is a concept, i.e., what the composition of the air would be if humans and their works were not on earth. We will never know the precise composition of unpolluted air because by the time we had the means and the desire to determine its composition, humans had been polluting the air for thousands of years. Now even at the most remote locations at sea, at the poles, and in the deserts and mountains, the air may be best described as dilute polluted air. It closely approximates unpolluted air, but differs from it to the extent that it contains vestiges of diffused and aged human-made pollution.

The real atmosphere is more than a dry mixture of permanent gases. It has other constituents—vapor of both water and organic liquids, and particulate matter held in suspension. Above their temperature of condensation, vapor molecules act just like permanent gas molecules in the air. The predominant vapor in the air is water vapor. Below its condensation temperature, if the air is saturated, water changes from vapor to liquid. We are all familiar with this phenomenon because it appears as fog or mist in the air and as condensed liquid water on windows and other cold surfaces exposed to air. The quantity of water vapor in the air varies greatly from almost complete dryness to supersaturation, i.e., between 0% and 4% by weight. If Table 2-1 is compiled on a wet air basis at a time when the water vapor concentration is 31,200 parts by volume per million parts by volume of wet air (Table 2-2), the concentration of condensable organic vapors is seen to be so low compared to that of water vapor that for all practical purposes the difference between wet air and dry air is its water vapor content.

Gaseous composition in Tables 2-1 and 2-2 is expressed as parts per million by volume—ppm (vol). (When a concentration is expressed simply

TABLE 2-1

The Gaseous Composition of Unpolluted Air (Dry Basis)

	ppm (vol)	$\mu g/m^3$
Nitrogen	780,000	8.95×10^8
Oxygen	209,400	2.74×10^8
Water	—	—
Argon	9,300	1.52×10^7
Carbon dioxide	315	5.67×10^5
Neon	18	1.49×10^4
Helium	5.2	8.50×10^2
Methane	1.0–1.2	$6.56–7.87 \times 10^2$
Krypton	1.0	3.43×10^3
Nitrous oxide	0.5	9.00×10^2
Hydrogen	0.5	4.13×10^1
Xenon	0.08	4.29×10^2
Organic vapors	ca. 0.02	—

TABLE 2-2

The Gaseous Composition of Unpolluted Air (Wet Basis)

	ppm (vol)	$\mu g/m^3$
Nitrogen	756,500	8.67×10^8
Oxygen	202,900	2.65×10^8
Water	31,200	2.30×10^7
Argon	9,000	1.47×10^7
Carbon dioxide	305	5.49×10^5
Neon	17.4	1.44×10^4
Helium	5.0	8.25×10^2
Methane	0.97–1.16	$6.35\text{–}7.63 \times 10^2$
Krypton	0.97	3.32×10^3
Nitrous oxide	0.49	8.73×10^2
Hydrogen	0.49	4.00×10^1
Xenon	0.08	4.17×10^2
Organic vapors	ca. 0.02	—

as ppm, one is in doubt as to whether a volume or weight basis is intended.) To avoid confusion caused by different units, air pollutant concentrations in this book are generally expressed as micrograms per cubic meter of air ($\mu g/m^3$) at 25°C and 760 mm Hg, i.e., in metric units. To convert from units of ppm (vol) to $\mu g/m^3$, it is assumed that the ideal gas law is accurate under ambient conditions. A generalized formula for the conversion at 25°C and 760 mm Hg is

$$\begin{aligned}1 \text{ ppm (vol) pollutant} &= \frac{1 \text{ liter pollutant}}{10^6 \text{ liter air}}\\ &= \frac{(1 \text{ liter}/22.4) \times \text{MW} \times 10^6\ \mu\text{g/gm}}{10^6 \text{ liters} \times 298°\text{K}/273°\text{K} \times 10^{-3}\ \text{m}^3/\text{liter}}\\ &= 40.9 \times \text{MW}\ \mu\text{g/m}^3 \qquad (2\text{-}1)\end{aligned}$$

where MW equals molecular weight. For convenience, conversion units for common pollutants are shown in Table 2-3.

A minor problem arises in regard to nitrogen oxides. It is common practice to add concentrations of nitrogen dioxide and nitric oxide in ppm (vol) and express the sum as "oxides of nitrogen." In metric units, conversion from ppm (vol) to $\mu g/m^3$ must be done separately for nitrogen dioxide and nitric oxide prior to addition.

III. PARTICULATE MATTER

Neither Table 2-1 nor Table 2-2 lists among the constituents of the air the suspended particulate matter that it always contains. The gases and vapors exist as individual molecules in random motion. Each gas or vapor

TABLE 2-3

Conversion Factors between Volume and Mass Units of Concentration (25°C, 760 mm Hg)

Pollutant	To convert from ppm (vol) to $\mu g/m^3$, multiply by:	To convert from $\mu g/m^3$ to ppm (vol), multiply by ($\times 10^{-3}$):
Ammonia (NH_3)	695	1.44
Carbon dioxide	1800	0.56
Carbon monoxide	1150	0.87
Chlorine	2900	0.34
Ethylene	1150	0.87
Hydrogen chloride	1490	0.67
Hydrogen fluoride	820	1.22
Hydrogen sulfide	1390	0.72
Methane (carbon)	655	1.53
Nitrogen dioxide	1880	0.53
Nitric oxide	1230	0.81
Ozone	1960	0.51
Peroxyacetylnitrate	4950	0.20
Sulfur dioxide	2620	0.38

exerts its proportionate partial pressure. The particles are aggregates of many molecules, sometimes of similar molecules, often of dissimilar ones. They age in the air by several processes. Some particles serve as nuclei upon which vapors condense. Some particles react chemically with atmospheric gases or vapors to form different compounds. When two particles collide in the air, they tend to adhere to each other because of attractive surface forces, thereby forming progressively larger and larger particles by agglomeration. The larger a particle becomes, the greater its weight and the greater its likelihood of falling to the ground rather than remaining airborne. The process by which particles fall out of the air to the ground is called *sedimentation*. Washout of particles by snowflakes, rain, hail, sleet, mist, or fog is a common form of agglomeration and sedimentation. Still other particles leave the air by impaction onto and retention by the solid surfaces of vegetation, soil, and buildings. The particulate mix in the atmosphere is dynamic, with continual injection into the air from sources of small particles; creation of particles in the air by vapor condensation or chemical reaction among gases and vapors; and removal of particles from the air by agglomeration, sedimentation, or impaction.

Before the advent of humans and their works, there must have been particles in the air from natural sources. These certainly included all the particulate forms of condensed water vapor; the condensed and reacted forms of natural organic vapors; salt particles resulting from the evaporation of water from sea spray; wind-borne pollen, fungi, molds, algae, yeasts, rusts, bacteria, and debris from live and decaying plant and animal life; particles eroded by the wind from beaches, desert, soil, and rock; particles from volcanic and other geothermal eruption and from forest fires started by lightning; and particles entering the troposphere from outer space. As mentioned earlier, the true natural background concentration will never be known because when it existed humans were not there to measure it, and by the time humans started measuring particulate matter levels in the air, they had already been polluting the atmosphere with particles resulting from their presence on earth for several million years. The best that can be done now is to assume that the particulate levels at remote places—the middle of the sea, the poles, and the mountaintops—approach the true background concentration. The very act of going to a remote location to make a measurement implies some change in the atmosphere of that remote location attributable to the means people used to travel and to maintain themselves while obtaining the measurements. Particulate matter is measured on a dry basis, thereby eliminating from the measurement not only water droplets and snowflakes but also all vapors, both aqueous and organic, that evaporate or are desiccated from the particulate matter during the drying process. Since different investigators and investigative processes employ different drying procedures and definitions of dryness, it is important to know the procedures and definition employed when comparing data.

There are ways of measuring particulate matter other than by weight per unit volume of air. They include a count of the total number of particles in a unit volume of air, a count of the number of particles of each size range, the weight of particles of each size range, and similar measures based on the surface area and volume of the particles rather than on their number or weight. Some particles in the air are so small that they cannot be seen by an optical microscope, individually weighing so little that their presence is masked in gravimetric analysis by the presence of a few large particles. The mass of a spherical particle is

$$w = \tfrac{4}{3}\pi p r^3 \qquad (2\text{-}2)$$

where w is the particle mass (gm), r is the particle radius (cm), and p is the particle density (gm/cm^3).

The size of small particles is measured in microns (μm). One micron is one-millionth of a meter or 10,000 Å (angstrom units)—the units used to measure the wavelength of light (visible light is between 3000 and 8000 Å) (Fig. 2-2) (2). Compare the weight of a 10-μm particle near the upper limit of those found suspended in the air and a 0.1-μm particle which is near the lower limit. If both particles have the same density (p), the smaller particle will have one-millionth the weight of the larger one. The usual gravimetric procedures can scarcely distinguish a 0.1-μm particle in the presence of a 10-μm particle. To measure the entire size range of particles in the atmosphere, several measurement techniques must therefore be combined, each most appropriate for its size range (Table 2-4). Thus the smallest particles—those only slightly larger than a gas molecule—are measured by the electric charge they carry and by electron microscopy. The next larger size range is measured by electron microscopy or by the ability of these particles to act as nuclei upon which water vapor can be condensed in a cloud chamber. (The water droplets are measured rather than the particles themselves.) The still larger size range is measured by electron or optical microscopy; and the largest size range is measured gravimetrically, either as suspended particles separated from the air by a sampling device or as sedimented particles falling out of the air into a receptacle.

By measuring each portion of the particle size spectrum by the most appropriate method, a composite diagram of the size distribution of the atmospheric aerosol can be produced. Figure 2-3 shows that there are separate size distributions with respect to the number, surface area, and volume (or mass) of the particles. The volume (mass) distribution is called *bimodal* because of its separate maxima at about 0.2 and 10 μm, which result from different mechanisms of particle formation. The mode with the 0.2-μm maximum results from coagulation and condensation formation mechanisms. These particles are created in the atmosphere by chemical reaction among gases and vapors. They are called *fine* particles to differentiate them from the particles in the 10-μm maximum mode, which are called *coarse*. These fine particles are primarily sulfates, nitrates, organics, ammonium, and lead compounds. The mode with the 10-μm maximum are particles introduced to the atmosphere as solids from the surface of the earth and the seas, plus particles from the coagulation–condensation mode which have grown larger and moved across the saddle between the modes into the larger size mode. These are primarily silicon, iron, aluminum, sea salt, and plant particles. Thus there is a dynamism that creates small particles, allows them to grow larger, and eventually allows the large particles to be scavenged from the atmosphere by sedimentation (in the absence of precipitation), plus washout and rainout when there is precipitation.

The majority of particles in the atmosphere are spherical in shape because they are formed by condensation or cooling processes or they contain core nuclei coated with liquid. Liquid surface tension draws the material in the particle into a spherical shape. Other important particle shapes exist in the atmosphere; e.g., asbestos is present as long fibers and fly ash can be irregular in shape.

The methods just noted tell something about the physical characteristics of atmospheric particulate matter but nothing about its chemical composition. One can seek this kind of information for either individual particles or all particles en masse. Analysis of particles en masse involves analysis of a mixture of particles of many different compounds. How much of

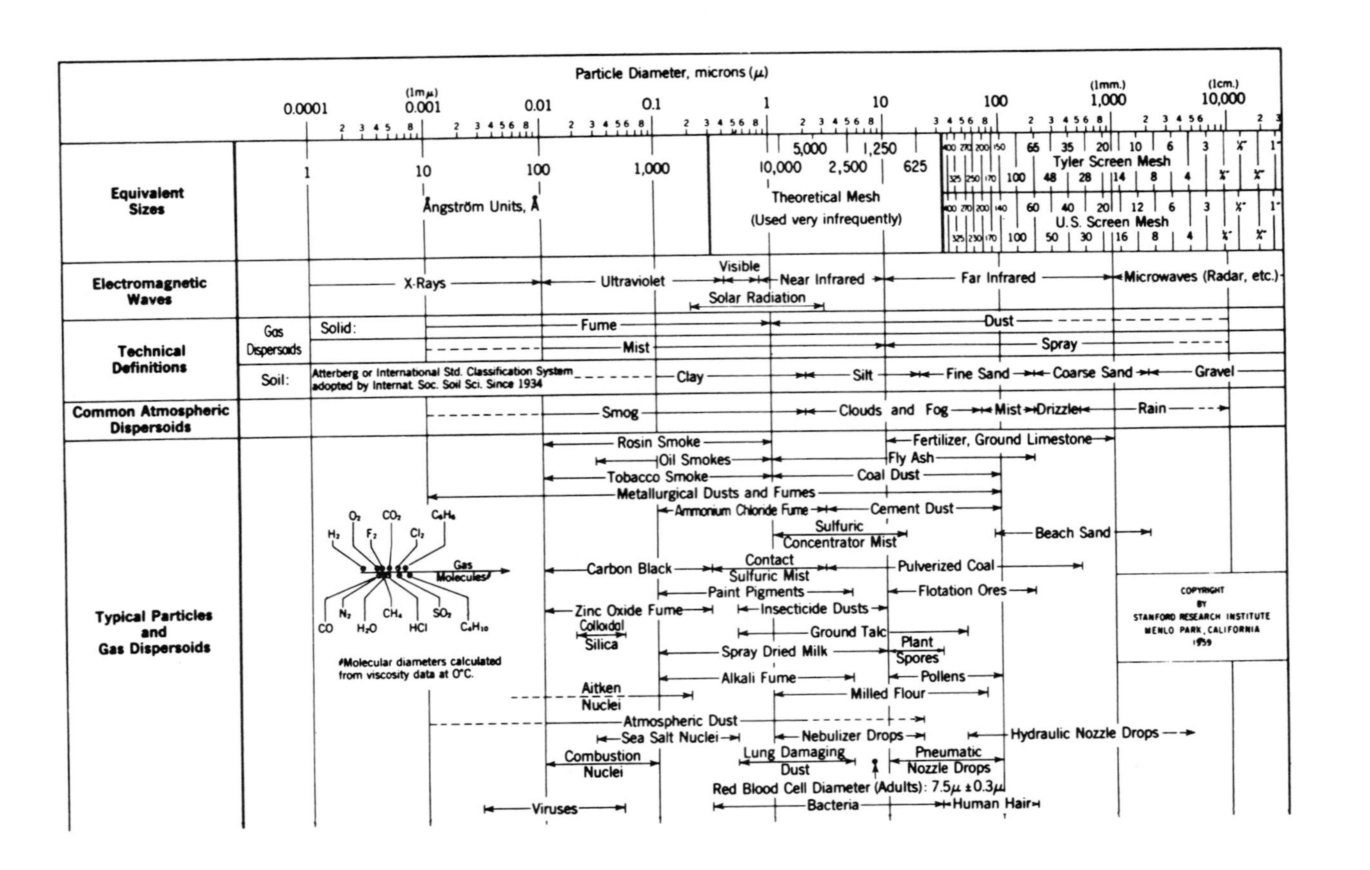
Particle Diameter, microns (μ)
0.0001
(1mμ) 0.001
0.01
0.1
1
10
100
(1mm.) 1,000
(1cm.) 10,000
Equivalent Sizes
Ångström Units, Å
1
10
100
1,000
10,000
5,000
2,500
1,250
625
Theoretical Mesh
(Used very infrequently)
Tyler Screen Mesh
U.S. Screen Mesh
Electromagnetic Waves
X-Rays
Ultraviolet
Visible
Near Infrared
Far Infrared
Microwaves (Radar, etc.)
Solar Radiation
Technical Definitions
Gas Dispersoids
Solid:
Fume
Dust
Mist
Spray
Soil:
Atterberg or International Std. Classification System adopted by Internat. Soc. Soil Sci. Since 1934
Clay
Silt
Fine Sand
Coarse Sand
Gravel
Common Atmospheric Dispersoids
Smog
Clouds and Fog
Mist
Drizzle
Rain
Typical Particles and Gas Dispersoids
Rosin Smoke
Oil Smokes
Tobacco Smoke
Metallurgical Dusts and Fumes
Ammonium Chloride Fume
Fertilizer, Ground Limestone
Fly Ash
Coal Dust
Cement Dust
Sulfuric Concentrator Mist
Beach Sand
Carbon Black
Contact Sulfuric Mist
Pulverized Coal
Paint Pigments
Flotation Ores
Zinc Oxide Fume
Insecticide Dusts
Colloidal Silica
Ground Talc
Spray Dried Milk
Plant Spores
Alkali Fume
Pollens
Aitken Nuclei
Milled Flour
Atmospheric Dust
Sea Salt Nuclei
Nebulizer Drops
Hydraulic Nozzle Drops
Combustion Nuclei
Lung Damaging Dust
Pneumatic Nozzle Drops
Red Blood Cell Diameter (Adults): 7.5μ ± 0.3μ
Viruses
Bacteria
Human Hair
H₂ O₂ F₂ CO₂ Cl₂ C₆H₆
Gas Molecules*
CO N₂ H₂O CH₄ HCl SO₂ C₄H₁₀
*Molecular diameters calculated from viscosity data at 0°C.
COPYRIGHT BY STANFORD RESEARCH INSTITUTE MENLO PARK, CALIFORNIA 1959

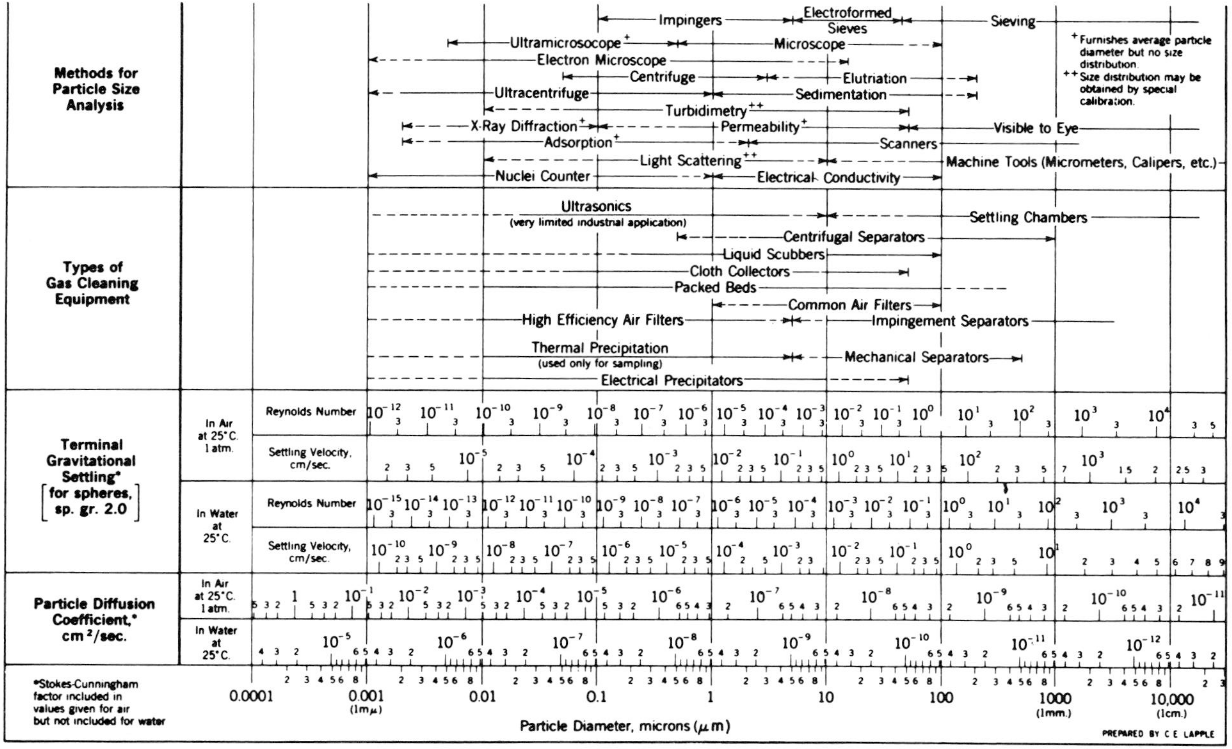

Methods for Particle Size Analysis

Impingers
Electroformed Sieves
Sieving
Ultramicroscope+
Microscope
Electron Microscope
Centrifuge
Elutriation
Ultracentrifuge
Sedimentation
Turbidimetry++
X-Ray Diffraction+
Permeability+
Visible to Eye
Adsorption+
Scanners
Light Scattering++
Machine Tools (Micrometers, Calipers, etc.)
Nuclei Counter
Electrical Conductivity

+ Furnishes average particle diameter but no size distribution.
++ Size distribution may be obtained by special calibration.

Types of Gas Cleaning Equipment

Ultrasonics
(very limited industrial application)
Settling Chambers
Centrifugal Separators
Liquid Scubbers
Cloth Collectors
Packed Beds
Common Air Filters
High Efficiency Air Filters
Impingement Separators
Thermal Precipitation
(used only for sampling)
Mechanical Separators
Electrical Precipitators

Terminal Gravitational Settling* [for spheres, sp. gr. 2.0]

In Air at 25°C. 1 atm.: Reynolds Number; Settling Velocity, cm/sec.

In Water at 25°C.: Reynolds Number; Settling Velocity, cm/sec.

Particle Diffusion Coefficient,* $cm^2/sec.$

In Air at 25°C. 1 atm.

In Water at 25°C.

*Stokes-Cunningham factor included in values given for air but not included for water

0.0001 | 0.001 (1mμ) | 0.01 | 0.1 | 1 | 10 | 100 | 1000 (1mm.) | 10,000 (1cm.)

Particle Diameter, microns (μm)

PREPARED BY C E LAPPLE

Fig. 2-2. Characteristics of particles and particle dispersoids. Reproduced by permission of SRI International, Menlo Park, California, 1959.

TABLE 2-4

Particle Size Ranges and Their Methods of Measurement

Particle size range (um)	Ions	Nuclei	Visability	Suspended or settleable; nonairborne	Dispersion aerosol	Condensation aerosol	Pollen and spores	Sedimentation, diffusion, and settling
10^{-4}–10^{-3}	Small	—	—	Suspended	—	Gas molecules	—	Diffusion
10^{-3}–10^{-2}	Intermediate and large	Aitken nuclei	Electron microscope	Suspended	—	Vapor molecules	—	Diffusion
10^{-2}–10^{-1}	Large	Aitken and condensation nuclei	Electron microscope	Suspended	—	Fume–mist	—	Diffusion
Air pollution 10^{-1}–10^{-0}	—	Condensation nuclei	Microscope: electron and optical	Suspended	Dust–mist	Fume–mist	—	Diffusion and sedimentation
10^{0}–10^{1}	—	—	Microscope: optical	Suspended and settleable	Dust–mist	Fume–mist	—	Sedimentation
10^{1}–10^{2}	—	—	Eye, sieves	Settleable	Dust–mist	Mist–fog	Pollen and spores	
10^{2}–10^{3}	—	—	Eye, sieves	Nonairborne	Dust–spray	Drizzle–rain	—	Sedimentation
10^{3}–10^{4}	—	—	Eye, sieves	Nonairborne	Sand–rocks	Rain	—	Sedimentation

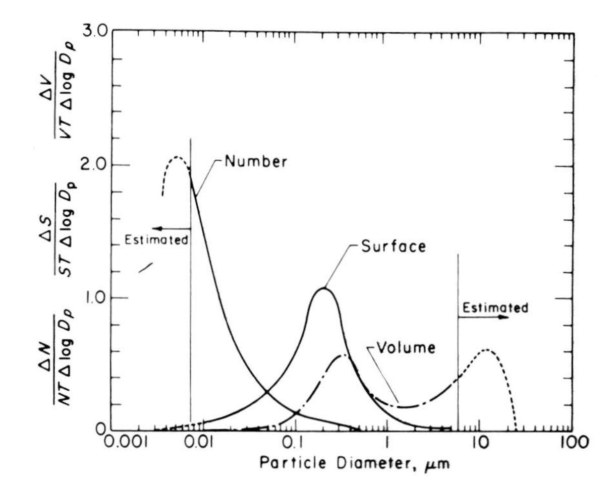

Fig. 2-3. Grand average number (N), surface area (S), and volume (V) distribution of Los Angeles smog. The linear ordinate normalized by total number (NT), area (ST), or volume (VT) is used so that the apparent area under the curves is proportional to the quantity in that size range. Source: Corn, M., Properties of non-viable particles in the air, *In* "Air Pollution," 3rd ed., Vol. I (A. C. Stern, ed.). Academic Press, New York, 1976, p. 123.

each element or radical, anion, or cation is present in the mixture can be determined. Specific organic compounds may be separated and identified. Individual particles may be analyzed by electron microscopy techniques.

Much of the concern about particulate matter in the atmosphere arises because particles of certain size ranges can be inhaled and retained by the human respiratory system. There is also concern because particulate matter in the atmosphere absorbs and scatters incoming solar radiation. For a detailed discussion of the human respiratory system and the defenses it provides against exposure of the lungs to particulate matter, see Chapter 7.

IV. CONCEPTS

A. Sources and Sinks

The places from which pollutants emanate are called *sources*. There are natural as well as anthropogenic sources of the permanent gases considered to be pollutants. These include plant and animal respiration and the decay of what was once living matter. Volcanoes and naturally caused forest fires are other natural sources. The places to which pollutants disappear from the air are called *sinks*. Sinks include the soil, vegetation, structures, and water bodies, particularly the oceans. The mechanisms whereby pollutants are removed from the atmosphere are called *scavenging mechanisms,* and the measure used for the aging of a pollutant is its *half-life*—the time it takes for half of the quantity of pollutant emanating from a source to disappear into its various sinks. Fortunately, most pollutants have a short enough half-life (i.e., days rather than decades) to prevent their accumulation in the air to the extent that they substantially alter the composition of unpolluted air shown in Table 2-1. Several gases do appear to be accumulating in the air to the extent that measurements have documented the increase in concentration from year to year. The best-known example is carbon dioxide (Fig. 2-4; see also Fig. 11-1). Other accumulating gases are nitrous oxide (N_2O), methane (CH_4), CFCs, and other halocarbons. All of these gases have complex roles in climate change processes, particularly global warming concerns. CFCs are chemically very stable compounds in the troposphere and have half-lives from tens of years to over 100 years. One of the sinks for CFCs is transport to the stratosphere, where shortwave UV radiation photodissociates the molecules, releasing chlorine (Cl) atoms. These Cl atoms are projected to reduce the steady-state stratospheric ozone concentration, in turn increasing the penetration of harmful UV radiation to the earth's surface.

Oxidation, either atmospheric or biological, is a prime removal mechanism for inorganic as well as organic gases. Inorganic gases, such as nitric oxide (NO), nitrogen dioxide (NO_2), hydrogen sulfide (H_2S), sulfur dioxide (SO_2), and sulfur trioxide (SO_3), may eventually form corresponding acids:

$$NO + \tfrac{1}{2}O_2 \rightarrow NO_2 \qquad (2\text{-}3)$$

$$4NO_2 + 2H_2O + O_2 \rightarrow 4HNO_3 \qquad (2\text{-}4)$$

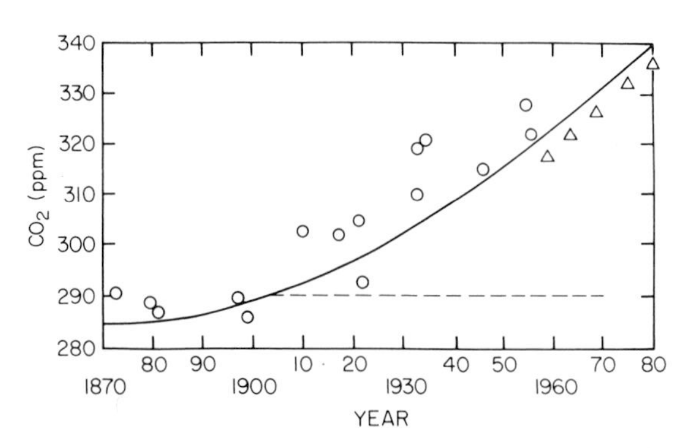

Fig. 2-4. Average CO_2 concentration: North Atlantic Region ○, Pacific Region △. (The dashed line is the nineteenth-century base value: 290 ppm.) Source: Combination of data from Callender, G. C., *Tellus,* **10,** 243 (1958), and Council on Environmental Quality, "Global Energy Futures and the Carbon Dioxide Problem." Superintendent of Documents, U.S. Government Printing Office, Washington, DC, 1981. (See also Fig. 11-1.)

$$H_2S + \tfrac{3}{2}O_2 \rightarrow SO_2 + H_2O \tag{2-5}$$

$$SO_2 + \tfrac{1}{2}O_2 \rightarrow SO_3 \tag{2-6}$$

$$SO_3 + H_2O \rightarrow H_2SO_4 \tag{2-7}$$

Oxidation of SO_2 is slow in a mixture of pure gases, but the rate is increased by light, NO_2, oxidants, and metallic oxides which act as catalysts for the reaction. The formed acids can react with particulate matter or ammonia to form salts.

B. Receptors

A *receptor* is something which is adversely affected by polluted air. A receptor may be a person or animal that breathes the air and whose health may be adversely affected thereby, or whose eyes may be irritated or whose skin made dirty. It may be a tree or plant that dies, or the growth yield or appearance of which is adversely affected. It may be some material such as paper, leather, cloth, metal, stone, or paint that is affected. Some properties of the atmosphere itself, such as its ability to transmit radiant energy, may be affected. Aquatic life in lakes and some soils are adversely affected by acidification via acidic deposition.

C. Transport and Diffusion

Transport is the mechanism that moves the pollution from a source to a receptor. The simplest source–receptor combination is that of an isolated point source and an isolated receptor. A point source may best be visualized as a chimney or stack emitting a pollutant into the air; the isolated point source might be the stack of a smelter standing by itself in the middle of a flat desert next to the body of ore it is smelting. The isolated receptor might be a resort hotel 5 miles distant on the edge of the desert. The effluent from the stack will flow directly from it to the receptor when the wind is along the line connecting them (Fig. 2-5). The wind is the means by which the pollution is transported from the source to the receptor. However, during its transit over the 5 miles between the source and the receptor, the plume does not remain a cylindrical tube of pollution of the same diameter as the interior of the stack from which it was emitted. Instead, as it travels over the 5-mile distance, turbulent eddies in the air and in the plume move parcels from the edges of the plume into the surrounding air and move parcels of surrounding air into the plume. If the wind speed is greater than the speed of ejection from the stack, the wind will stretch out the plume until the plume speed equals wind speed. These two processes—mixing by turbulence and stretch-out of the plume, plus a third one—meandering (which means that the plume may not follow a true

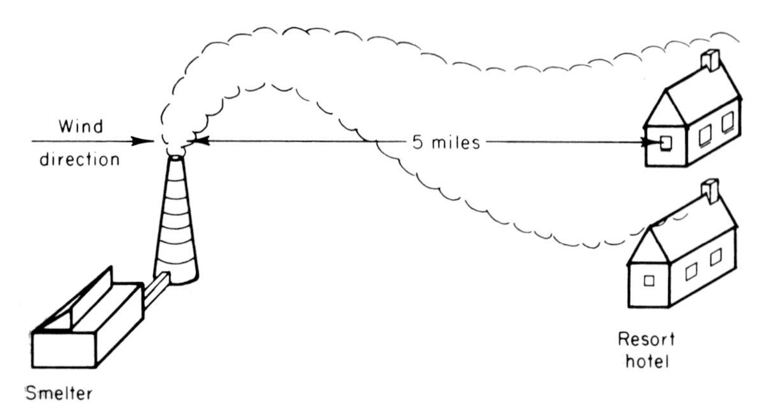

Fig. 2-5. Transport and diffusion from source to receptor.

straight line between the source and the receptor, but may meander somewhat about that line as wind direction fluctuates from its mean value over the time of transit between the two points)—tend to make the concentration of the plume as it arrives at the receptor less than its concentration on release from the stack. The sum of all these processes is called *diffusion*. The process of diffusion becomes increasingly complex as the number of sources and receptors increases; as sources and receptors begin to group together into towns and cities; as some of the sources and receptors move, e.g., vehicles; and, finally, as the weather and topography become more complex than a wind blowing in one direction over a flat desert for a prolonged period of time.

If the plume being transported is above the height where ground-based convective and turbulent processes will bring it down to the ground reasonably close to its origin, it may travel for hundreds of miles at that height before being brought to earth, by these processes, in a remote community. This is known as *long-range* or *long-distance transport*.

D. Significant Deterioration of Air Quality

It may be desirable to curtail transport of pollution to areas whose air is presently quite clean, even though, after such transport, the air quality of the area would be considerably cleaner than would be required by air quality standards. This concept is called prevention of significant deterioration of the air quality in such areas. It requires definition of how much deterioration can be considered insignificant.

E. Polluted Atmosphere

Sections II and III describe the types and form of gases, vapors, and particles in the atmosphere. Definitions of terms are offered at the beginning of this chapter. *When does air pollution exist?* This chapter presents some of the principles by which materials may be introduced into the atmosphere, moved about, and removed from the atmosphere. The definition of air pollutant or air pollution depends on the context of time, space, and impact for a particular set of circumstances. Smoke in a cave was a major problem for early humans but not one we consider today. Indoor air quality or pollution in our homes and workplaces are concerns of today. Elevated levels of ozone, organic gases, and other trace gases in our communities are the "air pollution" problems of today. Governments around the world have established and are continously evaluating the impact of elevated levels of trace gases and particulate material in the atmosphere. This process helps bring meaning to the definitions offered at the beginning of the chapter—"to defile, to make impure" the atmosphere.

REFERENCES

1. Turner, D. B. Meteorological fundamentals, *in* "Meteorological Aspects of Air Pollution—Course Manual. Air Pollution Training Institute, Office of Manpower Development, Office of Air Programs. United States Environmental Protection Agency, Research Triangle Park, NC.
2. Lapple, C. E., *Stanford Research Inst. J.* **5,** 95 (1961).

SUGGESTED READING

Bridgman, H. A., "Global Air Pollution: Problems for the 1990's." Belhaven, London, 1990.

Graedel, T. E., and Crutzen, P. J., "Atmospheric Change: An Earth System Perspective." W.H. Freeman, New York, 1993.

Seinfeld, J. H., "Atmospheric Chemistry and Physics of Air Pollution," Wiley, New York, 1986.

Warneck, P., "Chemistry of the Natural Atmosphere." Academic Press, San Diego, 1988.

QUESTIONS

1. Prepare a graph showing the conversion factor from ppm (vol) to $\mu g/m^3$ for compounds with molecular weights ranging from 10 to 200 at 25°C and 760 mm Hg as well as at 0°C and 760 mm Hg.
2. (a) Convert 0.2 ppm (vol) NO and 0.15 ppm (vol) NO_2 to $\mu g/m^3$ nitrogen oxides (NO_x) at 25°C and 760 mm Hg. (b) Convert 0.35 ppm (vol) NO_x to $\mu g/m^3$ at 25°C and 760 mm Hg.
3. Prepare a table showing the weight in grams and the surface area in m^3 of a 0.1-, 1.0-, 10.0-, and 100.0-μm-diameter spherical particle of unit density.
4. What is the settling velocity in cm/sec in air at 25°C and 1 atmosphere for a 100 mesh size spherical particle, i.e., one which just passes through the opening in the sieve (specific gravity = 2.0)?

5. How does the diameter of airborne pollen grains compare with the diameter of a human hair?
6. What are the principal chemical reactions that take place in the chemosphere to give it its name? How do they influence stratospheric and tropospheric chemical reactions?
7. What are the source and nature of the condensable organic vapors in unpolluted air?
8. Has the composition of the unpolluted air of the troposphere most probably always been the same as in Tables 2-1 and 2-2? Will Tables 2-1 and 2-2 most probably define unpolluted air in the year 2085? Discuss your answer.
9. Describe the apparatus and procedures used to measure atmospheric ions and nuclei.

3

Scales of the Air Pollution Problem

Problems of air pollution exist on all scales from extremely local to global. These are divided in this chapter into five different scales: local, urban, regional, continental, and global. The local scale includes up to about 5 km. The urban scale extends to the order of 50 km. The regional scale is from 50 to 500 km. Continental scales are from 500 to several thousand km. Of course, the global scale extends worldwide.

I. LOCAL

Local air pollution problems are usually characterized by one or several large emitters or a large number of relatively small emitters. The lower the release height of a source, the larger the potential impact for a given release. Carbon monoxide from motor vehicles will cause high concentrations near roadways. Any ground-level source, such as evaporation of volatile organic compounds from a waste treatment pond, will produce the highest concentrations near the source.

However, large sources that emit high above the ground through stacks, such as power plants or industrial sources, can cause local problems, especially under unstable meteorological conditions that cause portions of the plume to reach the ground in high concentrations.

There are many releases of pollutants from relatively short stacks or vents on the top of one- or two-story buildings. Under most conditions such releases are caught within the turbulent downwash downwind of the building. This allows high concentration to be brought to the ground. Many different pollutants can be released in this manner, including compounds and mixtures that can cause odors. The modeling of the transport and dispersion of pollutants on this scale is discussed in Chapter 20.

Usually the effects of accidental releases are confined to the local scale.

II. URBAN

There are two different types of air pollution problems in urban areas. One is the release of primary pollutants (those released directly from sources). The other is the formation of secondary pollutants (those that are formed through chemical reactions of the primary pollutants).

Air pollution problems can be caused by individual sources on the urban scale as well as the local scale. For pollutants that are relatively nonreactive, such as carbon monoxide and particulate matter, or relatively slowly reactive, such as sulfur dioxide, the contributions from individual sources add together to yield high concentrations. Since a major source of carbon monoxide is motor vehicles, "hot spots" of high concentration can occur especially near multilane intersections. The emissions are especially high from idling vehicles, and if high buildings surround the intersection, the volume in which the pollution is contained is severely restricted. The combination of these factors results in high concentrations.

Urban problems result from the formation of secondary pollutants. A major problem of many large metropolitan areas is the formation of ozone from photochemical reactions of oxides of nitrogen and various species of hydrocarbons. These reactions are catalyzed by the ultraviolet light in sunlight and are therefore called photochemical reactions. Many metropolitan areas are in nonattainment for ozone; that is, they are not meeting the air quality standards. The Clean Air Act Amendments (CAAA) of 1990 recognize this as a major problem and have classified the various metropolitan areas that are in nonattainment according to the severity of the problem for that area. The CAAA sets timetables for the various classifications for reaching attainment with the National Ambient Air Quality Standards (NAAQS). Oxides of nitrogen, principally nitric oxide, NO, but also nitrogen dioxide, NO_2, are emitted from automobiles and from combustion processes. Hydrocarbons are emitted from many different sources. The various species have widely varying reactivities. Determining the emissions of the various species from the many sources in order to conduct control programs presents formidable tasks. Atmospheric chemistry is discussed in Chapter 12.

III. REGIONAL

At least three types of problems contribute to air pollution problems on the regional scale. One is the carryover of urban oxidant problems to the regional scale. With the existence of major metropolitan areas in close proximity, the air from one metropolitan area, containing both secondary pollutants formed through reactions and primary pollutants, flows on to the adjacent metropolitan area. The pollutants from the second area are then added on top of the "background" from the first.

A second type of problem is the release of relatively slow-reacting primary air pollutants that undergo reactions and transformation during long transport times. These long transport times then result in transport distances over regional scales. The gas, sulfur dioxide, released primarily through combustion of fossil fuels, both coal and oil, is oxidized during long-distance transport to sulfur trioxide, SO_3, which reacts with water vapor to form sulfuric acid, which reacts with numerous compounds to form sulfates. These are fine (submicrometer) particulates.

Nitric oxide, NO, results from high-temperature combustion, both in stationary sources such as power plants or industrial plants in the production of process heat and in internal combustion engines in vehicles. The NO is oxidized in the atmosphere, usually rather slowly, or more rapidly if there is ozone present, to nitrogen dioxide, NO_2. NO_2 also reacts further with other constituents, forming nitrates, which is also in fine particulate form.

Both the sulfates and nitrates existing in the atmosphere as fine particulates, generally in the size range less than 1 μm, can then be removed from the atmosphere by several processes. One of these is the formation of clouds, with the particles serving as additional condensation nuclei. These are then deposited on the ground if the droplets grow to sufficient size to fall as raindrops (rainout). The other mechanism also involves rain, but is capture of the particles in air by raindrops falling through the air (washout). Both of these mechanisms contribute to "acid rain," which results in the sulfate and nitrate particles reaching lakes and streams and increasing their acidity.

A third type of regional problem is that of visibility. Visibility may be reduced by specific plumes or by the regional levels of particulate matter that produce various intensities of haze. The fine sulfate and nitrate particulates just discussed are largely responsible for reduction of visibility (see Chapter 10). This problem is of concern in locations of natural beauty, where it is desirable to keep scenic vistas as free of obstructions to the view as possible.

IV. CONTINENTAL

In a relatively small continental area such as Europe, there is not much difference between what would be considered the regional scale and the continental scale. However, on most other continents there would be a difference between what is considered regional and what continental. Perhaps of greatest concern on the continental scale is that the air pollution policies of a nation are likely to create impacts on neighboring nations. Acid rain in Scandanavia has been considered to have had impacts from Great Britain and Western Europe. Japan has considered that part of their air pollution problem, especially in the western part of the country, has origins in China and Korea. Cooperation in the examination of the North American acid rain problem has existed for a long time between Canada and the United States.

V. GLOBAL

The release of radioactivity from the accident at Chernobyl would be considered primarily a regional or continental problem. However, higher than usual levels of radioactivity were detected in the Pacific Northwest part of the United States soon after the accident. This indicates the long-range transport that occurred following this accident.

One air pollution problem of a global nature is the release of chlorofluorocarbons used as propellants in spray cans and in air conditioners and their effect on the ozone layer high in the atmosphere. (See Chapter 11.)

Some knowledge of the exchange processes between the stratosphere and the troposphere and between the Northern and Southern Hemispheres was learned in the late 1950s and early 1960s as a result of the injection of radioactive debris into the stratosphere from atomic bomb tests in the Pacific. The debris was injected primarily into the Northern Hemisphere in the stratosphere. The stratosphere is usually quite stable and resists vertical air exchange between layers. It was found that the exchange processes in the stratosphere between the Northern and Southern Hemispheres is quite slow. However, radioactivity did show up in the Southern Hemisphere within 3 years of the onset of the tests, although the levels remained much lower than those in the Northern Hemisphere. Similarly, the exchange processes between the troposphere and the stratosphere are quite slow. The main transfer from the troposphere into the stratosphere is injection through the tops of thunderstorms that occasionally penetrate the tropopause, the boundary between the troposphere and the stratosphere. Some transfer of stratospheric air downward also occurs through occasional gaps in the tropopause. Since the ozone layer is considerably above the stratosphere, the transfer of chlorofluorocarbons upward to the ozone layer is expected to take place very slowly. Thus there was a lag from the first release of these gases until an effect was seen. Similarly, on cessation of use of these materials worldwide, there will be another lag before an effect is noted (See Chapter 11.)

Another global problem is that generated by carbon dioxide, which is not normally considered an air pollutant. Because carbon dioxide is a "greenhouse gas," it absorbs and reradiates infrared radiation. At night the ground loses heat and cools, primarily through infrared radiation. But a portion of this radiation is intercepted by the carbon dioxide in the air and is reradiated both upward and downward. That which is radiated downward keeps the ground from cooling as rapidly. As the carbon dioxide concentration continues to increase, the earth's temperature is expected to increase.

A natural air pollution problem that can cause global effects is injection into the atmosphere of fine particulate debris by volcanoes. The addition of this particulate matter has caused some spectacular sunsets throughout the earth. If sufficient material is released, it can change the radiation balance. Blocking of the incoming solar radiation will reduce the normal degree of daytime warming of the earth's surface. A "mini-ice age" was caused in the mid-1800s, when a volcanic mountain in the Pacific erupted. The summer of that year was much cooler than usual and snow occurred in July in New England.

SUGGESTED READING

Critchfield, H. J., "General Climatology," 4th ed. Prentice-Hall, Englewood Cliffs, NJ, 1983.

Knap, A. H. (ed.), "The Long-Range Atmospheric Transport of Natural and Contaminant Substances." Kluwer Academic Press, Hingham, MA, 1989.

Kramer, M. L., and Porch, W. M., "Meteorological Aspects of Emergency Response." American Meteorological Society, Boston, 1990.

Landsberg, H. E., "The Urban Climate." Academic Press, New York, 1981.

QUESTIONS

1. What situation of emission has the potential to produce the greatest local problem?
2. What are the two types of air pollution problems found in urban areas?
3. What are the two primary gaseous pollutants that transform to fine-particle form during long-range transport?
4. What are major concerns on the continental scale?
5. What are two global issues?

4

Air Quality

The terms *ambient air, ambient air pollution, ambient levels, ambient concentrations, ambient air monitoring, ambient air quality,* etc. occur frequently in air pollution parlance. The intent is to distinguish pollution of the air outdoors by transport and diffusion by wind (i.e., ambient air pollution) from contamination of the air indoors by the same substances.

The air inside a factory building can be polluted by release of contaminants from industrial processes to the air of the workroom. This is a major cause of occupational disease. Prevention and control of such contamination are part of the practice of industrial hygiene. To prevent exposure of workers to such contamination, industrial hygienists use industrial ventilation systems that remove the contaminated air from the workroom and discharge it, either with or without treatment to remove the contaminants, to the ambient air outside the factory building.

The air inside a home, office, or public building is the subject of much interest and is referred to as indoor air pollution or indoor air quality (see Chapter 23). These interior spaces may be contaminated by such sources as fuel-fired cooking or space-heating ranges, ovens, or stoves that discharge their combustion products to the room; by solvents evaporated from inks, paints, adhesives, cleaners, or other products; by formaldehyde, radon, and other products emanating from building materials; and by other pollutant sources indoors (1). If some of these sources exist inside a building, the pollution level of the indoor air might be higher than that of the outside air. However, if none of these sources are inside the building, the pollution level inside would be expected to be lower than the ambient concentration outside because of the ability of the surfaces inside the building—walls, floors, ceilings, furniture, and fixtures—to adsorb or react with gaseous pollutants and attract and retain particulate pollutants, thereby partially removing them from the air breathed by occupants of the building. This adsorption and retention would occur even if doors and windows were open, but the difference between outdoor and indoor concentrations would be even greater if they were closed, in which case air could enter the building only by infiltration through cracks and walls.

Many materials used and dusts generated in buildings and other enclosed spaces are allergenic to their occupants. Occupants who do not smoke are exposed to tobacco and its associated gaseous and particulate emissions from those who do. This occurs to a much greater extent indoors than in the outdoor air. Many ordinances have been established to limit or prohibit smoking in public and work places. Attempts have been made to protect occupants of schoolrooms from infections and communicable diseases by using ultraviolet light or chemicals to disinfect the air. These attempts have been unsuccessful because disease transmission occurs instead outdoors and in unprotected rooms. There is, of course, a well-established technology for maintaining sterility in hospital operating rooms and for manufacturing operations in pharmaceutical and similar plants.

I. AVERAGING TIME

The variability inherent in the transport and diffusion process, the time variability of source strengths, and the scavenging and conversion mechanisms in the atmosphere, which cause pollutants to have an effective half-life, result in variability in the concentration of a pollutant arriving at a receptor. Thus, a continuous record of the concentration of a pollutant at a receptor, as measured by an instrument with rapid response, might look like Fig. 4-1(a). If, however, instead of measuring with a rapid-response instrument, the measurement at the receptor site was made with sampling and analytical procedures that integrated the concentration arriving at the receptor over various time periods, e.g., 15 minutes, 1 hour, or 6 hours, the resulting information would look variously like Figs. 4-1(b), (c), and (d), respectively. It should be noted that from the information in Fig. 4-1(a), it is possible to derive mathematically the information in Figs. 4-1(b), (c), and (d), and it is possible to derive the information in Figs. 4-1(c) and (d) from that in Fig. 4-1(b). The converse is not true. With only the information from Fig. 4-1(d) available, Figs. 4-l(a), (b), and (c) could never be constructed, nor could Figs. 4-1(a) and (b) be constructed from Fig. 4-1(c),

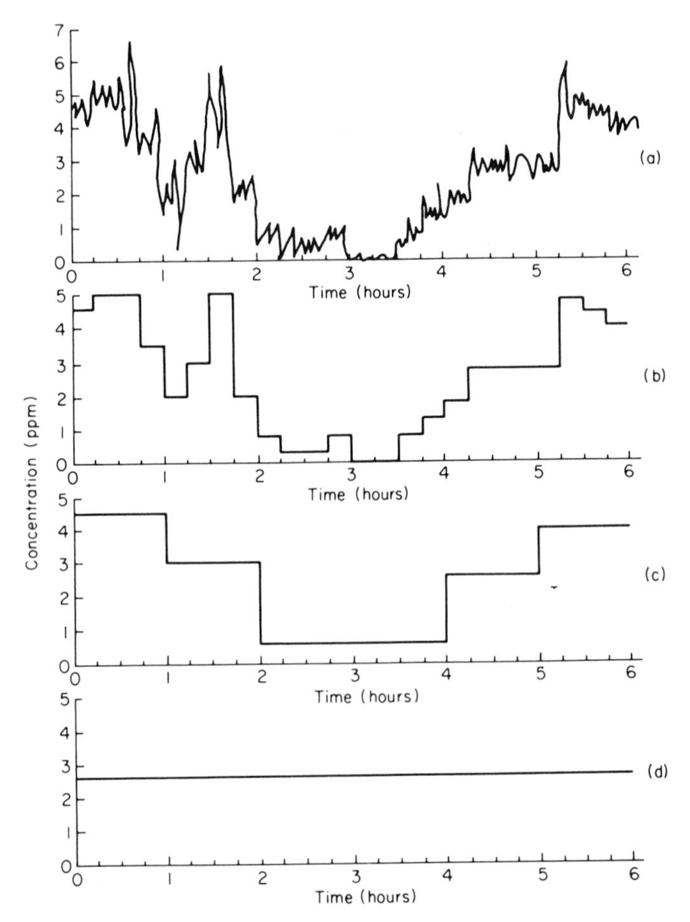

Fig. 4-1. The same atmosphere measured by (a) a rapid-response instrument and by sampling and analytical procedures that integrate the concentration arriving at the receptor over a time period of (b) 15 min, (c) 1 hr, and (d) 6 hr.

nor Fig. 4-1(a) from Fig. 4-1(b). In these examples the time intervals involved in Figs. 4-1(b), (c), and (d)—15 minutes, 1 hour, and 6 hours, respectively—are the *averaging times* of the measurement of pollutant exposure at the receptor.

The averaging time of the rapid-response record [Fig. 4-1(a)] is an inherent characteristic of the instrument and the data acquisition system. It can become almost an instantaneous record of concentration at the receptor. However, in most cases this is not desirable, because such an instantaneous record cannot be put to any practical air pollution control use. What such a record reveals is something of the turbulent structure of the atmosphere, and thus it has some utility in meteorological research. In communications science parlance, an instantaneous recording has too much *noise*. It is therefore necessary to filter or damp out the noise in order to extract the useful information about pollution concentration at the receptor that the signal is trying to reveal. This damping is achieved by building time lags into the response of the sampling, analysis, and recording systems (or into all three); by interrogating the instantaneous output of the analyzer at discrete time intervals, e.g., once every minute or once every 5 minutes, and recording only this extracted information; or by a combination of damping and periodic interrogation.

II. CYCLES

The most significant of the principal cyclic influences on variability of pollution concentration at a receptor is the diurnal cycle (Fig. 4-2). First, there is a diurnal pattern to source strength. In general, emissions from almost all categories of sources are less at night than during the day. Factories and businesses shut down or reduce activity at night. There is less automotive, aircraft, and railroad traffic, use of electricity, cooking, home heating, and refuse burning at night. Second, there is a diurnal pattern to transport and diffusion that will be discussed in detail later in this book.

The next significant cycle is the weekend–weekday cycle. This is entirely a source strength cycle associated with the change in the pattern of living on weekends as compared with weekdays.

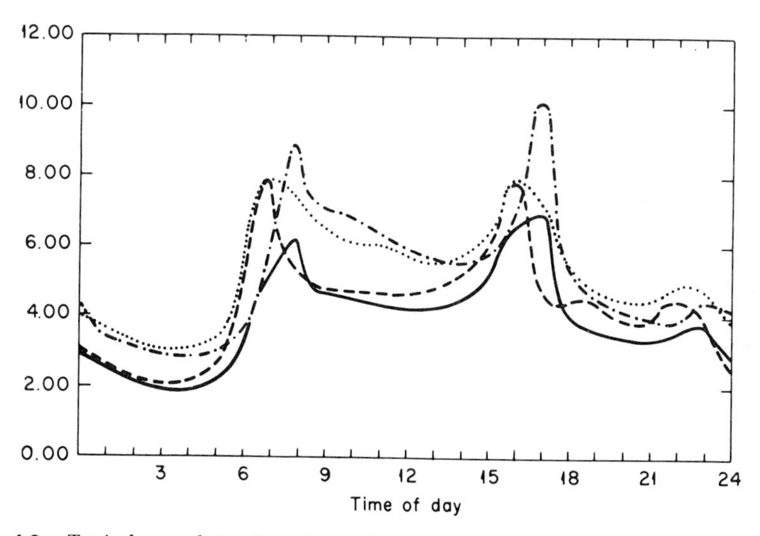

Fig. 4-2. Typical central city diurnal variation in carbon monoxide concentration, in a city in the United States. Spring (———), summer (- - - -), fall (······), winter (–·–·–).

Finally, there is the seasonal cycle associated with the difference in climate and weather over the four seasons—winter, spring, summer, and fall (Fig. 4-3). The climatic changes affect source strength, and the weather changes affect transport and diffusion.

On an annual basis, some year-to-year changes in source strength may be expected as a community, a region, a nation, or the world increases in population or changes its patterns of living. Source strength will be reduced if control efforts or changes in technology succeed in preventing more pollution emission than would have resulted from increases in population (Fig. 4-4). These changes are called *trends.* Although an annual trend in source strength is expected, none is expected in climate or weather, even though each year will have its own individuality with respect to its weather.

Other examples of trends come from Great Britain, where the emission of industrial smoke was reduced from 1.4 million tonnes per year in 1953 to 0.1 million tonnes per year in 1972; domestic smoke emission was reduced from 1.35 million tonnes per year in 1953 to 0.58 million tonnes per year in 1972; and the number of London fogs (smogs) capable of reducing visibility at 9 AM to less than 1 km was reduced from 59 per year in 1946 to 5 per year in 1976.

Annual trends in urban ozone are much more subtle because of the complex interaction among precursors (hydrocarbons and oxides of nitrogen) and meteorology (including solar radiation) (Fig. 4-5).

III. PRIMARY AND SECONDARY POLLUTANTS

A substantial portion of the gas and vapors emitted to the atmosphere in appreciable quantity from anthropogenic sources tends to be relatively simple in chemical structure: carbon dioxide, carbon monoxide, sulfur dioxide, and nitric oxide from combustion processes; hydrogen sulfide, ammonia, hydrogen chloride, and hydrogen fluoride from industrial processes. The solvents and gasoline fractions that evaporate are alkanes, alkenes, and aromatics with relatively simple structures. In addition, more complex

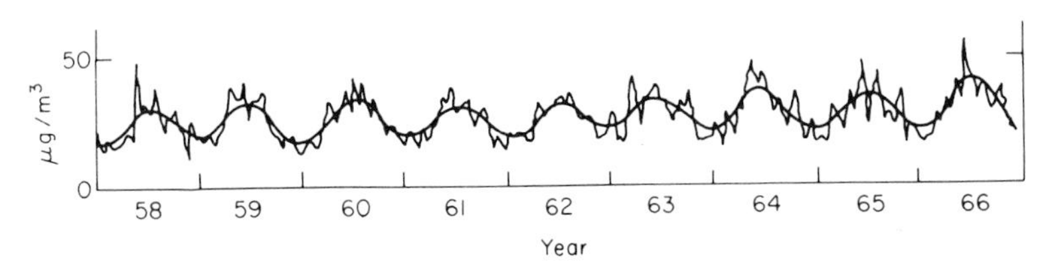

Fig. 4-3. Seasonal variation of suspended particulate matter concentration. Composite of 20 nonurban sites. United States. Source: Pirtas, R., and Levin, H. J., *J. Air Pollut. Control Assoc.* **21**(6), 329–333, 1971.

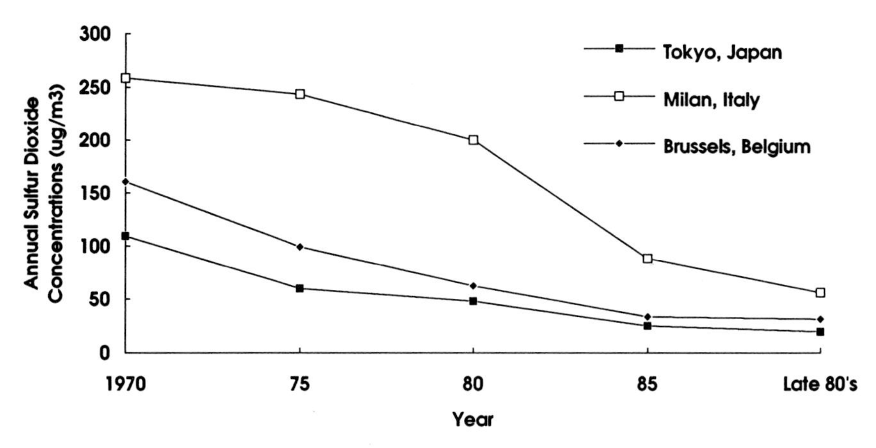

Fig. 4-4. Urban trends in annual sulfur dioxide concentrations. Source: U.S. Environmental Protection Agency, 1992.

molecules such as polynuclear aromatic hydrocarbons and dioxins are released from industrial processes and combustion sources and are referred to as toxic pollutants. Substances such as these, emitted directly from sources, are called *primary pollutants.* They are certainly not innocuous, as will be seen when their adverse effects are discussed in later chapters. However, the primary pollutants do not, of themselves, produce all of the adverse effects of air pollution. Chemical reactions may occur among the

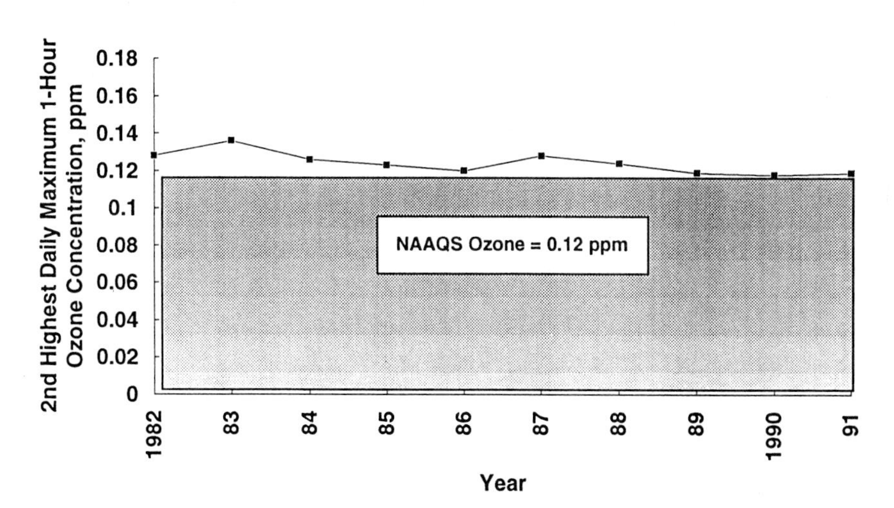

Fig. 4-5. Annual second highest daily maximum 1-hr ozone concentration at 495 sites in the United States for 1982–1991. Source: U.S. Environmental Protection Agency, 1992.

primary pollutants and the constituents of the unpolluted atmosphere (Fig. 4-6). The atmosphere may be viewed as a reaction vessel into which are poured reactable constituents and in which are produced a tremendous array of new chemical compounds, generated by gases and vapors reacting with each other and with the particles in the air. The pollutants manufactured in the air are called *secondary pollutants;* they are responsible for most of the smog, haze, and eye irritation and for many of the forms of plant and material damage attributed to air pollution. In air pollution parlance,

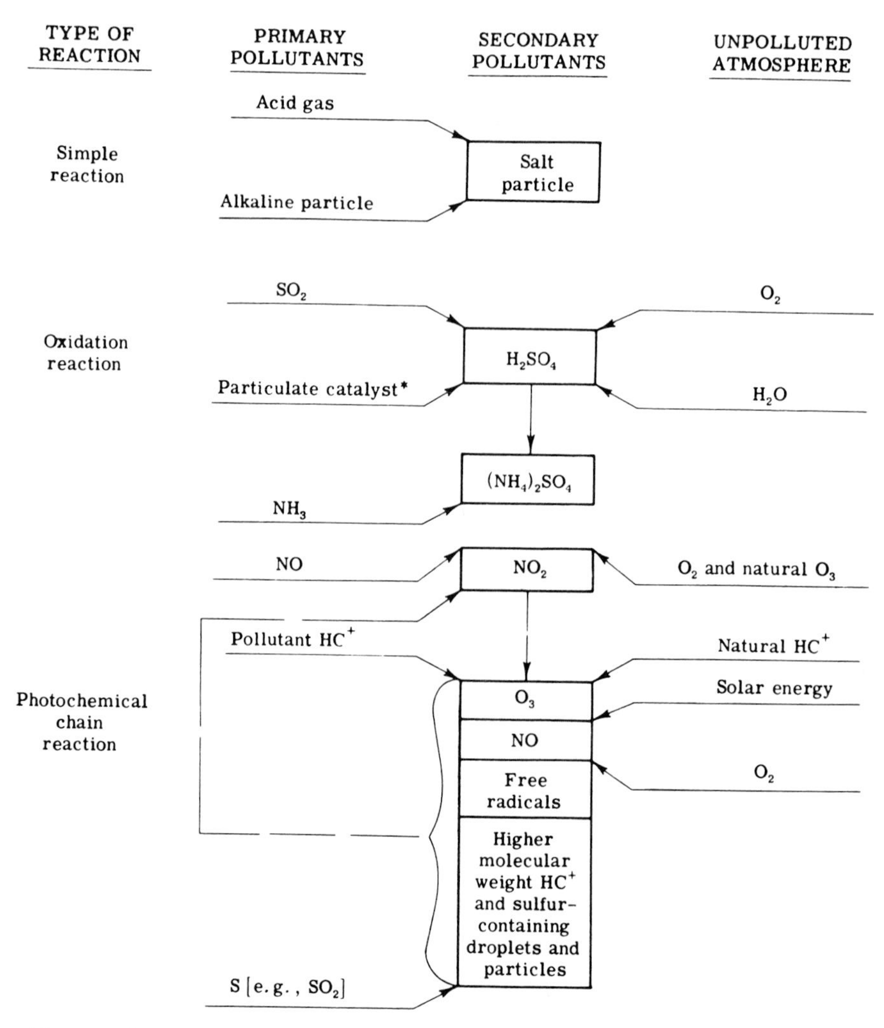

Fig. 4-6. Primary and secondary pollutants. *Reaction can occur without catalysis (HC^+, hydrocarbons).

the primary pollutants that react are termed the *precursors* of the secondary pollutants. With the knowledge that each secondary pollutant arises from specific chemical reactions involving specific primary reactants, we must control secondary pollutants by controlling how much of each primary pollutant is allowed to be emitted.

IV. MEASUREMENT SYSTEMS

Many methods of air quality measurement have inherent averaging times. In selecting methods for measuring air quality or assessing air pollution effects, this fact must be borne in mind (Table 4-1). Thus, an appropriate way to assess the influence of air pollution on metals is to expose identical specimens at different locations and compare their annual rates of corrosion among the several locations. Since soiling is mainly due to the sedimentation of particulate matter from the air, experience has shown that this can be conveniently measured by exposing open-topped receptacles to the atmosphere for a month and weighing the settled solids. Human health seems to be related to day-to-day variation in pollutant level. It is accepted practice the world over to assess suspended particulate matter levels in the air by a 24-hour filter sample, which in the United States is acquired by a high-volume sampler, known to workers in the field as a *hi-vol*.

Because a filter sample includes particles both larger and smaller than those retained in the human respiratory system (see Chapter 7, Section III), other types of samplers are used which allow measurement of the size ranges of particles retained in the respiratory system. Some of these are called *dichotomous samplers* because they allow separate measurement of the respirable and nonrespirable fractions of the total. *Size-selective samplers* rely on impactors, miniature cyclones, and other means. The United States has selected the size fraction below an aerodynamic diameter of 10 μm (PM_{10}) for compliance with the air quality standard for airborne particulate matter.

TABLE 4-1

Air Quality Measurement

Measure of averaging time	Cyclic factor measured	Measurement method with same averaging time	Effect with same averaging time
Year	Annual trend	Metal specimen	Corrosion
Month	Seasonal cycle	Dustfall	Soiling
Day	Weekly cycle	Hi-vol	Human health
Hour	Diurnal cycle	Sequential sampler	Vegetation damage
Minute	Turbulence	Continuous instrument	Irritation (odor)

Because an agricultural crop can be irreparably damaged by an excursion of the level of several gaseous pollutants lasting just a few hours, recording such an excursion requires a measuring procedure that will give hourly data. The least expensive device capable of doing this is the *sequential sampler*, which will allows a sequence of 1- or 2-hr samples day after day for as long as the bubblers in the sampler are routinely serviced and analyzed. As already discussed, the hourly data from sequential samplers can be combined to yield daily, monthly, and annual data.

None of the foregoing methods will tell the frequency or duration of exposure of any receptor to irritant or odorous gases when each such exposure may exceed the irritation or odor response threshold for only minutes or seconds. The only way that such an exposure can be measured instrumentally is by an essentially continuous monitoring instrument, the record from which will yield not only this kind of information but also all the information required to assess hourly, daily, monthly, and annual phenomena. Continuous monitoring techniques may be used at a particular location or involve remote sensing techniques.

V. AIR QUALITY LEVELS

A. Levels

Air quality levels vary between concentrations so low that they are less than the minimum detectable values of the instruments we use to measure them and maximum levels that are the highest concentrations ever measured. Table 4-2 gives data from monitoring sites reporting concentrations approaching maximum values for the four principal gaseous pollutants. Figure 4-7 gives national data (1982–1991) for CO in the United States. The mean chemical composition and atmospheric concentration of suspended particulate matter (total, coarse, and fine) measured in the United States in 1980 are shown in Table 4-3. The percentages do not add up to 100% because they exclude the oxygen (except for the nitrate and sulfate components), nitrogen (except for the nitrate component), hydrogen, and other components of the compounds of the listed elements in the form in which they actually exist in the atmosphere; for example, the most common form of particulate sulfur and sulfate in the atmosphere is $(NH_4)_2(SO_4)$. The table indicates that about 30% of the mass of particulate matter is in the fine fraction (<2.5 μm), 30% is in the coarse fraction ($>2.5 < 15$ μm), and 40% is coarser still—between 15 and ca. 50 μm.

Because there has been no recent analysis of the relationship between the concentration of total suspended particulate matter in the air of cities with populations of different sizes, we are forced to use data for the decade

TABLE 4-2

Air Pollutant Concentrations at United States Sites, 1980

			Maximum average concentrations in ppb, for different averaging times			
Gaseous pollutant	Monitoring site	% days[a]	1 hr	1 day	1 month[b]	1 year
Carbon monoxide	E. 45th St., New York, NY	95	30,896	11,713	6,014	5,217
Sulfur dioxide	Miami, AZ	99	2,537	228	27	13
Nitrogen dioxide	West St. and Capitol Ave., Hartford, CT	95	124	78	36	29
Ozone	N. Main St., Los Angeles, CA	97	290	72	38	23

[a] Percentage of days per year for which data were available for analysis.
[b] Four Weeks.
Source: U.S. Environmental Protection Agency, Research Triangle Park, NC.

1957–1967 to show this relationship (Table 4-4). The general rule that the larger the population base, the dirtier the air will be, still exists. The air of these cities can be expected to be cleaner in the 1990s and beyond than it was in the 1950s and 1960s because of the tremendous efforts made to clean up the air in the intervening decades.

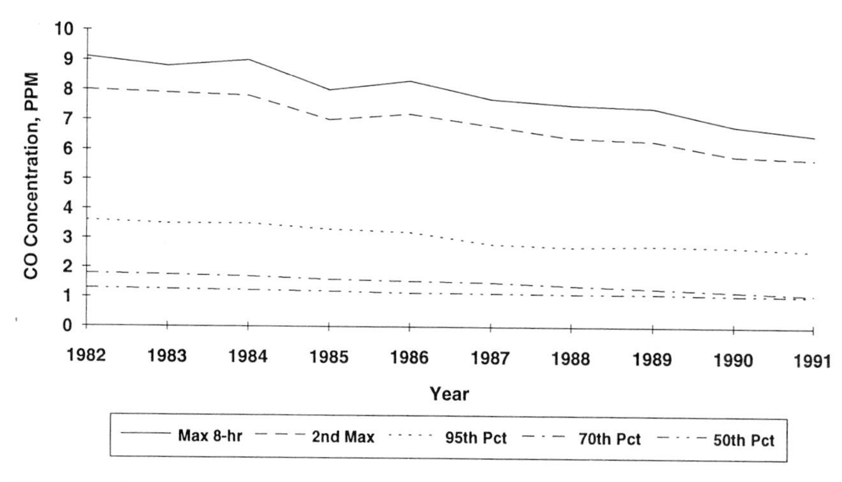

Fig. 4-7. Trend in carbon monoxide air quality indicators. Source: U.S. Environmental Protection Agency, 1992.

TABLE 4-3

Mean Chemical Composition and Atmospheric Concentrations of Suspended Particulate Matter Sampled by the United States Environmental Protection Agency's Inhalable Particle and National Air Surveillance Networks—$\mu g/m^3$ and Percentage of Total Mass Sampled, 1980

Type of sample:	Urban						Rural	
Number of samples	745				2255[a]		133[b]	
Particle size	Coarse, 15–2.5 μm		Fine, less than 2.5 μm		All less than ca. 50 μm			
	Mean value of 745 values	%	Mean value of 745 values	%	Mean value of 2255 values	%	Mean value of 133 values	%
All (total mass)	21.655	100.00	22.680	100.00	74.990	100.00	36.504[a]	100.00
Aluminum	1.797	8.30	0.353	1.56	—	—	—	—
Antimony	0.051	0.24	0.050	0.22	—	—	—	—
Arsenic	0.003	0.01	0.004	0.02	0.005[c]	0.01	0.003[d]	0.01
Barium	0.060	0.28	0.060	0.26	0.273	0.36	0.281	0.77
Beryllium[c]	—	—	—	—	(0.095)	—	(0.084)	—
Bromine	0.019	0.09	0.077	0.34	—	—	—	—
Cadmium	0.006	0.03	0.007	0.03	0.002	0.01	0.001	0.01
Calcium	1.503	6.94	0.340	1.50	—	—	—	—
Chlorine	0.440	2.03	0.155	0.68	—	—	—	—
Chromium	0.008	0.04	0.006	0.03	0.013[c]	0.02	0.015[d]	0.05
Cobalt	—	—	—	—	0.001[c]	0.01	0.001[d]	0.01

Copper	0.019	0.09	0.026	0.12	0.143	0.19	0.136	0.37
Iron	0.743	3.43	0.205	0.90	0.923	1.23	0.254	0.70
Lead	0.083	0.38	0.314	1.38	0.353	0.47	0.066	0.18
Manganese	0.021	0.10	0.013	0.06	0.031	0.04	0.008	0.02
Mercury	0.003	0.01	0.003	0.01	—	—	—	—
Molybdenum	—	—	—	—	0.002	0.01	0.001	0.01
Nickel	0.004	0.02	0.007	0.03	0.007	0.01	0.002	0.01
Phosphorus	0.056	0.26	0.021	0.09	—	—	—	—
Potassium	0.222	1.03	0.156	0.69	—	—	—	—
Selenium	0.001	0.01	0.002	0.01	—	—	—	—
Silicon	2.561	11.83	0.360	1.59	—	—	—	—
Strontium	0.246	0.21	0.051	0.22	—	—	—	—
Sulfur	0.339	1.56	2.056	9.07	—	—	—	—
Tin	0.006	0.03	0.006	0.03	—	—	—	—
Titanium	0.042	0.19	0.015	0.07	—	—	—	—
Vanadium	0.008	0.04	0.010	0.04	0.015	0.02	0.004	0.01
Zinc	0.038	0.18	0.067	0.30	0.147	0.20	0.114	0.31
Nitrate	0.699	3.23	1.071	4.72	4.647	6.20	2.341	6.41
Sulfate	0.706	3.26	5.30	23.37	10.811	14.42	8.675	23.77
Sum of percentages	—	43.82[f]	—	47.34[f]	—	23.20	—	32.64

[a] Except for arsenic, chromium, and cobalt where the number of samples was 1245.
[b] Except for arsenic, chromium, and cobalt where the number of samples was 30.
[c] Except for arsenic, chromium, and cobalt where the mean total mass was 76.647 $\mu g/m^3$.
[d] Except for arsenic, chromium, and cobalt where the mean total mass was 30.367 $\mu g/m^3$.
[e] Nanograms/m^3.
[f] Sulfur is counted twice—as sulfur and as sulfate. Some of this sulfur exists as sulfides, sulfites, and forms other than sulfate.

TABLE 4-4

Distribution of Cities by Population Class and Particulate Matter Concentration, 1957–1967

Population class	Average particulate matter concentration, $\mu m/m^3$										
	Less than 40	40–59	60–79	80–99	100–119	120–139	140–159	160–179	180–199	More than 200	Total
Over 3 million	—	—	—	—	—	—	1	—	1	—	2
1—3 million	—	—	—	—	—	—	2	1	—	—	3
0.7–1 million	—	—	1	—	2	—	4	—	—	—	7
400–700,000	—	—	—	4	5	6	1	1	1	—	18
100–400,000	—	3	7	30	24	17	12	3	2	1	99
50–100,000	—	2	20	28	16	12	6	5	1	3	93
25–50,000	—	5	24	12	12	10	2	1	2	3	71
10–25,000	—	7	18	19	9	5	2	3	1	—	64
Under 10,000	1	5	7	15	11	2	1	2	—	—	44
Total	1	22	77	108	79	52	31	16	8	7	401

Source: U.S. Environmental Protection Agency, "Air Quality Data from 1967" (Rev. 1971), Office of Air Programs, Pub. APTD 0471, Research Triangle Park, NC, 1971.

B. Display of Air Quality Data

A very useful format in which to display air quality data for analysis is that of Fig. 4-8, which has as its abscissa averaging time expressed in two different time units and, as its ordinate, concentration of the pollutant at the receptor. This type of chart is called an *arrowhead chart* and includes enough information to characterize fully the variability of concentration at the receptor.

To understand the meaning of the information given, let us concentrate on the data for 1-hr averaging time. In the course of a year there will be 8760 such values, one for each hour. If all 8760 are arrayed in decreasing value, there will be one maximum value and one minimum value. (For some pollutants the minimum value is indefinite if it is below the minimum detectable value of the analytical method or instrument employed.) In this array, the value 2628 from the maximum will be the value for which 30% of all values are greater and 70% are lower. Similarly, the value 876 from the maximum will be the one for which 10% of all values are greater and 90% are lower. The 1% value will be between the 87th and 88th values from the maximum, and the 0.1% value will lie between the 8th and 9th values in the array.

The 50% value, which is called the *median value*, is not necessarily the same as the average value, which is also called the *arithmetic mean value*. The arithmetic average value is obtained by adding all 8760 values and then dividing the total by 8760. The arithmetic average value obtained for other averaging times, e.g., by adding all 365 24-hr values and dividing

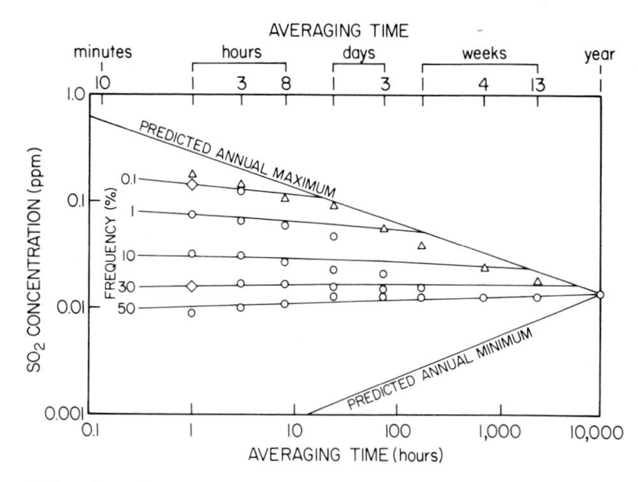

Fig. 4-8. Sulfur dioxide concentration versus averaging time and frequency for 1980 at U.S. National Aerometric Data Bank (NADB) Site 264280007 HO1, 8227 S. Broadway, St. Louis, Missouri. Source: Chart courtesy of Dr. Ralph Larsen, U.S. Environmental Protection Agency, Research Triangle Park, NC; see also Fig. 19-13.

the total by 365, will be the same and will equal the annual arithmetic average value. The median value will equal the arithmetic average value only if the distribution of all values allows this to occur. For example:

1 2 3 4 5 6 7 8 9 10 Median = 5.5 Mean = 5.5

1 3 3 3 5 6 6 6 9 13 Median = 5.5 Mean = 5.5

If air quality data at a receptor for any one averaging time are lognormally distributed, these data will plot as a straight line on log probability graph paper (Fig. 4-9) which bears a note $S_g = 2.35$. S_g is the standard geometric deviation about the geometric mean (the geometric mean is the Nth root of the product of the n values of the individual measurements).

$$S_g = \exp \sum_{i=1}^{n} \left[\frac{(\ln X_i - \overline{\ln X_i})^2}{n-1} \right]^{\frac{1}{2}}$$

where $X_i = X_1, X_2, \ldots X_n$ are the individual measurements and n is the number of measurements.

Further discussion of the significance of mean values and standard deviations can be found in Chapters 14 and 28 and in any textbook on statistics.

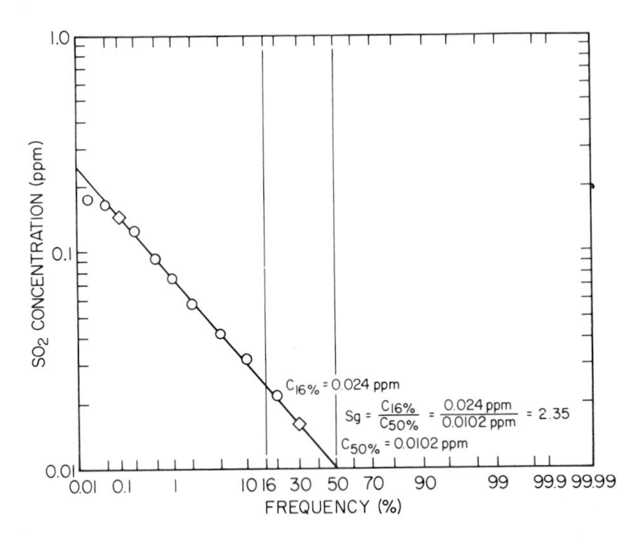

Fig. 4-9. Frequency of 1-hr average sulfur dioxide concentrations equal to or greater than stated values during 1980 at U.S. National Aerometric Data Bank (NADB) Site 264280007 HO1, 8227 S. Broadway, St. Louis. Missouri. Source: Chart courtesy of Dr. Ralph Larsen, U.S. Environmental Protection Agency, Research Triangle Park, NC.

C. Adverse Responses to Air Quality Levels

The objective of air pollution control is to prevent adverse responses by all receptor categories exposed to the atmosphere—human, animal, vegetable, and material. These adverse responses have characteristic response times—short-term (i.e., seconds or minutes), intermediate-term (i.e., hours or days), and long-term (i.e., months or years) (Table 4-5). For there to be no adverse responses, the pollutant concentration in the air must be lower than the concentration level at which these responses occur. Figure 4-9 illustrates the relationship between these concentration levels. This figure displays response curves, which remain on the concentration duration axes because they are characteristic of the receptors, not of the actual air quality to which the receptors are exposed. The odor response curve, e.g., to hydrogen sulfide, shows that a single inhalation requiring approximately 1 sec can establish the presence of the odor but that, due to odor fatigue, the ability to continue to recognize that odor can be lost in a matter of minutes. Nasopharyngeal and eye irritation, e.g., by ozone, is similarly subject to acclimatization due to tear and mucus production. The three visibility lines correlate with the concentration of suspended particulate matter in the air. Attack of metal, painted surfaces, or nylon hose is shown by a line starting at 1 sec and terminating in a matter of minutes (when the acidity of the droplet is neutralized by the material attacked).

TABLE 4-5

Examples of Receptor Category Characteristic Response Times

	Characteristic response times		
Receptor category	Short-term (seconds—minutes)	Intermediate-term (hours–days)	Long-term (months–years)
Human	Odor, visibility, nasopharyngeal and eye irritation	Acute respiratory disease	Chronic respiratory disease, lung cancer
Animal, vegetation	Field crop loss and ornamental plant damage	Field crop loss and ornamental plant damage	Fluorosis of livestock, decreased fruit and forest yield
Material	Acid droplet pitting, nylon hose destruction	Rubber cracking, silver tarnishing, paint blackening	Corrosion, soiling, materials deterioration

TABLE 4-6

Comparison of Pollutant Standard Index (PSI) Values, Pollutant Levels, and General Health Effects

	Pollutant level						Health	
PSI value	PM_{10} (24-hr) $\mu g/m^3$	SO_2 (24-hr) $\mu g/m^3$	CO (8-hr) mg/m^3	O_3 (1-hr) $\mu g/m^3$	NO_2 (1-hr) $\mu g/m^3$	Descriptor	Effects	Warning
400 and above	500 and above	2100 and above	46.0 and above	1000 and above	3000 and above	Hazardous	Premature death of ill and elderly. Healthy people will experience adverse symptoms that affect their normal activity.	All persons should remain indoors, keeping windows and doors closed. All persons should minimize physical exertion and avoid traffic.
300–399	420–500	1600–2099	34.0–45.9	800–1000	2260–2999	Hazardous	Premature onset of certain diseases in addition to significant aggravation of symptoms and decreased exercise tolerance in healthy persons.	Elderly and persons with existing diseases should stay indoors and avoid physical exertion. General population should avoid outdoor activity.

200–299	380–420	800–1599	17.0–33.9	400–800	1130–2259	Very unhealthy	Significant aggravation of symptoms and decreased exercise tolerance in persons with heart or lung disease, with widespread symptoms in the healthy population.	Elderly and persons with existing heart or lung disease should stay indoors and reduce physical activity.
100–199	150–380	365–799	10.0–16.9	235–400	NR	Unhealthy	Mild aggravation of symptoms in susceptible persons, with irritation symptoms in the healthy population.	Persons with existing heart or respiratory ailments should reduce physical exertion and outdoor activity.
50–99	50[a]–150	80[a]–364	5.0–9.9	120–235	NR	Moderate		
0–49	0–50	0–79	0–4.9	0–119	NR	Good		

NR = No index values reported at concentration levels below those specified by "alert level" criteria (Table 5-1).
[a] Annual primary National Ambient Air Quality Standard.
Source: 40 CFR, § 58, App. G, 1992.

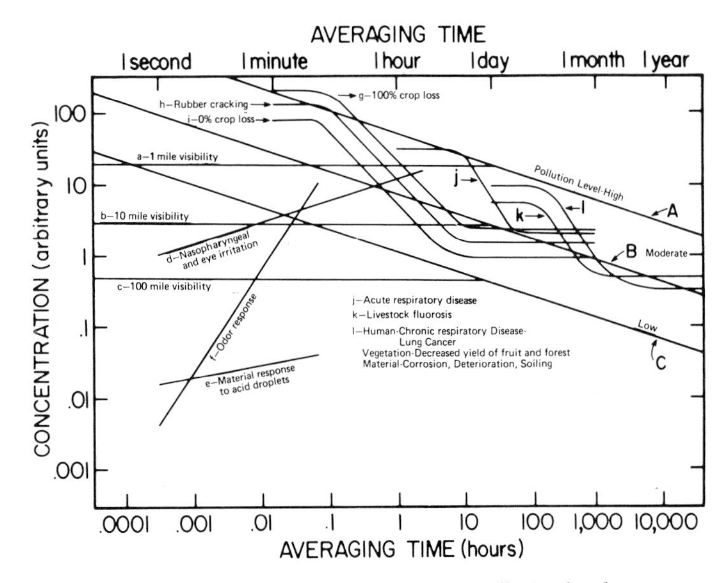

Fig. 4-10. Adverse responses to various pollution levels.

Vegetation damage can be measured biologically or socioeconomically. Using the latter measure, there is a 0% loss when there is no loss of the sale value of the crops or ornamental plants but a 100% loss if the crop is damaged to the extent that it cannot be sold. These responses are related to dose, i.e., concentration times duration of exposure, as shown by the percent loss curves on the chart. A number of manifestations of material damage, e.g., rubber cracking by ozone, require an exposure duration long enough for the adverse effects to be significant economically. That is, attack for just a few seconds or minutes will not affect the utility of the material for its intended use, but attack for a number of days will.

The biological response line for acute respiratory disease is a dose–response curve, which for a constant concentration becomes a duration–response curve. The shape of such a curve reflects the ability of the human body to cope with short-term, ambient concentration respiratory exposures and the overwhelming of the body's defenses by continued exposure.

Fluorosis of livestock is not induced until there has been a long enough period of deposition of a high enough ambient concentration of fluoride to increase the level of fluoride in the forage. Since the forage is either eaten by livestock or cut for hay at least once during the growing season, the duration of deposition ends after the growing season. The greater the duration of the season, the greater the time for deposition, hence the shape of the line labeled "fluorosis." Long-term vegetation responses—decreased yield of fruit and forest—and long-term material responses—corrosion, soiling, and material deterioration—are shown on the chart as having essentially the same response characteristics as human chronic respiratory disease and lung cancer.

The relationship of these response curves to ambient air quality is shown by lines A, B, and C, which represent the maximum or any other chose percentile line from a display such as Fig. 4-10, which shows actual air quality. Where the air quality is poor (line A), essentially all the adverse effects displayed will occur. Where the air quality is good (line C), most of the intermediate and long-term adverse effects displayed will not occur. Where the air quality is between good and poor, some of the intermediate and long-term adverse effects will occur, but in an attenuated form compared with those of poor air quality. These concepts will be referred to later in this text when air quality standards are discussed.

D. Air Quality Indexes

Air quality indexes have been devised for categorizing the air quality measurements of several individual pollutants by one composite number. The index used by the U.S. Environmental Protection Agency is called the *Pollutant Standards Index* (PSI) (Table 4-6).

REFERENCE

1. Samet, J. M., and Spengler, J. D. (eds), "Indoor Air Pollution: A Health Perspective." Johns Hopkins University Press, Baltimore, 1991.

SUGGESTED READING

Benedick, R. E., "Ozone Diplomacy: New Directions in Safeguarding the Planet." Harvard University Press, Cambridge, MA, 1991.

Brooks, B. O., and Davis, W. F., "Understanding Indoor Air Quality." CRC Press, Boca Raton, FL, 1992.

Council on Environmental Quality, Annual Reports. Washington, DC, 1984–1992.

QUESTIONS

1. How does the range of concentrations of air pollutants of concern to the industrial hygienist differ from that of concern to the air pollution specialist? To what extent are air sampling and analytical methods in factories and in the ambient air the same or different?

2. Using the data in Fig. 4-1, draw the variation in concentration over the 6-hr period as it would appear using sampling and analytical procedures which integrate the concentrations arriving at the receptor over 30 min and 2 hr, respectively.
3. Sketch the appearance of a stripchart record measuring one pollutant for a week to show the weekday–weekend cycle.
4. Draw a chart showing the most probable trend of the concentration of airborne particles of horse manure in the air of a large midwestern U.S. city from 1850 to 1950.
5. Describe an air quality measurement system used to assess the levels and types of aeroallergens.
6. Using the data of Fig. 4-8, determine the frequency with which a 30-min average SO_2 concentration of 0.2 ppm would probably have been exceeded at site 26428007 HO1 in St. Louis, Missouri in 1980.
7. Prepare a table describing air quality levels in your community, or in the nearest community to you that has such data available.
8. If you were to prepare a figure representing the relationship between air quality levels and the effects caused by these levels, what changes would you make in Fig. 4-10?
9. Discuss the extent and usefulness of dissemination by the media of the PSI values in the communities in which you have lived.

5

The Philosophy of Air Pollution Control

I. STRATEGY AND TACTICS—THE AIR POLLUTION SYSTEM

Since primary pollutants may have the dual role of causing adverse effects in their original unreacted form and of reacting chemically to form secondary pollutants, air pollution control consists mainly of reducing the emission of primary pollutants to the atmosphere. Air pollution control has two major aspects—strategic and tactical. The former is the long-term reduction of pollution levels at all scales of the problem from local to global. This aspect is called *strategic* in that long-term strategies must be developed. Goals can be set for air quality improvement 5, 10, or 15 yr ahead and plans made to achieve these improvements. There can be a regional strategy to effect planned reductions at the urban and local scales; a state or provincial strategy to achieve reductions at the state, provincial, urban, and local scales; and a national strategy to achieve them at national and lesser scales. The continental and global scales require an international strategy for which an effective instrumentality is being developed.

The other major aspect of air pollution reduction is the control of short-term episodes on the urban scale. This aspect is called *tactical* because, prior to an episode, a scenario of tactical maneuvers must be developed for application on very short notice to prevent an impending episode from becoming a disaster. Since an episode usually varies from a minimum of about 36 hr to a maximum of 3 or 4 days, temporary controls on emissions much more severe than are called for by the long-term strategic control scenario must be implemented rapidly and maintained for the duration of the episode. After the weather conditions that gave rise to the episode have passed, these temporary episode controls can be relaxed and controls can revert to those required for long-term strategic control.

The mechanisms by which a jurisdiction develops its air pollution control strategies and episode control tactics are outlined in Fig. 5-1. Most of the boxes in the figure have already been discussed—sources, pollutant emitted, transport and diffusion, atmospheric chemistry, pollutant half-life, air quality, and air pollution effects. To complete an analysis of the elements of the air pollution system, it is necessary to explain the several boxes not yet discussed.

II. EPISODE CONTROL

The distinguishing feature of an air pollution episode is its persistence for several days, allowing continued buildup of pollution levels. Consider the situation of the air pollution control officer who is expected to decide when to use the stringent control restrictions required by the episode control tactics scenario (Fig. 5-2 and Table 5-1). If these restrictions are imposed and the episode does not mature, i.e., the weather improves and blows away the pollution without allowing it to accumulate for another 24 hours or more, the officer will have required for naught a very large expenditure by the community and a serious disruption of the community's normal activities. Also, part of the officer's credibility in the community will be destroyed. If this happens more than once, the officer will be accused of crying wolf, and when a real episode occurs the warnings will be unheeded. If, however, the reverse situation occurs—i.e., the restrictions are not invoked and an episode does occur—there can be illness or possibly deaths in the community that could have been averted.

In deciding whether or not to initiate episode emergency plans, the control officer cannot rely solely on measurements from air quality monitoring stations, because even if pollutant concentrations rise toward acute levels over the preceding hours, these readings give no information on whether they will rise or fall during the succeeding hours.

The only way to avert this dilemma is for the community to develop and utilize its capability of forecasting the advent and persistence of the stagnation conditions during which an episode occurs and its capability of computing pollution concentration buildup under stagnation conditions. The details of how these forecasts and computations are made are discussed

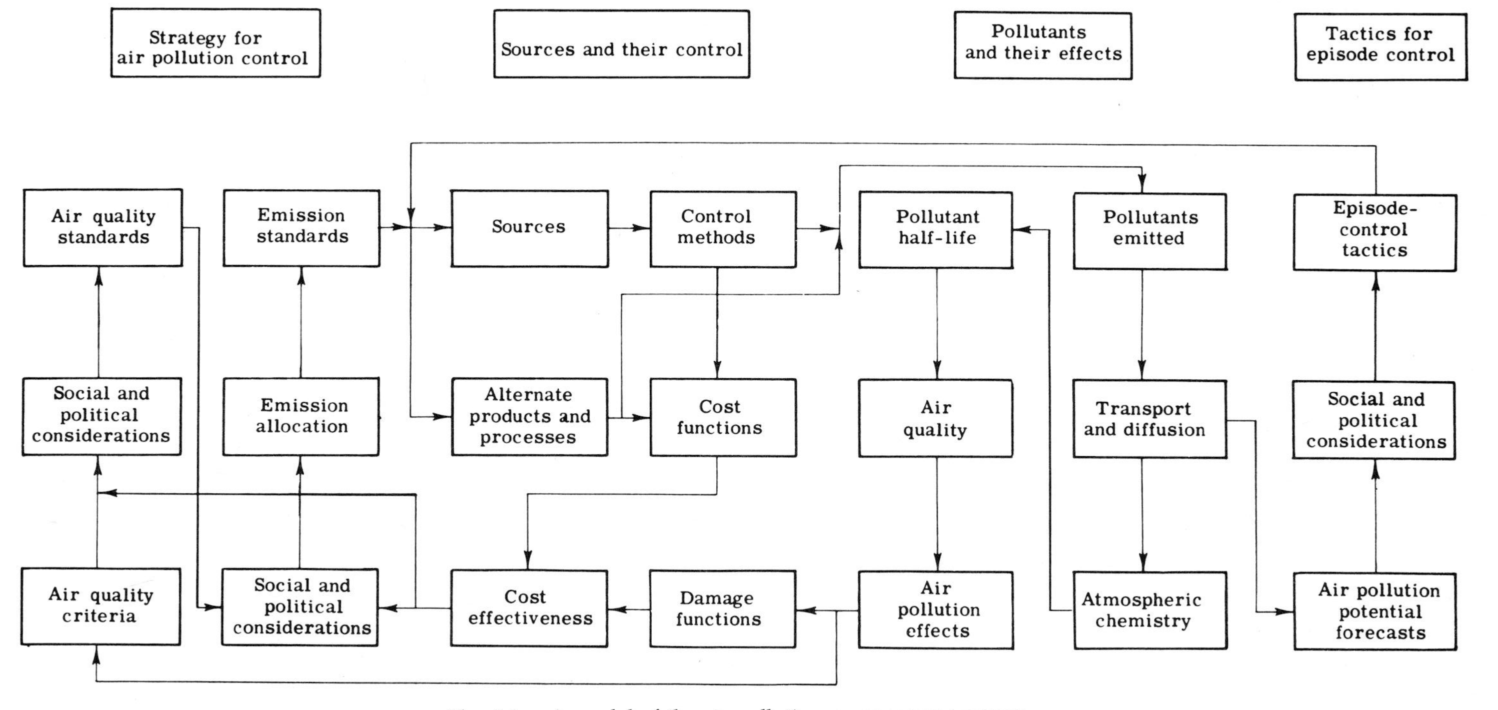

Fig. 5-1. A model of the air pollution management system.

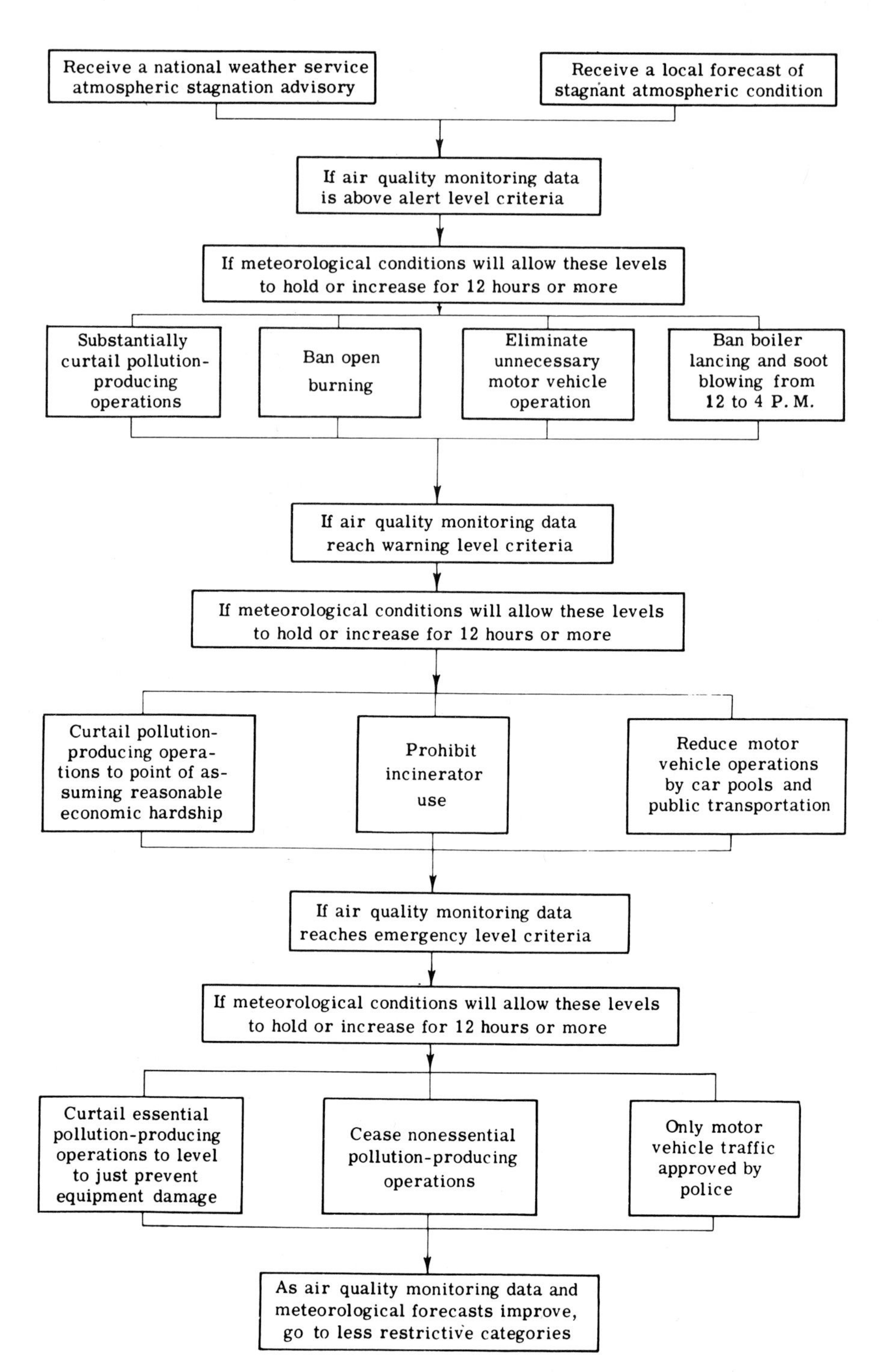

Fig. 5-2. Air pollution episode control scenario. *See Table 5-1.

TABLE 5-1

United States Alert, Warning, and Emergency Level Criteria[a, b, c]

Level	Pollutant	Criteria
Alert level criteria		
	SO_2	800 $\mu g/m^3$ (0.3 ppm), 24-hr average
	PM_{10}	350 $\mu g/m^3$, 24-hr average
	CO	17mg/m^3 (15 ppm), 8-hr average
	Ozone	400 $\mu g/m^3$ (0.2 ppm), 1-hr average
	NO_2	1130 $\mu g/m^3$ (0.6 ppm), 1-hr average; 282 $\mu g/m^3$ (0.15 ppm), 24-hr average
Warning level criteria		
	SO_2	1600 $\mu g/m^3$ (0.6 ppm), 24-hr average
	PM_{10}	420 $\mu g/m^3$, 24-hr average
	CO	34 mg/m^3 (30 ppm), 8-hr average
	Ozone	800 $\mu g/m^3$ (0.4 ppm), 1-hr average
	NO_2	2260 $\mu g/m^3$ (1.2 ppm), 1-hr average; 565 $\mu g/m^3$ (0.15 ppm), 24-hr average
Emergency level criteria		
	SO_2	2100 $\mu g/m^3$ (0.8 ppm), 24-hr average
	PM_{10}	500 $\mu g/m^3$, 24-hr average
	CO	46 mg/m^3 (15 ppm), 8-hr average
	Ozone	1000 $\mu g/m^3$ (0.1 ppm), 1-hr average
	NO_2	3000 $\mu g/m^3$ (0.6 ppm), 1-hr average; 750 $\mu g/m^3$ (0.15 ppm), 24-hr average

[a] 1992 Code of Federal Regulations, Title 40-Protection of Environment, Chapter 1—Environmental Protection Agency, Appendix L, Example Regulations for Prevention of Air Pollution Emergency Episodes, 1.1 Episode Criteria, pp. 841–842.

[b] There is no criterion for lead, due to the chronic nature of the health effects of concern.

[c] Note: Append to each entry: Meterological conditions are such that pollutant concentrations can be expected to remain at the above levels for 12 or more hours or increase, or in the case of ozone the situation is likely to reoccur within the next 24 hours unless control actions are taken.

in Chapter 19, but at this point the foregoing discussion should explain the reason for the box in Fig. 5-1 labeled Air Pollution Potential Forecasts. The connecting box marked Social and Political Considerations provides a place in the system for the public debates, hearings, and action processes necessary to decide, well in advance of an episode, what control tactics to use and when to call an end to the emergency. The public needs to be involved because alternatives have to be written into the scenario concerning where, when, and in what order to impose restrictions on sources. This should be done in advance and should be well publicized, because during the episode there is no time for public debate. In any systems analysis, the system must form a closed loop with feedback to keep the system under control. It will be noted that the system for tactical episode control is closed by the line connecting Episode-Control Tactics to Sources, which means that the episode control tactics are to limit sources severely during the episode. Since it takes hours before emergency plans can be put into effect and their impact on pollution levels felt, it is possible that by the time the community responds, the situation has disappeared. Experience has shown that the time for community response is slowed by the need to write orders to close down sources and to respond in court to requests for judicial relief from such orders. To circumvent the former type of delay, orders can be written in blank in advance. To circumvent the latter type of delay, the agency's legal counsel must move as rapidly as the counsel seeking such relief.

III. AIR QUALITY MANAGEMENT CONTROL STRATEGY

Now let us consider the system for long-range strategy for air pollution control. The elements in this system that have not yet been discussed include several listed in Fig. 5-1 under Sources and Their Control and all those listed under Strategy for Air Pollution Control. Control of sources is effected in several ways. We can (1) use devices to remove all or part of the pollutant from the gases discharged to the atmosphere, (2) change the raw materials used in the pollution-producing process, or (3) change the operation of the process so as to decrease Pollutants Emitted. These are Control Methods (Table 5-2). Such control methods have a cost associated with them and are the Cost Functions that appear in the system. There is always the option of seeking Alternate Products or Processes which will provide the same utility to the public but with less Pollutants Emitted. Such products and processes have their own Cost Functions.

Just as it costs money to control pollution, it also costs the public money not to control pollution. All the adverse Air Pollution Effects represent economic burdens on the public for which an attempt can be made to assign dollar values, i.e., the cost to the public of damage to vegetation, materials,

TABLE 5-2

Control Methods

I. Applicable to all emissions
  A. Decrease or eliminate production of emission
    1. Change specification of product
    2. Change design of product
    3. Change process temperature, pressure, or cycle
    4. Change specification of materials
    5. Change the product
  B. Confine the emissions
    1. Enclose the source of emissions
    2. Capture the emissions in an industrial exhaust system
    3. Prevent drafts
  C. Separate the contaminant from effluent gas stream
    1. Scrub with liquid

II. Applicable specifically to particulate matter emissions
  A. Decrease or eliminate particulate matter production
    1. Change to process that does not require blasting, blending, buffing, calcining, chipping, crushing, drilling, drying, grinding, milling, polishing, pulverizing, sanding, sawing, spraying, tumbling, etc.
    2. Change from solid to liquid or gaseous material
    3. Change from dry to wet solid material
    4. Change particle size of solid material
    5. Change to process that does not require particulate material
  B. Separate the contaminant from effluent gas stream
    1. Gravity separator
    2. Centrifugal separator
    3. Filter
    4. Electrostatic precipitator

III. Applicable specifically to gaseous emissions
  A. Decrease or eliminate gas or vapor production
    1. Change to process that does not require annealing, baking, boiling, burning, casting, coating, cooking, dehydrating, dipping, distilling, expelling, galvanizing, melting, pickling, plating, quenching, reducing, rendering, roasting, smelting, etc.
    2. Change from liquid or gaseous to solid material
    3. Change to process that does not require gaseous or liquid material
  B. Burn the contaminant to CO_2 and H_2O
    1. Incinerator
    2. Catalytic burner
  C. Adsorb the contaminant
    1. Activated carbon

structures, animals, the atmosphere, and human health. These costs are called Damage Functions. To the extent that there is knowledge of Cost Functions and Damage Functions, the Cost Effectiveness of control methods and strategies can be determined by their interrelationship. Cost Effectiveness is an estimate of how many dollars worth of damage can be averted per dollar expended for control. It gives information on how to economically optimize an attack on pollution, but it gives no information on the reduction in pollution required to achieve acceptable public health and well-being. However, when these goals can be achieved by different control alternatives, it behooves us to utilize the alternatives that show the greatest cost effectiveness.

To determine what pollution concentrations in air are compatible with acceptable public health and well-being, use is made of Air Quality Criteria, which are statements of the air pollution effects associated with various air quality levels. It is inconceivable that any jurisdiction would accept levels of pollution it recognizes as damaging to health. However, the question of what constitutes damage to health is judgmental and therefore debatable. The question of what damage to well-being is acceptable is even more judgmental and debatable. Because they are debatable, the same Social and Political Considerations come into the decision-making process as in the previously discussed case of arriving at episode control tactics. Cost Effectiveness is not a factor in the acceptability of damage to health, but it is a factor in determining acceptable damage to public well-being. Some jurisdictions may opt for a pollution level that allows some damage to vegetation, animals, materials, structures, and the atmosphere as long as they are assured that there will be no damage to their constituents' health. The concentration level the jurisdiction selects by this process is called an Air Quality Standard. This is the level the jurisdiction says it wishes to maintain.

Adoption of air quality standards by a jurisdiction produces no air pollution control. Control is produced by the limitation of emission from sources, which, in turn, is achieved by the adoption and enforcement of Emission Standards. However, before emission standards are adopted, the jurisdiction must make some social and political decisions on which of several philosophies of emission standard development are to be utilized and which of the several responsible groups in the jurisdiction should bear the brunt of the control effort—its homeowners, landlords, industries, or institutions. This latter type of decision making is called Emission Allocation. It will be seen in Fig. 5-1 that the system for strategic control is closed by the line connecting Emission Standards and Sources, which means that long-range pollution control strategy consists of applying emission limitation to sources.

IV. ALTERNATIVE CONTROL STRATEGIES

There are several different strategies for air pollution control. The strategy just discussed and shown in Fig. 5-1 is called the *air quality management* strategy. It is distinguished from other strategies by its primary reliance on the development and promulgation of ambient air quality standards.

This is the strategy in use in the United States. The second principal strategy is the *emission standard* strategy, also known as the *best practicable means of control* approach. In this strategy, neither air quality criteria nor ambient air quality standards are developed and promulgated. Either an emission standard is developed and promulgated or an emission limit on sources is determined on a case-by-case basis, representing the best practicable means for controlling emissions from those sources. This is the strategy in use in Great Britain. A third strategy controls pollution by adopting financial incentives (Table 5-3). This is usually but not necessarily in addition to the promulgation of air quality standards. Among the countries that have adopted tax, fee, or fine schedules on a national basis are Czechoslovakia, Hungary, Japan, The Netherlands, and Norway. There is additional discussion of this strategy in Chapter 27, Section III. A fourth strategy seeks to

TABLE 5-3

Financial Incentives to Supplement or Replace Regulation

Taxes	
Sales taxes	On fuel, fuel additives, ingredients in fuel, pollution-producing equipment
Ultimate disposal taxes	On automobiles or other objects requiring ultimate disposal
Land use taxes	For pollution-producing activities
Tax remission	
Corporate income tax	For investment in or operation of pollution control equipment; accelerated write-off of pollution control equipment
Property taxes	For pollution control equipment
Fines, effluent charges, and fees	
Fines	For violation of regulations
Effluent charges	For permission to emit excessive quantities of pollution; paid after emission
Fees	For permission to emit excessive quantities of pollution, paid before emission
Subsidies	
Direct	Governmental production (e.g., nuclear fuel)
Grants-in-aid	For pollution control installations
Indirect (low-interest bonds and loans)	For pollution-control process or equipment development
Import restraints	
Duties and quotas	On materials, fuels, and pollution control-producing apparatus
Domestic production restraints	
Quotas, land and offshore use restraints	On material and fuels

Source: From Stern, A. C., Heath, M. S., and Hufschmidt, M. M., A critical review of the role of fiscal policies and taxation in air pollution control, Proceedings of the Third International Clean Air Congress, Verein Deutscher Ingeniuere, Dusseldorf, pp. D-10–12, 1974.

maximize cost effectiveness and is called the *cost–benefit* strategy. These strategies may result in lower emissions from existing processes or promote process modifications which reduce pollution generation.

V. ECONOMIC CONSIDERATIONS

The situation with regard to economic considerations has been so well stated in the First Report of the British Royal Commission on Environmental Pollution (1) that this section contains an extensive quotation from that report.

> Our survey of the activities of the Government, industry and voluntary bodies in the control of pollution discloses several issues which need further enquiry. The first and most difficult of these is how to balance the considerations which determine the levels of public and private expenditure on pollution control. Some forms of pollution bear more heavily on society than others; some forms are cheaper than others to control; and the public are more willing to pay for some forms of pollution control than for others. There are also short and long-term considerations: in the short-term the incidence of pollution control on individual industries or categories of labor may be heavy; but . . . what may appear to be the cheapest policy in the short-term may prove in the long-term to have been a false economy.
>
> While the broad outlines of a general policy for protecting the environment are not difficult to discern, the economic information needed to make a proper assessment of the considerations referred to in the preceding paragraph . . . seems to us to be seriously deficient. This is in striking contrast with the position regarding the scientific and technical data where, as our survey has shown, a considerable amount of information is already available and various bodies are trying to fill in the main gaps. The scientific and technical information is invaluable, and in many cases may be adequate for reaching satisfactory decisions, but much of it could be wasted if it were not supported by some economic indication of priorities and of the best means of dealing with specific kinds of pollution.
>
> So, where possible, we need an economic framework to aid decision making about pollution, which would match the scientific and technical framework we already have. This economic framework should include estimates of the way in which the costs of pollution, including disamenity costs, vary with levels of pollution; the extent to which different elements contribute to the costs; how variations in production and consumption affect the costs; and what it would cost to abate pollution in different ways and by different amounts. There may well be cases where most of the costs and benefits of abatement can be assessed in terms of money. Many of the estimates are likely to be speculative, but this is no reason for not making a start. There are other cases where most of the costs and benefits cannot be given a monetary value. In these cases decisions about pollution abatement must not await the results of a full economic calculation: they will have to be based largely on subjective judgments anyway. Even so, these subjective judgments should be supported by as much quantitative information as possible, just as decisions about health and education are supported by extensive statistical data. Further, even if decisions to abate pollution are not based on rigorous economic criteria, it is still desirable to find the most economic way of achieving the abatement.

As air pollution management moves forward, economics has a major role in reducing pollution. Multimedia considerations are forcing a blend of traditional emission reduction approaches and innovative methods for waste minimization. These efforts are directed toward full cost accounting of the life cycle of products and residuals from the manufacturing, use, and ultimate disposal of materials.

REFERENCE

1. Royal Commission on Environmental Pollution. First Report, Command 4585, H.M. Stationery Office, London, 1971.

SUGGESTED READING

Ashby, E., and Anderson, M., "The Politics of Clean Air." Oxford Univ. Press (Clarendon), New York, 1981.

Cohen, R. E., "Washington at Work: Back Rooms and Clean Air." Macmillian, New York, 1992.

Crandall, R. W., "Controlling Industrial Pollution: The Economics and Politics of Clean Air." Brookings Institution, Washington, DC, 1983.

Hoberg, G., "Pluralism by Design: Environmental Policy and the American Regulatory State." Praeger, New York, 1992.

QUESTIONS

1. Explain why certain important long-range air pollution control strategies will not suffice for short-term episode control, and vice versa.
2. Develop an episode control scenario for a single large coal-fired steam electric generating station.
3. In Fig. 5-1, the words "Social and political considerations" appear several times. Discuss these considerations for the various contexts involved.
4. Discuss the relative importance of air quality criteria and cost effectiveness in the setting of air quality standards.
5. The quotation in Section V contains the words "what may appear to be the cheapest policy in the short-term may prove in the long-term to have been a false economy." Give some examples of this.
6. Draw a simplified version of Fig. 5-1 with fewer than 10 boxes.
7. How would one go about developing an air pollution damage function for human health?
8. The Organization for Economic Cooperation and Development (OECD) has been a proponent of the "polluter pays" principle. What is the principle and how can it be implemented?
9. Study Table 5-2 and determine whether there are any control methods that you believe should be added or deleted.

6

Sources of Air Pollution

I. GENERAL

The sources of air pollution are nearly as numerous as the grains of sand. In fact, the grains of sand themselves are air pollutants when the wind entrains them and they become airborne. We would class them as a natural air pollutant, which implies that such pollution has always been with us. Natural sources of air pollution are defined as sources not caused by people in their activities.

Consider the case in which someone has removed the ground cover and left a layer of exposed soil. Later the wind picks up some of this soil and transports it a considerable distance to deposit it at another point, where it affects other people. Would this be classed as a natural pollutant or an anthropogenic pollutant? We might call it natural pollution if the time span between when the ground cover was removed and when the material became airborne was long enough. How long would be long enough? The answers to such questions are not as simple as they first appear. This is one of the reasons why pollution problems require careful study and analysis before a decision to control them at a certain level can be made.

A. Natural Sources

An erupting volcano emits particulate matter. Pollutant gases such as SO_2, H_2S, and methane are also emitted. The emission from an eruption may be of such magnitude as to harm the environment for a considerable distance from the volcanic source. Clouds of volcanic particulate matter and gases have remained airborne for very long periods of time. The eruption of Mt. St. Helens in the state of Washington is a classic example of volcanic activity. Figure 6-1 is a photograph of Mt. St. Helens during the destructive eruption of May 18, 1980.

Accidental fires in forests and on the prairies are usually classified as natural sources even though they may have been originally ignited by human activities. In many cases foresters intentionally set fires in forest lands to burn off the residue, but lightning setting off a fire in a large section of forest land could only be classed as natural. A large uncontrolled forest fire, as shown in Fig. 6-2, is a frightening thing to behold. Such a fire emits large quantities of pollutants in the form of smoke, unburned hydrocarbons, carbon monoxide, carbon dioxide, oxides of nitrogen, and ash. Forest fires in the Pacific Northwest of the United States have been observed to emit a plume which caused reduction in visibility and sunlight as far away as 350 km from the actual fire.

Dust storms that entrain large amounts of particulate matter are a common natural source of air pollution in many parts of the world. Even a relatively small dust storm can result in suspended particulate matter read-

Fig. 6-1. Mt. St. Helens during the eruption of May 1980. Source: Photo by C. Rosenfeld, Oregon Air National Guard.

Fig. 6-2. Uncontrolled forest fire. Source: Information and Education Section, Oregon Department of Forestry.

ings one or two orders of magnitude above ambient air quality standards. Visibility reduction during major dust storms is frequently the cause of severe highway accidents and can even affect air travel. The particulate matter transferred by dust storms from the desert to urban areas causes problems to householders, industry, and automobiles. The materials removed by the air cleaner of an automobile are primarily natural pollutants such as road dust and similar entrained material.

The oceans of the world are an important natural source of pollutant material. The ocean is continually emitting aerosols to the atmosphere, in the form of salt particles, which are corrosive to metals and paints. The action of waves on rocks reduces them to sand, which may eventually become airborne. Even the shells washed up on the beach are eroded by wave and tidal action until they are reduced to such a small size that they too may become airborne.

An extensive source of natural pollutants is the plants and trees of the earth. Even though these green plants play a large part in the conversion of carbon dioxide to oxygen through photosynthesis, they are still the major source of hydrocarbons on the planet. The familiar blue haze over forested areas is nearly all from the atmospheric reactions of the volatile organics given off by the trees of the forest (1). Another air pollutant problem, which can be attributed to plant life, is the pollens which cause respiratory distress and allergic reactions in humans.

Other natural sources, such as alkaline and saltwater lakes, are usually quite local in their effect on the environment. Sulfurous gases from hot springs also fall into this category in that the odor is extremely strong when close to the source but disappears a few kilometers away.

B. Anthropogenic Sources

1. Industrial Sources

The reliance of modern people on industry to produce their needs has resulted in transfer of the pollution sources from the individual to industry. A soap factory will probably not emit as much pollution as did the sum total of all the home soap-cooking kettles it replaces, but the factory is a source that all soap consumers can point to and demand that it be cleaned up.

A great deal of industrial pollution comes from manufacturing products from raw materials—(1) iron from ore, (2) lumber from trees, (3) gasoline from crude oil, and (4) stone from quarries. Each of these manufacturing processes produces a product, along with several waste products which we term pollutants. Occasionally, part or all of the polluting material can be recovered and converted into a usable product.

Industrial pollution is also emitted by industries that convert products to other products—(1) automobile bodies from steel, (2) furniture from lumber, (3) paint from solids and solvents, and (4) asphaltic paving from rock and oil.

Industrial sources are stationary, and each emits relatively consistent qualities and quantities of pollutants. A paper mill, for example, will be in the same place tomorrow that it is today, emitting the same quantity of the same kinds of pollutants unless a major process change is made. Control of industrial sources can usually be accomplished by applying known technology. The most effective regulatory control is that which is applied uniformly within all segments of industries in a given region, e.g., "Emission from all asphalt plant dryers in this region shall not exceed 230 mg of particulate matter per standard dry cubic meter of air."

2. Utilities

The utilities in our modern society are so much a part of our lives that it is hard to imagine how we survived without them. An electric power plant generates electricity to heat and light our homes in addition to providing power for the television, refrigerator, and electric toothbrush. When our homes were heated with wood fires, home-made candles were used for light, there was no television, and food was stored in a cellar, the total of the air pollution generated by all the individual sources probably exceeded that of the modern generating stations supplying today's energy. It is easy for citizens to point out the utility as an air pollution source without connecting their own use of the power to the pollution from the utility. Figure 6-3 illustrates a modern electric power plant.

Utilities are in the business of converting energy from one form to another and transporting that energy. If a large steam generating plant, producing 2000 MW, burns a million kilograms per hour of 4% ash coal, it must somehow dispose of 40,000 kg of ash per hour. Some will be removed from the furnaces by the ash-handling systems, but some will go up the stack with the flue gases. If 50% of the ash enters the stack and the fly ash collection system is 99% efficient, 200 kg of ash per hour will be emitted to the atmosphere. For a typical generating plant, the gaseous emissions would include 341,000 kg of oxides of sulfur per day and 185,000 kg of oxides of nitrogen per day. If this is judged as excessive pollution, the management decision can be to (1) purchase lower-ash or lower-sulfur coal, (2) change the furnace so that more ash goes to the ash pit and less goes up the stack, or (3) install more efficient air pollution control equipment. In any case, the cost of operation will be increased and this increase will be passed on to the consumer.

Another type of utility that is a serious air pollution source is the one that handles the wastes of modern society. An overloaded, poorly designed,

Fig. 6-3. Coal-fired electric generating plant.

or poorly operated sewage treatment plant can cause an air pollution problem which will arouse citizens to demand immediate action. A burning dump is a sure source of public complaint, even though it may be explained to the same public that it is the "cheapest" way to dispose of their solid waste. The public has shown its willingness to ban burning dumps and pay the additional cost of adequate waste disposal facilities to have a pollution-free environment.

3. Personal Sources

Even though society has moved toward centralized industries and utilities, we still have many personal sources of air pollution for which we alone can answer—(1) automobiles, (2) home furnaces, (3) home fireplaces and stoves, (4) backyard barbecue grills, and (5) open burning of refuse and leaves. Figure 6-4 illustrates the personal emissions of a typical U.S. family.

The energy release and air pollution emissions from personal sources in the United States are greater than those from industry and utilities combined. In any major city in the United States, the mass of pollutants emitted

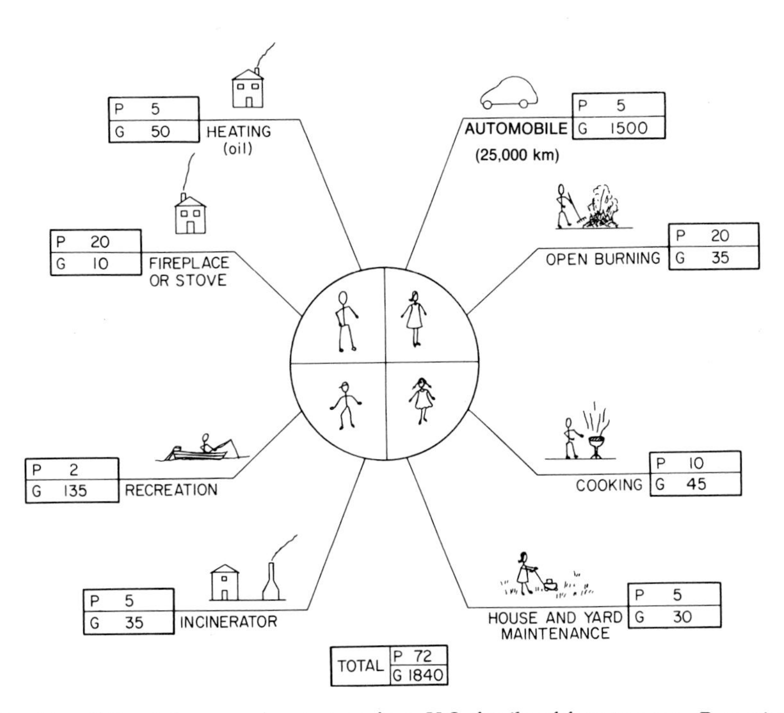

Fig. 6-4. Estimated personal emissions from U.S. family of four persons. P, particulate matter in kilograms per year; G, gases in kilograms per year.

by the vast numbers of private automobiles exceeds that from any other source.

Control of these personal sources of pollution takes the form of (1) regulation (fireplaces and stoves may be used only when atmospheric mixing is favorable), (2) change of lifestyle (sell the automobile and ride public transportation), (3) change from a more polluting to a less polluting source (convert the furnace to natural gas), or (4) change the form of pollution (instead of burning leaves, haul them to the city dump). Whatever method is used for control of pollution from personal sources, it will probably be difficult and unpopular to enforce. It is difficult to get citizens to believe that their new, highly advertised, shiny, unpaid-for automobiles are as serious a pollution problem as the smoking factory stack on the horizon. It is also a very ineffective argument to point out that the workers at that factory put more pollution into the air each day by driving their automobiles to and from work than does the factory with its visible plume of smoke.

II. COMBUSTION

Combustion is the most widely used, and yet one of the least understood, chemical reactions at our disposal. *Combustion* is defined as the rapid union of a substance with oxygen accompanied by the evolution of light and heat (2).

We use combustion primarily for heat by changing the potential chemical energy of the fuel to thermal energy. We do this in a fossil fuel-fired power plant, a home furnace, or an automobile engine. We also use combustion as a means of destruction for our unwanted materials. We reduce the volume of a solid waste by burning the combustibles in an incinerator. We subject combustible gases, with undesirable properties such as odors, to a high temperature in an afterburner system to convert them to less objectionable gases.

The simple combustion equations are very familiar:

$$C + O_2 \rightarrow CO_2 \tag{6-1}$$

$$2H_2 + O_2 \rightarrow 2H_2O \tag{6-2}$$

They produce the products carbon dioxide and water, which are odorless and invisible.

The problems with the combustion reaction occur because the process also produces many other products, most of which are termed *air pollutants*. These can be carbon monoxide, carbon dioxide, oxides of sulfur, oxides of nitrogen, smoke, fly ash, metals, metal oxides, metal salts, aldehydes, ketones, acids, polynuclear hydrocarbons, and many others. Only in the past few decades have combustion engineers become concerned about these relatively small quantities of materials emitted from the combustion process. An automotive engineer, for example, was not overly concerned about the 1% of carbon monoxide in the exhaust of the gasoline engine. By getting this 1% to burn to carbon dioxide inside the combustion chamber, the engineer could expect an increase in gasoline mileage of something less than one-half of 1%. This 1% of carbon monoxide, however, is 10,000 ppm by volume, and a number of such magnitude cannot be ignored by an engineer dealing with air pollution problems.

Combustion is extremely complicated but is generally considered to be a free radical chain reaction. Several reasons exist to support the free radical mechanism. (1) Simple calculations of the heats of disassociation and formation for the molecules involved do not agree with the experimental values obtained for heats of combustion. (2) A great variety of end products may be found in the exhaust from a combustion reaction. Many complicated organic molecules have been identified in the effluent from a system burning pure methane with pure oxygen. (3) Inhibitors, such as tetraethyl lead, can greatly change the rate of reaction (3).

When visualizing a combustion process, it is useful to think of it in terms of the three Ts: time, temperature, and turbulence. Time for combustion to occur is necessary. A combustion process that is just initiated, and suddenly has its reactants discharged to a chilled environment, will not go to completion and will emit excessive pollutants. A high enough temperature must exist for the combustion reaction to be initiated. Combustion is an exothermic reaction (it gives off heat), but it also requires energy to be initiated. This is illustrated in Fig. 6-5.

Turbulence is necessary to ensure that the reacting fuel and oxygen molecules in the combustion process are in intimate contact at the proper

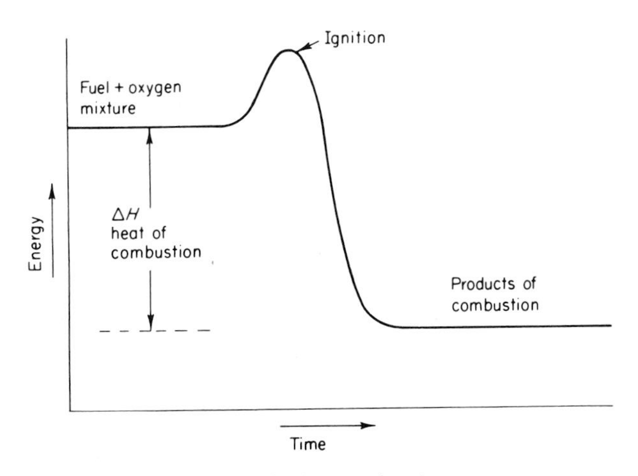

Fig. 6-5. Energies involved in combustion processes.

instant, when the temperature is high enough to cause the reaction to begin.

The physical state of the fuel for a combustion process dictates the type of system to be used for burning. A fuel may be composed of volatile material, fixed carbon, or both. The volatile material burns as a gas and exhibits a visible flame, whereas the fixed carbon burns without a visible flame in a solid form. If a fuel is in the gaseous state, such as natural gas, it is very reactive and can be fired with a simple burner.

If a fuel is in the liquid state, such as fuel oil, most of it must be vaporized to the gaseous state before combustion occurs. This vaporization can be accomplished by supplying heat from an outside source, but usually the liquid fuel is first atomized and then the finely divided fuel particles are sprayed into a hot combustion chamber to accomplish the gasification.

With a solid fuel, such as coal or wood, a series of steps are involved in combustion. These steps occur in a definite order, and the combustion device must be designed with these steps in mind. Figure 6-6 shows what happens to a typical solid fuel during the combustion process.

The cycle of operation of the combustion source is very important as far as emissions are concerned. A steady process, such as a large steam boiler, operates with a fairly uniform load and a continuous fuel flow. The effluent gases, along with any air pollutants, are discharged steadily and continually from the stack. An automobile engine, on the other hand, is a series of intermittent sources. The emissions from the automotive engine will be vastly different from those from the boiler in terms of both quantity and quality. A four-cylinder automotive engine operating at 2500 rpm has 5000 separate combustion processes started and completed each minute of its

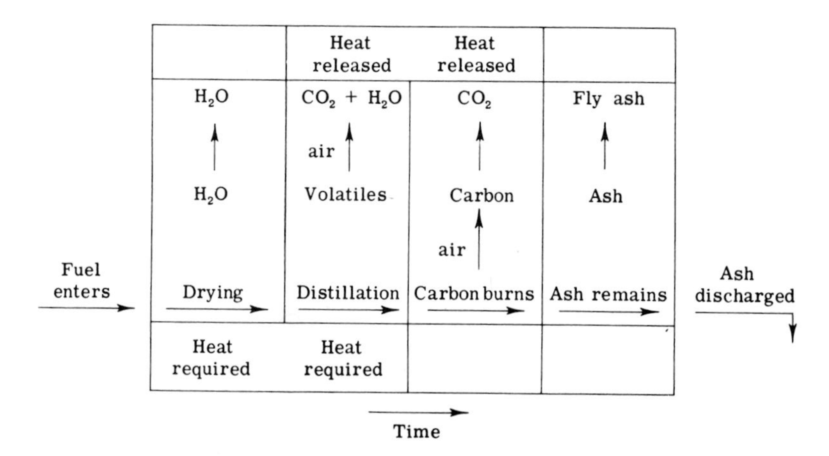

Fig. 6-6. Solid fuel combustion schematic.

operation. Each of these lasts about 1/100 of a second from beginning to end.

The emissions from combustion processes may be predicted to some extent if the variables of the processes are completely defined. Figure 6-7 indicates how the emissions from a combustion source would be expected to vary with the temperature of the reaction. No absolute values are shown, as these will vary greatly with fuel type, independent variables of the combustion process, etc.

A comparison of typical emissions from various common combustion sources may be seen in Table 6-1.

III. STATIONARY SOURCES

Emissions from industrial processes are varied and often complex (4). These emissions can be controlled by applying the best available technology. The emissions may vary slightly from one facility to another, using apparently similar equipment and processes, but in spite of this slight variation, similar control technology is usually applied (5). For example, a

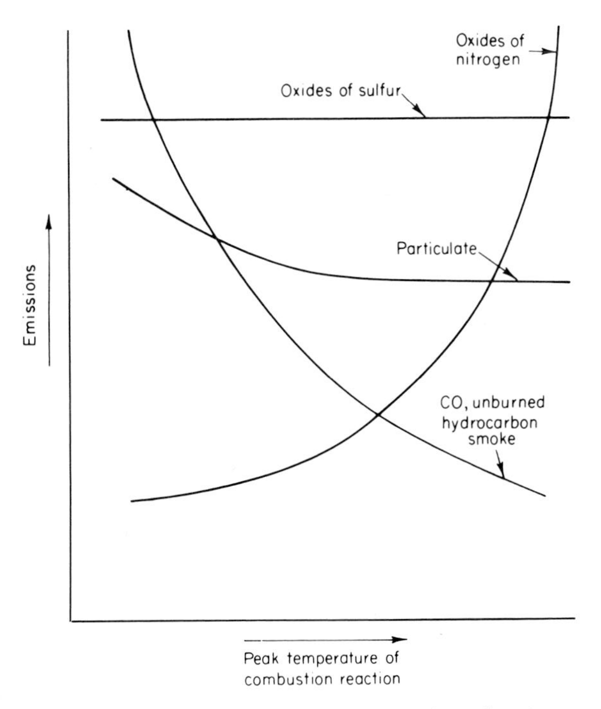

Fig. 6-7. Combustion emissions as a function of peak combustion temperatures.

TABLE 6-1

Comparison of Combustion Pollutants[a]

Contaminant	Power plant emission (gm/kg fuel)			Refuse burning emission (gm/kg refuse)		Uncontrolled automotive emission (gm/kg fuel)	
	Coal	Oil	Gas	Open burning	Multiple chamber	Gasoline	Diesel
Carbon monoxide	Nil	Nil	Nil	50.0	Nil	165.0	Nil
Oxides of sulfur (SO_2)	(20)x	(20)x	(16)x	1.5	1.0	0.8	7.5
Oxides of nitrogen (NO_2)	0.43	0.68	0.16	2.0	1.0	16.5	16.5
Aldehydes and ketones	Nil	0.003	0.001	3.0	0.5	0.8	1.6
Total hydrocarbons	0.43	0.05	0.005	7.5	0.5	33.0	30.0
Total particulate	(75)y	(2.8)y	Nil	11	11	0.05	18.0

[a] x = percentage of sulfur in fuel; y = percentage of ash in fuel.

method used to control the emissions from steel mill X may be applied to control similar emissions at plant Y. It should not be necessary for plant Y to spend excessive amounts for research and development if plant X has a system that is operating satisfactorily. The solution to the problem is often to look for a similar industrial process, with similar emissions, and find the type of control system used.

A. Chemical and Allied Products

The emissions from a chemical process can be related to the specific process. A plant manufacturing a resin might be expected to emit not only the resin being manufactured but also some of the raw material and some other products which may or may not resemble the resin. A plant manufacturing sulfuric acid can be expected to emit sulfuric acid fumes and SO_2. A plant manufacturing soap products could be expected to emit a variety of odors. Depending on the process, the emissions could be any one or a combination of dust, aerosols, fumes, or gases. The emissions may or may not be odorous or toxic. Some of the primary emissions might be innocuous but later react in the atmosphere to form an undesirable secondary pollutant. A flowchart and material balance sheet for the particular process are very helpful in understanding and analyzing any process and its emissions (6).

In any discussion of the importance of emissions from a particular process for an area, several factors must be considered—(1) the percentage of the total emissions of the area that the particular process emits, (2) the degree of toxicity of the emissions, and (3) the obvious characteristics of the source (which can be related to either sight or smell).

B. Resins and Plastics

Resins are solid or semisolid, water-insoluble, organic substances with little or no tendency to crystallize. They are the basic components of plastics and are also used for coatings on paper, particleboard, and other surfaces that require a decorative, protective, or special-purpose finish. The common characteristic of resins is that heat is used in their manufacture and application, and gases are exhausted from these processes. Some of the gases that are economically recoverable may be condensed, but a large portion are lost to the atmosphere. One operation, coating a porous paper with a resin to form battery separators, was emitting to the atmosphere about 85% of the resin purchased. This resin left the stacks of the plant as a blue haze, and the odor was routinely detected more than 2 km away. Since most resins and their by-products have low odor thresholds, disagreeable odor is the most common complaint against any operation using them.

C. Varnish and Paints

In the manufacture of varnish, heat is necessary for formulation and purification. The same may be true of operations preparing paints, shellac, inks, and other protective or decorative coatings. The compounds emitted to the atmosphere are gases, some with extremely low odor thresholds. Acrolein, with an odor threshold of about 4000 $\mu g/m^3$, and reduced sulfur compounds, with odor thresholds of 2 $\mu g/m^3$, are both possible emissions from varnish cooking operations. The atmospheric emissions from varnish cooking appear to have little or no recovery value, whereas some of the solvents used in paint preparation are routinely condensed for recovery and return to the process. If a paint finish is baked to harden the surface by removal of organic solvents, the solvents must either be recovered, destroyed, or emitted to the atmosphere. The last course, emission to the atmosphere, is undesirable and may be prohibited by the air pollution control agency.

D. Acid Manufacture

Acids are used as basic raw materials for many chemical processes and manufacturing operations. Figure 6-8 illustrates an acid plant with its flow diagram. Sulfuric acid is one of the major inorganic chemicals in modern industry. The atmospheric discharges from a sulfuric acid plant can be expected to contain gases including SO_2 and aerosol mists, containing SO_3 and H_2SO_4, in the submicron to 10-μm size range. The aerosol mists are particularly damaging to paint, vegetation, metal, and synthetic fibers.

Other processes producing acids, such as nitric, acetic, and phosphoric acids, can be expected to produce acid mists from the processes themselves as well as various toxic and nontoxic gases. The particular process must be thoroughly studied to obtain a complete listing of all the specific emissions.

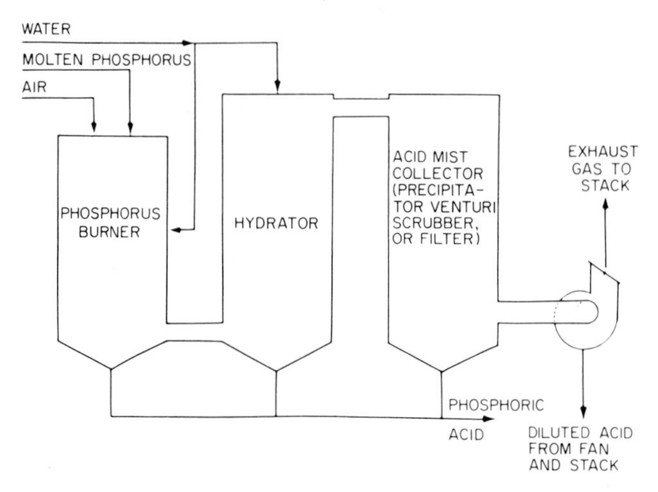

Fig. 6-8. Flow diagram for a phosphoric acid plant.

E. Soaps and Detergents

Soaps are made by reacting fats or oils with a base. Soaps are produced in a number of grades and types. They may be liquid, solid, granules, or powder. The air pollution problems of soap manufacture are primarily odors from the chemicals, greases, fats, and oils, although particulate emissions may occur during drying and handling operations. Detergents are manufactured from base stocks similar to those used in petroleum refineries, so the air pollution problems are similar to those of refineries.

F. Phosphate Fertilizers

Phosphate fertilizers are prepared by beneficiation of phosphate rock to remove its impurities, followed by drying and grinding. The PO_4 in the rock may then be reacted with sulfuric acid to produce normal superphosphate fertilizer. Over 100 plants operating in the United States produce approximately a billion kilograms of phosphate fertilizer per year. Figure 6-9 is a flow diagram for a normal superphosphate plant which notes the pollutants emitted. The particulate and gaseous fluoride emissions cause greatest concern near phosphate fertilizer plants.

G. Other Inorganic Chemicals

Production of the large quantities of inorganic chemicals necessary for modern industrial processes can result in air pollutant emissions as undesirable by-products. Table 6-2 lists some of the more common inorganic chemi-

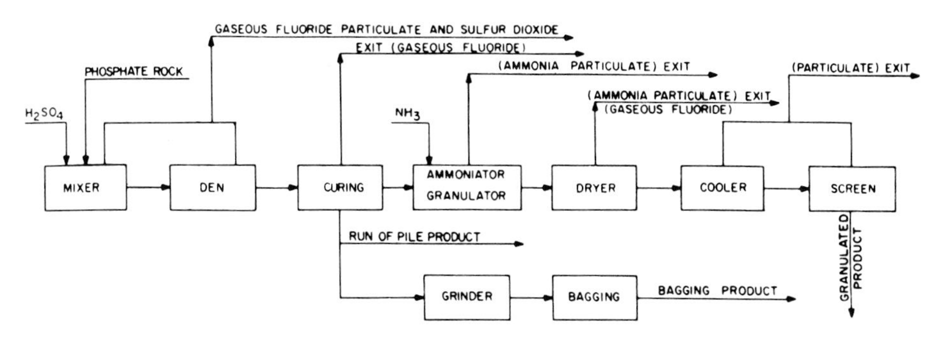

Fig. 6-9. Flow diagram for a normal superphosphate plant.

cals produced, along with the major atmospheric emissions from the specific process (6).

H. Petroleum and Coal

Petroleum and coal supply the majority of the energy in all industrial countries. This fact gives an indication of the vast quantities of materials handled and also hints at the magnitude of the air pollution problems associated with obtaining the resource, transporting it, refining it, and transporting it again. The emission problems from burning fossil fuel have been previously discussed.

1. *Petroleum*

Petroleum products are obtained from crude oil. In the process of getting the crude oil from the ground to the refinery, many possibilities for emission of hydrocarbon and reduced sulfur gaseous emissions occur. In many cases, these operations take place in relatively remote regions and affect only those employed by the industry, so that little or no control is attempted.

TABLE 6-2

Miscellaneous Inorganic Chemicals and Associated Air Pollution Emissions

Inorganic chemical produced	Major air pollution emissions
Calcium oxide (lime)	Lime dust
Sodium carbonate (soda ash)	Ammonia—soda ash dust
Sodium hydroxide (caustic soda)	Ammonia—caustic dust and mist
Ammonium nitrate	Ammonia—nitric oxides
Chlorine	Chlorine gas
Bromine	Chlorine gas

As shown in Fig. 6-10, an uncontrolled petroleum refinery is a potential source of large tonnages of atmospheric emissions. All refineries are odorous, the degree being a matter of the housekeeping practices around the refinery. Since refineries are essentially closed processes, emissions are not normally considered a part of the operation. Refineries do need pressure relief systems and vents, and emissions from them are possible. Most refineries use very strict control measures for economic as well as regulatory reasons. The recovery of 1 or 2% of a refinery throughput which was previously lost to the atmosphere can easily pay for the cost of the control equipment. The expense of the catalyst charge in some crackers and regenerators requires that the best possible control equipment be used to prevent catalyst emissions to the atmosphere.

Potential air pollutants from a petroleum refinery could include (1) hydrocarbons from all systems, leaks, loading, and sampling; (2) sulfur oxides from boilers, treaters, and regenerators; (3) carbon monoxide from regenerators and incinerators; (4) nitrogen oxides from combustion sources and regenerators; (5) odors from air and steam blowing, condensers, drains, and vessels; and (6) particulate matter from boilers, crackers, regenerators, coking, and incinerators.

Fig. 6-10. Uncontrolled petroleum refinery.

Loading facilities must be designed to recover all vapors generated during filling of tank trucks or tanker ships. Otherwise these vapors will be lost to the atmosphere. Since they may be both odorous and photochemically reactive, serious air pollution problems could result. The collected vapors must be returned to the process or disposed of by some means.

2. *Coal*

The air pollution problems associated with combustion of coal are of major concern. These problems generally occur away from the coal mine. The problems of atmospheric emissions due to mining, cleaning, handling, and transportation of coal from the mine to the user are of lesser significance as far as the overall air pollution problems are concerned. Whenever coal is handled, particulate emission becomes a problem. The emissions can be either coal dust or inorganic inclusions. Control of these emissions can be relatively expensive if the coal storage and transfer facilities are located near residential areas.

I. Primary Metals Industry

Metallurgical equipment has long been an obvious source of air pollution. The effluents from metallurgical furnaces are submicron-size dusts and fumes and hence are highly visible. The emissions from associated coke ovens are not only visible but odorous as well.

1. *Ferrous Metals*

Iron and steel industries have been concerned with emissions from their furnaces and cupolas since the industry started. Pressures for control have forced the companies to such a low level of permissible emissions that some of the older operations have been closed rather than spend the money to comply. The companies controlling these operations have not gone out of business but rather have opened a new, controlled plant to replace each old plant. Table 6-3 illustrates the changes in the steelmaking processes that have occurred in the United States.

Air-polluting emissions from steelmaking furnaces include metal oxides, smoke, fumes, and dusts to make up the visible aerosol plume. They also may include gases, both organic and inorganic. If steel scrap is melted, the charge may contain appreciable amounts of oil, grease, and other combustibles that further add to the organic gas and smoke loadings. If the ore used has appreciable fluoride concentrations, the emission of both gaseous and particulate fluorides can be a serious problem.

Emissions from foundry cupolas are relatively small but still significant, in some areas. An uncontrolled 2-m cupola can be expected to emit up to 50 kg of dust, fumes, smoke, and oil vapor per hour. Carbon monoxide, oxides of nitrogen, and organic gases may also be expected. Control is

TABLE 6-3

Changes in Steel-Making Processes in the United States

	Production by specific process (%)				
Year	Bessemer	Open hearth	Electric	Basic oxygen furnace	Total
1920	21	78	1	0	100
1940	6	92	2	0	100
1960	2	89	7	2	100
1970	1	36	14	48	100
1980	1	22	31	46	100
1990	0	4	37	59	100

possible, but the cost of the control may be prohibitive for the small foundry which only has one or two heats per week.

2. *Nonferrous Metals*

Around the turn of the century, one of the most obvious effects of industry on the environment was the complete destruction of vegetation downwind from copper, lead, and zinc smelters. This problem was caused by the smelting of the metallic sulfide ores. As the metal was released in the smelting process, huge quantities of sulfur were oxidized to SO_2, which was toxic to much of the vegetation fumigated by the plume. Present smelting systems go to great expense to prevent the uncontrolled release of SO_2, but in many areas the recovery of the ecosystem will take years and possibly centuries.

Early aluminum reduction plants were responsible for air pollution because of the fluoride emissions from their operations. Fluoride emissions can cause severe damage to vegetation and to animals feeding on such vegetation. The end result was an area surrounding the plant devoid of vegetation. Such scenes are reminiscent of those downwind from some of the uncontrolled copper smelters. New aluminum reduction plants are going to considerable expense to control fluoride emissions. Some of the older plants are finding that the cost of control will exceed the original capital investment in the entire facility. Where the problem is serious, control agencies have developed extensive sampling networks to monitor emissions from the plant of concern.

Emissions from other nonferrous metal facilities are primarily metal fumes or metal oxides of extremely small diameter. Zinc oxide fumes vary from 0.03 to 0.3 μm and are toxic. Lead and lead oxide fumes are extremely toxic and have been extensively studied. Arsenic, cadmium, bismuth, and other trace metals can be emitted from many metallurgical processes.

J. Stone and Clay Products

The industries which produce and handle various stone products emit considerable amounts of particulate matter at every stage of the operation. These particulates may include fine mineral dusts of a size to cause damage to the lungs. The threshold values for such dusts have been set quite low to prevent disabling diseases for the worker.

In the production of clay, talc, cement, chalk, etc., an emission of particulate matter will usually accompany each process. These processes may involve grinding, drying, and sieving, which can be enclosed and controlled to prevent the emission of particles. In many cases, the recovered particles can be returned to the process for a net economic gain.

During the manufacture of glass, considerable dust, with particles averaging about 300 μm in size, will be emitted. Some dusts may also be emitted from the handling of the raw materials involved. Control of this dust to prevent a nuisance problem outside the plant is a necessity. When glass is blown or formed into the finished product, smoke and gases can be released from the contact of the molten glass with lubricated molds. These emissions are quite dense but of a relatively short duration.

K. Forest Products Industry

1. *Wood Processing*

Trees are classified as a renewable resource which is being utilized in most portions of the world on a sustained yield basis. A properly managed forest will produce wood for lumber, fiber, and chemicals forever. Harvesting this resource can generate considerable dust and other particulates. Transportation over unpaved roads causes excessive dust generation. The cultural practice of burning the residue left after a timber harvest, called *slash burning*, is still practiced in some areas and is a major source of smoke, gaseous, and particulate air pollution in the localities downwind from the fire. Visibility reduction from such burning can be a serious problem.

Processing the harvested timber into the finished product may involve sawing, peeling, planing, sanding, and drying operations, which can release considerable amounts of wood fiber and lesser amounts of gaseous material to the atmosphere. Control of wood fiber emissions from the pneumatic transport and storage systems can be a major problem of considerable expense for a plywood mill or a particleboard plant.

2. *Pulp and Paper*

Pulp and paper manufacture is increasing in the world at an exponential rate. The demand for paper will continue as new uses are found for this product. Since most paper is manufactured from wood or wood residue, it provides an excellent use for this renewable resource.

The most widely used pulping process is the kraft process, as shown in Fig. 6-11, which results in recovery and regeneration of the chemicals. This occurs in the recovery furnace, which operates with both oxidizing and reducing zones. Emissions from such recovery furnaces include particulate matter, very odorous reduced sulfur compounds, and oxides of sulfur. If extensive and expensive control is not exercised over the kraft pulp process, the odors and aerosol emissions will affect a wide area. Odor complaints have been reported over 100 km away from these plants. A properly controlled and operated kraft plant will handle huge amounts of material and produce millions of kilograms of finished products per day, with little or no complaint regarding odor or particulate emissions.

L. Noxious Trades

As the name implies, these operations, if uncontrolled, can cause a serious air pollution problem. The main problem is the odors associated with the process. Examples of such industries are tanning works, rendering plants, and many of the food processing plants such as fish meal plants. In most cases, the emissions of particulates and gases from such plants are not of concern, only the odors. Requiring these industries to locate away from the business or residential areas is no longer acceptable as a means of control.

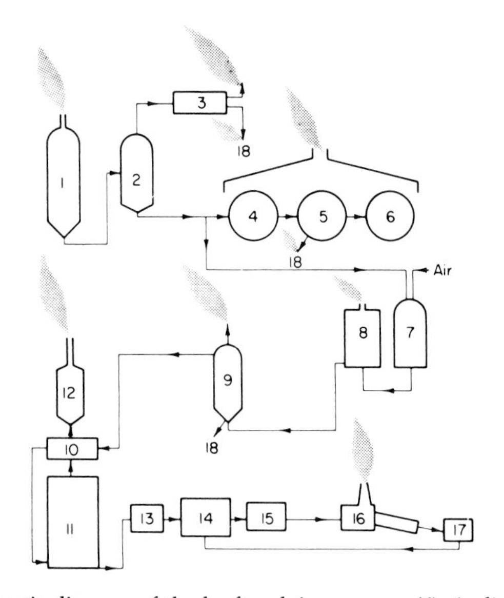

Fig. 6-11. Schematic diagram of the kraft pulping process (6). 1, digester; 2, blow tank; 3, blow heat recovery; 4, washers; 5, screens; 6, dryers; 7, oxidation tower; 8, foam tank; 9, multiple effect evaporator; 10, direct evaporator; 11, recovery furnace; 12, electrostatic precipitator; 13, dissolver, 14, causticizer; 15, mud filter; 16, lime kiln; 17, slaker; 18, sewer.

IV. MOBILE SOURCES

A mobile source of air pollution can be defined as one capable of moving from one place to another under its own power. According to this definition, an automobile is a mobile source and a portable asphalt batching plant is not. Generally, mobile sources imply transportation, but sources such as construction equipment, gasoline-powered lawn mowers, and gasoline-powered tools are included in this category.

Mobile sources therefore consist of many different types of vehicles, powered by engines using different cycles, fueled by a variety of products, and emitting varying amounts of both simple and complex pollutants. Table 6-4 includes the more common mobile sources.

The predominant mobile air pollution source in all industrialized countries of the world is the automobile, powered by a four-stroke cycle (Otto cycle) engine and using gasoline as the fuel. In the United States, over 85 million automobiles were in use in 1990. If the 15 million gasoline-powered trucks and buses and the 4 million motorcycles are included, the United States total exceeds 100 million vehicles. The engine used to power these millions of vehicles has been said to be the most highly engineered machine of the century. When one considers the present reliability, cost, and life expectancy of the internal combustion engine, it is not difficult to see why it has remained so popular. A modern automotive engine traveling 100,000 km will have about 2.5×10^8 power cycles.

The emissions from a gasoline-powered vehicle come from many sources. Figure 6-12 illustrates what might be expected from an uncontrolled (1960 model) automobile and a controlled (1983 or later model) automobile if it complies with the 1983 federal standards (7). With most of today's automobiles using unleaded gasoline, lead emissions are no longer a major concern.

TABLE 6-4

Emissions from Mobile Sources

Power plant type	Fuel	Major emissions	Vehicle type
Otto cycle	Gasoline	HC, CO, CO_2, NO_x	Auto, truck, bus, aircraft, marine, motorcycle, tractor
Two-stroke cycle	Gasoline	HC, CO, CO_2, NO_x, particulate	Motorcycle, outboard motor
Diesel	Diesel oil	NO_x, particulate, SO_x, CO_2	Auto, truck, bus, railroad, marine, tractor
Gas turbine (jet)	Turbine	NO_x, particulate, CO_2	Aircraft, marine, railroad
Steam	Oil, coal	NO_x, SO_x, particulate, CO_2	Marine

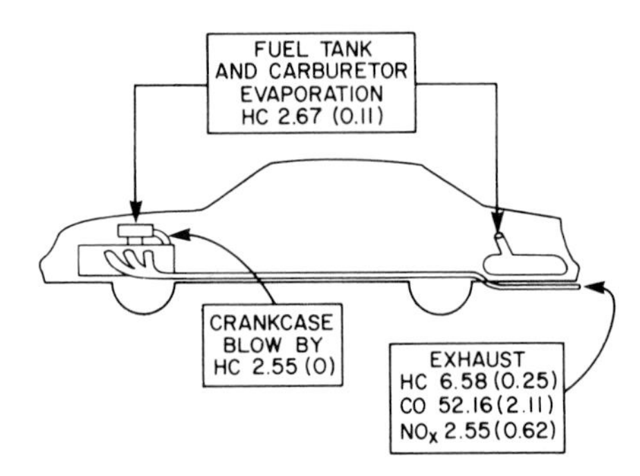

Fig. 6-12. Emissions from uncontrolled automobiles (and those meeting U.S. Environmental Protection Agency standards) in grams per kilometer.

V. EMISSION INVENTORY

An *emission inventory* is a list of the amount of pollutants from all sources entering the air in a given time period. The boundaries of the area are fixed (8).

The tables of emission inventory are very useful to control agencies as well as planning and zoning agencies. They can point out the major sources whose control can lead to a considerable reduction of pollution in the area. They can be used with appropriate mathematical models to determine the degree of overall control necessary to meet ambient air quality standards. They can be used to indicate the type of sampling network and the locations of individual sampling stations if the areas chosen are small enough. For example, if an area uses very small amounts of sulfur-bearing fuels, establishing an extensive SO_2 monitoring network in the area would not be an optimum use of public funds. Emission inventories can be used for publicity and political purposes: "If natural gas cannot meet the demands of our area, we will have to burn more high-sulfur fuel, and the SO_2 emissions will increase by 8 tons per year."

The method used to develop the emission inventory does have some elements of error, but the other two alternatives are expensive and subject to their own errors. The first alternative would be to monitor continually every major source in the area. The second method would be to monitor continually the pollutants in the ambient air at many points and apply appropriate diffusion equations to calculate the emissions. In practice, the most informative system would be a combination of all three, knowledgeably applied.

The U.S. Clean Air Act Amendments of 1990 (10) strengthened the emission inventory requirements for plans and permits in nonattainment areas. The amendments state:

> "INVENTORY.—Such plan provisions shall include a comprehensive, accurate, current inventory of actual emissions from all sources of the relevant pollutant or

pollutants in such area, including such periodic revisions as the Administrator may determine necessary to assure that the requirements of this part are met.

"IDENTIFICATION AND QUANTIFICATION.—Such plan provisions shall expressly identify and quantify the emissions, if any, of any such pollutant or pollutants which will be allowed, from the construction and operation of major new or modified stationary sources in each such area. The plan shall demonstrate to the satisfaction of the Administrator that the emissions quantified for this purpose will be consistent with the achievement of reasonable further progress and will not interfere with the attainment of the applicable national ambient air quality standard by the applicable attainment date.

A. Inventory Techniques

To develop an emission inventory for an area, one must (1) list the types of sources for the area, such as cupolas, automobiles, and home fireplaces; (2) determine the type of air pollutant emission from each of the listed sources, such as particulates and SO_2; (3) examine the literature (9) to find valid emission factors for each of the pollutants of concern (e.g., "particulate emissions for open burning of tree limbs and brush are 10 kg per ton of residue consumed"); (4) through an actual count, or by means of some estimating technique, determine the number and size of specific sources in the area (the number of steelmaking furnaces can be counted, but the number of home fireplaces will probably have to be estimated); and (5) multiply the appropriate numbers from (3) and (4) to obtain the total emissions and then sum the similar emissions to obtain the total for the area.

A typical example will illustrate the procedure. Suppose we wish to determine the amount of carbon monoxide from oil furnaces emitted per day, during the heating season, in a small city of 50,000 population:

1. The source is oil furnaces within the boundary area of the city.
2. The pollutant of concern is carbon monoxide.
3. Emission factors for carbon monoxide are listed in various ways (9) (240 gm per 1000 liters of fuel oil, 50 gm per day per burner, $1\frac{1}{2}\%$ by volume of exhaust gas, etc.). For this example, use 240 gm per 1000 liters of fuel oil.
4. Fuel oil sales figures, obtained from the local dealers association, average 40,000 liters per day.
5. $\dfrac{250 \text{ gm CO}}{1000 \text{ liters}} \times \dfrac{40{,}000 \text{ liters}}{\text{day}} = 9.6 \text{ kg CO/day}$

B. Emission Factors

Valid emission factors for each source of pollution are the key to the emission inventory. It is not uncommon to find emission factors differing by 50%, depending on the researcher, variables at the time of emission measurement, etc. Since it is possible to reduce the estimating errors in the inventory to ± 10% by proper statistical sampling techniques, an emission factor error of 50% can be overwhelming. It must also be realized that an uncontrolled source will emit at least 10 times the amount of pollutants released from one operating properly with air pollution control equipment installed.

Actual emission data are available from many handbooks, government publications, and literature searches of appropriate research papers and journals. It is always wise to verify the data, if possible, as to the validity of the source and the reasonableness of the final number. Some emission factors, which have been in use for years, were only rough estimates proposed by someone years ago to establish the order of magnitude of the particular source.

Emission factors must be also critically examined to determine the tests from which they were obtained. For example, carbon monoxide from an automobile will vary with the load, engine speed, displacement, ambient temperature, coolant temperature, ignition timing, carburetor adjustment, engine condition, etc. However, in order to evaluate the overall emission of carbon monoxide to an area, we must settle on an average value that we can multiply by the number of cars, or kilometers driven per year, to determine the total carbon monoxide released to the area.

C. Data Gathering

To compile the emission inventory requires a determination of the number and types of units of interest in the study area. It would be of interest, for example, to know the number of automobiles in the area and the number of kilometers each was driven per year. This figure would require considerable time and expense to obtain. Instead, it can be closely approximated by determining the liters of gasoline sold in the area during the year. Since a tax is collected on all gasoline sold for highway use, these figures can be obtained from the tax collection office.

Data regarding emissions are available from many sources. Sometimes the same item may be checked by asking two or more agencies for the same information. An example of this would be to check the liters of gasoline sold in a county by asking both the tax office and the gasoline dealers association. Sources of information for an emission inventory include (1) city, county, and state planning commissions; (2) city, county, and state chambers of commerce; (3) city, county, and state industrial development commissions; (4) census bureaus; (5) national associations such as coal associations; (6) local associations such as the County Coal Dealers Association; (7) individual dealers or distributors of oil, gasoline, coal, etc.; (8) local utility companies; (9) local fire and building departments; (10) data gathered by air pollution control agencies through surveys, sampling, etc.; (11) traffic maps; and (12) insurance maps.

TABLE 6-5

Thermal Conversion Factors for Fuels

Fuel	Joule × 10^6
Bituminous coal	30.48 per kg
Anthracite coal	29.55 per kg
Wood	20.62 per kg
Distillate fuel oil	38.46 per kg
Residual fuel oil	41.78 per liter
Natural gas	39.08 per m^3
Manufactured gas	20.47 per m^3

D. Data Reduction and Compilation

The final emission inventory can be prepared on a computer. This will enable the information to be stored on magnetic tape or disk so that it can be updated rapidly and economically as new data or new sources appear. The computer program can be written so that changes can easily be made. There will be times when major changes occur and the inventory must be completely changed. Imagine the change that would take place when natural gas first becomes available in a commercial–residential area which previously used oil and coal for heating.

To determine emission data, as well as the effect that fuel changes would produce, it is necessary to use the appropriate thermal conversion factor from one fuel to another. Table 6-5 lists these factors for fuels in common use.

A major change in the emissions for an area will occur if control equipment is installed. This can be shown in the emission inventory to illustrate the effect on the community.

By keeping the emission inventory current and updating it at least yearly as fuel uses change, industrial and population changes occur, and control equipment is added, a realistic record for the area is obtained.

REFERENCES

1. Tombach, I., "A Critical Review of Methods for Determining Visual Range in Pristine and Near-Pristine Areas," Proceedings of the Annual Meeting of the Air Pollution Control Association, Pittsburgh, 1978.
2. Singer, J. G. (ed.), "Combustion, Fossil Power Systems," 3rd ed. Combustion Engineering, Windsor, CT, 1981.
3. Popovich, M., and Hering, C., "Fuels and Lubricants." Oregon State University Press, Corvallis, 1978.
4. Straus, W., "Industrial Gas Cleaning," 2nd ed. Pergamon, Oxford, 1975.
5. Buonicore, A. J., and Davis, W. T. (eds.), "Air Pollution Engineering Manual." Van Nostrand Reinhold, New York, 1992.

6. Stern, A. C. (ed.), "Air Pollution," 3rd ed., Vol. IV. Academic Press, New York, 1977.
7. Elston, J. C., *J. Air Pollut. Control Assoc.* **31**(5), 524–547 (1981).
8. Stern, A. C. (ed.), "Air Pollution," 3rd ed., Vol. III. Academic Press, New York, 1977.
9. "Compilation of Air Pollutant Emission Factors," AP-42 and Supplements. U.S. Environmental Protection Agency, Research Triangle Park, NC, 1973–1992.
10. Public Law 101-549, 101st Congress—November 15, 1990, An Act to Amend the Clean Air Act to provide for attainment and maintenance of health protective national ambient air quality standards, and for other purposes.

SUGGESTED READING

Faith, W. L., and Atkission, A. A., "Air Pollution," 2nd ed. Wiley, New York, 1972.

"U.S. Interagency Team Proposes Program to Quantify Effects of Kuwait Oil Fires," *J. Air Waste Management Assoc.* **41**,(6), June 1991.

Wark, K., and Warner, C. F., "Air Pollution: Its Origin and Control." IEPA, Dun-Donnelley, New York, 1976.

"Wood Heating and Air Quality," 1981 Annual Report, Oregon Department of Environmental Quality, June 1982.

QUESTIONS

1. Calculate the heat generated by dissociation and formation as one molecular weight of methane, CH_4, burns to carbon dioxide and water. How does this heating value compare to the tabular heating value for methane?
2. Many control districts have banned the use of private backyard incinerators. Would you expect a noticeable increase in air quality as a result of this action?
3. Show a free radical reaction which results in ethane in the effluent of a combustion process burning pure methane with pure oxygen.
4. A power plant burns oil that is 4% ash and 3% sulfur. At 50% excess air, what particulate (mg/m^3) would you expect?
5. Many control districts have very tight controls over petroleum refineries. Suppose these refineries produce 100 million liters of products per day and required air pollution control devices to recover all of the 2% which was previously lost. What are the savings in dollars per year at an average product cost of 10 cents per liter? How does this compare to the estimate that the refineries spent $400 million for control equipment over a 10-year period?
6. Suppose a 40,000-liter gasoline tank is filled with liquid gasoline with an average vapor pressure of 20 mm Hg. At 50% saturation, what weight of gasoline would escape to the atmosphere during filling?
7. If a major freeway with four lanes of traffic in one direction passes four cars per second at 100 km per hour during the rush period, and each car carries two people, how often would a commuter train of five cars carrying 100 passengers per car have to be operated to handle the same load? Assume the train would also operate at 100 km per hour.
8. An automobile traveling 50 km per hour emits 0.1% CO from the exhaust. If the exhaust rate is 80 m^3 per minute, what is the CO emission in grams per kilometer?
9. List the following in increasing amounts from the exhaust of an idling automobile: O_2, NO_x, SO_x, N_2, unburned hydrocarbons, CO_2, and CO.

Part II

The Effects of Air Pollution

7

Effects on Health and Human Welfare

I. AIR–WATER–SOIL INTERACTIONS

The harmful effects of air pollutants on human beings have been the major reason for efforts to understand and control their sources. During the past two decades, research on acidic deposition on water-based ecosystems has helped to reemphasize the importance of air pollutants in other receptors, such as soil-based ecosystems (1). When discussing the impact of air pollutants on ecosystems, the matter of scale becomes important. We will discuss three examples of elements which interact with air, water, and soil media on different geographic scales. These are the carbon cycle on a global scale, the sulfur cycle on a regional scale, and the fluoride cycle on a local scale.

A. The Carbon Cycle: Global Scale

Human interaction with the global cycle is most evident in the movement of the element carbon. The burning of biomass, coal, oil, and natural gas to generate heat and electricity has released carbon to the atmosphere and oceans in the forms of CO_2 and carbonate. Because of the relatively slow reactions and removal rates of CO_2, its concentration has been increasing steadily since the beginning of the Industrial Revolution (Fig. 2-4).

In its natural cycle, CO_2 enters the global atmosphere from vegetative decay and atmospheric oxidation of methane and is removed from the atmosphere by photosynthesis and solution by water bodies. These natural sources and sinks of CO_2 have balanced over thousands of years to result in an atmospheric CO_2 concentration of about 200–250 ppm by volume. Over the past 200 years, however, the burning of fossil fuels has caused a steady increase in the atmospheric CO_2 concentration to its current value of ~360 ppm by volume, with projected concentrations over the next 50 years ranging up to 400–600 ppm by volume worldwide (2). In raising the concentration of CO_2, humans are clearly interacting with nature on a global scale, producing a potential for atmospheric warming and subsequent changes in ocean depths and agricultural zones. This topic is currently subject to considerable research. Current research topics include further development of radiative–convective models, determination of global temperature trends, measurement of changes in polar ice packs, and refinement of the global carbon cycle.

B. The Sulfur Cycle: Regional Scale

Human production of sulfur from fossil fuel and ore smelting has caused an observable impact on the regional scale (hundreds of kilometers). Considerable evidence suggests that long-range transport of SO_2 occurs in the troposphere. In transit, quantities of SO_2 are converted to sulfate, with eventual deposition by dry or wet processes on the surface far from the original source of SO_2. Sulfate deposition plays the principal role in acid deposition which results in lowering the pH of freshwater lakes and alters the composition of some soils. These changes affect the viability of some plant and aquatic species. The long-range transport of SO_2 and the presence of sulfates as fine particulate matter play a significant role in reduction of visibility in the atmosphere.

C. The Fluoride Cycle: Local Scale

The movement of fluoride through the atmosphere and into a food chain illustrates an air–water interaction at the local scale (<100 km) (3). Industrial sources of fluoride include phosphate fertilizer, aluminum, and glass manufacturing plants. Domestic livestock in the vicinity of substantial fluoride sources are exposed to fluoride by ingestion of forage crops. Fluoride released into the air by industry is deposited and accumulated in vegetation. Its concentration is sufficient to cause damage to the teeth and bone structure of the animals that consume the crops.

The atmospheric movement of pollutants from sources to receptors is only one form of translocation. A second one involves our attempt to control air pollutants at the source. The control of particulate matter by wet or dry scrubbing techniques yields large quantities of waste materials—often toxic—which are subsequently taken to landfills. If these wastes are not properly stored, they can be released to soil or water systems. The prime examples involve the disposal of toxic materials in dump sites or landfills. The Resource Conservation and Recovery Act of 1976 and subsequent revisions are examples of legislation to ensure proper management of solid waste disposal and to minimize damage to areas near landfills (4).

II. TOTAL BODY BURDEN

The presence of air pollutants in the surrounding ambient air is only one aspect of determining the impact on human beings. An air pollution instrument can measure the ambient concentration of a pollutant gas, which may or may not be related to its interaction with individuals. More detailed information about where and for how long we are breathing an air pollutant provides additional information about our actual exposure. Finally, how an air pollutant interacts with the human body provides the most useful information about the dose to a target organ or bodily system.

The human body and other biological systems have a tremendous capacity to take in all types of chemicals and either utilize them to support some bodily function or eliminate them. As analytical capabilities have improved, lower and lower concentrations of chemicals have been observed in various parts of the body. Some of these chemicals enter the body by inhalation.

The concept of *total body burden* refers to the way a trace material accumulates in the human system. The components of the body that can store these materials are the blood, urine, soft tissue, hair, teeth, and bone. The blood and urine allow more rapid removal of trace materials than the soft tissue, hair, and bone (5). Accumulation results when trace materials are stored more rapidly than they can be eliminated. It can be reversed when the source of the material is reduced. The body may eliminate the trace material over a period of a few hours to days, or may take much longer—often years.

The effect of accumulation in various systems depends greatly on the quantity of pollutants involved. Many pollutants can be detected at concentrations lower than those necessary to affect human health. For pollutants which are eliminated slowly, individuals can be monitored over long periods of time to detect trends in body burden; the results of these analyses can then be related to total pollutant exposure. Following are two examples of air pollutants that contribute to the total body burden for lead and carbon monoxide.

A. Lead and the Human Body

The major sources of airborne lead are leaded gasoline, incineration of solid wastes and discarded oil, and certain manufacturing processes (6). The populations most sensitive to lead exposure are young children and fetuses. Lead can degrade renal function, impair hemoglobin synthesis, and alter the nervous system. The neurobehavioral impairment of children's intellectual development is a major concern for lead exposure. There are two routes for the entry of lead—inhalation and ingestion. The importance of each depends on the circumstances. As noted in Chapter 20, the U.S. national ambient air quality standard for lead is based on the ingestion route, which accounts for 80% of the allowed body burden, with only the remaining 20% permissible via inhalation. Inhalation results in primary exposure to airborne lead, whereas ingestion may result in secondary exposure via contamination of the ingested material by atmospheric lead. When lead is inhaled, some of it is absorbed directly into the bloodstream and a fraction into the gastrointestinal tract through lung clearance mechanisms that result in swallowing of mucus.

The absorption, distribution, and accumulation of lead in the human body may be represented by a three-part model (6). The first part consists of red blood cells, which move the lead to the other two parts, soft tissue and bone. The blood cells and soft tissue, represented by the liver and kidney, constitute the mobile part of the lead body burden, which can fluctuate depending on the length of exposure to the pollutant. Lead accumulation over a long period of time occurs in the bones, which store up to 95% of the total body burden. However, the lead in soft tissue represents a potentially greater toxicological hazard and is the more important component of the lead body burden. Lead measured in the urine has been found to be a good index of the amount of mobile lead in the body. The majority of lead is eliminated from the body in the urine and feces, with smaller amounts removed by sweat, hair, and nails.

B. Carbon Monoxide and the Human Body

The second example of an air pollutant that affects the total body burden is carbon monoxide (CO). In addition to CO in ambient air, there are other sources for inhalation. People who smoke have an elevated CO body burden compared to nonsmokers. Individuals indoors may be exposed to elevated levels of CO from incomplete combustion in heating or cooking stoves. CO gas enters the human body by inhalation and is absorbed directly into the bloodstream; the total body burden resides in the circulatory system. The human body also produces CO by breakdown of hemoglobin. Hemoglobin breakdown gives every individual a baseline level of CO in the circulatory system. As the result of these factors, the body burden can fluctuate over a time scale of hours.

In the normal interaction between the respiratory and circulatory systems, O_2 is moved into the body for use in biochemical oxidation and CO_2, a waste product, is removed. Hemoglobin molecules in the blood play an important role in both processes. Hemoglobin combines with O_2 and CO_2 as these gases are moved between the lung and the cells. The stability of the hemoglobin–O_2 and hemoglobin–CO_2 complex is sufficiently strong to transport the gases in the circulatory system but not strong enough to prevent the release of CO_2 at the lung and O_2 where it is needed at the cellular level. CO interferes with this normal interaction by forming a much more stable complex with hemoglobin (COHb) (7). This process reduces the number of hemoglobin molecules available to maintain the necessary transport of O_2 and CO_2.

The baseline level of COHb is ~0.5% for most individuals. Upon exposure to elevated levels of atmospheric CO, the percentage of COHb will increase in a very predictable manner. Analytical techniques are available to measure COHb from <0.1 to $>80\%$ in the bloodstream, providing a very rapid method for determining the total body burden. If elevated levels of CO are reduced, the percentage of COHb will decrease over a period of time.

At low levels of COHb (0.5–2.0%) the body burden is measurable, but research has not shown any substantive effects at these low levels. When COHb increases to higher levels the body burden of CO is elevated, producing adverse effects on the cardiovascular system and reducing physical endurance.

III. THE HUMAN RESPIRATORY SYSTEM

The primary function of the human respiratory system is to deliver O_2 to the bloodstream and remove CO_2 from the body. These two processes occur concurrently as the breathing cycle is repeated. Air containing O_2 flows into the nose and/or mouth and down through the upper airway to the alveolar region, where O_2 diffuses across the lung wall to the bloodstream. The counterflow involves transfer of CO_2 from the blood to the alveolar region and then up the airways and out the nose. Because of the extensive interaction of the respiratory system with the surrounding atmosphere, air pollutants or trace gases other than N_2 and O_2 can be delivered to the respiratory system.

The anatomy of the respiratory system is shown in Fig. 7-1. This system may be divided into three regions—the nasal, tracheobronchial, and pulmonary. The nasal region is composed of the nose and mouth cavities and the throat. The tracheobronchial region begins with the trachea and extends through the bronchial tubes to the alveolar sacs. The pulmonary region is composed of the terminal bronchi and alveolar sacs, where gas exchange with the circulatory system occurs. Figure 7-1 illustrates the continued

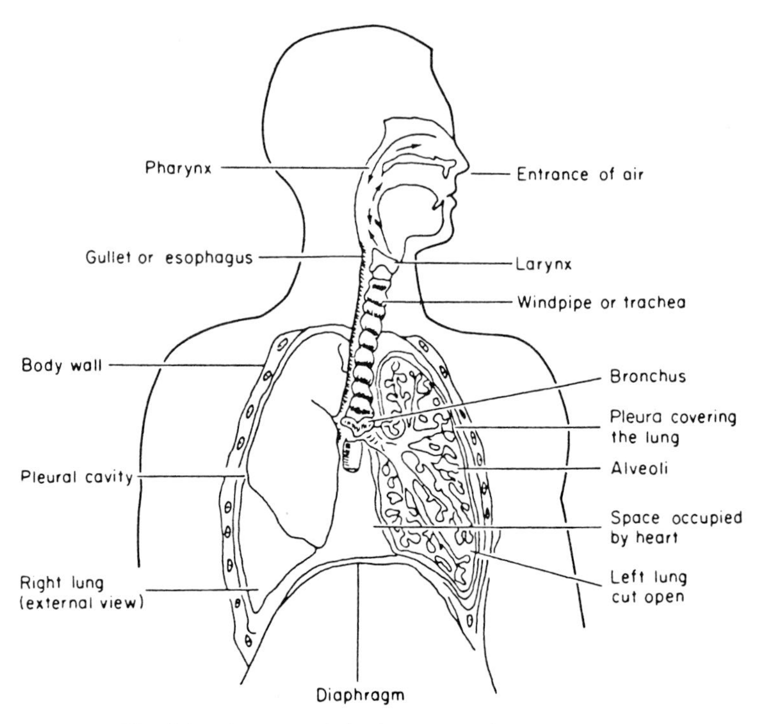

Fig. 7-1. Anatomy of the human respiratory system.

bifurcation of the trachea to form many branching pathways of increasingly smaller diameter by which air moves to the pulmonary region. The trachea branches into the right and left bronchi. Each bronchus divides and subdivides at least 20 times; the smallest units, bronchioles, are located deep in the lungs. The bronchioles end in about 3 million air sacs, the alveoli.

The behavior of particles and gases in the respiratory system is greatly influenced by the region of the lung in which they are located (8). Air passes through the upper region and is humidified and brought to body temperature by gaining or losing heat. After the air is channeled through the trachea to the first bronchi, the flow is divided at each subsequent bronchial bifurcation until very little apparent flow is occurring within the alveolar sacs. Mass transfer is controlled by molecular diffusion in this final region. Because of the very different flows in the various sections of the respiratory region, particles suspended in air and gaseous air pollutants are treated differently in the lung.

A. Particle and Gas Behavior in the Lung

Particle behavior in the lung is dependent on the aerodynamic characteristics of particles in flow streams. In contrast, the major factor for gases is the solubility of the gaseous molecules in the linings of the different regions of the respiratory system. The aerodynamic properties of particles are related to their size, shape, and density. The behavior of a chain type or fiber may also be dependent on its orientation to the direction of flow. The deposition of particles in different regions of the respiratory system depends on their size. The nasal openings permit very large dust particles to enter the nasal region, along with much finer airborne particulate matter. Particles in the atmosphere can range from less than 0.01 μm to more than 50 μm in diameter.

The relationship between the aerodynamic size of particles and the regions where they are deposited is shown in Fig. 7-2 (9). Larger particles are deposited in the nasal region by impaction on the hairs of the nose or at the bends of the nasal passages. Smaller particles pass through the nasal region and are deposited in the tracheobronchial and pulmonary regions. Particles are removed by impacts with the walls of the bronchi when they are unable to follow the gaseous streamline flow through subsequent bifurcations of the bronchial tree. As the airflow decreases near the terminal bronchi, the smallest particles are removed by Brownian motion, which pushes them to the alveolar membrane.

B. Removal of Deposited Particles from the Respiratory System

The respiratory system has several mechanisms for removing deposited particles (8). The walls of the nasal and tracheobronchial regions are coated with a mucous fluid. Nose blowing, sneezing, coughing, and swallowing

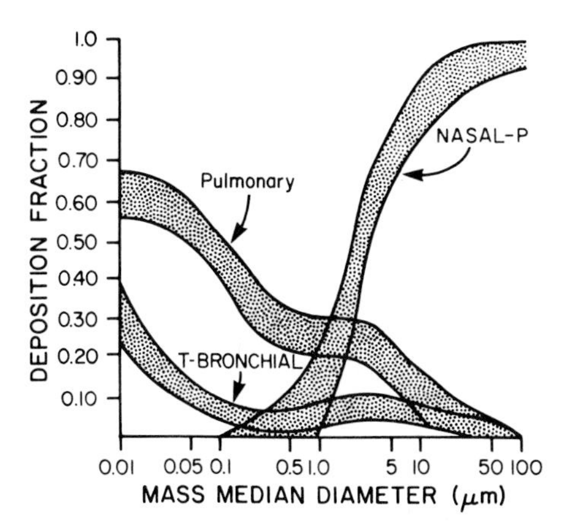

Fig. 7-2. Particle deposition as a function of particle diameter in various regions of the lung. The nasopharyngeal region consists of the nose and throat; the tracheobronchial region consists of the windpipe and large airways; and the pulmonary region consists of the small bronchi and the alveolar sacs. Source: Task Group on Lung Dynamics, *Health Phys.* **12,** 173 (1966).

help remove particles from the upper airways. The tracheobronchial walls have fiber cilia which sweep the mucous fluid upward, transporting particles to the top of the trachea, where they are swallowed. This mechanism is often referred to as the *mucociliary escalator*. In the pulmonary region of the respiratory system, foreign particles can move across the epithelial lining of the alveolar sac to the lymph or blood systems, or they may be engulfed by scavenger cells called *alveolar macrophages*. The macrophages can move to the mucociliary escalator for removal.

For gases, solubility controls removal from the airstream. Highly soluble gases such as SO_2 are absorbed in the upper airways, whereas less soluble gases such as NO_2 and O_3 may penetrate to the pulmonary region. Irritant gases are thought to stimulate neuroreceptors in the respiratory walls and cause a variety of responses, including sneezing, coughing, bronchoconstriction, and rapid, shallow breathing. The dissolved gas may be eliminated by biochemical processes or may diffuse to the circulatory system.

IV. IMPACT OF AIR POLLUTION ON HUMANS

The impact of air pollution on human beings has been the major force motivating efforts to control it. Most persons do not have the luxury of choosing the air they breathe. Working adults can make some choices in the selection of their occupation and the place where they live and work, but children and the nonworking elderly often cannot. The receptor population in an urban location includes a wide spectrum of demographic traits with respect to age, sex, and health status. Within this group, certain sensitive subpopulations have been identified: (1) very young children, whose respiratory and circulatory systems are still undergoing maturation; (2) the elderly, whose respiratory and circulatory systems function poorly; and (3) persons who have preexisting diseases such as asthma, emphysema, and heart disease. These subpopulations have been found to exhibit more adverse responses from exposure to air pollutants than the general population (10).

Air pollution principally affects the respiratory, circulatory, and olfactory systems. The respiratory system is the principal route of entry for air pollutants, some of which may alter the function of the lungs.

Health effects data come from three types of studies: clinical, epidemiological, and toxicological. Clinical and epidemiological studies focus on human subjects, whereas toxicological studies are conducted on animals or simpler cellular systems. Ethical considerations limit human exposure to low levels of air pollutants which do not have irreversible effects. Table 7-1 lists the advantages and disadvantages of each type of experimental information.

TABLE 7-1

Three Disciplinary Approaches for Obtaining Health Information

Discipline	Population	Strengths	Weaknesses
Epidemiology	Communities	Natural exposure	Difficulty of quantifying exposure
	Diseased groups	No extrapolations	Many covariates
		Susceptible groups	Minimal dose–response data
		Long-term, low-level effects	Association vs. causation
Clinical studies	Experimental	Controlled exposure	Artificial exposure
	Diseased subjects	Few covariates	Acute effects only
		Vulnerable persons	Hazards
		Cause–effect	Public acceptance
Toxicology	Animals	Maximal dose–response data	Realistic models of human disease?
	Cells	Rapid acquisition of data	Threshold of human response?
	Biochemical systems	Cause–effect	Extrapolation
		Mechanism of response	

In general, clinical studies provide evidence on the effects of air pollutants under reproducible laboratory conditions. The exposure level may be accurately determined. The physiological effect may be quantified, and the health status of the subject is well known. This type of study can determine the presence or absence of various endpoints for a given sample group exposed to short-term, low-level concentrations of various air pollutants.

The fact that the subjects are exposed to the actual pollutants existing in their community is both the greatest strength and the greatest weakness of epidemiological studies. The strength is the real-world condition of the exposure and the subjects; the weakness is the difficulty in quantifying the relationship between exposure and subsequent effects. In the future, the development of biomarkers may provide a better indication of target dose.

The effects attributed to air pollutants range from mild eye irritation to mortality. In most cases, the effect is to aggravate preexisting diseases or to degrade the health status, making persons more susceptible to infection or development of a chronic respiratory disease. Some of the effects associated with specific pollutants are listed in Table 7-2. Further information is available in the U.S. Environmental Protection Agency criteria documents summarized in Chapter 22.

TABLE 7-2

Specific Air Pollutants and Associated Health Effects

Pollutant	Effects
CO	Reduction in the ability of the circulatory system to transport O_2
	Impairment of performance on tasks requiring vigilance
	Aggravation of cardiovascular disease.
NO_2	Increased susceptibility to respiratory pathogens
O_3	Decrement in pulmonary function
	Coughing, chest discomfort
	Increased asthma attacks
Lead	Neurocognitive and neuromotor impairment
	Heme synethesis and hematologic alterations
Peroxyacyl nitrates, aldehydes	Eye irritation
SO_2/particulate matter	Increased prevalence of chronic respiratory disease
	Increased risk of acute respiratory disease

V. IMPACT OF ODOR ON HUMANS

Odors are perceived via the olfactory system, which is composed of two organs in the nose: the olfactory epithelium, a very small area in the nasal system, and the trigeminal nerve endings, which are much more widely distributed in the nasal cavity (11). The olfactory epithelium is extremely sensitive, and humans often sniff to bring more odorant in contact with this area. The trigeminal nerves initiate protective reflexes, such as sneezing or interruption of inhalation, with exposure to noxious odorants.

The health effects of odors are extremely hard to quantify, yet people have reported nausea, vomiting, and headache; induction of shallow breathing and coughing; upsetting of sleep, stomach, and appetite; irritation of the eyes, nose, and throat; destruction of the sense of well-being and the enjoyment of food, home, and the external environment; disturbance; annoyance; and depression (11). Research under controlled conditions has qualitatively revealed changes in respiratory and cardiovascular systems. The difficulty has been in establishing the relationship between the intensity or duration of the exposure and the magnitude of the effects on these systems.

REFERENCES

1. Cowling, E. B., *Environ. Sci. Technol.* **16,** 110A–123A (1982).
2. Fantechi, R., and Ghazi, A. (eds.), "Carbon Dioxide and Other Greenhouse Gases: Climatic and Associated Impacts." Kluwer Academic Publishers, Boston, 1989.

3. International Symposium on Fluorides, "Fluorides: Effects on Vegetation, Animals and Humans." Paragon Press, Salt Lake City, UT, 1983.
4. Code of Federal Regulations, Protection of Environment Title 40, Subchapter I, Parts 240–272. U.S. Government Printing Office, Washington, DC, 1992.
5. Lee, D. H. K. (ed.), "Handbook of Physiology," Vol. 9, "Reactions to Environmental Agents." American Physiological Society, Bethesda, MD, 1977.
6. U.S. Environmental Protection Agency, "Air Quality Criteria for Lead," EPA 600/8-83-018F. Research Triangle Park, NC, June 1986.
7. National Research Council, "Carbon Monoxide." National Academy of Sciences. Washington, DC, 1977.
8. American Lung Association, "Health Effects of Air Pollution." New York, 1978.
9. Task Group on Lung Dynamics, *Health Phys.* **12,** 173 (1966).
10. Shy, C., *Am. J. Epidemiol.* **110,** 661–671 (1979).
11. National Research Council, "Odors from Stationary and Mobile Sources." National Academy of Sciences, Washington, DC, 1979.

SUGGESTED READING

Nagy, G. Z., The odor impact model. *J. Air Waste Manage. Assoc.* **42,** 1567 (1992).

National Research Council, "Human exposure assessment for airborne pollutants: advances and opportunities," National Academy Press, Washington, DC, 1991.

Tomatis, L. (ed.), "Air Pollution and Human Cancer." Springer-Verlag, New York, 1990.

QUESTIONS

1. By extrapolating from Fig. 2-4, what will be the concentration of CO_2 in the year 2050? How does this compare with the concentration in 1980?
2. What factors influence the accumulation of a chemical in the human body?
3. Explain why the inhalation route for lead is considered an important hazard when it accounts for only about 20% of the potential allowable body burden.
4. (a) Explain how CO interacts with the circulatory system. (b) Why are individuals with heart disease at greater risk when exposed to elevated CO levels?
5. Describe normal lung function.
6. How is particle deposition and removal from the lung influenced by the size of the particles?
7. How do exposure time and type of population influence the air quality standards established for the community and the workplace?
8. Compare the strengths and weaknesses of health effects information obtained from epidemiological, clinical, and toxicological studies.

8

Effects on Vegetation and Animals

I. INJURY VERSUS DAMAGE

The U.S. Department of Agriculture makes a distinction between air pollution damage and air pollution injury. *Injury* is considered to be any observable alteration in the plant when exposed to air pollution. *Damage* is defined as an economic or aesthetic loss due to interference with the intended use of a plant. This distinction indicates that injury by air pollution does not necessarily result in damage because any given injury may not prevent the plant from being used as intended, e.g., marketed.

Vegetation reacts with air pollution over a wide range of pollutant concentrations and environmental conditions. Many factors influence the outcome, including plant species, age, nutrient balance, soil conditions, temperature, humidity, and sunlight (1). Any type of observable effect due to exposure can be termed plant injury. A schematic diagram of the potential levels of injury with increasing exposure to air pollution is presented in Fig. 8-1. At low levels of exposure for a given species and pollutant, no significant effects may be observed. However, as the exposure level increases, a series of potential injuries may occur, including biochemical alterations, physiological response, visible symptoms, and eventual death.

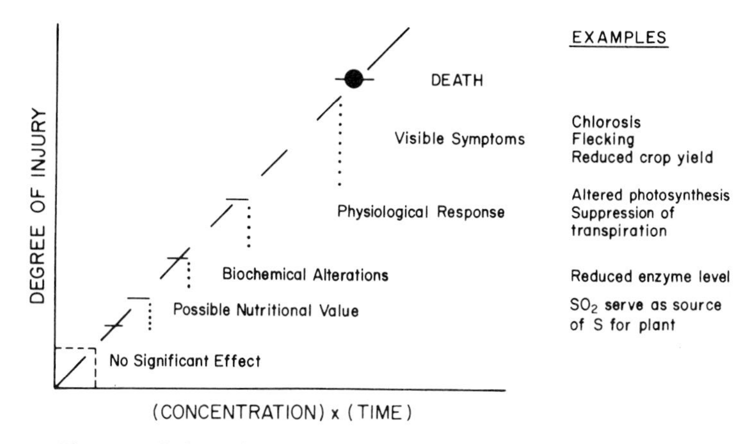

Fig. 8-1. Biological response spectrum for plants and air pollution.

Air pollutants may enter plant systems by either a primary or a secondary pathway. The primary pathway is analogous to human inhalation. Figure 8-2 shows the cross section of a leaf. Both of the outer surfaces are covered by a layer of epidermal cells, which help in moisture retention. Between the epidermal layers are the mesophyll cells—the spongy and palisade parenchyma. The leaf has a vascular bundle which carries water, minerals, and carbohydrates throughout the plant. Two important features shown in Fig. 8-2 are the openings in the epidermal layers called *stomates*, which are controlled by guard cells which can open and close, and air spaces in the interior of the leaf.

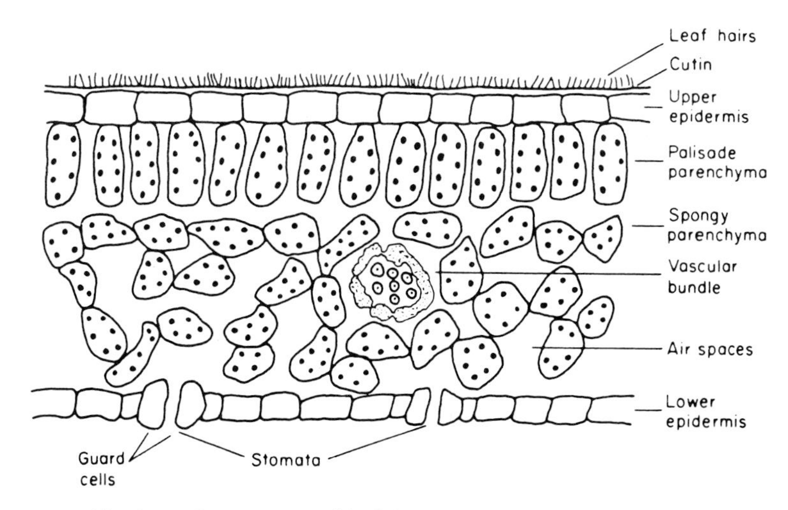

Fig. 8-2. Cross section of leaf showing various components.

The leaf structure has several important functions, three of which are photosynthesis, transpiration, and respiration (2). Photosynthesis is accomplished by chloroplasts in the leaf, which combine water and CO_2 in the presence of sunlight to form sugars and release O_2. This process is shown in Eq. (8-1).

$$6CO_2 + 6H_2O \rightarrow C_6H_{12}O_6 + 6O_2 \tag{8-1}$$

Transpiration is the movement of water from the root system up to the leaves and its subsequent evaporation to the atmosphere. This process moves nutrients throughout the plant and cools the plant. Respiration is a heat-producing process resulting from the oxidation of carbohydrates by O_2 to form CO_2 and H_2O, as shown in Eq. (8-2).

$$C_6H_{12}O_6 + 6O_2 \rightarrow 6CO_2 + 6H_2O \tag{8-2}$$

These three functions involve the movement of O_2, CO_2, and H_2O through the epidermal layers of the leaf. The analogy to human inhalation is obvious. With the diffusion of gases into and out of the leaf, pollutant gases have a direct pathway to the cellular system of the leaf structure. Direct deposition of particulate matter also occurs on the outer surfaces of the leaves.

The indirect pathway by which air pollutants interact with plants is through the root system. Deposition of air pollutants on soils and surface waters can cause alteration of the nutrient content of the soil in the vicinity of the plant. This change in soil condition can lead to indirect or secondary effects of air pollutants on vegetation and plants.

Injury to plants and vegetation is caused by a variety of factors, of which air pollution is only one. Drought, too much water, heat and cold, hail, insects, animals, disease, and poor soil conditions are some of the other causes of plant injury and possible plant damage (3). Estimates suggest that less than 5% of total crop losses are related to air pollution. Air pollution has a much greater impact on some geographic areas and crops than others. Crop failure can be caused by fumigation from a local air pollution source or by more widespread and more frequent exposure to adverse levels of pollution.

The subtle interaction of air pollutants with these other stressors to plants and vegetation is the subject of ongoing research. For some plant systems, exposure to air pollutants may induce biochemical modifications which interfere with the water balance in plants, thereby reducing their ability to tolerate drought conditions.

II. EFFECTS ON VEGETATION AND CROPS

The effects of air pollution on plants range from subtle to catastrophic, as shown in Fig. 8-1. Historically, these effects have been classified as visible symptoms and nonvisible or subtle effects (4). Visible symptoms are deviations from the normal healthy appearance of the leaves. For broadleaf plants, a healthy leaf has good color, with a normal cell structure in the various layers. Deviations from this healthy appearance include tissue collapse and various degrees of loss of color. Extensive tissue collapse or necrosis results from injury to the spongy or palisade cells in the interior of the leaf. The leaf is severely discolored and loses structural integrity. Dead tissue may fall out of the leaf, leaving holes in the structure. Less dramatic discolorations are caused by a reduction in the number of chloroplasts, a symptom referred to as *chlorosis*. Injury to the outer or epidermal layer is referred to as *glazing* or *silvering* of the leaf surface. When the pattern is spotty, the terms *flecking* and *stippling* are used to describe the injury.

Other forms of visible injury are related to various physiological alterations. Air pollution injury can cause early senescence or leaf drop. Stems and leaf structure may be elongated or misshapen. Ornamentals and fruit trees can also show visible injury to the blooms of the fruit, which can result in decreased yield.

The nonvisual or subtle effects of air pollutants involve reduced plant growth and alteration of physiological and biochemical processes, as well as changes in the reproductive cycle. Reduction in crop yield can occur without the presence of visible symptoms. This type of injury is often related to low-level, long-term chronic exposure to air pollution. Studies have shown that field plantings exposed to filtered and unfiltered ambient air have produced different yields when no visible symptoms were present (5). Reduction in total biomass can lead to economic loss for forage crops or hay.

Physiological or biochemical changes have been observed in plants exposed to air pollutants, including alterations in net photosynthesis, stomate response, and metabolic activity. Such exposure studies have been conducted under controlled laboratory conditions. An understanding of the processes involved will help to identify the cause of reduction in yield.

Laboratory studies have also investigated the interaction of air pollutants and the reproductive cycle of certain plants. Subtle changes in reproduction in a few susceptible species can render them unable to survive and prosper in a given ecosystem.

The major air pollutants which are phytotoxic to plants are ozone, sulfur dioxide, nitrogen dioxide, fluorides, and peroxyacyl nitrate (2). Table 8-1 lists some of the types of plants injured by exposure to these pollutants. The effects range from slight reduction in yield to extensive visible injury, depending on the level and duration of exposure. Examples of the distinction between air pollution injury and damage are also given in Table 8-1. Visible markings on plants or crops such as lettuce, tobacco, and orchids caused by air pollution translate into direct economic loss, i.e., damage. In contrast, visible markings on the leaves of grapes, potatoes, or corn

TABLE 8-1

Examples of Types of Injury and Air Pollution

Pollutant	Symptoms	Maturity of leaf affected	Part of leaf affected	Injury threshold		
				ppm (vol)	$\mu g/m^3$	Sustained exposure
Sulfur dioxide	Bleached spots, bleached areas between veins, chlorosis; insect injury, winter and drought conditions may cause similar markings	Middle-aged leaves most sensitive; oldest least sensitive	Mesophyll cells	0.3	785	8 hr
Ozone	Flecking, stippling, bleached spotting, pigmentation; conifer needle tips become brown and necrotic	Oldest leaves most sensitive; youngest least sensitive	Palisade or spongy parenchyma in leaves with no palisade	0.03	59	4 hr
Peroxyacetyl-nitrate (PAN)	Glazing, silvering, or bronzing on lower surface of leaves	Youngest leaves most sensitive	Spongy cells	0.01	50	6 hr
Nitrogen dioxide	Irregular, white or brown collapsed lesions on intercostal tissue and near leaf margin	Middle-aged leaves most sensitive	Mesophyll cells	2.5	4700	4 hr
Hydrogen fluoride	Tip and margin burns, dwarfing, leaf abscission; narrow brown-red band separates necrotic from green tissue; fungal disease, cold and high temperatures, drought, and wind may produce similar markings; suture red spot on peach fruit	Youngest leaves most sensitive	Epidermis and mesophyll cells	0.1 (ppb)	0.08	5 weeks

Ethylene	Sepal withering, leaf abnormalities; flower dropping, and failure of leaf to open properly; abscission; water stress may produce similar markings	Young leaves recover; older leaves do not recover fully	All	0.05	58	6 hr
Chlorine	Bleaching between veins, tip and margin burn, leaf abscission; marking often similar to that of ozone	Mature leaves most sensitive	Epidermis and mesophyll cells	0.10	290	2 hr
Ammonia	"Cooked" green appearance becoming brown or green on drying; overall blackening on some species	Mature leaves most sensitive	Complete tissue	~20	~14,000	4 hr
Hydrogen chloride	Acid-type necrotic lesion; tip burn on fir needles; leaf margin necrosis on broad leaves	Oldest leaves most sensitive	Epidermis and mesophyll cells	~5–10	~11,200	2 hr
Mercury	Chlorosis and abscission; brown spotting; yellowing of veins	Oldest leaves most sensitive	Epidermis and mesophyll cells	<1	<8,200	1–2 days
Hydrogen sulfide	Basal and marginal scorching	Youngest leaves most affected		20	28,000	5 hr
2,4-Dichlorophenoxyacetic acid (2–4D)	Scalloped margins, swollen stems, yellow-green mottling or stippling, suture red spot (2,4,5-T); epinasty	Youngest leaves most affected	Epidermis	<1	<9,050	2 hr
Sulfuric acid	Necrotic spots on upper surface similar to those caused by caustic or acidic compounds; high humidity needed	All	All	—	—	—

caused by air pollution will not result in a determination of damage if there is no loss in yield. Individual circumstances determine whether air pollution damage has occurred.

The costs of air pollution damage are difficult to estimate. However, estimates indicate crop losses of $1 billion to $5 billion for the United States (6). When compared to the crop losses due to all causes, this percentage is small. However, for particular crops in specific locations, the economic loss can be very high. Certain portions of the Los Angeles, California, basin are no longer suitable for lettuce crops because they are subject to photochemical smog. This forces producers either to move to other locations or to plant other crops that are less susceptible to air pollution damage. Concern has been expressed regarding the future impact of air pollution on the much larger Imperial Valley of California, which produces up to 50% of certain vegetables for the entire United States.

III. EFFECTS ON FORESTS

Approximately 1.95×10^{10} km^2 of the earth's surface has at least 20% or more crown tree cover, representing about one-third of the total land area (7). Several different types of forest ecosystems can be defined based on their location and the species present. The largest in area are tropical forest systems, followed by temperate forests, rain forests, and tidal zone systems. The temperate forest systems are located in the latitudes where the greatest industrialization is occurring and have the most opportunity to interact with pollutants in the atmosphere. The impact of air pollution on forest ecosystems ranges from beneficial to detrimental. Smith (8, 9) classified the relationship of air pollutants with forests into three categories: low dose (I), intermediate dose (II), and high dose (III). With this classification scheme, seemingly contradictory statements on the impact of air pollution on forests can be understood.

A. Low-Dose Levels

Under low-dose conditions, forest ecosystems act as sinks for atmospheric pollutants and in some instances as sources. As indicated in Chapter 7, the atmosphere, lithosphere, and oceans are involved in cycling carbon, nitrogen, oxygen, sulfur, and other elements through each subsystem with different time scales. Under low-dose conditions, forest and other biomass systems have been utilizing chemical compounds present in the atmosphere and releasing others to the atmosphere for thousands of years. Industrialization has increased the concentrations of NO_2, SO_2, and CO_2 in the "clean background" atmosphere, and certain types of interactions with forest systems can be defined.

Forests can act as sources of some of the trace gases in the atmosphere, such as hydrocarbons, hydrogen sulfide, NO_x, and NH_3. Forests have been identified as emitters of terpene hydrocarbons. In 1960, Went (10) estimated that hydrocarbon releases to the atmosphere were on the order of 108 tons per year. Later work by Rasmussen (11) suggested that the release of terpenes from forest systems is 2×10^8 tons of reactive materials per year on a global basis. This is several times the anthropogenic input. Yet, it is important to remember that forest emissions are much more widely dispersed and less concentrated than anthropogenic emissions. Table 8-2 shows terpene emissions from different types of forest systems in the United States.

Forest systems also act as sources of CO_2 when controlled or uncontrolled burning and decay of litter occur. In addition, release of ethylene occurs during the flowering of various species. One additional form of emission to the atmosphere is the release of pollen grains. Pollen is essential to the reproductive cycle of most forest systems but becomes a human health hazard for individuals susceptible to hay fever. The contribution of sulfur from forests in the form of dimethyl sulfide is considered to be about 10–25% of the total amount released by soils and vegetation (12).

Trees and soils of forests act as sources of NH_3 and oxides of nitrogen. Ammonia is formed in the soil by several types of bacteria and fungi. The volatilization of ammonia and its subsequent release to the atmosphere are dependent on temperature and the pH of the soil. Fertilizers are used as a tool in forest management. The volatilization of applied fertilizers may become a source of ammonia to the atmosphere, especially from the use of urea.

Nitrogen oxides are formed at various stages of the biological denitrification process. This process starts with nitrate; as the nitrate is reduced through various steps, NO_2, NO, N_2O, and N_2 can be formed and, depending on the conditions, released to the atmosphere.

The interactions of air pollutants with forests at low-dose concentrations result in imperceptible effects on the natural biological cycles of these species. In some instances, these interactions may be beneficial to the forest ecosystem. Forests, as well as other natural systems, act as sinks for the removal of trace gases from the atmosphere.

B. Intermediate-Dose Levels

The second level of interaction, the intermediate-dose level, can result in measurable effects on forest ecosystems. These effects consist of a reduction in forest growth, change in forest species, and susceptibility to forest pests. Both laboratory investigations and field studies show SO_2 to be an inhibitor of forest growth. When various saplings have been exposed to SO_2 in the laboratory, they show reduction in growth compared with unexposed

TABLE 8-2

Composition of U.S. Forest-Type Groups by Foliar Terpene Emissions

	Percent of total U.S. forest area	Percent α pinene emitters	Percent of isoprene emitters
Eastern type group			
Softwood types			
Loblolly-shortleaf pine	11	~100	Some from oak and sweetgum associates
Longleaf-slash pine	5	~100	Some from oak and sweetgum associates
Spruce-fir	4	~75	25% from spruce, which also emits α pinene
White-red-jack pine	2	~90	10% from aspen trees
Subtotal	22%	~91%	~9%
Hardwood types			
Oak-hickory	23	~10	70%, diluted by hickory, maple & black walnut
Oak-gum cypress	7	~50	50% from plurality of oak, cottonwood and willow
Oak-pine	5	~30	60%, diluted by black gum & hickory associates
Maple-beech-birch	6	~15	Terpene foliates are hemlock and white pine
Aspen-birch	5	~20	60%, diluted by birch, α pinene source balsam fir and balsam poplar
Elm-ash-cottonwood	4	—	30% from cottonwood, sycamore, willow
Subtotal	50%	~21%	~45%
Total	72%	—	—
Western type groups			
Softwoods			
Douglas fir	7	~100	—
Ponderosa pine	7	~100	5% from aspen associates
Lodgepole pine	3	~90	10% from Engelmann spruce and aspen
Fir-spruce	3	~100	40% from spruce trees
Hemlock-Sitka spruce	2	~100	25% from Sitka spruce
White pine	1	~100	5% from Englemann spruce
Larch	1	~100	—
Redwood	0.5	~100	—
Subtotal	24.5%	~98%	~12%
Hardwoods	2	—	~100% from aspen trees
Total	26.5%		

Source: Rasmussen, R. A., *J. Air Pollut. Contrl. Assoc.* **22,** 537–543 (1972).

saplings (13). Various field investigations of forest systems in the vicinity of large point sources show the effects of elevated SO_2 levels on the trees closer to the source. For example, Linzon (14) found that SO_2 from the Sudbury, Ontario (Canada), smelter caused a reduction in forest growth over a very large area, with the closer-in trees severely defoliated, damaged, and killed.

The effect of photochemical oxidants, mainly O_3 and peroxyacyl nitrate (PAN), on the forests located in the San Bernardino Mountains northeast of Los Angeles, California, has been to change the forest composition and to alter the susceptibility of forest species to pests. This area has been subjected to increasing levels of oxidant since the 1950s (Fig. 8-3). During the late 1960s and early 1970s, changes in the composition and aesthetic quality of the forest were observed (15).

During this period, the photochemical problem was expanding to a wider geographical region; and photochemical oxidant was transported to the San Bernardino Mountains with increasing frequency and at higher concentrations. The receptor forest system has been described as a mixed conifer system containing ponderosa pine, Jeffrey pine, white fir, and cedar, along with deciduous black oak. The damage to the ponderosa pine ranged from no visible injury to death. As the trees came under increased stress due to exposure to oxidant, they became more susceptible to pine beetle, which ultimately caused their death. The ponderosa pine appears to be more

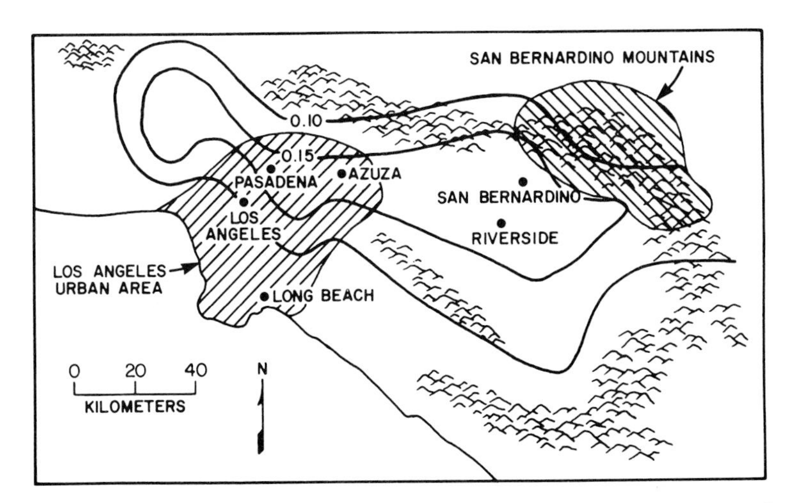

Fig. 8-3. Relationship between Los Angeles Basin's urban sources of photochemical smog and the San Bernardino Mountains, where ozone damage has occurred to the ponderosa pines. The solid lines are the average daily 1-hr maximum dose of ozone (ppm), July–September 1975–1977. Source: Adapted from Davidson, A., Ozone trends in the south coast air basin of California, *in* "Ozone/Oxidants: Interaction with the Total Environment." Air Pollution Control Association, Pittsburgh, 1979, pp. 433–450.

susceptible than the other members of this forest system, and continued exposure to photochemical oxidant may very well shift it from the dominant species to a minor one.

The interactions in the intermediate-dose category may result in effects on the reproduction cycle of species, the utilization of nutrients, the production of biomass, and the susceptibility to disease.

C. High-Dose Levels

The third category for interactions is high dose (III). The effects produced by this level of interaction can be seen by the casual observer. The result of high-dose exposure is destruction or severe injury of the forest system. High-dose conditions are almost always associated with point source emissions. The pollutants most often involved are SO_2 and hydrogen fluoride. Historically, the most harmful sources of pollution for surrounding forest ecosystems have been smelters and aluminum reduction plants.

One example of high-dose interaction is the impact of a smelter on the surrounding area in Wawa, Ontario, Canada. This smelter began operating about 1940. Gordon and Gorham (16) documented the damage in the prevailing downwind northeast sector for a distance of 60 km. They analyzed vegetative plots and established four zones of impact in the downwind direction: Within 8 km of the plant, damage was classified as "very severe" where no trees or shrubs survived; "severe damage" occurred at ~17 km, where no tree canopy existed; "considerable damage" existed at ~25 km, where some tree canopy remained, but with high tree mortality; and "moderate damage" was found at ~35 km, where a tree canopy existed but was put under stress and where the number of ground flora species was still reduced.

This type of severe air pollution damage has occurred several times in the past. If care is not taken, additional examples will be documented in the future.

D. Acid Deposition

Acid deposition refers to the transport of acid constituents from the atmosphere to the earth's surface. This process includes dry deposition of SO_2, NO_2, HNO_3, and particulate sulfate matter and wet deposition ("acid rain") to surfaces. This process is widespread and alters distribution of plant and aquatic species, soil composition, pH of water, and nutrient content, depending on the circumstances.

The impact of acid deposition on forests depends on the quantity of acidic components received by the forest system, the species present, and the soil composition. Numerous studies have shown that widespread areas in the eastern portion of North America and parts of Europe are being altered by acid deposition. Decreased pH in some lakes and streams in the affected areas was observed in the 1960s (17) and further evidence shows this trend.

When a forest system is subjected to acid deposition, the foliar canopy can initially provide some neutralizing capacity. If the quantity of acid components is too high, this limited neutralizing capacity is overcome. As the acid components reach the forest floor, the soil composition determines their impact. The soil composition may have sufficient buffering capacity to neutralize the acid components. However, alteration of soil pH can result in mobilization or leaching of important minerals in the soil. In some instances, trace metals such as Ca or Mg may be removed from the soil, altering the Al tolerance for trees.

This interaction between airborne acid components and the tree–soil system may alter the ability of the trees to tolerate other environmental stressors such as drought, insects, and other air pollutants like ozone. In Germany, considerable attention is focused the role of ozone and acid deposition as a cause of forest damage. Forest damage is a complex problem involving the interaction of acid deposition, other air pollutants, forestry practices, and naturally occurring soil conditions.

IV. EFFECTS ON ANIMALS

Acid deposition and the alteration of the pH of aquatic systems has led to the acidification of lakes and ponds in various locations in the world. Low-pH conditions result in lakes which contain no fish species.

Heavy metals on or in vegetation and water have been and continue to be toxic to animals and fish. Arsenic and lead from smelters, molybdenum from steel plants, and mercury from chlorine-caustic plants are major offenders. Poisoning of aquatic life by mercury is relatively new, whereas the toxic effects of the other metals have been largely eliminated by proper control of industrial emissions. Gaseous (and particulate) fluorides have caused injury and damage to a wide variety of animals—domestic and wild—as well as to fish. Accidental effects resulting from insecticides and nerve gas have been reported.

Autopsies of animals in the Meuse Valley, Donora, and London episodes described in Chapter 16, Section III, revealed evidence of pulmonary edema. Breathing toxic pollutants is not, however, the major form of pollutant intake for cattle; ingestion of pollution-contaminated feeds is the primary mode.

In the case of animals we are concerned primarily with a two-step process: accumulation of airborne contaminants on or in vegetation or forage that serves as their feed and subsequent effects of the ingested herbage on animals. In addition to pollution-affected vegetation, carnivores (humans included) consume small animals that may have ingested exotic chemicals including pesticides, herbicides, fungicides, and antibiotics. Increasing environmental concern has pointed out the importance of the complete food chain for the physical and mental well-being of human beings.

A. Heavy Metal Effects

One of the earliest cattle problems involved widespread poisoning of cattle by arsenic at the turn of the century. Abnormal intake of arsenic results in severe colic (salivation, thirst, vomiting), diarrhea, bloody feces, and a garliclike odor on the breath; cirrhosis of the liver and spleen as well as reproductive effects may be noted. Arsenic trioxide in the feed must be approximately 10 mg/kg body weight for these effects to occur.

Cattle feeding on herbage containing 25–50 mg/kg (ppm wt) lead develop excitable jerking of muscles, frothing at the mouth, grinding of teeth, and paralysis of the larynx muscles; a "roaring" noise is caused by the paralysis of the muscles in the throat and neck.

Symptoms of molybdenum poisoning in cattle include emaciation, diarrhea, anemia, stiffness, and fading of hair color. Vegetation containing 230 mg/kg of this substance affects cattle.

Mercury in fish has been found in waters in the United States and Canada. Mercury in the waters is converted into methyl mercury by aquatic vegetation. Small fish consume such vegetation and in turn are eaten by larger fish and eventually by humans; food with more than 0.5 ppm of mercury (0.5 mg/kg) cannot be sold in the United States for human consumption.

B. Gaseous and Particulate Effects

Periodically, accidental emissions of a dangerous chemical affect animal well-being. During nerve gas experimentation in a desolate area in Utah, a high-speed airplane accidentally dropped several hundred gallons of nerve gas. As a result of the discharge, 6200 sheep were killed. Considering the large number of exotic chemicals being manufactured, such unfortunate accidents may be anticipated in the future.

Fluoride emissions from industries producing phosphate fertilizers or phosphate derivatives have caused damage to cattle throughout the world; phosphate rock, the raw material, can contain up to 4% fluoride, part of which is discharged to air (and waters) during processing. In Polk and Hillsborough counties of Florida, the cattle population decreased by 30,000 between 1953 and 1960 as a result of fluoride emissions. Since 1950, research has greatly increased our knowledge of the effect of fluorides on animals; standards and guides for diagnosing and evaluating fluorosis in cattle have been compiled.

Chronic fluoride toxicity (fluorosis) is the type most frequently observed in cattle. The primary effects of fluorides in cattle are seen in the teeth and bones. Excessive intake weakens the enamel of developing teeth; the initially dulled erupted teeth can develop into soft teeth, with uneven wearing of molar teeth. Characteristic osteofluorotic bone lesions develop, causing intermittent lameness and stiffness in the animal. Fluoride content of the bone increases with dosage despite excretion in urine and feces. Secondary symptoms include reduced lactation, nonpliable skin, and dry, rough hair coat. As shown in Fig. 8-4, the fluoride ingestion level correlates with the fluoride content of bones and urine as well as incisor teeth classification (18).

Tolerance of animals for fluorides varies, dairy cattle being most sensitive and poultry least (Table 8-3). Fluorosis of animals in contaminated areas can be avoided by keeping the intake levels below those listed by incorporating clean feeds with those high in fluorides. It has also been determined

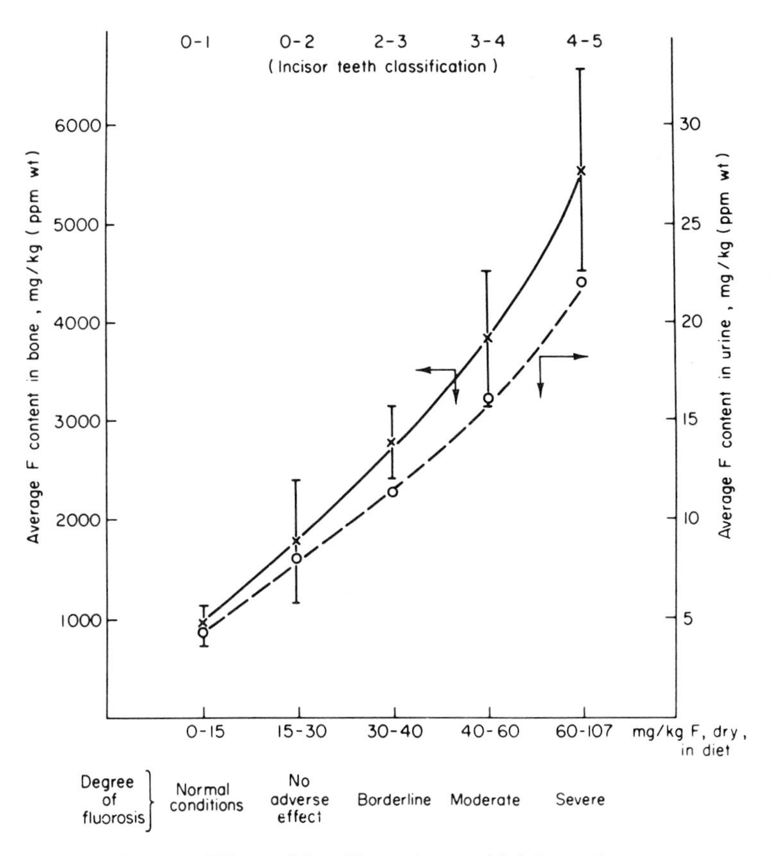

Fig. 8-4. Effects of fluoride on 4-year-old dairy cattle.

TABLE 8-3

Fluoride Tolerance of Animals (ppm wt in Ration, Dry)[a]

Species	Breeding or lactating animals (ppm)[b]	Finishing animals to be sold for slaughter with average feeding period (ppm)[b]
Dairy, beef heifers	30	100
Dairy cows	30	100
Beef cows	40	100
Steers	—	100
Sheep	50	160
Horse	60	—
Swine	70	—
Turkeys	—	100
Chickens	—	150

[a] Data based on soluble fluoride; increased values for insoluble fluoride compounds.
[b] 1 ppm wt = 1 mg/kg.

that increased consumption of aluminum and calcium salts can reduce the toxicity of fluorides in animals.

REFERENCES

1. Levitt, J., "Responses of Plants to Environmental Stresses." Academic Press, New York, 1972.
2. Hindawi I. I., "Air Pollution Injury to Vegetation, AP-71." United States Department of Health, Education, and Welfare, Raleigh, NC, 1970.
3. Treshow, M., "Environment and Plant Response." McGraw-Hill, New York, 1970.
4. Heck, W. W., and Brandt, C. S., Effects on vegetation, *in* "Air Pollution," 3rd ed., Vol. 111 (A. C. Stern, ed.), pp. 157–229. Academic Press, New York, 1977.
5. Hileman, B., *Environ. Sci. Technol.* **16,** 495A–499A (1982).
6. Heck, W. W., Heagle, A. S., and Shriner, D. S., Effects on vegetation: native, crops, forests, *in* "Air Pollution VI" (A. C. Stern, ed.), pp. 247–350. Academic Press, Orlando, FL, 1986.
7. "McGraw-Hill Encyclopedia of Science & Technology," Vol. 5, p. 658. McGraw-Hill, New York, 1982.
8. Smith, W. H., *Environ. Pollut.* **6,** 111–129 (1974).
9. Smith, W. H., "Air Pollution and Forests: Interaction between Air Contaminants and Forest Ecosystems," 2nd ed. Springer-Verlag, New York, 1990.
10. Went, F. W., *Nature (London)* **187,** 641–643 (1960).
11. Rasmussen, R. A., *J. Air Pollut. Control Assoc.* **22,** 537–543 (1972).
12. Bremmer, J. M., and Steele, C. G., *Adv. Microbiol. Ecol.* **2,** 155–201 (1978).
13. Keller, T., *Environ. Pollut.* **16,** 243–247 (1978).
14. Linzon, S. N., *J. Air Pollut. Control Assoc.* **21,** 81–86 (1971).

15. Miller, P. R., and Elderman, M. J. (eds.), "Photochemical Oxidant Air Pollution Effects of a Mixed Conifer System." EPA-600/3-77-104. U.S. Environmental Protection Agency, Corvallis, OR, 1977.
16. Gordon, A. G., and Gorham, E., *Can. J. Bot.* **41,** 1063–1078 (1973).
17. Oden, S., "The Acidification of Air and Precipitation and Its Consequences in the Natural Environment." Ecology Committee, Bulletin No. 1, Swedish National Science Research Council, Stockholm, 1967.
18. Shupe, J. L., *Am. Ind. Hyg. Assoc. J.* **31,** 240–247 (1970).

SUGGESTED READING

"Acidic Precipitation," Vols. 1–5. Advances in Environmental Sciences, Springer-Verlag, London, 1989–90.

Georgii, H. W., "Atmospheric Pollutants in Forest Areas: Their Deposition and Interception." Kluwer Academic Publishers, Boston, 1986.

Legge, A. H., and Krupa, S. V., "Air Pollutants and Their Effects on the Terrestrial Ecosystem." Wiley, New York, 1986.

Mellanby, K. (ed.), "Air Pollution, Acid Rain, and the Environment." Elsevier Science Publishers, Essex, England, 1988.

National Research Council, "Biologic Markers of Air Pollution Stress and Damage in Forests." National Academy Press, Washington, DC, 1989.

Smith, W. H., "Air Pollution and Forests: Interactions between Air Contaminants and Forest Ecosystems," 2nd ed. Springer-Verlag, New York, 1990.

QUESTIONS

1. Distinguish between air pollution damage and injury.
2. Why is it difficult to prove that effects on plants in the field observed visually were cause by exposure to air pollution?
3. What functions do the stomates serve in gas exchange with the atmosphere?
4. Why is air pollution damage important when estimates suggest that it accounts for less than 5% of total crop losses in the United States?
5. List examples of air pollution effects on plants that cannot be detected by visual symptoms.
6. What types of trace gases are released to the atmosphere by forest ecosystems?
7. How have ozone and insects interacted to damage trees in the San Bernardino Mountain National Forest of California?
8. Why are animals used in research on air pollution effects?
9. Calculate the daily fluoride intake of a dairy animal from (a) air and (b) food and water, based on the conditions below and assuming 100% retention of the fluoride:

   Animal breathing rate: 30 kg air per day containing 6 μg fluoride per cubic meter of air (STP)

   Animal food and water intake:

   Herbage 10 kg containing 200 mg/kg of fluoride

   Water 5 kg containing 1 mg/kg of fluoride

9

Effects on Materials and Structures

I. EFFECTS ON METALS

The principal effects of air pollutants on metals are corrosion of the surface, with eventual loss of material from the surface, and alteration in the electrical properties of the metals. Metals are divided into two categories—ferrous and nonferrous. Ferrous metals contain iron and include various types of steel. Nonferrous metals, such as zinc, aluminum, copper, and silver, do not contain iron.

Three factors influence the rate of corrosion of metals—moisture, type of pollutant, and temperature. A study by Hudson (1) confirms these three factors. Steel samples were exposed for 1 year at 20 locations throughout the world. Samples at dry or cold locations had the lowest rate of corrosion, samples in the tropics and marine environments were intermediate, and samples in polluted industrial locations had the highest rate of corrosion. Corrosion values at an industrial site in England were 100 times higher than those found in an arid African location.

The role of moisture in corrosion of metals and other surfaces is twofold: surface wetness acts as a solvent for containments and for metals is a medium for electrolysis. The presence of sulfate and chloride ions accelerates the corrosion of metals. Metal surfaces can by wetted repeatedly over a period of time as the humidity fluctuates.

Several studies have been conducted in urban areas to relate air pollution exposure and metal corrosion. In Tulsa, Oklahoma, wrought iron disks were exposed in various locations (2). Using weight change as a measure of air pollution corrosion, the results indicated higher corrosion rates near industrial sectors containing an oil refinery and fertilizer and sulfuric acid manufacturing facilities. Upham (3) conducted a metal corrosion investigation in Chicago. Steel plates were exposed at 20 locations, and SO_2 concentrations were also measured. Figure 9-1 shows the relationship between weight loss during 3-, 6-, and 12-month exposure periods and the mean SO_2 concentration. Corrosion was also found to be higher in downtown locations than in suburban areas. Nonferrous metals are also subject to corrosion, but to a lesser degree than ferrous metals. Table 9-1 compares the weight loss of several nonferrous metals over a 20-year period (4). The results vary depending on the type of exposure present.

Zinc is often used as a protective coating over iron to form galvanized iron. In industrial settings exposed to SO_2 and humidity, this zinc coating is subject to sufficient corrosion to destroy its protective capacity. Haynie and Upham (5) used their results from a zinc corrosion study to predict the useful life of a zinc-coated galvanized sheet in different environmental settings. Table 9-2 shows the predicted useful life as a function of SO_2 concentration.

Aluminum appears to be resistant to corrosion from SO_2 at ambient air concentrations. Aluminum alloys tend to form a protective surface film

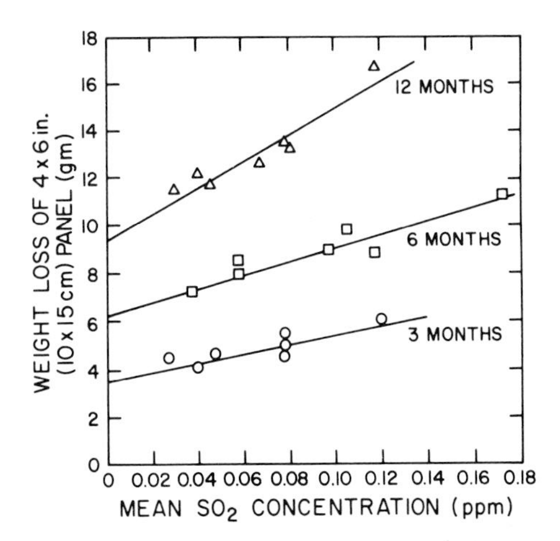

Fig. 9-1. Relationship between corrosion of mild steel and corresponding mean sulfur dioxide concentration at seven Chicago, Illinois sites. Source: Upham, J. B., *J. Air Pollut. Control Assoc.* **17,** 400 (1967).

TABLE 9-1

Weight Loss of Metal Panels[a] after 20 Years' Exposure in Various Atmospheres (ca. 1930–1954)[b]

City	Exposure classification	Average loss in weight, %					
		Commercial copper (99.9% + Cu)	Commercial aluminum (99% + Al)	Brass (85% Cu, 15% Zn)	Nickel (99% + Ni)	Commercial lead (99.92% Pb, 0.06% Cu)	Commercial zinc (99% Zn, 0.85% Pb)
Altoona, PA	Industrial	6.1	—	8.5	25.2	1.8	30.7
New York, NY	Industrial	6.4	3.4	8.7	16.6	—	25.1
La Jolla, CA	Seacoast	5.4	2.6	1.3	0.6	2.1	6.9
Key West, FL	Seacoast	2.4	—	2.5	0.5	—	2.9
State College, PA	Rural	1.9	0.4	2.0	1.0	1.4	5.0
Phoenix, AZ	Rural	0.6	0.3	0.5	0.2	0.4	0.8

[a] Panels—9 × 12 × 0.035 in (22.86 × 30.48 × 0.089 cm).

[b] Data from H. R. Copson, Report of ASTM Subcommittee VI, of Committee B-3 on Atmospheric Corrosion, Am. Soc. Test Mater., Spec. Tech. Publ. 175, 1955. Used by permission of the American Society for Testing and Materials, Philadelphia.

Source: Yocom, J. E., and Upham, J. B., Effects on Economic Materials and Structures, *in* "Air Pollution," 3rd ed., Vol. I (A. C. Stern, ed.), p. 80. Academic Press, New York, 1977.

TABLE 9-2

Predicted Useful Life of Galvanized Sheet Steel with a 53-μm Coating at an Average Relative Humidity of 65%[a]

SO_2 concentration, $\mu g/m^2$	Type of environment	Useful life (years)		
		Predicted best estimate	Predicted range	Observed range
13	Rural	244	41.0	30–35
130	Urban	24	16.0–49.0	
260	Semi-industrial	12	10.0–16.0	15–20
520	Industrial	6	5.5–7.0	
1,040	Heavy industrial	3	2.9–3.5	3–5

[a] Source: Yocom, J. E., and Upham, J. B., Effects of Economic Materials and Structures, *in* "Air Pollution," 3rd ed., Vol. I (A. C. Stern, ed.), p. 80. Academic Press, New York, 1977.

that limits further corrosion upon exposure to SO_2. Laboratory studies at higher concentrations (280 ppm) show corrosion of aluminum at higher humidities (>70%), with the formation of a white powder of aluminum sulfate.

Copper and silver are used extensively in the electronics industry because of their excellent electrical conductivity. These metals tend to form a protective surface coating which inhibits further corrosion. When exposed to H_2S, a sulfide coating forms, increasing the resistance across contacts on electrical switches (6).

II. EFFECTS ON STONE

The primary concern in regard to air pollution is the soiling and deterioration of limestone, which is widely used as a building material and for marble statuary. Figure 9-2 shows the long-term effects of urban air pollution on the appearance of stone masonry. Many buildings in older cities have been exposed to urban smoke, SO_2, and CO_2 for decades. The surfaces have become soiled and are subjected to chemical attack by acid gases. Exterior building surfaces are also subjected to a wet–dry cycle from rain and elevated humidity. SO_2 and moisture react with limestone ($CaCO_3$) to form calcium sulfate ($CaSO_4$) and gypsum ($CaSO_4 \cdot 2H_2O$). These two sulfates are fairly soluble in water, causing deterioration in blocks and in the mortar used to hold the blocks together. The soluble calcium sulfates can penetrate into the pores of the limestone and recrystallize and expand, causing further deterioration of the stone. CO_2 in the presence of moisture forms carbonic acid. This acid converts the limestone into bicarbonate, which is also water soluble and can be leached away by rain. This type of mechanism is present in the deterioration of marble statues.

Fig. 9-2. Old post office building being cleaned in St. Louis, Missouri, 1963. Source: Photo by H. Neff Jenkins.

III. EFFECTS ON FABRICS AND DYES

The major effects of air pollution on fabrics are soiling and loss of tensile strength. Sulfur oxides are considered to cause the greatest loss of tensile strength. The most widely publicized example of this type of problem has been damage to women's nylon hose by air pollution, described in newspaper accounts. The mechanism is not understood, but it is postulated that fine droplets of sulfuric acid aerosol deposit on the very thin nylon fibers, causing them to fail under tension. Cellulose fibers are also weakened by sulfur dioxide. Cotton, linen, hemp, and rayon are subject to damage from SO_2 exposure.

Brysson and co-workers (7) conducted a study in St. Louis, Missouri, on the effects of urban air pollution on the tensile strength of cotton duck material. Samples were exposed at seven locations for up to 1 year. Figure 9-3 shows the relationship between tensile strength and pollutant exposure. For two levels of ambient air exposure, the materials exhibited less than one-half their initial tensile strength when exposed to air pollution for 1 year.

Particulate matter contributes to the soiling of fabrics. The increased frequency of washing to remove dirt results in more wear on the fabric, causing it to deteriorate in the cleaning process.

In addition to air pollution damage to fabrics, the dyes used to color fabrics have been subject to fading caused by exposure to air pollutants. Since the early 1900s, fading of textile dyes has been a continuing problem. The composition of dyes has been altered several times to meet demands of new fabrics and to "solve" the fading problem. Before World War I, dyes used on wool contained free or substituted amino acid groups, which were found to be susceptible to exposure to nitrogen dioxide.

When cellulose acetate rayon was introduced in the mid-1920s, old dye technology was replaced with new chemicals called dispersive dyes. Not long after their initial use, fading of blue, green, and violet shades began to be observed in material exposed to nitrogen oxides. The fabric was marked by a reddening discoloration. Laboratory studies duplicating am-

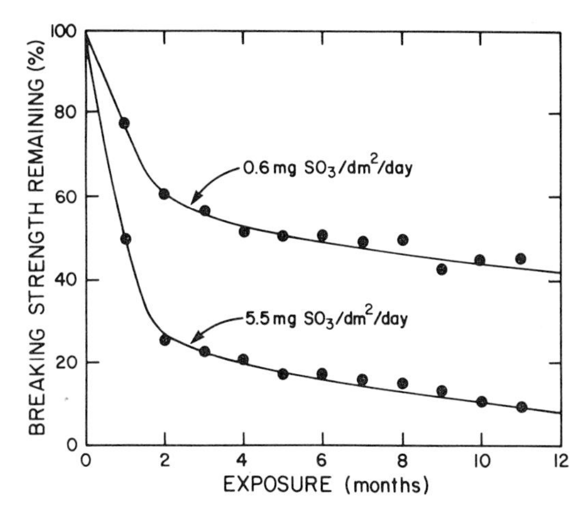

Fig. 9-3. Effects of sulfation and time on tensile strength of cotton duck. Source: Brysson, R. J., Trask, B. J., Upham, J. B., and Booras, S. A. *J. Air Pollut. Control Assoc.* **17,** 294 (1967).

bient air levels of nitrogen dioxide and humidity reproduced these effects (8).

Ozone has also been found to cause fading of material. This was discovered when white fabrics developed a yellow discoloration (9, 10), leading researchers to investigate the effects of ozone on other chemicals added to the material, including optical brighteners, antistatic and soil-release finishes, and softeners. A very complex process was occurring where the dyes were migrating to the permanent-press-finish materials, e.g., softeners. Softeners have been found to be good absorbers of gases. Fading results from the combination of dye and absorbed nitrogen dioxide and ozone. This combination with high relative humidities has caused color fading in numerous types of material and dye combinations. Although some progress has been made in the development of fade-resistant dyes, they are more expensive and have poorer dyeing properties.

IV. EFFECTS ON LEATHER, PAPER, PAINT, AND GLASS

Sulfur dioxide affects the composition of leather and paper, causing significant deterioration. The major concern is the destruction of leatherbound books in the libraries of the world. SO_2 is absorbed by leather and converted to sulfuric acid, which attacks the structure of the leather. Initially, the edges of the exposed back of the book begin to crack at the hinges. As the cracks expand, more leather is exposed and the cracks widen, with the back eventually falling off the book. Preventive measures now include storage in sulfur dioxide–free air.

The cellulose fiber in paper is attacked and weakened by sulfur dioxide. Paper made before about 1750 is not significantly affected by sulfur dioxide (11). At about that time, the manufacture of paper changed to a chemical treatment process that broke down the wood fiber more rapidly. It is thought that this process introduces trace quantities of metals, which catalyze the conversion of sulfur dioxide to sulfuric acid. Sulfuric acid causes the paper to become brittle and more subject to cracking and tearing. New papers have become available to minimize the interaction with SO_2.

Paints are designed to decorate and protect surfaces. During normal wear, paint chalks moderately to clean the surface continuously. A hardened paint surface resists sorption by gases, although the presence of relatively high concentrations of 2620 to 5240 $\mu g/m^3$ SO_2 (1–2 ppm) increases the drying time of newly painted surfaces. Hydrogen sulfide reacts with lead base pigments

$$Pb^{2+} + H_2S \rightarrow PbS + 2H^+ \quad (9\text{-}1)$$

to blacken white and light-tinted paints. Wohlers and Feldstein (12) concluded that lead base paints could discolor surfaces in several hours at a concentration of 70 $\mu g/m^3$ H_2S (0.05 ppm). In time the black lead sulfide oxidizes to the original color. Paints pigmented with titanium or zinc do not form a black precipitate. Alkyd or vinyl vehicles and pigments contain no heavy metal salts for reaction with H_2S. Painted surfaces are also dirtied by particulate matter. Contaminating dirt can readily become attached to wet or tacky paint, where it is held tenaciously and forms focal points for gaseous sorption for further attack. Dirt that collects on roofs or in gutters, blinds, screens, windowsills, or other protuberances is eventually washed over external surfaces to mar decorative effects.

Paints and coatings for automobiles have not been immune to damage by air pollution. Wolff and co-workers (13) found that damage to automobile finishes was the result of scarring by calcium sulfate crystals formed when sulfuric acid in rain or dew reacted with dry deposited calcium.

Glass is normally considered a very stable material. However, there is growing evidence that SO_2 air pollution may be accelerating the deterioration of medieval glass. A corrosion surface forms on these glass surfaces and the sulfate present helps prolong surface wetness. This condition is conducive to further attack and degradation of the glass surface (14).

V. EFFECTS ON RUBBER

Although it was known for some time that ozone cracks rubber products under tension, the problem was not related to air pollution. During the early 1940s, it was discovered that rubber tires stored in warehouses in Los Angeles, California, developed serious cracks. Intensified research soon identified the causative agent as ozone that resulted from atmospheric reaction between sunlight (3000–4600 Å), oxides of nitrogen, and specific types of organic compounds, i.e., photochemical air pollution.

Natural rubber is composed of polymerized isoprene units. When rubber is under tension, ozone attacks the carbon–carbon double bond, breaking the bond. The broken bond leaves adjacent $C=C$ bonds under additional stress, eventually breaking and placing still more stress on surrounding $C=C$ bonds. This "domino" effect can be discerned from the structural formulas in Fig. 9-4. The number of cracks and the depth of the cracks in rubber under tension are related to ambient concentrations of ozone.

Rubber products may be protected against ozone attack by the use of a highly saturated rubber molecule, the use of a wax inhibitor which will "bloom" to the surface, and the use of paper or plastic wrappings to protect the surface. Despite these efforts, rubber products still crack more on the West Coast than on the East Coast of the United States.

Natural rubber

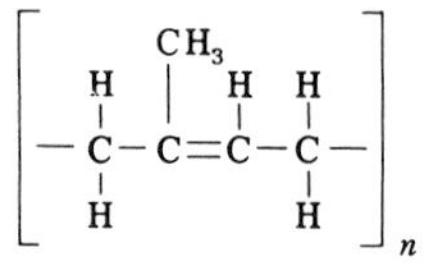

Ozone attacks C=C bond.
Natural rubber has the formula $(C_5H_8)_n$.

Butadiene-styrene rubber

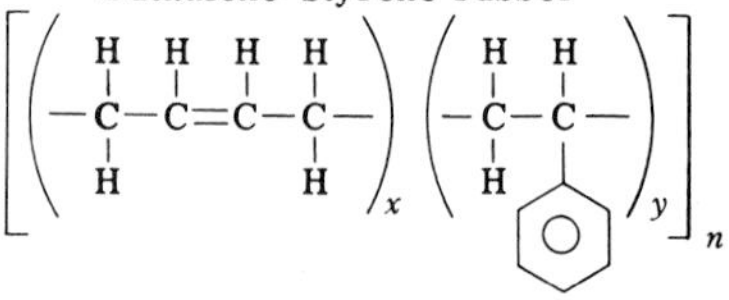

This synthetic rubber shows about same low resistance to ozone as natural rubber.

Polychloroprene rubber

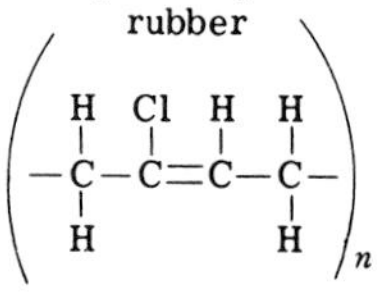

Although this rubber is unsaturated, the chlorine atom near the C=C makes molecule more resistant to ozone.

Isobutylene-diolefin rubber

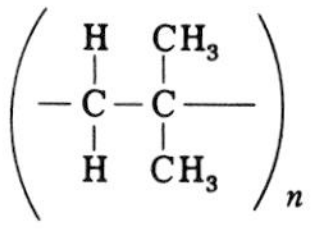

n + a few percent dienes

Since this rubber contains few C=C bonds, it is relatively resistant to ozone.

Silicon rubber

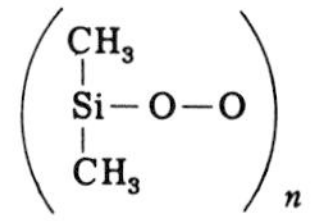

This synthetic rubber contains no C=C bond and hence is resistant to ozone.

Fig. 9-4. Susceptibility of natural and synthetic rubbers to attack by ozone.

REFERENCES

1. Hudson, J. D., *J. Iron Steel Inst. London* **148,** 161 (1943).
2. Galegar, W. C., and McCaldin, R. O., *Am. Ind. Hyg. Assoc. J.* **22,** 187 (1961).
3. Upham, J. B., *J. Air Pollut. Control Assoc.* **17,** 400 (1967).
4. Yocom, J. E., and Upham, J. B., Effects on economic materials and structures, *in* "Air Pollution," 3rd ed., Vol. II (A. C. Stern, ed.). Academic Press, New York, 1977, p. 80.
5. Haynie, F. H., and Upham, J. B., *Mater. Prot. Performance* **9,** 35–40 (1970).
6. Leach, R. H., Corrosion in Liquid Media, the Atmosphere, and Gases—Silver and Silver Alloys, *in* "Corrosion Handbook," Sect. 2 (H. H. Uhlig, ed.). Wiley, New York, 1948, p. 319.
7. Brysson, R. J., Trask, B. J., Upham, J. B., and Booras, S. G., *J. Air Pollut. Control Assoc.* **17,** 294 (1967).

8. Salvin, V. S., *Am. Dyest. Rep.* **53,** 33–41 (1964).
9. McLendon, V., and Richardson, F., *Am. Dyest. Rep.* **54,** 305–311 (1965).
10. Salvin, V. S., "Proceeding of Annual Conference, American Society for Quality Control, Textile and Needle Trades Division," Vol. **16,** pp. 56–64 (1969).
11. Langwell, W. H., *Proc. R. Inst. Gr. Brit.* **37** (Part II, No. 166), 210 (1958).
12. Wohlers, H. C., and Feldstein, M., *J. Air Pollut. Control Assoc.* **16,** 19–21 (1966).
13. Wolff, G. T., Rodgers, W. R., Collins, D. C., Verma, M. H., Wong, C. A., *J. Air Waste Manage. Assoc.* **40,** 1638–1648 (1990).
14. Newton, R. G., "The Deterioration and Conservation of Painted Glass: A Critical Bibliography." Oxford University Press, Oxford, 1982.

SUGGESTED READING

Butlin, R. N., The effects of pollutants on materials, *in* "Energy and the Environment." Royal Society of Chemistry, Cambridge, 1990.

Liu, B., and Yu, E. S., "Air Pollution Damage Functions and Regional Damage Estimates." Technomic, Westport, CT, 1978.

Yocum, J. E., and Upham J. B., Effects on Economic Materials and Structures, *in* "Air Pollution," 3rd ed, Vol. II (A. C. Stern, ed.). Academic Press, New York, 1977.

QUESTIONS

1. Assuming that a relationship exists among corrosion, population, and sulfur dioxide, why might one expect this interdependence?
2. Compare the solubilities in water of calcium carbonate, calcium sulfite, calcium sulfate, magnesium sulfate, and dolomite.
3. Describe possible mechanisms for the deterioration of marble statuary.
4. Explain why soiling and corrosion are hidden costs of air pollution.
5. Describe possible preventive actions to limit deterioration of books and other print material in libraries.
6. Explain the role of moisture in corrosion of materials.
7. In your location, determine whether sulfur dioxide, ozone, or particulate matter contributes to soiling or corrosion problems.
8. Describe why some types of synthetic rubber are less susceptible to ozone attack than natural rubber.

10

Effects on the Atmosphere, Soil, and Water Bodies

I. THE PHYSICS OF VISIBILITY

Impairment of visibility involves degradation of the ability to perceive the environment. Several factors are involved in determining visibility in the atmosphere (Fig. 10-1): the optical characteristics of the illumination source, the viewed targets, the intervening atmosphere, and the characteristics of the observer's eyesight (1).

In order to see an object, an observer must be able to detect the contrast between the object and its surroundings. If this contrast decreases, it is more difficult to observe the object. In the atmosphere, visibility can decrease for a number of reasons. For example, we may be farther away from the object (e.g., an airplane can move away from us); the sun's angle may change with the time of day; and if air pollution increases, the contrast may decrease, reducing our ability to see the object.

Objects close to us are easily perceived, but as we attempt to detect objects farther and farther away from us, the contrast between the object and the background decreases. The lowest limit of contrast for human observers is called the *threshold contrast* and is important because this value influences the maximum distance at which we can see various objects.

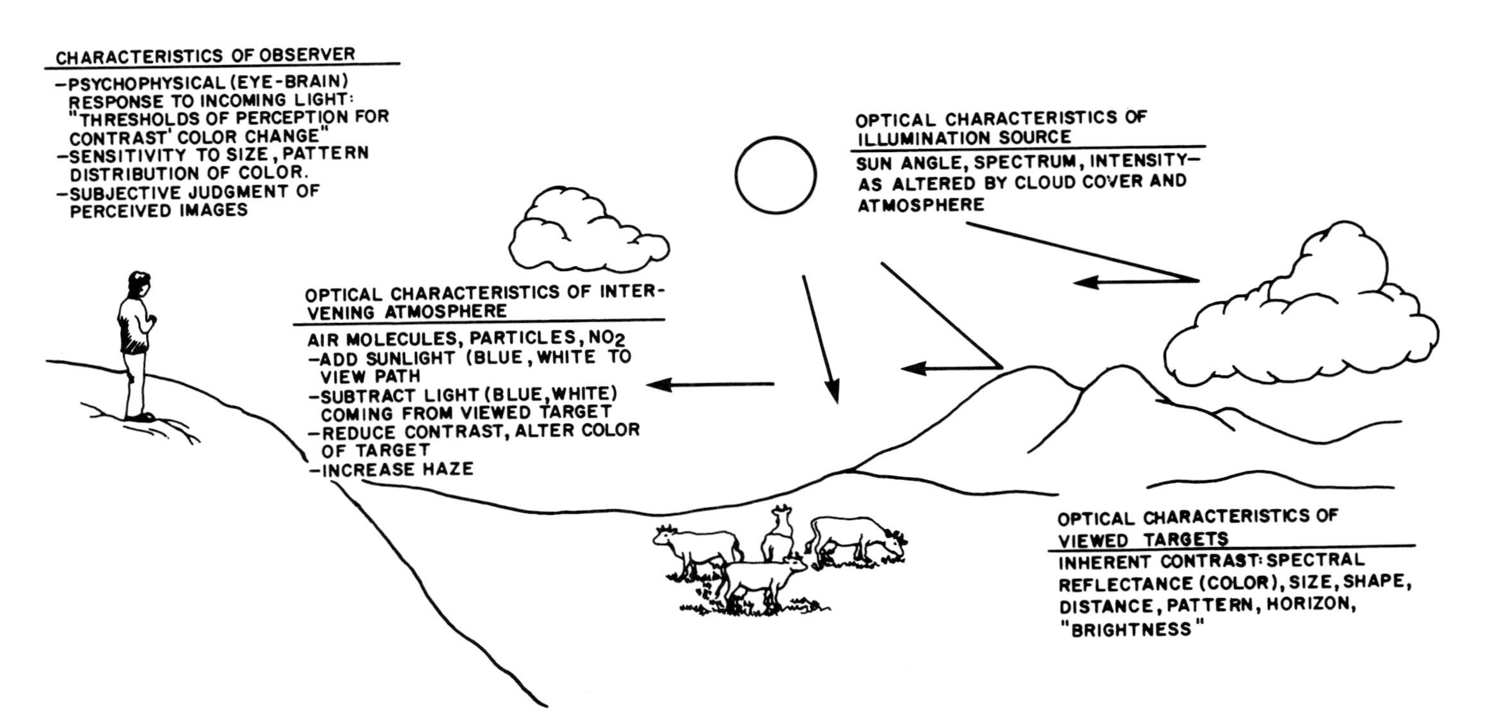

Fig. 10-1. Factors determining visibility in the atmosphere. Source: U.S. Environmental Protection Agency, "Protecting Visibility," EPA-450/5-79-008. Office of Air Quality Planning and Standards, Research Triangle Park, NC, 1979.

Thus, it is closely related to our understanding of good versus bad visibility for a particular set of environmental conditions.

Threshold contrast is illustrated in Fig. 10-2. I is the intensity of light received by the eye from the object, and $I + \Delta I$ represents the intensity coming from the surroundings. The threshold contrast can be as low as 0.018–0.03 and the object can still be perceptible. Other factors, such as the physical size of the visual image on the retina of the eye and the brain's response to the color of the object, influence the perception of contrast.

Let's consider the influence of gases and particles on the optical properties of the atmosphere. Reduction in visibility is caused by the following interactions in the atmosphere: light scattering by gaseous molecules and particles, and light absorption by gases and particles (2).

Light-scattering processes involve the interaction of light with gases or particles in such a manner that the direction or frequency of the light is altered. Absorption processes occur when the electromagnetic radiation interacts with gases or particles and is transferred internally to the gas or particle.

Light scattering by gaseous molecules is wavelength dependent and is the reason why the sky is blue. This process is dominant in atmospheres that are relatively free of aerosols or light-absorbing gases. Light scattering by particles is the most important cause of visibility reduction. This phenomenon is dependent on the size of the particles suspended in the atmosphere.

Light absorption by gases in the lower troposphere is limited to the absorption characteristics of nitrogen dioxide. This compound absorbs the shorter, or blue, wavelengths of visible light, causing us to observe the red wavelengths. We therefore perceive a yellow to reddish-brown tint in atmospheres containing quantities of NO_2. Light absorption by particles is related principally to carbonaceous or black soot in the atmosphere. Other

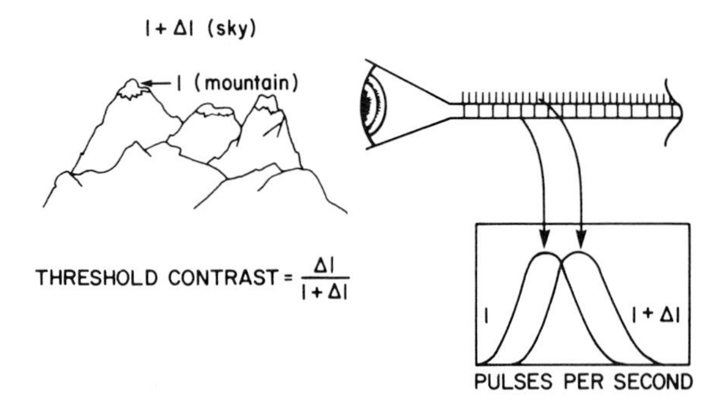

Fig. 10-2. Threshold contrast in distinguishing an object from its surroundings. The eye responds to an increment in light intensity by increasing the number of signals (pulses) sent to the brain. The detection of threshold contrast involves the ability to discriminate between the target (I) and the brighter background ($I + \Delta I$). Source: Gregory, R. L., Eye and Brain: "The Psychology of Seeing." Weidenfeld and Nicolson, London, 1977.

types of fine particles such as sulfates, although not good light absorbers, are very efficient at scattering light.

A. Light Extinction in the Atmosphere

The interaction of light in the atmosphere is described mathematically in Eq. (10-1):

$$-\,dI = b_{\text{ext}}\, I\, dx \tag{10-1}$$

where $-dI$ is the decrease in intensity, b_{ext} the extinction coefficient, I the original intensity of the beam of light, and dx the length of the path traveled by the beam of light.

Figure 10-3(a) shows a beam of light transmitted through the atmosphere. The intensity of the beam $I(x)$ decreases with the distance from the illumination source as the light is absorbed or scattered out of the beam. For a short period, this decrease is proportional to the intensity of the beam and the length of the interval at that point. Here b_{ext} is the extinction or attenuation coefficient and is a function of the degree of scattering and absorption of the particles and gases which are present in the beam path.

Figure 10-3(b) illustrates a slightly more complicated case, but one more applicable to atmospheric visibility. In this example, the observer still de-

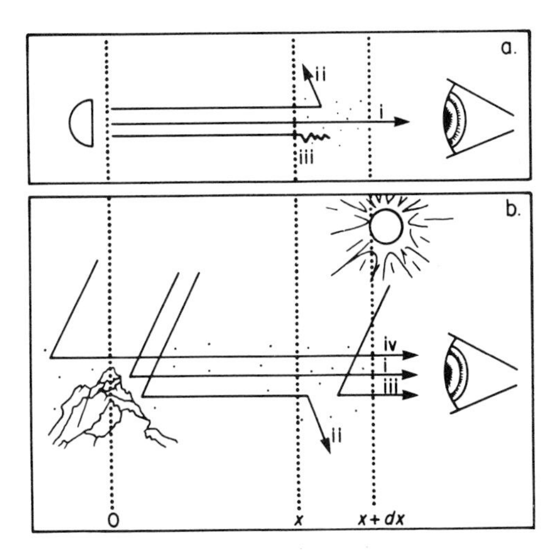

Fig. 10-3. (a) A diagram of extinction of light from a source such as an electric light in a reflector, illustrating (i) transmitted, (ii) scattered, and (iii) absorbed light. (b) A diagram of daylight visibility, illustrating (i) residual light from a target reaching an observer, (ii) light from a target scattered out of an observer's line of sight, (iii) air light from the intervening atmosphere, and (iv) air light constituting horizon sky. Source: U.S. Environmental Protection Agency, "Protecting Visibility," EPA-450/5-79-008. Office of Air Quality Planning and Standards, Research Triangle Park, NC, 1979.

pends on the ability to perceive light rays emanating from the target object and on the scattering and absorption of those rays out of the beam. In addition, however, the observer must contend with additional light scattered into the line of sight from other angles. This extraneous light is sometimes called *air light*. Equation (10-1) is modified to account for this phenomenon by adding a term to represent this background intensity.

$$-dI = -dI(\text{extinction}) + dI(\text{air light}) \tag{10-2}$$

This air light term contributes to the reduced visibility we call *atmospheric haze*.

A simplified relationship developed by Koschmieder which relates the visual range and the extinction coefficient is given by Eq. (10-3),

$$L_v = 3.92/b_{ext} \tag{10-3}$$

where L_v is the distance at which a black object is just barely visible (3). Equation (10-3) is based on the following assumptions:

1. The background behind the target is uniform.
2. The object is black.
3. An observer can detect a contrast of 0.02.
4. The ratio of air light to extinction is constant over the path of sight.

While the Koschmieder relationship is useful as a first approximation for determining visual range, many situations exist in which the results are only qualitative.

The extinction coefficient b_{ext} is dependent on the presence of gases and molecules that scatter and absorb light in the atmosphere. The extinction coefficient may be considered as the sum of the air and pollutant scattering and absorption interactions, as shown in Eq. (10-4):

$$b_{ext} = b_{rg} + b_{ag} + b_{scat} + b_{ap} \tag{10-4}$$

where b_{rg} is scattering by gaseous molecules (Rayleigh scattering), b_{ag} absorption by NO_2 gas, b_{scat} scattering by particles, and b_{ap} absorption by particles. These various extinction components are a function of wavelength. As extinction increases, visibility decreases.

The Rayleigh scattering extinction coefficient for particle-free air is 0.012 km^{-1} for "green" light ($\gamma = 0.05\ \mu m$) at sea level (4). This permits a visual range of ~320 km. The particle-free, or Rayleigh scattering, case represents the best visibility possible with the current atmosphere on earth.

The absorption spectrum of NO_2 shows significant absorption in the visible region (see Fig. 10-4) (5). As a strong absorber in the blue region, NO_2 can color plumes red, brown, or yellow. Figure 10-5 shows a comparison of extinction coefficients of 0.1 ppm NO_2 and Rayleigh scattering by air (6). In urban areas, some discoloration can be due to areawide NO_2 pollution. In rural areas, the biggest problem with NO_2 is that in coherent plumes from power plants, it contributes to the discoloration of the plume.

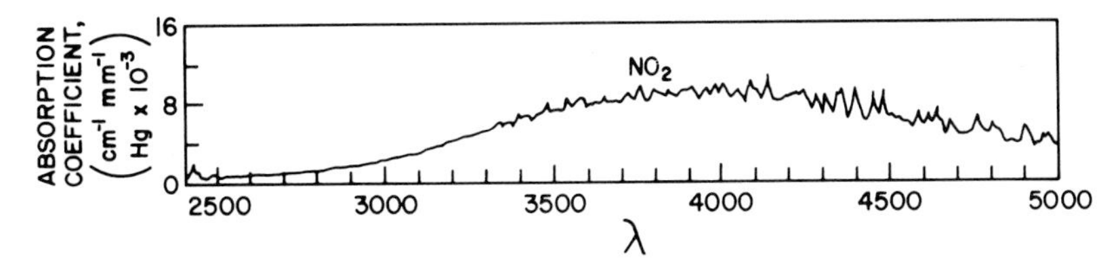

Fig. 10-4. Absorption spectrum of NO_2. Source: Hall, T. C., Jr., and Blacet, F. E., *J. Chem. Phys.* **20,** 1745 (1952).

Suspended particles are the most important factor in visibility reduction. In most instances, the visual quality of air is controlled by particle scattering and is characterized by the extinction coefficient b_{scat}. The size of particles plays a crucial role in their interaction with light. Other factors are the refractive index and shape of the particles, although their effect is harder to measure and is less well understood. If we could establish these properties, we could calculate the amount of light scattering and absorption. Alternatively, the extinction coefficient associated with an aerosol can be measured directly.

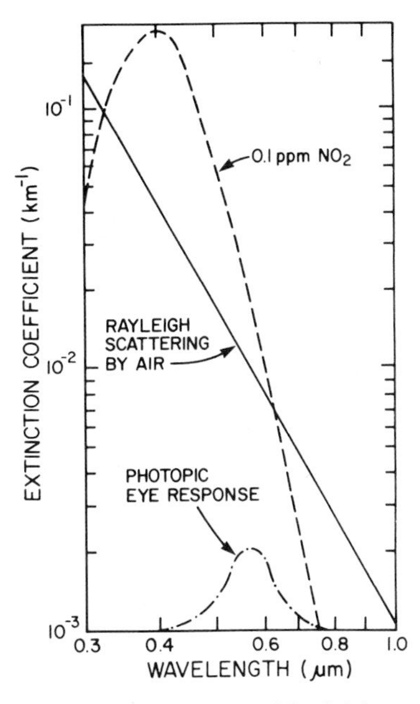

Fig. 10-5. Comparison of b_{ext} for 0.1 ppm NO_2 and Rayleigh scattering by air. The photopic eye response represents the range of wavelengths over which the eye detects light. Source: Husar, R., White, W. H., Paterson, D. E., and Trijonis, J., "Visibility Impairment in the Atmosphere," Draft report prepared for the U.S. Environmental Protection Agency under Contract No. 68022515, Task Order No. 28.

Light and suspended particles interact in the four basic ways shown in Fig. 10-6: refraction, diffraction, phase shift, and absorption. For particles with a diameter of 0.1–1.0 μm, scattering and absorption can be calculated by using the Mie equations (7). Figure 10-7 shows the relative scattering and absorption efficiency per unit volume of particle for a typical aerosol containing some light-absorbing soot (8). This clearly shows the importance of atmospheric particles in the diameter range 0.1 to 1.0 μm as efficient light-scattering centers. With particles of larger and smaller diameters, scattering decreases. Absorption generally contributes less to the extinction coefficient than does the scattering processes. Atmospheric particles of different chemical composition have different refractive indices, resulting in different scattering efficiencies. Figure 10-8 shows the scattering-to-mass ratio for four different materials (9). Clearly, carbon or soot aerosols, and aerosols of the same diameter with water content, scatter with different efficiencies at the same diameter.

Visibility is also affected by alteration of particle size due to hydroscopic particle growth, which is a function of relative humidity. In Los Angeles, California, the air, principally of marine origin, has numerous sea salt particles. Visibility is noticeably reduced when humidity exceeds about 67%. In a study of visibility related to both relative humidity and origin of

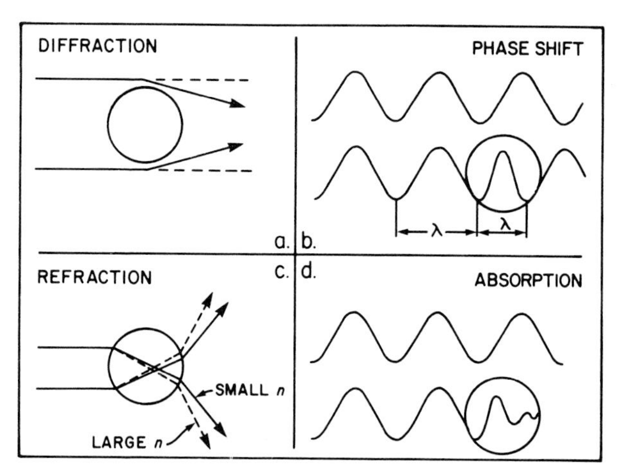

Fig. 10-6. Four forms of particle light interaction. Light scattering by coarse particles (>2 μm) is the combined effect of diffraction and refraction. (a) Diffraction is an edge effect whereby the light is bent to fill in the shadow behind the particle. (b) The speed of a wavefront entering a particle with refractive index $n > 1$ (for water, $n = 1.33$) is reduced. (c) Refraction produces a lens effect. The angular dispersion resulting from bending incoming rays increases with n. (d) For absorbing media, the refracted wave intensity decays within the particle. When the particle size is comparable to the wavelength of light (0.1–1.0 μm), these interactions (a–d) are complex and enhanced. Source: U.S. Environmental Protection Agency, "Protecting Visibility," EPA-450/5-79-008. Office of Air Quality Planning Standards, Research Triangle Park, NC, 1979.

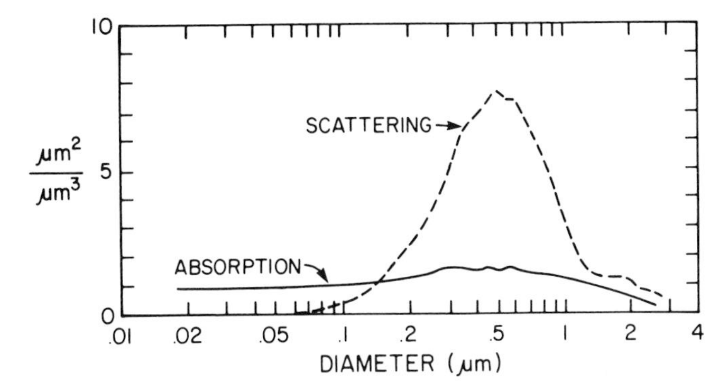

Fig. 10-7. Scattering and absorption cross-section per unit volume as a function of particle diameter. Source: Charlson, R. J., Waggoner, A. P., and Thielke, H. F., "Visibility Protection for Class I Areas. The Technical Basis," Report to the Council on Environmental Quality, Washington, DC, 1978.

air (maritime or continental), Buma (10) found that at a set relative humidity, continental air reduced visibility below 7 km more often than did air of maritime origin. This effect is presumably due to numerous hygroscopic aerosols from air pollution sources. Some materials, such as sulfuric acid mist, exhibit hygroscopic growth at humidity as low as 30%.

B. Turbidity

The attenuation of solar radiation has been studied by McCormick and his associates (11, 12) using the Voltz sun photometer, which makes 'mea-

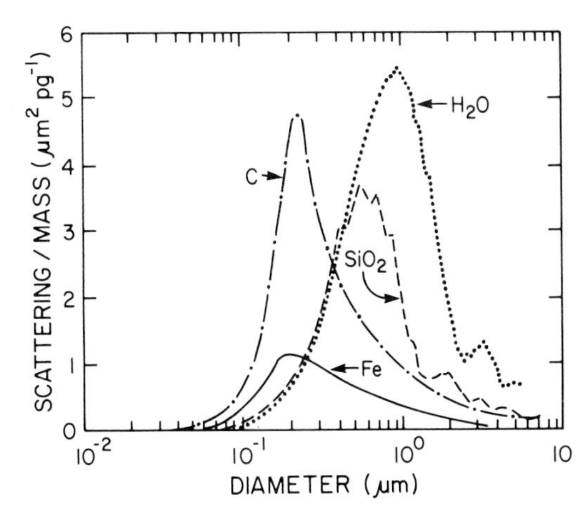

Fig. 10-8. Single particle scattering to mass ratio for particles of four different compositions. Carbon particles are also very efficient absorbers of light. Source: U.S. Environmental Protection Agency, "Protecting Visibility," EPA-450/5-79-008, Office of Air Quality Planning Standards, Research Triangle Park, NC, 1979.

surements at a wavelength of 0.5 μm. The ratio of the incident solar transmissivity to the extraterrestial solar intensity can be as high as 0.5 in clean atmospheres but can drop to 0.2–0.3 in polluted areas, indicating a decrease of 50% in ground-level solar intensity. The turbidity coefficient can also be derived from these measurements and used to approximate the aerosol loading of the atmosphere. By assuming a particle size distribution in the size range 0.1–10.0 μm and a particle density, the total number of particles can be estimated. The mass loading per cubic meter can also be approximated. Because of the reasonable cost and simplicity of the sun photometer, it is useful for making comparative measurements around the world.

C. Precipitation

Pollution can cause opposite effects in relation to precipitation. Addition of a few particles that act as ice nuclei can cause ice particles to grow at the expense of supercooled water droplets, producing particles large enough to fall as precipitation. An example of this is commercial cloud seeding with silver iodide particles released from aircraft to induce rain. If too many particles are added, none of them grow sufficiently to cause precipitation. Therefore, the effects of pollution on precipitation are complex.

II. FORMATION OF ATMOSPHERIC HAZE

Atmospheric haze is the condition of reduced visibility caused by the presence of fine particles or NO_2 in the atmosphere. The particles must be 0.1–1.0 μm in diameter, the size range in which light scattering occurs. The source of these particles may be natural or anthropogenic.

Atmospheric haze has been observed in both the western and eastern portions of the United States. Typical visual ranges in the East are <15 miles and in the Southwest >50 miles. The desire to protect visual air quality in the United States is focused on the national parks in the West. The ability to see vistas over 50–100 km in these locations makes them particularly vulnerable to atmospheric haze. This phenomenon is generally associated with diffuse or widespread atmospheric degradation as opposed to individual plumes.

The major component of atmospheric haze is sulfate particulate matter (particularly ammonium sulfate), along with varying amounts of nitrate particulate matter, which in some areas can equal the sulfate. Other components include graphitic material, fine fly ash, and organic aerosols.

The sources of particulate matter in the atmosphere can be primary, directly injected into the atmosphere, or secondary, formed in the atmosphere by gas-to-particle conversion processes (13). The primary sources of fine particles are combustion processes, e.g., power plants and diesel engines. Power plants with advanced control technology still emit substantial numbers and masses of fine particles with diameters <1.0 μm. The composition of these particles includes soot or carbonaceous material, trace metals, V_2O_5, and sulfates. In addition, large quantities of NO_2 and SO_2 are released to the atmosphere.

A. Particle Formation in the Atmosphere

The secondary source of fine particles in the atmosphere is gas-to-particle conversion processes, considered to be the more important source of particles contributing to atmospheric haze. In gas-to-particle conversion, gaseous molecules become transformed to liquid or solid particles. This phase transformation can occur by three processes: absortion, nucleation, and condensation. *Absorption* is the process by which a gas goes into solution in a liquid phase. Absorption of a specific gas is dependent on the solubility of the gas in a particular liquid, e.g., SO_2 in liquid H_2O droplets. *Nucleation* and *condensation* are terms associated with aerosol dynamics.

Nucleation is the growth of clusters of molecules that become a thermodynamically stable nucleus. This process is dependent on the vapor pressure of the condensable species. The molecular clusters undergo growth when the saturation ratio, S, is greater than 1, where *saturation ratio* is defined as the actual pressure of the gas divided by its equilibrium vapor pressure. $S > 1$ is referred to as a *supersaturated condition* (14).

The size at which a cluster may be thermodynamically stable is influenced by the Kelvin effect. The equilibrium vapor pressure of a component increases as the droplet size decreases. Vapor pressure is determined by the energy necessary to separate a single molecule from the surrounding molecules in the liquid. As the curvature of the droplet's surface increases, fewer neighboring molecules will be able to bind a particular molecule to the liquid phase, thus increasing the probility of a molecule escaping the liquid's surface. Thus, smaller droplets will have a higher equilibrium vapor pressure. This would affect the minimum size necessary for a thermodynamically stable cluster, suggesting that components with lower equilibrium saturation vapor pressures will form stable clusters at smaller diameters.

Condensation is the result of collisions between a gaseous molecule and an existing aerosol droplet when supersaturation exists. Condensation occurs at much lower values of supersaturation than nucleation. Thus, when particles already exist in sufficient quantities, condensation will be the dominant process occurring to relieve the supersaturated condition of the vapor-phase material.

A simple model for the formation and growth of an aerosol at ambient conditions involves the formation of a gas product by the appropriate chemical oxidation reactions in the gas phase. This product must have a sufficiently low vapor pressure for the gas-phase concentration of the oxidized product to exceed its saturation vapor pressure. When this condition occurs, nucleation and condensation may proceed, relieving supersaturation. These processes result in the transfer of mass to the condensed phase. Aerosol growth in size occurs while condensation is proceeding.

Coagulation, i.e., the process by which discrete particles come in contact with each other in the air and remain joined together by surface forces, represents another way in which aerosol diameter will increase. However, it does not alter the mass of material in the coagulated particle.

The clearest example of this working model of homogeneous gas-to-particle conversion is sulfuric acid aerosol formation. Sulfuric acid (H_2SO_4) has an extremely low saturation vapor pressure. Oxidation of relatively small amounts of sulfur dioxide (SO_2) can result in a gas-phase concentration of H_2SO_4 that exceeds its equilibrium vapor pressure in the ambient atmosphere, with the subsequent formation of sulfuric acid aerosol. In contrast, nitric acid (HNO_3) has a much higher saturation vapor pressure. Therefore, the gas-phase concentration of HNO_3 is not high enough to permit nucleation of nitric acid aerosol in typical atmospheric systems.

Atmospheric haze can occur over regions of several thousand square kilometers, caused by the oxidation of widespread SO_2 and NO_2 to sulfate and nitrate in relatively slow-moving air masses. In the eastern United States, large air masses associated with slow- moving or stagnating anticyclones have become sufficiently contaminated to be called *hazy blobs*. These blobs have been tracked by satellites as they develop and move across the country (15).

The evolution of regional hazy air masses has been documented in several case studies. The development of one such system is shown in Fig. 10-9. During a 10-day period in the summer of 1975, a large region of the eastern United States had decreased visibility associated with the presence of fine particles in the atmosphere. The phenomenon occurred in association with a slow-moving high-pressure system. Because it seldom rains during the passage of these systems, the fine particles may have stayed airborne for a longer period of time than usual.

III. EFFECTS OF ATMOSPHERIC HAZE

The United States Clean Air Act of 1977 set as a national goal the prevention of any future degradation and the reduction of any existing impairment of visibility in mandatory class I federal areas caused by anthropogenic air pollution. The Clean Air Act Amendments of 1990 reinforce the support of these goals. (See Chapter 22 for a discussion of federal classes of areas.) These areas include most of the major national parks, such as the Grand Canyon, Yosemite, and Zion Park. This portion of the Clean Air Act ad-

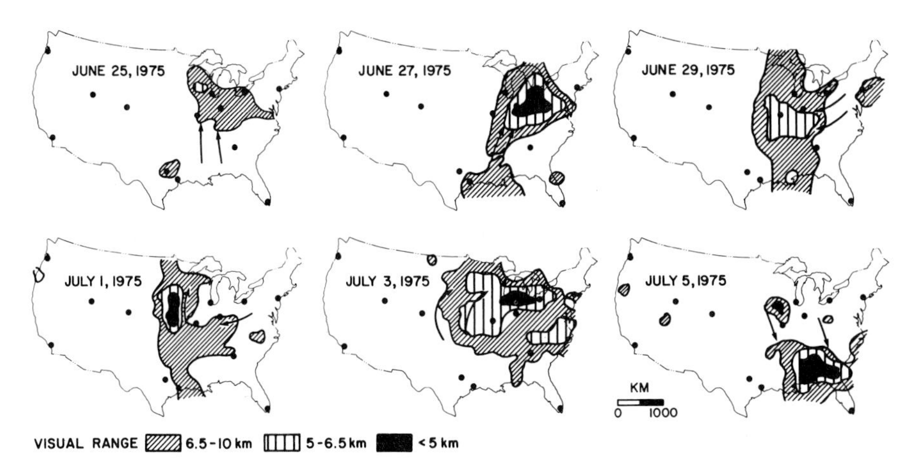

Fig. 10-9. The evolution and transport of a large hazy air mass. Contour maps of noon visibility for June 25–July 5, 1975. Source: Lyons, W. A., and Husar, R. B., *Mon. Weather Rev.* **104,** 1623–1626 (1976).

dresses the problem of visibility degradation by atmospheric haze of anthropogenic origin. This legislation recognizes atmospheric haze as a cause of degradation in visual air quality.

All nations contain areas of exceptional scenic beauty. The value of these areas is largely determined by society. Many nations, determined to protect these areas, have established parks or preserves where only limited development can occur, in many instances limited to facilities such as food and lodging for visitors to the area.

Grand Canyon National Park in the southwestern United States is a prime example of an area of natural beauty. This park is ~250 km long, varying in width up to ~45 km. The actual canyon is ~10 km at its widest point, with the Colorado River running in the bottom of the canyon, 1600 m below the edge of the outer rim. Visitors go to parks for many reasons, such as hiking, camping, wildlife, and the enjoyment of solitude, but the overwhelming majority visit the Grand Canyon to enjoy the magnificent views from its rim. These views have detail in the foreground (0–5 km), with colored layers of rock strata on canyon walls perhaps 5–25 km distant and in the far background (25–50 km) additional geologic features which contribute to the viewers' appreciation of the scene. To enjoy these views, one must have good visibility over the entire path length from the details in the foreground to the objects in the distant background.

A survey by national park personnel indicates that large areas of the United States are subject to varying degrees of visibility degradation (1). The middle portion of the eastern half of the country and the Florida Gulf Coast are subject to widespread hazy air masses associated with stagnation conditions. Large portions of the western half of the country are subject to atmospheric haze problems associated with power plants, urban plumes, and agricultural activities.

Average airport visibilities over the eastern half of the United States have been determined over a period of approximately 25 years (1948–1974) (6). Although seasonal variations occur, the long-term trend has been decreased visual air quality over the time period.

IV. VISIBILITY

Holzworth and Maga (16) developed a technique for examining the trend in visibility and analyzed data for several California airports. Bakersfield's visibility deteriorated over the period 1948–1957, and Sacramento's visibility decreased over the period 1935–1958. Los Angeles had decreasing visibility from 1932 to 1947, with little change over the period 1948–1959.

Holzworth (17) reported on the frequency of visibility of less than 7 miles for 28 cities. Two periods of record were compared for each city. There were increases in low visibility in only 26% of the comparisons from the early period (around 1930–1940) to a later period (around the mid-1950s).

Miller et al. (18), using analyses for Akron, Ohio; Lexington, Kentucky; and Memphis, Tennessee, concluded that "summer daytime visibilities were significantly lower during the period 1966–69 than visibilities for the preceding 4-year period."

Faulkenberry and Craig (19), in examining the trends at three Oregon cities, utilized a modification of the Holzworth–Maga technique by which a single statistic can be calculated for each year, indicating the probability of observing better visibility at Salem, Oregon, with no trends at Portland and Eugene, Oregon over the period 1950–1971.

Arizona has traditionally been a large copper-producing state. SO_x emissions from copper smelters near Phoenix and Tucson are shown in Fig. 10-10 (1). Phoenix is located 100 km from the nearest smelter, and Tucson is 60 km from the nearest smelter. The improvement in visibility in the 1967–1968 period was due to a decrease in SO_x emissions when there was 9-month shutdown caused by a strike. Improvement in visibility in the mid-1970 was the result of better control technology and process changes.

Zannetti et al. (20) did an analysis of visual range in the eastern United States again showing the importance of humidity but also showing the importance of air mass type, which is usually related to its direction of origin.

Mathai (21) summarized the specialty conference on atmospheric visibility. With the exception of water content of particles and the measurement of organic species, analytical laboratory techniques are readily available for particle analysis. Regulatory approaches to mitigate existing visibility impairment and to prevent further impairment are being formulated. A

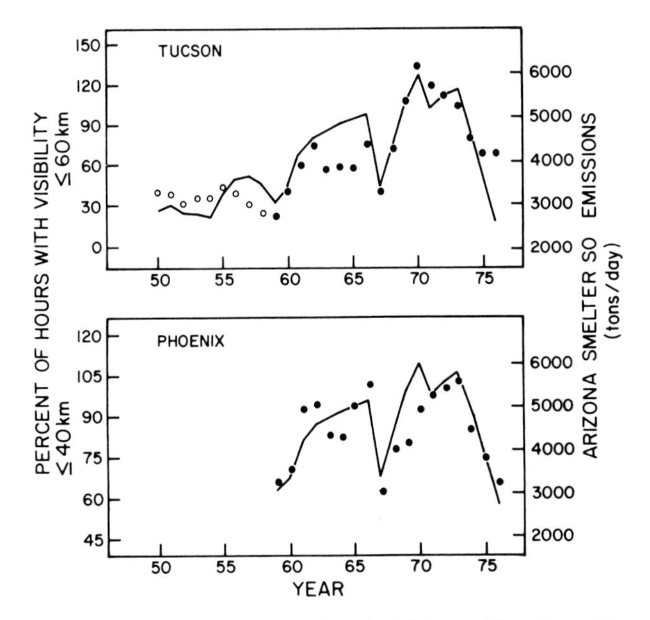

Fig. 10-10. Historic trends in hours of reduced visibility at Phoenix and Tucson, Arizona, compared to trends in SO_x emissions from Arizona smelters. The solid lines (————) represent yearly SO_x emissions. The dots (. . .) represent yearly percentages of hours of reduced visibility. Note the open dots used to represent different locations in Tucson before 1958. Source: U.S. Environmental Protection Agency, "Protecting Visibility," EPA-450/5-79-008. Office of Air Quality Planning Standards, Research Triangle Park, NC, 1979.

significant problem for regulation is the lack of proven techniques to quantify the contributions due to various sources.

V. ACIDIC DEPOSITION

Acid rain is the popular term for a very complex environmental problem. Over the past 25 years, evidence has accumulated on changes in aquatic life and soil pH in Scandinavia, Canada, and the northeastern United States. Many believe that these changes are caused by acidic deposition traceable to pollutant acid precursors that result from the burning of fossil fuels. Acid rain is only one component of *acidic deposition,* a more appropriate description of this phenomenon. Acidic deposition is the combined total of wet and dry deposition, with wet acidic deposition being commonly referred to as acid rain.

Acidity is defined in terms of the pH scale, where pH is the negative logarithm of the hydrogen ion $[H^+]$ concentration.

$$pH = -\log [H^+] \tag{10-5}$$

In the simplest case, CO_2 dissolves in raindrops, forming carbonic acid. At a temperature of 20°C, the raindrops have a pH of 5.6, the value often labeled as that of clean or natural rainwater. It represents the baseline for comparing the pH of rainwater which may be altered by SO_2 or NO_x oxidation products. Figure 10-11 illustrates the pH scale with the pH of common items and the pH range observed in rainwater. The pH of rainwater can vary from 5.6 due to the presence of H_2SO_4 and HNO_3 dissolved or formed in the droplets. These strong acids dissociate and release hydrogen ions, resulting in more acidic droplets. Basic compounds can also influence the pH. Calcium (Ca^{2+}), magnesium (Mg^{2+}), and ammonium (NH_4^+) ions help to neutralize the rain droplet and shift the overall H^+ toward the basic end of the scale. The overall pH of any given droplet is a combination of the effects of carbonic acid, sulfuric and nitric acids, and any neutralizers such as ammonia.

The principal elements of acidic deposition are shown in Fig. 10-12. Dry deposition occurs when it is not raining. Gaseous SO_2, NO_2, and HNO_3 and acid aerosols are deposited when they come in contact with and stick to the surfaces of water bodies, vegetation, soil, and other materials. If the surfaces are moist or liquid, the gases can go directly into solution; the acids formed are identical to those that fall in the form of acid rain. SO_2 and NO_2 can undergo oxidation, forming acids in the liquid surfaces if oxidizers are present. During cloud formation, when rain droplets are created, fine particles or acid droplets can act as seed nuclei for water to condense. This is one process by which sulfuric acid is incorporated in the droplets. While the droplets are in the cloud, additional gaseous SO_2 and NO_2 impinge on them and are absorbed. These absorbed gases can be oxidized by dissolved H_2O_2 or other oxidizers, lowering the pH of the raindrop. As the raindrop falls beneath the cloud, additional acidic gases and aerosol particles may be incorporated in it, also affecting its pH.

The United States has established a National Acid Deposition Program (NADP), and Canada has established the CANSAP program, which consists

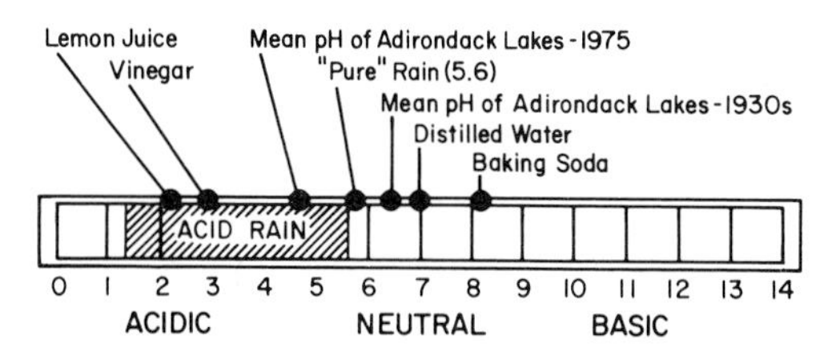

Fig. 10-11. The pH scale is a measure of hydrogen ion concentration. The pH of common substances is shown with various values along the scale. The Adirondack Lakes are located in the state of New York and are considered to be receptors of acidic deposition. Source: U.S. Environmental Protection Agency, "Acid Rain—Research Summary," EPA-600/8-79-028, Cincinnati, 1979.

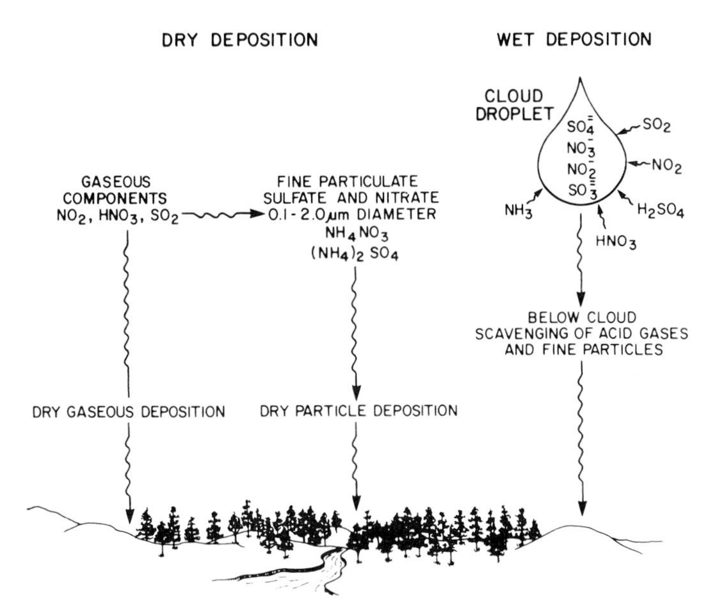

Fig. 10-12. Atmospheric processes involved in acidic deposition. The two principal deposition pathways are dry deposition (nonrain events) and wet deposition (rain events).

of sampling networks and organizational and support structures to obtain quantitative information on the spatial and temporal distribution patterns of acid deposition (22, 23). The lowest rainwater pH isopleths are associated with the regions of highest SO_2 emissions. Although there is considerable controversy over the quality and strength of the link between SO_2 and NO_x emissions from stationary sources and subsequent acid deposition hundreds of kilometers downwind, the National Research Council has concluded that a 50% reduction in the emissions of sulfur and nitrogen gases will produce about a 50% reduction in the acids deposited on the land and water downwind of the emission source. They also state that current meteorological models of atmospheric transport cannot identify specific sources with acid deposition at a particular downwind location (24).

A field study, the Eulerian Model Evaluation Field Study (EMEFS) (25) has been used to evaluate two models: the Acidic Deposition and Oxidant Model (ADOM) (26) and the Regional Acid Deposition Model (RADM) (27). For both models the calculated values, such as air concentrations, are volume averages over grid cells which are 127 km on a side for the ADOM and 80 km for the RADM. These are compared with point measurements at a possible 97 locations. The ADOM tended to overestimate ground-level SO_2 and underestimate ground-level sulfate. Two factors not included in the model that may contribute to these results are consideration of conversion during surface fog and conversion in nonprecipitating clouds. The RADM also overestimated SO_2 and underestimated aerosol sulfate.

In the eastern United States, acid rain consists of ~65% sulfuric acid, ~30% nitric acid, and ~5% other acids. In the West, windblown alkaline dusts buffer the acidity in rains occurring over many rural areas, whereas in urban areas 80% of the acidity is due to nitric acid (28). Average pH in rainfall over the eastern United States for the period April 1979–March 1980 was less than 5.0, with some areas less than pH 4.2 (29). The lowest annual pH recorded was 3.78 at De Bilt, The Netherlands, in 1967, and the lowest in an individual rainfall was 2.4 at Pitlochry, Scotland, on April 10, 1974 (30).

One of the major effects of acidic deposition is felt by aquatic ecosystems in mountainous terrain, where considerable precipitation occurs due to orographic lifting. The maximum effect is felt where there is little buffering of the acid by soil or rock structures and where steep lakeshore slopes allow little time for precipitation to remain on the ground surface before entering the lake. Maximum fish kills occur in the early spring due to the "acid shock" of the first meltwater, which releases the pollution accumulated in the winter snowpack. This first melt may be 5–10 times more acidic than rainfall.

Although the same measurement techniques for rainfall acidity have not been used over a long period of time and sampling has been carried out at relatively few locations, the trend between 1955–1956 and 1975–1976 was for the area with a pH of less than 4.6 to expand greatly over the eastern United States. The largest increases occurred over the southeastern United States, where industrialization grew rapidly during the period. The last several decades have also seen an increased area of lower pH over northern Europe.

VI. EFFECTS OF ACIDIC DEPOSITION

Land, vegetation, and bodies of water are the surfaces on which acidic deposition accumulates. Bodies of fresh water represent the smallest proportion of the earth's surface area available for acidic deposition. Yet, the best-known effect is acidification of freshwater aquatic systems.

Consider a lake with a small watershed in a forest ecosystem. The forest and vegetation can be considered as an acid concentrator. SO_2, NO_2, and acid aerosol are deposited on vegetation surfaces during dry periods and rainfalls; they are washed to the soil floor by low-pH rainwater. Much of the acidity is neutralized by dissolving and mobilizing minerals in the soil. Aluminum, calcium, magnesium, sodium, and potassium are leached from the soil into surface waters. The ability of soils to tolerate acidic deposition is very dependent on the alkalinity of the soil. The soil structure in the northeastern United States and eastern Canada is quite varied, but much of the area is covered with thin soils with a relatively limited neutralizing capacity. In watersheds with this type of soil, lakes and streams are susceptible to low pH and elevated levels of aluminum. This combination has been found to be very toxic to some species of fish. When the pH drops to ~5, many species of fish are no longer to reproduce and survive. In Sweden, thousands of lakes are no longer able to support fish. In the United States the number of polluted lakes is much smaller, but many more may be pushed into that condition by continued acidic deposition. In Canada, damage to aquatic systems and forest ecosystems is a matter of considerable concern.

Aquatic systems in areas of large snowfall accumulation are subjected to a pH surge during the spring thaw. Acidic deposition is immobilized in the snowpack, and when warm springtime temperatures cause melting, the melted snow flows into streams and lakes, potentially overloading the buffering capacity of the aquatic system.

A second area of concern is reduced tree growth in forests. As acidic deposition moves through forest soil, the leaching process removes nutrients. If the soil base is thin or contains barely adequate amounts of nutrients to support a particular mix of species, the continued loss of a portion of the soil minerals may cause a reduction in future tree growth rates or a change in the types of trees able to survive in a given location.

REFERENCES

1. U.S. Environmental Protection Agency, "Protecting Visibility," EPA-450/5-79-008. Office of Air Quality Planning and Standards, Research Triangle Park, NC, 1979.
2. Friedlander, S. K., "Smoke, Dust and Haze." Wiley, New York, 1977.
3. Middleton, W. E. K., "Vision through the Atmosphere." University of Toronto Press, Toronto, 1952.
4. Van De Hulst, H. C., Scattering in the atmosphere of earth and planets, *in* "The Atmospheres of the Earth and Planets," (G. P. Kuiper, ed.), pp. 49–111. University of Chicago Press, Chicago, 1949.
5. Hall, T. C., Jr., and Blacet, F. E., *J. Chem. Phys.* **20,** 1745 (1952).
6. Husar, R., White, W. H., Paterson, D. E., and Trijonis, J., "Visibility Impairment in the Atmosphere," Draft report prepared for the U.S. Environmental Protection Agency under contract 68022515, Task Order No. 28.
7. Twomey, S., "Atmospheric Aerosols." Elsevier, North-Holland, New York, 1977.
8. Charlson, R. J., Waggoner, A. P., and Thielke, H. F., "Visibility Protection for Class I Areas. The Technical Basis." Report to the Council of Environmental Quality, Washington, DC, 1978.
9. Faxvog, F. R., *Appl. Opt.* **14,** 269–270 (1975).
10. Buma, T. J., *Bull. Am. Meteorol. Soc.* **41,** 357–360 (1960).
11. McCormick, R. A., and Baulch, D. M., *J. Air Pollut. Control Assoc.* **12,** 492–496 (1962).

12. McCormick, R. A., and Kurfis, K. R., *Q. J. R. Meteorol. Soc.* **92,** 392–396 (1966).
13. National Research Council, "Airborne Particles." University Park Press, Baltimore, MD, 1979.
14. Reiss, H., *Ind. Eng. Chem.* **44,** 1284–1288 (1952).
15. Lyons, W. A., and Husar, R. B., *Mon. Weather Rev.* **104,** 1623–1626 (1976).
16. Holzworth, G. C., and Maga, J. A., *J. Air Pollut. Control Assoc.* **10,** 430–435 (1960).
17. Holzworth, G. C., "Some Effects of Air Pollution on Visibility in and Near Cities," Sanitary Engineering Center Technical Report A62-5, Department of Health, Education and Welfare. United States Public Health Service, Cincinnati, OH, 1962.
18. Miller, M. E., Canfield, N. L., Ritter, T. A., and Weaver, C. R., *Mon. Weather Rev.* **100,** 65–71 (1972).
19. Faulkenberry, D. G., and Craig, C. D., "Visibility Trends in the Willamette Valley, 1950–71." Third Symposium on Atmospheric Turbulence, Diffusion, and Air Quality. American Meteorological Society, Boston, 1976.
20. Zannetti, P., Tombach, I. H., and Cvencek, S. J., An analysis of visual range in the eastern United States under different meteorological regimes. *J. Air Pollut. Control Assoc.* **39,** 200–203 (1989).
21. Mathai, C. V., *J. Air Waste Manage. Assoc.* **40,** 1486–1494 (1990).
22. Interagency Task Force on Acid Precipitation. National Acid Precipitation Assessment Plan. NTIS, PB82-244 617, 1982.
23. Whelpdale, D. M., and Barrie, L. A., *J. Air Water Soil Pollut.* **14,** 133–157 (1982).
24. National Research Council, "Acid Deposition: Atmospheric Processes in Eastern North America." National Academy Press, Washington, DC, 1983.
25. Hansen, D. A., Puckett, K. J., Jansen, J. J., Lusis, M., and Vickery, J. S., The Eulerian Model Evaluation Field Study (EMEFS). Paper 5.1, pp 58–62, *in* "Preprints, Seventh Joint Conference on Applications of Air Pollution Meteorology with AWMA," Jan. 14–18, 1991, New Orleans. American Meteorological Society, Boston, 1991.
26. Fung, C., Bloxam, R., Misra, P. K., and Wong, S., Understanding the performance of a comprehensive model. Paper N2.9, pp 46–49, *in* "Preprints, Seventh Joint Conference on Applications of Air Pollution Meteorology with AWMA," Jan. 14–18, 1991, New Orleans. American Meteorological Society, Boston, 1991.
27. Barchet, W. R., Dennis, R. L., and Seilkop, S. K., Evaluation of RADM using surface data from the Eulerian model evaluation field study. Paper 5.2, pp 63–66, *in* "Preprints, Seventh Joint Conference on Applications of Air Pollution Meteorology with AWMA," Jan. 14–18, 1991, New Orleans. American Meteorological Society, Boston, 1991.
28. Nicholas, G., and Boyd, R. R., *Dames and Moore Eng. Bull.* **58,** 4–12 (1981).
29. La Bastille, A., *Natl. Geogr.* **160,** 652–681 (1981).
30. Likens, G. E., Wright, R. F., Galloway, J. N., and Butler, T. J., *Sci. Am.* **241,** Jan. 43–51 (1979).

SUGGESTED READING

Chang, J. S., Brost, R. A., Isaksen, I. S. A., Madronich, S., Middleton, P., Stockwell, W. R., and Walcek, C. J., A three-dimensional Eulerian acid deposition model: physical concepts and formulation. *J. Geophys. Res.* **92,** 14681–14700 (1987).

Hidy, G. M., Mueller, P. K., Grosjean, D., Appel, B. R., and Wesolowski, J. J. (eds.), "Advances in Environmental Science and Technology," Vol. 9, "The Character and Origins of

Smog Aerosols. A Digest of Results from the California Aerosol Characterization Experiment (ACHEX)." Wiley, New York, 1980.

Keith, L. H., "Energy and Environmental Chemistry, " Vol. II, "Acid Rain." Ann Arbor Science Publishers, Ann Arbor, MI, 1982.

Mathai, C. V. (ed.), "Visibility and Fine Particles," TR-17. Air Waste Management Association, Pittsburgh, PA, 1990.

Suffet, I. H. (ed.), "Fate of Pollutants in the Air and Water Environments, "Part 2," Chemical and Biological Fate of Pollutants in the Environment." Wiley, New York, 1977.

U.S. Environmental Protection Agency, "Protecting Visibility," EPA-450/5-79-008. Office of Air Quality Planning and Standards, Research Triangle Park, NC, 1979.

QUESTIONS

1. Define threshold contrast.
2. List the four components which contribute to the extinction coefficient b_{ext}. Describe the circumstances in which each component would dominate extinction.
3. Derive the Koschmieder relationship from Eq. (10-1).
4. Compare visibility measurements at a nearby airport with those of particle-free clean air.
5. Explain why stringent emission standards for particulate matter based on mass/heat input will do little to improve visual air quality.
6. Explain the differences in visual air quality between the western and eastern portions of the United States.
7. Compare the wavelengths of visible light with the range of particle diameters which most efficiently scatter light.
8. Describe the impact of future visibility degradation on your area (e.g., on specific areas of scenic attraction).
9. In the last decade, where have the greatest increases in rainfall acidity occurred in the United States? What is the suspected reason?

11

Long-Term Effects on the Planet

I. GLOBAL WARMING

Warming on the global scale is expected to occur as a result of the increase of carbon dioxide, CO_2, and other greenhouse gases (those that absorb and reradiate portions of the infrared radiation from the earth). What is debatable is the amount of warming that will occur by a particular point in time. The CO_2 concentration has increased by about 25% since 1850 (1). This is due to both combustion of fossil fuels and deforestation, which decreases the surface area available for photosynthesis and the resulting breakdown of CO_2 to oxygen and water vapor. Measurements of CO_2 for a period of more than 25 years at the Mauna Loa observatory in Hawaii show the rather dramatic increase (2) (Fig. 11-1). Other greenhouse gases such as methane and chlorofluorocarbons have increased by much higher factors.

Firm evidence for the amount of warming taking place in terms of actual temperature measurements has been complicated primarily by the magnitudes of natural climatic variations that occur. A summary of the available measurements shown by Kellogg (3) is given in Fig. 11-2. Other factors contributing to observing trends are the length of temperature records; the lack of representative measurements over large portions of the earth,

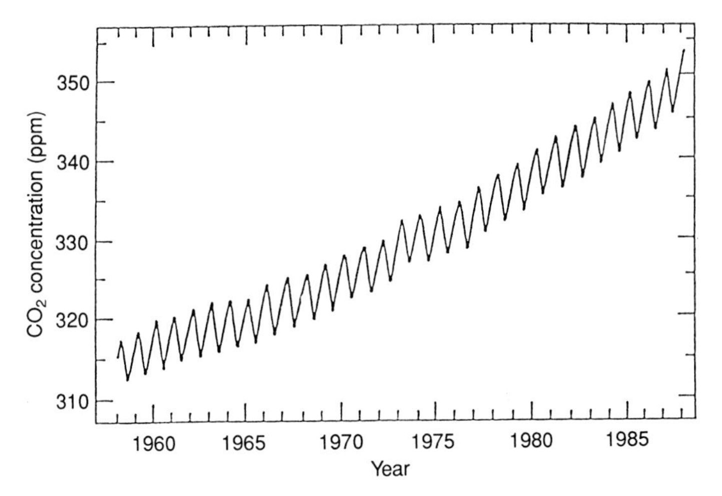

Fig. 11-1. Mean monthly concentrations of atmospheric CO_2 at Mauna Loa. The yearly oscillation is explained mainly by the annual cycle of photosynthesis and respiration of plants in the Northern Hemisphere. Source: Lindzen (2).

primarily the oceans and polar regions; and the urban sprawl toward locations at which temperature measurements are made, such as airports.

Although there has been a warming trend over the past 100 years, it is not necessarily due to the greenhouse effect. The concern of the scientific community about accelerating changes in the next 40 to 50 years is based not only on the recent observations of temperature compared with past observations, but also on the physical principles related to the greenhouse effect.

Global climate models have been used to estimate the effects in terms of temperature changes. Considerable difficulties are encountered in at least two areas. One is the difficulty in accounting properly for moisture changes including cloud formation. An important mechanism of heat transfer is through water vapor and water droplets. Of course, cloud cover alters the radiational heating at any given time. The second difficulty is accounting for solar radiation variability and the occasional injection of fine particulate matter into the atmosphere by volcanic activity, both of which alter the amount of solar radiation reaching the ground. Results from most model attempts suggest that global average surface temperatures will increase on the order of 1.5°C to 4.5°C over the next century.

The greatest concern about global warming is the regional and seasonal effects that will result. Of considerable significance could be changes in the patterns of precipitation in the agricultural and forested regions during the growing seasons in particular.

The Intergovernmental Panel on Climate Change (IPCC) was established in 1988 by the World Meteorological Organization (WMO) and United Nations Environment Program (UNEP). This panel gave its official report

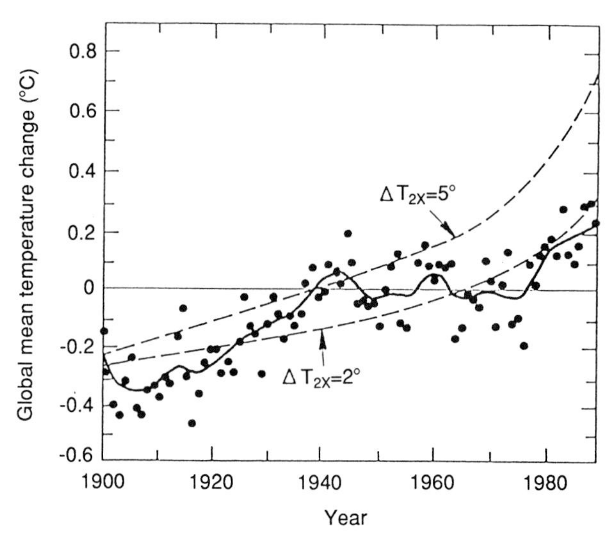

Fig. 11-2. Combined land–air and sea surface temperatures from 1900, relative to 1951–1980 average (solid line and dots), adapted from Folland *et al.* (4). The land–air temperatures were derived by Jones (5) and the sea surface temperatures by the UK Meteorological Office and Farmer et al. (6). The smoothed curve was obtained by a low-pass binomial filter operating on the annual data (shown by dots), passing fluctuations having a period of 290 years or more almost unattenuated. The dashed lines are calculated global temperature changes from the mean of the period 1861–1900, using the climate model of Wigley and Raper (7), and observed concentrations of greenhouse gases, adapted from Wigley and Barnett (8). The upper curve assumes an equilibrium temperature increase for a doubling of greenhouse gases of 5 K, and the lower one assumes it to be 2 K; both curves are based on an ocean vertical diffusion coefficient (K) of 0.63 $cm^2 sec^{-1}$ and temperature of the water in sinking regions (π) the same as the global mean. Source: Kellogg (3).

in Geneva in November 1990 at the Second World Climate Conference. This group predicted that if no significant actions are taken to curtail consumption of fossil fuel worldwide, the global mean temperature will increase at a rate of 0.2 to 0.5 K per decade over the next century (9). This is at a rate faster than seen over the past 10,000 years or longer.

Among the groups using advanced climate system models are the following five: National Center for Atmospheric Research (NCAR), NOAA Geophysical Fluid Dynamics Laboratory (GFDL), NASA Goddard Institute for Space Studies (GISS), United Kingdom Meteorological Office (UKMO), and Oregon State University (OSU). Proper simulation of cloudiness is difficult with the models. As part of a study of sensitivity to the inclusion of cloudiness, each of 14 models was run with clear skies and then with their simulation of cloudiness (10). A climate sensitivity parameter (CSP) was determined for each model. If the ratio of the CSP with clouds included to the CSP with clear skies was 1.0, the clouds had no feedback effect on temperature, but if the ratio was greater than 1, the cloud feedback was positive (to increase temperature). For the foregoing five models the range of ratios was from unity (no feedback) to 1.55 (a fairly strong positive feedback). All of these models consider two cloud types, stratiform and convective, but there are differences in the way these are calculated. It can be concluded that considerable work will be needed before the treatment of clouds can be considered satisfactory (3). Some arguments have been made that cloudiness may provide a negative feedback to temperature increases (2), that is, cause a decrease in temperature. These results indicate that this is not likely.

Penner (11) has pointed out that short-lifetime constituents of the atmosphere such as nitrogen oxides, carbon monoxide, and nonmethane hydrocarbons may also play roles related to global warming because of their chemical relations to the longer-lived greenhouse gases. Also, SO_2 with a very short life interacts with ozone and other constituents to be converted to particulate sulfate, which has effects on cloud droplet formation.

II. OZONE HOLES

During the mid-1980s, each September scientists began to observe a decrease in ozone in the stratosphere over Antarctica. These observations are referred to as "ozone holes." In order to understand ozone holes, one needs to know how and why ozone is present in the earth's stratosphere.

Stratospheric ozone is in a dynamic equilibrium with a balance between the chemical processes of formation and destruction. The primary components in this balance are ultraviolet (UV) solar radiation, oxygen molecules (O_2), and oxygen atoms (O) and may be represented by the following reactions:

$$O_2 + h\nu \rightarrow O + O \quad (11\text{-}1)$$

$$O + O_2 + M \rightarrow O_3 + M \quad (11\text{-}2)$$

$$O_3 + h\nu \rightarrow O_2 + O \quad (11\text{-}3)$$

where $h\nu$ represents a photon with energy dependent on the frequency of light, ν, and M is a molecule of oxygen or nitrogen. The cycle starts with the photodissociation of O2 to form atomic oxygen O (Eq. 11-1). O atoms react with O_2 in the presence of a third molecule (O_2 or N_2) to form O_3 (Eq. 11-2). Ozone absorbs UV radiation and can undergo photodissociation to complete the cycle of formation and destruction (Eq. 11-3). At a given altitude and latitude a dynamic equilibrium exists with a corresponding steady-state ozone concentration. This interaction of UV radiation with oxygen and ozone prevents the penetration of shortwave UV to the earth's surface. Stratospheric ozone thus provides a UV shield for human life and biological processes at the earth's surface.

In 1975, Rowland and Molina (12) postulated that chlorofluorocarbons (CFCs) could modify the steady- state concentrations of stratospheric ozone. CFCs are chemically very stable compounds and have been used for over 50 years as refrigerants, aerosol propellants, foam blowing agents, cleaning agents, and fire suppressants. Because of their stability in the troposphere, they remain in the troposphere for long periods of time, providing the opportunity for a portion of these chemicals to diffuse into the stratosphere. Rowland and Molina suggested that CFCs in the stratosphere would upset the balance represented by Eqs. (11-2) to (11-3). In the stratosphere, CFCs would be exposed to shortwave UV radiation with wavelengths λ <220 nm and undergo photodissociation, releasing chlorine atoms (C1), and Cl would interfere with the ozone balance in the following manner:

$$CCl_3F + h\nu \rightarrow CCl_2F + Cl \quad \text{(CFC role in formation of Cl)} \tag{11-4}$$

$$Cl + O_3 \rightarrow ClO + O_2 \tag{11-5}$$

$$ClO + O \rightarrow Cl + O_2 \tag{11-6}$$

$$\text{Net:} \quad O + O_3 \rightarrow O_2 + O_2$$

The chlorine atoms would provide another destruction pathway for ozone in addition to (11-3), shifting the steady-state ozone to a lower value. Because of the catalytic nature of Eqs. (11-5) and (11-6), one chlorine atom destroys many ozone molecules.

The discovery of ozone holes over Antarctica in the mid-1980s was strong observational evidence to support the Rowland and Molina hypothesis. The atmosphere over the south pole is complex because of the long periods of total darkness and sunlight and the presence of a polar vortex and polar stratospheric clouds. However, researchers have found evidence to support the role of ClO in the rapid depletion of stratospheric ozone over the south pole. Figure 11-3 shows the profile of ozone and ClO measured at an altitude of 18 km on an aircraft flight from southern Chile toward the south pole on September 21, 1987. One month earlier the ozone levels were fairly uniform around 2 ppm (vol).

Ozone holes are considered by many as a harbinger of atmospheric modification. Investigators have found a similar but less intense annual decrease in ozone over the Arctic region of the globe. Additional studies are providing evidence for stratospheric ozone depletion over the northern temperate regions of the globe. These observations have prompted a worldwide phaseout of the manufacture and use of chlorofluorocarbons and halogens over the next decade. These chemicals will be present at elevated levels for many years to come because of their stability.

CFCs represent only one class of chemicals being released to the atmosphere which have long-term effects. Replacement chemicals will be re-

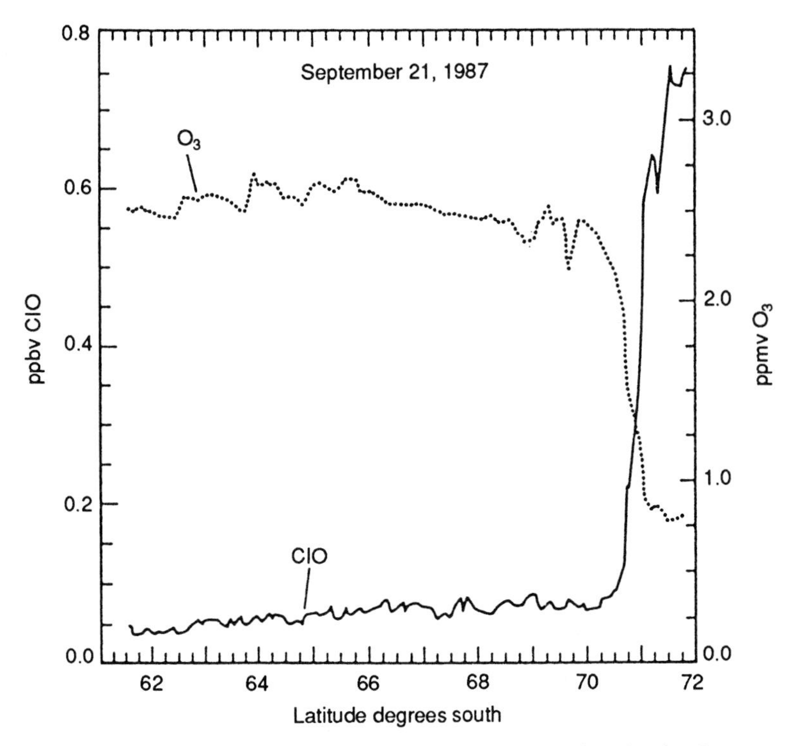

Fig. 11-3. Stratospheric ozone and ClO concentrations at an altitude of 18 km measured by aircraft flying south over Antarctica on September 27, 1987. The dramatic decrease in ozone at a latitude of 71 degrees is attributed to the role of ClO in catalytic destruction of ozone. Adapted from Anderson et al. (13).

viewed for potential adverse effects on the atmosphere. In addition, other greenhouse gases will be subject to investigation and questioning of their role in global warming scenarios.

REFERENCES

1. Schneider, S. H., *Bull. Am. Meteorol. Soc.* **71,** 1292–1304 (1990).
2. Lindzen, R., *Bull. Am. Meteorol. Soc.* **71,** 288–299 (1990).
3. Kellogg, W. W., *Bull. Am. Meteorol. Soc.* **72,** 499–511 (1991).
4. Folland, C. K., Karl, T., and Vinnikov, K. Ya., Observed climate variations and change, *in* "Climate Change: The IPCC Scientific Assessment," Working Group 1 Report (J. T. Houghton, G. J. Jenkins, and J. J. Ephramus, eds.), pp 195–238. Cambridge University Press, Cambridge, 1990.
5. Jones, P. D., *J. Climatol.* **1,** 654–660 (1988).

6. Farmer, G., Wigley, T. M. L., Jones, P. D., and Salmon, M., "Documenting and Explaining Recent Global-Mean Temperature Changes." Climatic Research Unit, Norwich, Final Report to NERC, UK, Contract GR3/6565, 1989.
7. Wigley, T. M. L., and Raper, S. C. B., *Nature* **330,** 127–131 (1987).
8. Wigley, T. M. L., and Barnett, T. P., Detection of the greenhouse effect in observations, *in* "Climate Change: The IPCC Scientific Assessment," Working Group 1 Report (J. T. Houghton, G. J. Jenkins, and J. J. Ephramus, eds.), pp. 239–256. Cambridge University Press, Cambridge, 1990.
9. World Meteorological Organization, "WMO and Global Warming," WMO No. 741. Geneva, Switzerland, 1990.
10. Cess, R. D., Potter, G. L., Blanchet, J. P., Boer, G. J., Ghan, S. J., Kiehl, J. T., Le Treut, H., Li, Z.-X., Liang, X.-Z., Mitchell, J. F. B., Morcrette, J.-J., Randall, D. A., Riches, M. R., Roeckner, E., Schlese, U., Slingo, A., Taylor, K. E., Washington, W. M., Wetherald, R. T., and Yagai, I., *Science,* **245,** 513–516 (1989).
11. Penner, J. E., *J. Air Waste Manage. Assoc.* **40,** 456–461 (1990).
12. Rowland, F. S., and Molina, M. J., *Rev. Geophys. Space Phys.* **13,** 1–35 (1975).
13. Anderson, J. G., Brune, W. H., and Proffitt, M. H., *J. Geophys. Res.* **94,** 465–479 (1989).

SUGGESTED READING

American Meteorological Society, *Bull. Am. Meteorol. Soc.* **72,** 57–59 (1991).

QUESTIONS

1. What are the difficulties in the use of climate models to estimate effects of increased CO_2?
2. What is the primary reason for the increase of CO_2 in the atmosphere?
3. What is the importance of having ozone in the stratosphere?
4. What is the effect of chlorofluorocarbons entering the stratosphere?

Part III

The Measurement and Monitoring of Air Pollution

12

Atmospheric Chemistry

Atmospheric chemistry encompasses all of the chemical transformations occurring in the various atmospheric layers from the troposphere to beyond the stratosphere. Air pollution chemistry represents the subset of these atmospheric chemical processes which have a direct impact on human beings, vegetation, and surface water bodies. Classification of atmospheric chemical processes as either human-made (anthropogenic) or natural is useful but not precise. For example, the trace gases nitric oxide (NO) and sulfur dioxide (SO_2) have both anthropogenic and natural sources, and their atmospheric behavior is independent of their source. A vivid example was the 1980 Mt. St. Helen's volcanic eruption in Washington, a gigantic point source for SO_2 and particulate matter in the atmosphere. This natural source was of such magnitude as to become first a regional air pollution problem and subsequently a global atmospheric chemical problem.

I. TYPES OF ATMOSPHERIC CHEMICAL TRANSFORMATIONS

The chemical transformations occurring in the atmosphere are best characterized as oxidation processes. Reactions involving compounds of carbon (C), nitrogen (N), and sulfur (S) are of most interest. The chemical processes in the troposphere involve oxidation of hydrocarbons, NO, and SO_2 to form oxygenated products such as aldehydes, nitrogen dioxide (NO_2), and sulfuric acid (H_2SO_4). These oxygenated species become the secondary products formed in the atmosphere from the primary emissions of anthropogenic or natural sources (Fig. 12-1).

Solar radiation influences the chemical processes in the atmosphere by interacting with molecules that act as photoacceptors. Free radicals are formed by the photodissociation of certain types of molecules. Free radicals are neutral fragments of stable molecules and are very reactive. Examples are O, atomic oxygen; H, atomic hydrogen; OH, the hydroxyl radical; and HO_2, the hydroperoxy radical. In areas with photochemical smog, the principal photoacceptors are aldehydes, NO_2, nitrous acid (HNO_2), and ozone. The photodissociation process is energy dependent, and only photons with sufficient energy are capable of causing photodissociation. The wavelength dependence of solar radiation is discussed in Chapter 17.

The reactivity of chemical compounds will differ because of their structure and molecular weight. Hydrocarbon compounds have been ranked according to their rate of reaction with various types of oxidizing species such as OH, NO_3, and O_3 (1). The role of hydrocarbons, along with oxides of nitrogen, in the formation of ozone is very complex. Ozone formation is a function of the mixture of hydrocarbons present and the concentration of NO_x, $[NO_x]$ ($= [NO] + [NO_2]$). The concept of an incremental reactivity scale permits accessing the increment of ozone formation per incremental change in a single hydrocarbon component (2). Incremental reactivity is determined by calculating the ozone formation potential in a baseline scenario using a simple mixture of hydrocarbons representing an urban atmosphere. Then for each hydrocarbon species of interest, the ozone formation is recalculated with incremental hydrocarbons added to the mixture. From this approach, the $\Delta[O_3]/\Delta[HC]$ values represent the impact of a specific hydrocarbon on urban photochemical smog formation.

The vapor pressure of a compound is important in determining the upper limit of its concentration in the atmosphere. High vapor pressures will permit higher concentrations than low vapor pressures. Examples of organic compounds are methane and benzo[*a*]pyrene. Methane, with a relatively high vapor pressure, is always present as a gas in the atmosphere; in contrast, benzo[*a*]pyrene, with a relatively low vapor pressure, is ad-

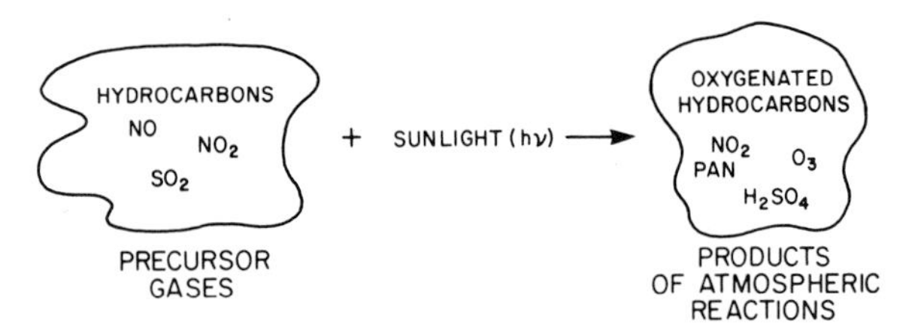

Fig. 12-1. Precursor–product relationship of atmospheric chemical reactions.

sorbed on particulate matter and is therefore not present as a gas. Vapor pressure also affects the rate of evaporation of organic compounds into the atmosphere and the conversion of atmospheric gases to particulate matter, e.g., SO_2 to the aerosol H_2SO_4 (3).

Atmospheric chemical reactions are classified as either photochemical or thermal. Photochemical reactions are the interactions of photons with species which result in the formation of products. These products may undergo further chemical reaction. These subsequent chemical reactions are called *thermal* or *dark* reactions.

Finally, atmospheric chemical transformations are classified in terms of whether they occur as a gas (homogeneous), on a surface, or in a liquid droplet (heterogeneous). An example of the last is the oxidation of dissolved sulfur dioxide in a liquid droplet. Thus, chemical transformations can occur in the gas phase, forming secondary products such as NO_2 and O_3; in the liquid phase, such as SO_2 oxidation in liquid droplets or water films; and as gas-to-particle conversion, in which the oxidized product condenses to form an aerosol.

II. ROLE OF SOLAR RADIATION IN ATMOSPHERIC CHEMISTRY

The time required for atmospheric chemical processes to occur is dependent on chemical kinetics. Many of the air quality problems of major metropolitan areas can develop in just a few days. Most gas-phase chemical reactions in the atmosphere involve the collision of two or three molecules, with subsequent rearrangement of their chemical bonds to form molecules by combination of their atoms. Consider the simple case of a bimolecular reaction of the following type:

$$\text{B} + \text{C} \rightarrow \text{products} \tag{12-1}$$

$$\text{Rate of reaction} = k[\text{B}][\text{C}] \tag{12-2}$$

where $k = A \exp[-E_a/RT]$ and k, the rate constant, is dependent on the frequency factor (A), the temperature (T), the activation energy of the reaction (E_a) and the ideal gas constant (R). The frequency factor, A, is of the same order of magnitude for most gas reactions. For $T = 298$ K the rate of reaction is strongly dependent on the activation energy E_a as shown in Table 12-1 (4). When E_a is > 30 kJ/mol, the rates become very small, limiting the overall rate of reaction. Table 12-2 contains the activation energies for bimolecular collisions of different molecular species. For the first three reactions between molecular species, E_a is > 100 kJ, but for the last three reactions E_a < 10 kJ. The last three reactions involve the participation of free radical or atomic species. The activation energies of reactions involving atomic or free radicals are very small, permitting chemical transformations on a short time scale.

TABLE 12-1

Values of $\exp(-E_a/RT)$ as a Function of the Activation Energy for $T = 298K$[a]

E_a	$\exp(-E_a/RT)$
1.0	0.67
3.0	0.30
10.0	0.0177
30.0	5.51×10^{-6}
100.0	2.95×10^{-18}
300.0	2.56×10^{-53}

[a] In SI units, $R = 8.31434\ kJ^{-1}\ mol^{-1}$.

III. GAS-PHASE CHEMICAL REACTION PATHWAYS

The complexity of the atmospheric chemical reactions occurring in major metropolitan areas can be staggering. Urban atmospheres are characterized as complex mixtures of hydrocarbons and oxides of sulfur and nitrogen. Table 12-3 show the hydrocarbons identified in the urban air of St. Petersburg, Florida (5). The interactions among this large number of compounds can be understood by studying simpler systems. Figure 12-2 shows the diurnal patterns of NO, NO_2, and O_3 for St. Louis, Missouri (6). These diurnal patterns are interrelated. The concentration profiles of Fig. 12-2 are the result of a combination of atmospheric chemical and meteorological processes. To uncouple this combination of factors, laboratory (smog chamber) studies such as those of the propene-NO_x system (Fig. 12-3) have been undertaken (7). These profiles show chemical transformations separated from meteorological processes.

Similar chemical steps occur in the ambient air and in laboratory smog chamber simulations. Initially, hydrocarbons and nitric oxide are oxidized

TABLE 12-2

Activation Energies for Atmospheric Reactions

Reaction	E_a (kJ/mol)
$N_2 + O_2 \rightarrow N_2O + O$	538
$CO + O_2 \rightarrow CO_2 + O$	251
$SO_2 + NO_2 \rightarrow SO_3 + NO$	106
$O + H_2S \rightarrow OH + HS$	6.3
$O + NO_2 \rightarrow NO + O_2$	<1
$HO_2 + NO \rightarrow NO_2 + OH$	<1

Source: Campbell, I. M., "Energy and the Atmosphere," pp. 212–213. Wiley, New York, 1977.

TABLE 12-3

Hydrocarbon Compounds Identified in Ambient Air Samples from St. Petersburg, Florida

Acetaldehyde	*m*-Ethyltoluene	Methylcyclohexane	Propene
Acetylene	*o*-Ethyltoluene	3-Methylhexane	*n*-Propylbenzene
1,3-Butadiene	*p*-Ethyltoluene	2-Methylpentane	Toluene
n-Butane	*n*-Heptane	Nonane	2,2,4-Trimethylpentane
trans-2-Butene	Isobutane	*n*-Pentane	*m*-Xylene
Cyclopentane	Isobutylene	1-Pentene	*o*-Xylene
n-Decane	Isopentane	*cis*-2-Pentene	*p*-Xylene
2,3-Dimethylpentane	Isopropyl benzene	*trans*-2-Pentene	1,2,4-Trimethylbenzene
Ethane	Limonene	*alpha*-Pinene	1,3,5-Trimethylbenzene
Ethylbenzene	Methane	*beta*-Pinene	
Ethylene	2-Methyl-1-butene	Propane	

Source: Lonneman, W. A., Seila, R. L., and Bufalini, J. J., Environ. *Sci. Technol.* **12,** 459–463 (1978).

to form nitrogen dioxide, ozone, and other oxidation products such as peroxyacyl nitrate (PAN) and aldehydes. The complete process is very complicated, with many reaction steps.

The principal components of atmospheric chemical processes are hydrocarbons, oxides of nitrogen, oxides of sulfur, oxygenated hydrocarbons, ozone, and free radical intermediates. Solar radiation plays a crucial role in the generation of free radicals, whereas water vapor and temperature can influence particular chemical pathways. Table 12-4 lists a few of the components of each of these classes. Although more extensive tabulations may be found in "Atmospheric Chemical Compounds" (8), those listed in

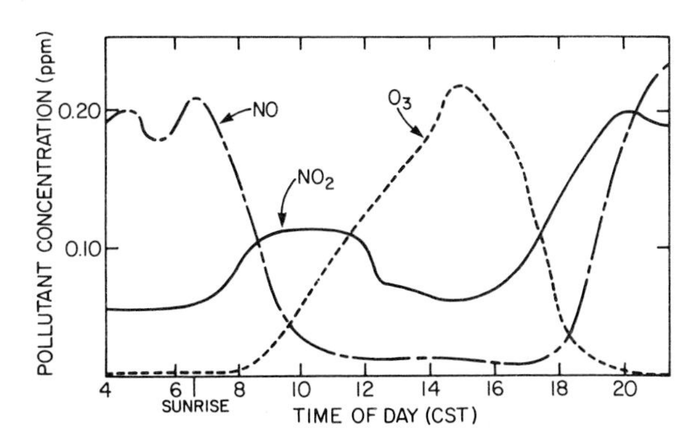

Fig. 12-2. NO-NO_2-O_3 ambient concentration profiles from average of four Regional Air Monitoring Stations (RAPS) in downtown St. Louis, Missouri (USA) on October 1, 1976. Source: RAPS, Data obtained from the 1976 data file for the Regional Air Pollution Study Program. U.S. Environmental Protection Agency, Research Triangle Park, NC, 1976.

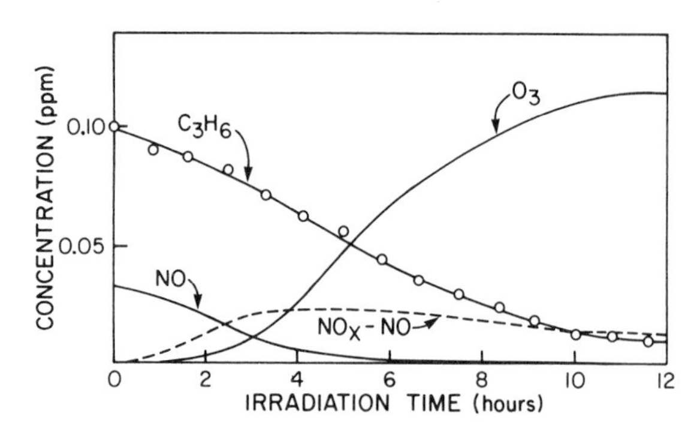

Fig. 12-3. Concentration versus time profiles of propene, NO, NO_x-NO, and O_3 from smog chamber irradiation; $k_1 = 0.16\ \text{min}^{-1}$. Source: Akimoto, H., Sakamaki, F., Hoshino, M., Inoue, G., and Oduda, M., *Environ. Sci. Technol.* **13,** 53–58 (1979).

Table 12-4 are sufficient for an understanding of smog chemistry. The major undesirable components of photochemical smog are NO_2, O_3, SO_2, H_2SO_4, PAN, and aldehydes. Air quality standards have been established in several countries for SO_2, NO_2, and O_3; H_2SO_4 contributes to acidic deposition and reduction in visibility; and PAN and aldehydes can cause eye irritation and plant damage if their concentrations are sufficiently high.

A. Photoabsorption of Solar Radiation

Solar radiation initiates the formation of free radicals. According to the elementary quantum theory of atoms and molecules, the internal energy of molecules is composed of electronic energy states. Molecules interact with solar radiation by absorbing photons. This absorption process causes the molecule to undergo a transition from the ground electronic state to an excited state. The change in energy between the two states corresponds to a quantum or photon of solar radiation. The frequencies ν of absorption are expressed by Planck's law:

$$E = h\nu = hc/\lambda \tag{12-3}$$

where h is Planck's constant, c is the speed of light, and ν and λ are the frequency and wavelength of the light of the photon, respectively. The spectrum of solar radiation in the lower troposphere starts at ~295 nm and increases. Photons of shorter wavelength and higher energy are absorbed in the upper atmosphere and therefore do not reach the lower troposphere.

Molecules and atoms interact with photons of solar radiation under certain conditions to absorb photons of light of various wavelengths. Figure 10-4 shows the absorption spectrum of NO_2 as a function of the wavelength of light from 240 to 500 nm. This molecule absorbs solar radiation from

TABLE 12-4

Classes and Examples of Atmospheric Compounds

Hydrocarbons	Oxygenated hydrocarbons
Alkenes	Aldehydes
Ethene C_2H_4	Formaldehyde HCHO
Propene C_3H_6	Acetylaldehyde CH_3CHO
trans-2-Butene	Other aldehydes RCHO
Alkanes	Acids
Methane CH_4	Formic acid HCOOH
Ethane C_2H_6	Acetic acid CH_3COOH
Alkynes	Alcohols
Acetylene C_2H_2	Methanol CH_3OH
Aromatics	
Toluene C_6H_6	
m-Xylene C_6H_{10}	
Oxides of nitrogen	Oxides of sulfur
Nitric oxide NO	Sulfur dioxide SO_2
Nitrogen dioxide NO_2	Sulfur trioxide SO_3
Nitrous acid HNO_2	Sulfuric acid H_2SO_4
Nitric acid HNO_3	Ammonium bisulfate $(NH_4)HSO_4$
Nitrogen trioxide NO_3	Ammonium sulfate $(NH_4)_2SO_4$
Dinitrogen pentoxide N_2O_5	
Ammonium nitrate NH_4NO_3	
Free radicals	Oxidants
Atomic oxygen O	PAN $CH_3COO_2NO_2$
Atomic hydrogen H	Ozone O_3
Hydroxyl OH	
Hydroperoxyl HO_2	
Acyl RCO	
Peroxyacyl $RCOO_2$	

295 nm through the visible region. The absorption of photons at these different wavelengths causes the NO_2 molecule to enter an excited state. For longer wavelengths, transitions only in the rotational–vibrational states occur, whereas for shorter wavelengths changes in electronic states may occur. The process of photoabsorption for NO_2 is expressed as

$$NO_2 + h\nu \rightarrow NO_2^* \quad (12\text{-}4)$$

where $h\nu$ represents the photon of solar radiation of energy and NO_2^* is the NO_2 molecule in the excited state. The excited NO_2 molecule can follow several pathways:

Fluorescence $$NO_2^* \rightarrow NO_2 + h\nu \quad (12\text{-}5)$$

Collisional deactivation where X_2 is N_2 or O_2 $$NO_2^* + X_2 \rightarrow NO_2 + X_2 \quad (12\text{-}6)$$

Direct reaction $$NO_2^* + O_2 \rightarrow NO_3 + O \quad (12\text{-}7)$$

Photodissociation $$NO_2^* \rightarrow NO + O \quad (12\text{-}8)$$

In the case of NO_2, for each photon absorbed below 400 nm, photodissociation occurs. For other photoabsorbers, HNO_2 and aldehydes, the photodissociation process leads to the formation of free radicals.

B. Nitric Oxide, Nitrogen Dioxide, and Ozone Cycles

Three relatively simple reactions can describe the interrelationships among these components.

$$NO_2 + h\nu \rightarrow NO + O \quad (12\text{-}9)$$

$$O + O_2 + M \rightarrow O_3 + M \quad (12\text{-}10)$$

$$NO + O_3 \rightarrow NO_2 + O_2 \quad (12\text{-}11)$$

Reaction (12-9) shows the photochemical dissociation of NO_2. Reaction (12-10) shows the formation of ozone from the combination of O and molecular O_2 where M is any third-body molecule (principally N_2 and O_2 in the atmosphere). Reaction (12-11) shows the oxidation of NO by O_3 to form NO_2 and molecular oxygen. These three reactions represent a cyclic pathway (Fig. 12-4) driven by photons represented by $h\nu$. Throughout the daytime period, the flux of solar radiation changes with the movement of the sun. However, over short time periods (~10 min) the flux may be considered constant, in which case the rate of reaction (12-9) may be expressed as

$$\text{Rate} = k_1[NO_2] \quad (12\text{-}12)$$

where k_1 is a function of time of day. Expressions for the time rate of change for each of the components may be written. If this cycle reaches a steady

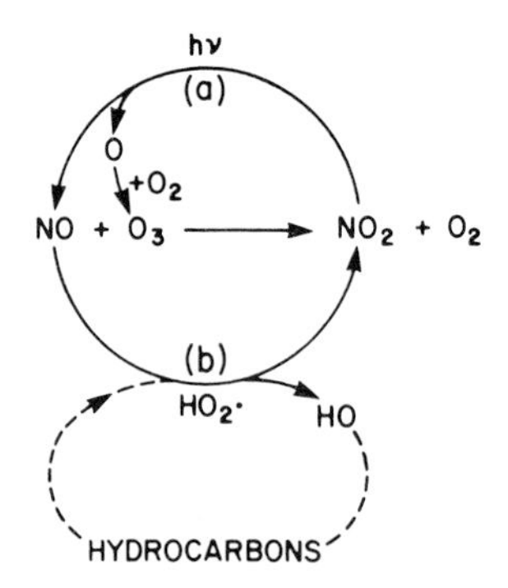

Fig. 12-4. Photochemical cycle of NO, NO_2, O_3, and free radicals.

state, the change in concentration with time no longer occurs, so that $d[\text{conc}]/dt$ is equal to zero.

$$d[NO]/dt = -k_1[NO_2] + k_3[NO][O_3] \quad (12\text{-}13)$$

$$d[O]/dt = k_1[NO_2] - k_2[O][O_2][M] \quad (12\text{-}14)$$

$$d[O_3]/dt = k_2[O][O_2][M] - k_3[NO][O_3] \quad (12\text{-}15)$$

From Eq. (12-13), it is possible to obtain an expression for the relationship of NO, NO_2, and O_3:

$$d[NO]/dt = O; \qquad k_1[NO_2] = k_3[NO][O_3] \quad (12\text{-}16)$$

$$[O_3] = k_1[NO_2]/k_3[NO] \quad (12\text{-}17)$$

Equation (12-17) is called the *photostationary state expression* for ozone. Upon examination, one sees that the concentration of ozone is dependent on the ratio NO_2/NO for any value of k_1. The maximum value of k_1 is dependent on the latitude, time of year, and time of day. In the United States, the range of k_1 is from 0 to 0.55 min^{-1}. Table 12-5 illustrates the importance of the NO_2/NO ratio with respect to how much ozone is required for the photostationary state to exist. The conclusion to be drawn from this table is that most of the NO must be converted to NO_2 before O_3 will build up in the atmosphere. This is also seen in the diurnal ambient air patterns shown in Fig. 12-2 and the smog chamber simulations shown in Fig. 12-3. It is apparent that without hydrocarbons, the NO is not converted to NO_2 efficiently enough to permit the buildup of O_3 to levels observed in urban areas.

The cycle represented by Eqs. (12-9), (12-10), and (12-11) is illustrated by the upper loop (a) in Fig. 12-4. In this figure, the photolysis of NO_2 by a photon forms an NO and an O_3 molecule. If no other chemical reaction is occurring, these two species react to form NO_2, which can start the cycle over again. In order for the O_3 concentration to build up, oxidizers other than O_3 must participate in the oxidation of NO to form NO_2. This will

TABLE 12-5

$[O_3]$ Predicted from Photostationary State Approximation as a Function of Initial $[NO_2]$[a]

$[NO_2]^o$ (ppm)	$[NO_2]_{final}$ (ppm)	$[O_3]_{final}$ (ppm)	$[NO_2]/[NO]$
0.1	0.064	0.036	1.78
0.2	0.145	0.055	2.64
0.3	0.231	0.069	3.35
0.4	0.319	0.081	3.94
0.5	0.408	0.092	4.43

[a] $k_1 = 0.5\ min^{-1}$; $k_3 = 24.2\ ppm^{-1}\ min^{-1}$.

permit the NO_2/NO ratio to build up and steady-state O_3 concentrations as represented by Eq. (12-17) to achieve typical ambient values. The other oxidizers in the atmosphere are free radicals. In the lower loop (b) of Fig. 12-4, a second pathway for NO oxidation is shown, with free radicals participating. These free radicals are derived from the participation of hydrocarbons in atmospheric chemical reactions.

C. Role of Hydrocarbons

The important hydrocarbon classes are alkanes, alkenes, aromatics, and oxygenates. The first three classes are generally released to the atmosphere, whereas the fourth class, the oxygenates, is generally formed in the atmosphere. Propene will be used to illustrate the types of reactions that take place with alkenes. Propene reactions are initiated by a chemical reaction of OH or O_3 with the carbon–carbon double bond. The chemical steps that follow result in the formation of free radicals of several different types which can undergo reaction with O_2, NO, SO_2, and NO_2 to promote the formation of photochemical smog products.

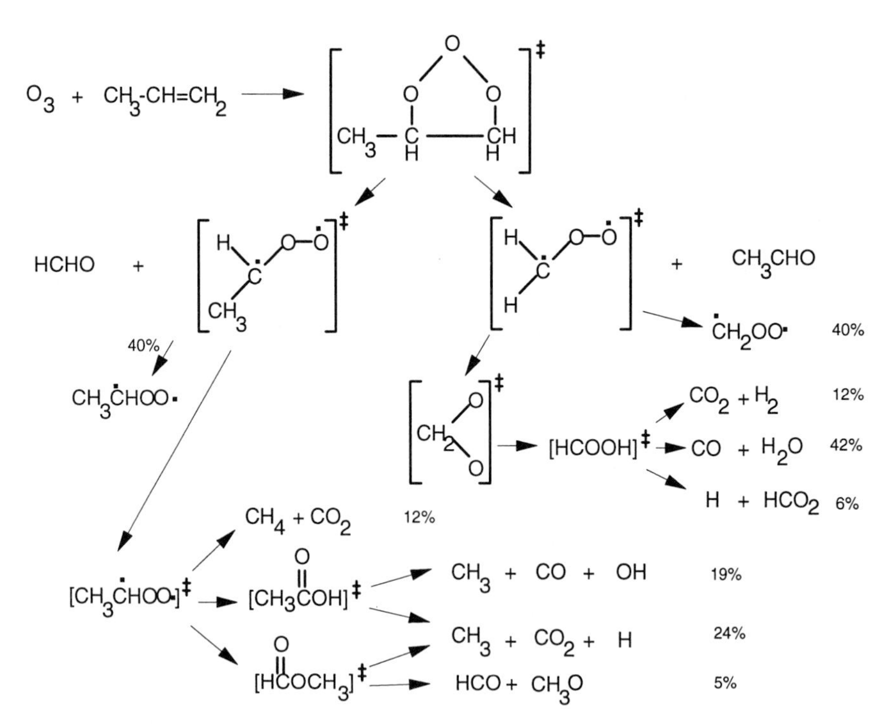

Fig. 12-5. Ozone–propene reaction pathways showing oxidation products.

1. *Ozone Reaction with Propene*

A schematic diagram of the O_3 reaction with propene (Fig. 12-5) is based on the work of Atkinson and Lloyd (9). The molozonide formed by addition of ozone to the double bond decomposes to form an aldehyde and an energy-rich (‡) biradical. In the case of propene, two sets of products are formed. Along the pathway on the right, approximately 40% of the biradicals $(H\dot{C}HOO\cdot)$‡ form a thermalized biradical $(\dot{C}H_2OO\cdot)$.* The remainder undergo rearrangement to form energy-rich acetic acid (HCOOH)‡, which subsequently decomposes to form H_2O, CO, CO_2, H_2, H, and HCO_2 radicals with percentages assigned to each pathway. The larger biradical $(CH_3\dot{C}HOO\cdot)$‡ follows a slightly different pathway. Approximately 40% forms a thermalized biradical $(CH_3\dot{C}HOO\cdot)$. Of the remaining 60%, a portion decomposes to CH_4 and CO_2 and two additional energy-rich species (CH_3COOH)‡ and $(CHOOCH_3)$‡. These two unstable species decompose as shown to form CH_3, OH, H, HCO, CH_3O, CO, and CO_2.

Alkyl radicals, R, react very rapidly with O_2 to form alkylperoxy radicals. H reacts to form the hydroperoxy radical HO_2. Alkoxy radicals, RO, react with O_2 to form HO_2 and R′CHO, where R′ contains one less carbon. This formation of an aldehyde from an alkoxy radical ultimately leads to the process of hydrocarbon chain shortening or clipping upon subsequent reaction of the aldehyde. This aldehyde can undergo photodecomposition forming R, H, and CO; or, after OH attack, forming CH(O)OO, the peroxyacyl radical.

2. *Hydroxyl Radical Addition to Propene*

As shown in Fig. 12-6, hydroxyl radicals primarily add to either of the carbon atoms which form the double bond. The remaining carbon atom has an unpaired electron which combines with molecular oxygen, forming an RO_2 radical. There are two types of RO_2 radicals labeled C_3OHO_2 in Fig. 12-6. Each of these RO_2 radicals reacts with NO to form NO_2, and an alkoxy radical reacts with O_2 to form formaldehyde, acetaldehyde, and HO_2.

3. *Aldehyde Photolysis and Reactions*

Aldehydes undergo two primary reactions: photolysis and reaction with OH radicals. These reactions lead to formation of CO, H, and R radicals.

4. *Radical Reactions with Nitric Oxide and Nitrogen Dioxide*

Alkylperoxy (RO_2) and peroxyacyl (RC(O)OO) radicals react with NO to form NO_2. The alkylperoxy radicals (RO_2) react with NO_2 to form pernitric acid-type compounds, which decompose thermally as the temperature increases. The peroxyacyl radical reacts with NO_2 to form PAN-type compounds, which also decompose thermally.

* The dots represent unpaired electrons.

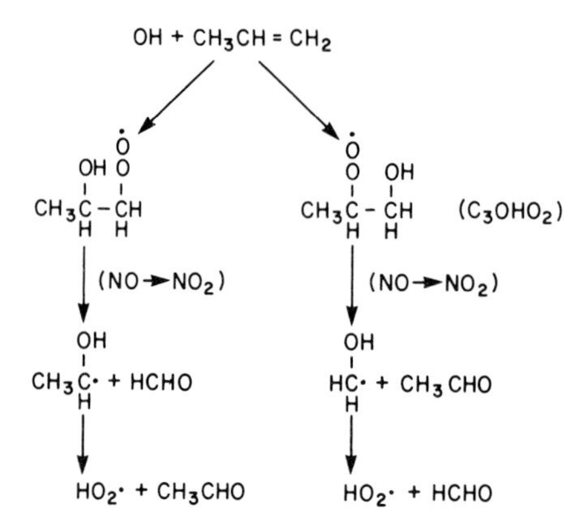

Fig. 12-6. OH–propene reaction pathways showing oxidation products.

5. *Radical Oxidation of Sulfur Dioxide*

Sulfur dioxide is oxidized in the atmosphere eventually to form sulfate compounds. The oxidation process includes both homogeneous and heterogeneous pathways. The free radicals produced from the degradation of hydrocarbons can and do react with SO_2 in the gas phase. Both OH and HO_2 oxidize SO_2 to reactive intermediates such as HSO_3 and SO_3 (10). These intermediates combine rapidly with water vapor in the atmosphere to form sulfuric acid aerosol. This type of process is dependent on atmospheric conditions. In urban areas with existing photochemical smog problems, the homogeneous oxidation of SO_2 by free radicals is probably dominant during the daytime.

IV. HETEROGENEOUS REACTIONS

Heterogeneous reactions are defined as those involving the gas–liquid or gas–solid phases. The chemistry of NO_2 and SO_2 has a heterogeneous component in the atmosphere. Heterogeneous reactions involve the dissolving of NO_2 and SO_2 in water droplets, with subsequent chemical reactions occurring to form HNO_3 and H_2SO_4 in the liquid phase. The heterogeneous oxidation of SO_2 in liquid droplets and water films is also a major pathway for conversion to sulfate in wet plumes and during humid or foggy conditions.

V. SCAVENGING AND REMOVAL FROM THE ATMOSPHERE

The atmosphere is a dynamic system, with gases and particulate matter entering, undergoing transformation, and leaving. Atmospheric chemical processes of oxidation transform gases into more highly oxidized products, e.g., NO to NO_2 to HNO_3, hydrocarbons to aldehydes, and SO_2 to sulfate particles. The removal of material from the atmosphere involves two processes: wet and dry deposition. The water solubility of gases influences the extent of removal by wet versus dry deposition. Gases such as SO_2 and NO_2 are sufficiently soluble to dissolve in water associated with in-cloud formation of rain droplets. These soluble gases may be removed by wet deposition of liquid droplets in the form of rain or fog. Less soluble gases such as O_3 and hydrocarbon vapors are removed by transport to the surface of the earth, where they diffuse to vegetation, materials, or water bodies (see Chapter 10).

REFERENCES

1. Pitts, J. N., Jr., Winer, A. M., Darnall, K. R., Lloyd, A. C., and Doyle, G. J., *in* "Proceedings, International Conference on Photochemical Oxidant Pollution and Its Control," Vol. II (B. Dimitriades, ed.), EPA-600/3-77-OOlb, pp. 687–707. U.S. Environmental Protection Agency, Research Triangle Park, NC, 1977.
2. Carter, W. P. L., "Development of Ozone Reactivity Scales for Volatile Organic Compounds," EPA 600/3-91-050. U.S. Environmental Protection Agency, August 1991.
3. National Research Council, "Ozone and Other Photochemical Oxidants." National Academy of Sciences, Washington, DC, 1977.
4. Campbell, I. M., "Energy and the Atmosphere." Wiley, New York, 1977.
5. Lonneman, W. A., Seila, R. L., and Bufalini, J. J., *Environ. Sci. Technol.* **12,** 459–463 (1978).
6. Data obtained from the 1976 data file of the Regional Air Pollution Study Program. U.S. Environmental Protection Agency, Research Triangle Park, NC, 1976.
7. Akimoto, H., Sakamaki, F., Hoshino, M., Inoue, G., and Oduda, M., *Environ. Sci. Technol.* **13,** 53–58 (1979).
8. Graedel, T. E., Hawkins, D. T., and Claxton, L. D., "Atmospheric Chemical Compounds: Sources, Occurrence, and Bioassay," Academic Press, Orlando, FL, 1986.
9. Atkinson, R., and Lloyd, A. C., *J. Phys. Chem. Ref. Data* **13,** 315–444 (1984).
10. Calvert, J. G., Su, F., Bottenheim, J. W., and Strausz, O. P., *Atmos. Environ.* **12,** 197–226 (1978).

SUGGESTED READING

National Research Council, "Rethinking the Ozone Problem in Urban and Regional Air Pollution." National Academy Press, Washington, DC, 1991.

Sloane, C. S., and Tesche, T. W., "Atmospheric Chemistry: Models and Predictions for Climate and Air Quality." Lewis Publishers, Chelsea, MI, 1991.

Warneck, P., "Chemistry of the Natural Atmosphere." Academic Press, San Diego, 1988.

QUESTIONS

1. What wavelength band of solar radiation leads to photodissociation for nitrogen dioxide? What determines the lower limit?

2. Equation (12-17) describes the steady-state $[O_3]$ in the presence of solar radiation. If k_1 = 0.3 min^{-1} and k_3 = 30 ppm^{-1} min^{-1}, which sets of conditions are consistent with $[O_3]$ = 0.1 ppm?

[NO] (ppm)	$[NO_2]$ (ppm)
0.005	0.050
0.125	0.125
0.047	0.47
0.47	0.047
0.001	0.010

3. What does the answer to question 2 indicate about the functional dependence of ozone concentrations on the absolute magnitude of $[NO_x]$ = [NO] + $[NO_2]$?
4. How are free radicals formed, and why are they so reactive?
5. From Fig. 12-6, how many molecules of NO can be oxidized to NO_2 by the reaction of one OH free radical with one propene molecule?
6. List various classification approaches used to account for hydrocarbon reactivity.
7. Classify the compounds in Table 12-4 into precursors (reactants) and products of atmospheric reaction processes.
8. List two photoacceptors in addition to nitrogen dioxide that provide an initial source of free radicals.
9. Describe the role of hydrocarbons in photochemical oxidant formation.

13

Ambient Air Sampling

I. ELEMENTS OF A SAMPLING SYSTEM

The principal requirement of a sampling system is to obtain a sample that is representative of the atmosphere at a particular place and time and that can be evaluated as a mass or volume concentration. Remote monitoring techniques are discussed in Chapter 15. The sampling system should not alter the chemical or physical characteristics of the sample in an undesirable manner. The major components of most sampling systems are an inlet manifold, an air mover, a collection medium, and a flow measurement device.

The inlet manifold transports material from the ambient atmosphere to the collection medium or analytical device, preferably in an unaltered condition. The inlet opening may be designed for a specific purpose. All inlets for ambient sampling must be rainproof. Inlet manifolds are made out of glass, Teflon, stainless steel, or other inert materials and permit the remaining components of the system to be located at a distance from the sample manifold inlet. The air mover provides the force to create a vacuum or lower pressure at the end of the sampling system. In most instances, air movers are pumps. The collection medium for a sampling system may be a liquid or solid sorbent for dissolving gases, a filter surface for collecting particles, or a chamber to contain an aliquot of air for analysis. The flow device measures the volume of air associated with the sampling system. Examples of flow devices are mass flow meters, rotameters, and critical orifices.

Sampling systems can take several forms and may not necessarily have all four components (Fig. 13-1). Figure 13-1(a) is typical of many extractive sampling techniques in practice, e.g., SO_2 in liquid sorbents and polynuclear aromatic hydrocarbons on solid sorbents. Figure 13-1(b) is used for "open-face" filter collection, in which the filter is directly exposed to the atmosphere being sampled. Figure 13-1(c) is an evacuated container used to collect an aliquot of air or gas to be transported to the laboratory for chemical analysis; e.g., polished stainless steel canisters are used to collect ambient hydrocarbons for air toxic analysis. Figure 13-1(d) is the basis for many of the automated continuous analyzers, which combine the sampling and analytical processes in one piece of equipment, e.g., continuous ambient air monitors for SO_2, O_3, and NO_x.

Regardless of the configuration or the specific material sampled, several characteristics are important for all ambient air sampling systems. These are collection efficiency, sample stability, recovery, minimal interference, and an understanding of the mechanism of collection. Ideally, the first three would be 100% and there would be no interference or change in the material when collected.

One example is sampling for SO_2. Liquid sorbents for SO_2 depend on the solubility of SO_2 in the liquid collection medium. Certain liquids at the correct pH are capable of removing ambient concentrations of SO_2 with 100% efficiency until the characteristics of the solution are altered so that no more SO_2 may be dissolved in the volume of liquid provided. Under these circumstances, sampling is 100% efficient for a limited total mass of SO_2 transferred to the solution, and the technique is acceptable as long as sampling does not continue beyond the time that the sampling solution is saturated (1). A second example is the use of solid sorbents such as Tenax

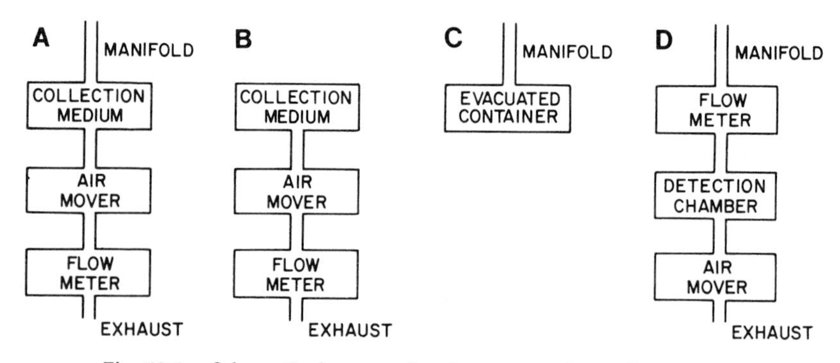

Fig. 13-1. Schematic diagram of various types of sampling systems.

for volatile hydrocarbons by the physical adsorption of the individual hydrocarbon molecules on active sites of the sorbent (2). Collection efficiency drops drastically when the active sites become saturated.

Sample stability becomes increasingly important as the time between sampling and analysis increases. Effects of temperature, trace contaminants, and chemical reactions can cause the collected species to be lost from the collection medium or to undergo a transformation that will prevent its recovery. Nearly 100% recovery is also required because a variable recovery rate will prevent quantification of the analysis. Interference should be minimal and, if present, well understood.

II. SAMPLING SYSTEMS FOR GASEOUS POLLUTANTS

Gaseous pollutants are generally collected by the sampling systems shown in Fig. 13-1(a–d). The sampling manifold's only function is to transport the gas from the manifold inlet to the collection medium in an unaltered state. The manifold must be made of nonreactive material. Tests of material for manifold construction can be made for specific gases to be sampled. In most cases, glass or Teflon will not adsorb or react with the gases. No condensation should be allowed to occur in the sampling manifold.

The volume of the manifold and the sampling flow rate determine the time required for the gas to move from the inlet to the collection medium. This residence time can be minimized to decrease the loss of reactive species in the manifold by keeping the manifold as short as possible.

The collection medium for gases can be liquid or solid sorbents, an evacuated flask, or a cryogenic trap. Liquid collection systems take the form of bubblers which are designed to maximize the gas–liquid interface. Each design is an attempt to optimize gas flow rate and collection efficiency. Higher flow rates permit shorter sampling times. However, excessive flow rates cause the collection efficiency to drop below 100%.

A. Extractive Sampling

When bubbler systems are used for collection, the gaseous species generally undergoes hydration or reaction with water to form anions or cations. For example, when SO_2 and NH_3 are absorbed in bubblers they form HSO_3^- and NH_4^+, and the analytical techniques for measurement actually detect these ions. Table 13-1 gives examples of gases which may be sampled with bubbler systems.

Bubblers are more often utilized for sampling programs that do not require a large number of samples or frequent sampling. The advantages of these types of sampling systems are low cost and portability. The disadvantages are the high degree of skill and careful handling needed to ensure

TABLE 13-1

Collection of Gases by Absorption

Gas	Sampler	Sorption medium	Air flow (liters/m)	Minimum sample (liters)	Collection efficiency	Analysis	Interferences
Ammonia	Midget impinger	25 ml 0.1 *N* sulfuric acid	1–3	10		Nessler reagent	—
	Petri bubbler	10 ml of above	1–3	10	+95	Nessler reagent	—
Benzene	Glass bead column	5 ml nitrating acid	0.25	3–5	+95	Butanone method	Other aromatic hydrocarbons
Carbon dioxide	Fritted bubbler	10 ml 0.1 *N* barium hydroxide	1	10–15	60–80	Titration with 0.05 *N* oxalic acid	Other acids
Ethyl benzene	Fritted bubbler or midget impinger	15 ml spectrograde isooctane	1	20	+90	Alcohol extraction, ultraviolet analysis	Other aromatic hydrocarbons
Formaldehyde	Fritted bubbler	10 ml 1% sodium bisulfite	1–3	25	+95	Liberated sulfite titrated, 0.01 *N* iodine	Methyl ketones
Hydrochloric acid	Fritted bubbler	0.005 *N* sodium hydroxide	10	100	+95	Titration with 0.01 *N* silver nitrate	Other chlorides
Hydrogen sulfide	Midget impinger	15 ml 5% cadmium sulfate	1–2	20	+95	Add 0.05 *N* iodine, 6 *N* sulfuric acid, back-titrate 0.01 *N* sodium thiosulfate	Mercaptans, carbon disulfide, organic sulfur compounds
Lead, tetraethyl, tetramethyl	Dreschel-type scrubber	100 ml 0.1 *M* iodine monochloride in 0.3 *N* hydrochloric acid	1.8–2.9	50–75	100	Dithizone	Bismuth, thallium, stannous tin
Mercury, diethyl and dimethyl	Midget impinger	15 ml of above	1.9	50–75	91–95	Same as above	Same as above

	Midget impinger	10 ml 0.1 *M* iodine monochloride in .3 *N* hydrochloric acid	1–1.5	100	91–100	Dithizone	Copper
Nickel carbonyl	Midget impinger	15 ml 3% hydrochloric acid	2.8	50–90	+90	Complex with alpha-furil-dioxime	—
Nitrogen dioxide	Fritted bubbler (60–70 μm pore size)	20–30 ml Saltzman reagent[a]	0.4	Sample until color appears; probably 10 ml of air	94–99	Reacts with absorbing solution	Ozone in fivefold excess peroxyacyl nitrate
Ozone	Midget impinger	1% potassium iodide in 1 *N* potassium hydroxide	1	25	+95	Measures color of iodine liberated	Other oxidizing agents
Phosphine	Fritted bubbler	15 ml 0.5% silver diethyl dithiocarbamate in pyridine	0.5	5	86	Complexes with absorbing solution	Arsine, stibine, hydrogen sulfide
Styrene	Fritted midget impinger	15 ml spectrograde isooctane	1	20	+90	Ultraviolet analysis	Other aromatic hydrocarbons
Sulfur dioxide	Midget impinger, fritted rubber	10 ml sodium tetrachloromer-curate	2–3	2	99	Reaction of dichloro-sulfitomercurate and formalde-hydepararosani-line	Nitrogen dioxide,[b] hydrogen sulfide[c]

(*Continued*)

TABLE 13-1 (*Continued*)

Gas	Sampler	Sorption medium	Air flow (liters/m)	Minimum sample (liters)	Collection efficiency	Analysis	Interferences
Toluene diisocyanate	Midget impinger	15 ml Marcali solution	1	25	95	Diazotization and coupling reaction	Materials containing reactive hydrogen attached to oxygen (phenol); certain other diamines
Vinyl acetate	Fritted midget impinger and simple midget impinger in series	Toluene	1.5	15	+99 (84 with fritted bubbler only)	Gas chromatography	Other substances with same retention time on column

[a] 5 gm sulfanilic; 140 ml glacial acetic acid; 20 ml 0.1% aqueous *N*-(1-naphthyl) ethylene diamine.
[b] Add sulfamic acid after sampling.
[c] Filter or centrifuge any precipitate.
Source: Pagnotto, L. D., and Keenan, R. G., "Sampling and Analysis of Gases and Vapors," *in* "The Industrial Environment—Its Evaluation and Control." United States Department of Health, Education and Welfare, United States Government Printing Office, Washington, D.C., 1973, pp. 167–179.

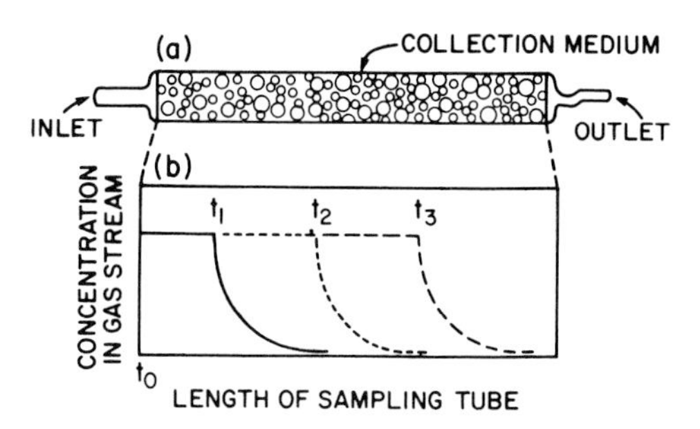

Fig. 13.2 Solid sorbent collection tube. (a) The tube is packed with a granular medium. (b) As the hydrocarbon-containing air is passed through the collection tube at t_1, t_2, and t_3, the collection medium becomes saturated at increasing lengths along the tube.

quality results. Solid sorbents such as Tenax, XAD, and activated carbon (charcoal) are used to sample hydrocarbon gases by trapping the species on the active sites of the surface of the sorbent. Figure 13-2 illustrates the loading of active sites with increasing sample time. It is critical that the breakthrough sampling volume, the amount of air passing through the tube that saturates its absorptive capacity, not be exceeded. The breakthrough volume is dependent on the concentration of the gas being sampled and the absorptive capacity of the sorbent. This means that the user must have an estimate of the upper limit of concentration for the gas being sampled.

Once the sample has been collected on the solid sorbent, the tube is sealed and transported to the analytical laboratory. To recover the sorbed gas, two techniques may be used. The tube may be heated while an inert gas is flowing through it. At a sufficiently high temperature, the absorbed molecules are desorbed and carried out of the tube with the inert gas stream. The gas stream may then be passed through a preconcentration trap for injection into a gas chromatograph for chemical analysis. The second technique is liquid extraction of the sorbent and subsequent liquid chromatography. Sometimes a derivitization step is necessary to convert the collected material chemically into compounds which will pass through the column more easily, e.g., conversion of carboxylic acids to methyl esters. Solid sorbents have increased our ability to measure hydrocarbon species under a variety of field conditions. However, this technique requires great skill and sophisticated equipment to obtain accurate results. Care must be taken to minimize problems of contamination of the collection medium, sample instability on the sorbent, and incomplete recovery of the sorbed gases.

Special techniques are employed to sample for gases and particulate matter simultaneously (3). Sampling systems have been developed which permit the removal of gas-phase molecules from a moving airstream by diffusion to a coated surface and permit the passage of particulate matter downstream for collection on a filter or other medium. These diffusion denuders are used to sample for SO_2 or acid gases in the presence of particulate matter. This type of sampling has been developed to minimize the interference of gases in particulate sampling and vice versa.

The third technique, shown in Fig. 13-1(c), involves collection of an aliquot of air in its gaseous state for transport back to the analytical laboratory. Use of a preevacuated flask permits the collection of a gas sample in a specially polished stainless steel container. By use of pressure–volume relationships, it is possible to remove a known volume from the tank for subsequent chemical analysis. Another means of collecting gaseous samples is the collapsible bag. Bags made of polymer films can be used for collection and transport of samples. The air may be pumped into the bag by an inert pump such as one using flexible metal bellows, or the air may be sucked into the bag by placing the bag in an airtight container which is then evacuated. This forces the bag to expand, drawing in the ambient air sample.

B. *In Situ* Sampling and Analysis

The fourth sampling technique involves a combination of sampling and analysis. The analytical technique is incorporated in a continuous monitoring instrument placed at the sampling location. Most often, the monitoring equipment is located inside a shelter such as a trailer or a small building, with the ambient air drawn to the monitor through a sampling manifold. The monitor then extracts a small fraction of air from the manifold for analysis by an automated technique, which may be continuous or discrete. Instrument manufacturers have developed automated *in situ* monitors for several air pollutants, including SO_2, NO, NO_2, O_3, and CO.

III. SAMPLING SYSTEMS FOR PARTICULATE POLLUTANTS AND PM_{10}

Sampling for particles in the atmosphere involves a different set of parameters from those used for gases. Particles are inherently larger than the molecules of N_2 and O_2 in the surrounding air and therefore behave differently with increasing diameter. When one is sampling for particulate matter in the atmosphere, three types of information are of interest—the mass concentration, size, and chemical composition of the particles. Particle size is important in determining adverse effects and atmospheric removal processes. The U.S. Environmental Protection Agency has specified a PM_{10} sampling method for compliance monitoring for the National Ambient Air Quality Standards (NAAQS) for particulate matter. This technique must be able to sample particulate matter with an aerodynamic diameter less than 10 μm with a prescribed efficiency.

Particles in the atmosphere come from different sources, e.g., combustion, windblown dust, and gas-to-particle conversion processes (see Chapter 6). Figure 2-2 illustrates the wide range of particle diameters potentially present in the ambient atmosphere. A typical size distribution of ambient particles is shown in Fig. 2-3. The distribution of number, surface, and mass can occur over different diameters for the same aerosol. Variation in chemical composition as a function of particle diameter has also been observed, as shown in Table 4-3.

The major purpose of ambient particulate sampling is to obtain mass concentration and chemical composition data, preferably as a function of particle diameter. This information is valuable for a variety of problems: effects on human health, identification of particulate matter sources, understanding of atmospheric haze, and particle removal processes.

The primary approach is to separate the particles from a known volume of air and subject them to weight determination and chemical analysis. The principal methods for extracting particles from an airstream are filtration and impaction. All sampling techniques must be concerned with the behavior of particles in a moving airstream. The difference between sampling for gases and sampling for particles begins at the inlet of the sampling manifold and is due to the discrete mass associated with individual particles.

Behavior of Particles at Sampling Inlets

Sampling errors may occur at the inlet, and particles may be lost in the sampling manifold while being transported to the collection surface. Figure

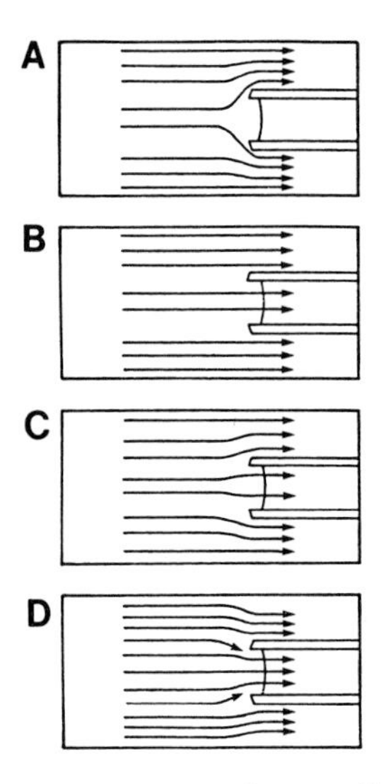

Fig. 13-3. The streamline flow patterns around a sampling inlet in a uniform-flow field.

13-3 illustrates the flow patterns around a sampling inlet in a uniform flow field. Figure 13-3(a) shows that when no air is permitted to flow into the inlet, the streamline flow moves around the edges of the inlet. As the flow rate through the inlet increases, more and more of the streamlines are attracted to the inlet. Figure 13-3(b) is called the *isokinetic condition,* in which the sampling flow rate is equal to the flow field rate. An example is an inlet with its opening into the wind pulling air at the wind speed. When one is sampling for gases, this is not a serious constraint because the composition of the gas will be the same under all inlet flow rates; i.e., there is no fractionation of the air sample by different gaseous molecules.

Particle-containing air streams present a different situation. Figure 13-3(b), the isokinetic case, is the ideal case. The ideal sample inlet would always face into the wind and sample at the same rate as the instantaneous wind velocity (an impossibility). Under isokinetic sampling conditions, parallel air streams flow into the sample inlet, carrying with them particles of all diameters capable of being carried by the stream flow. When the sampling rate is lower than the flow field (Fig. 13-3c), the streamlines start to diverge around the edges of the inlet and the larger particles with more inertia are unable to follow the streamlines and are captured by the sampling inlet. The opposite happens when the sampling rate is higher than the flow field. The inlet captures more streamlines, but the larger particles near the edges of the inlet may be unable to follow the streamline flow and escape collection by the inlet. The inlet may be designed for particle size fractionation; e.g., a PM_{10} inlet will exclude particles larger than 10 μm aerodynamic diameter.

These inertial effects become less important for particles with diameters less than 5 μm and for low wind velocities, but for samplers attempting to collect particles above 5 μm, the inlet design and flow rates become important parameters. In addition, the wind speed has a much greater impact on sampling errors associated with particles more than 5 μm in diameter (4).

After the great effort taken to get a representative sample into the sampling manifold inlet, care must be taken to move the particles to the collection medium in an unaltered form. Potential problems arise from too long or too twisted manifold systems. Gravitational settling in the manifold will remove a fraction of the very large particles. Larger particles are also subject to loss by impaction on walls at bends in a manifold. Particles may also be subject to electrostatic forces which will cause them to migrate to the walls of nonconducting manifolds. Other problems include condensation or agglomeration during transit time in the manifold. These constraints require sampling manifolds for particles to be as short and have as few bends as possible.

The collection technique involves the removal of particles from the air stream. The two principal methods are filtration and impaction. Filtration consists of collecting particles on a filter surface by three processes—direct interception, inertial impaction, and diffusion (5). Filtration attempts to remove a very high percentage of the mass and number of particles by these three processes. Any size classification is done by a preclassifier, such as an impactor, before the particle stream reaches the surface of the filter.

IV. STATIC SAMPLING SYSTEMS

Static sampling systems are defined as those that do not have an active air-moving component, such as the pump, to pull a sample to the collection medium. This type of sampling system has been used for over 100 years. Examples include the lead peroxide candle used to detect the presence of SO_2 in the atmosphere and the dust-fall bucket and trays or slides coated with a viscous material used to detect particulate matter. This type of system suffers from inability to quantify the amount of pollutant present over a short period of time, i.e., less than 1 week. The potentially desirable characteristics of a static sampling system have led to further developments in this type of technology to provide quantitative information on pollutant concentrations over a fixed period of time. Static sampling systems have been developed for use in the occupational environment and are also used to measure the exposure levels in the general community, e.g., radon gas in residences.

The advantages of static sampling systems are their portability, convenience, reliability, and low cost. The systems are lightweight and can be attached directly to individuals. Nonstatic sampling systems can, of course, also be attached to individuals, but are less convenient because the person must carry a battery-powered pump and its batteries. Static sampling systems are very reliable, and the materials used limit the costs to acceptable levels.

Two principles are utilized in the design of static samplers—diffusion and permeation (6, 7). Samplers based on the diffusion principle depend on the molecular interactions of N_2, O_2, and trace pollutant gases. If a concentration gradient can be established for the trace pollutant gas, under certain conditions the movement of the gas will be proportional to the concentration gradient (Fick's law of diffusion), and a sampler can be designed to take advantage of this technique. Figure 13-4 illustrates this principle. The sampler has a well-defined inlet, generally with a cylindrical shape, through which the pollutant gas must diffuse. At the end of the tube, a collection medium removes the pollutant gas for subsequent analysis and

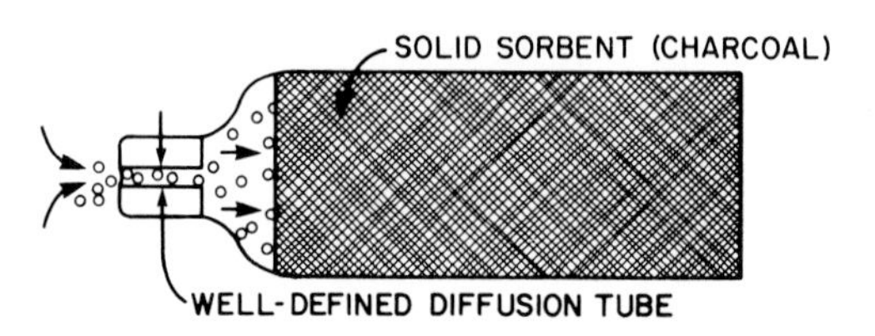

Fig. 13-4. Static sampler based on the diffusion principle.

maintains a concentration gradient between the inlet of the tube and the collection medium. The mathematical relationship (Fick's law) describing this type of passive sampler is given by Eq. (13-1):

$$R = -DA\ (dC/dx) \tag{13-1}$$

where R is the rate of transport by diffusion in moles per second, D the diffusion coefficient in square centimeters per second, A the cross-sectional area of the diffusion path in square centimeters, C the concentration of species in moles per cubic centimeter, and x the path length in centimeters.

The ability of gases to permeate through various polymers at a fixed rate depending on a concentration gradient has been used to create static samplers. This principle was originally developed to provide a standard calibration source of trace gas by putting that gas in a polymer tube under pressure and letting the material diffuse or permeate through the wall to the open atmosphere. Permeation samplers operate in the reverse direction. Figure 13-5 illustrates this type of system. A thin film membrane is open to the atmosphere on one side and to a collection medium on the other. A pollutant gas in the atmosphere diffuses through the membrane and is collected in the medium. The mathematical relationship for a permeation sampler is given by Eq. (13-2):

$$k = Ct/m \tag{13-2}$$

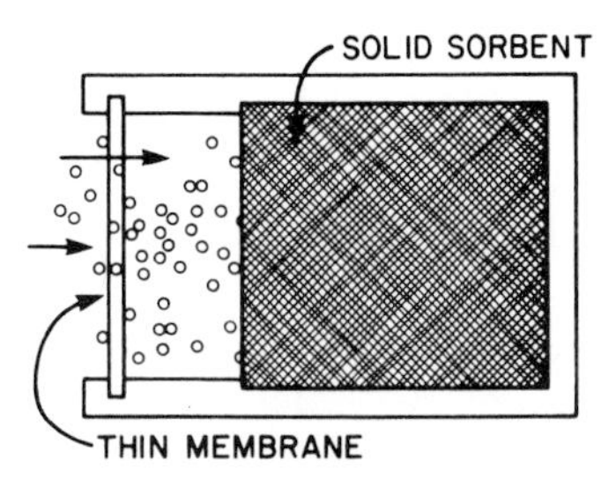

Fig. 13-5. Static sampler based on the permeation principle.

where k is the permeation constant, C the concentration of gas in parts per million, t the time of exposure, and m the amount of gas absorbed in micrograms.

Permeation systems can be calibrated in the laboratory and then used in the field for sample collection for a fixed period of time, e.g., 8 hr or 7 days. The sampler is returned to the laboratory for analysis. These systems can be made for specific compounds by selecting the appropriate collection medium and the polymer membrane (Table 13-2).

V. SAMPLER SITING REQUIREMENTS

Sampling site selection is dependent on the purpose or use of the results of the monitoring program. Sampling activities are typically undertaken to determine the ambient air quality for compliance with air quality standards, for evaluation of the impact of a new air pollution source during the preconstruction phase, for hazard evaluation associated with accidental spills of chemicals, for human exposure monitoring, and for research on atmospheric chemical and physical processes. The results of ambient air monitoring can be used to judge the effectiveness of the air quality management approach to air pollution problems. The fundamental reason for controlling

TABLE 13-2

Permeation Samplers for Selected Gases

Gas	Membrane	Sorber	Sensitivity
Chlorine	Dimethyl silicone (DMS) (single-backed)	Buffered (pH 7) Fluorescein, 0.005% NaBr (0.31%)	0.013 ppm (8-hr exposure)
Sulfur dioxide	DMS (single-backed)	Tetrachloromercurate (II)	0.01 ppm (8-hr exposure)
Vinyl chloride	DMS (single-backed)	Activated charcoal (CS_2 desorption)	0.02 ppm (linear to 50 ppm+)
Alkyl lead	DMS (unbacked)	Silica gel (ICl desorption)	0.2 μg
Benzene	Silicon polycarbonate	Activated charcoal (CS_2 desorption)	0.02 ppm (8-hr exposure)
Ammonia	Vinyl silicone	0.6% boric acid	0.4 ppm (8-hr exposure)
Hydrogen sulfide	DMS (single-backed)	0.02 *N* NaOH, 0.003 *M* EDTA	0.01 ppm
Hydrogen cyanide	DMS (single-backed)	0.01 *N* NaOH	0.01 ppm (8-hr exposure)

Source: West, P. W., *Am. Lab.* **12,** 35–39 (1980).

air pollution sources is to limit the buildup of contaminants in the atmosphere so that adverse effects are not observed. This suggests that sampling sites should be selected to measure pollutant levels close to or representative of exposed populations of people, plants, trees, materials, structures, etc. Generally, sites in air quality networks are near ground level, typically 3 m aboveground, and are located so as not to be unduly dominated by a nearby source such as a roadway. Sampling sites require electrical power and adequate protection (which may be as simple as a fence). A shelter, such as a small building, may be necessary. Permanent sites require adequate heating and air conditioning to provide a stable operating environment for the sampling and monitoring equipment.

VI. SAMPLING FOR AIR TOXICS

Public awareness of the release of chemicals into the atmosphere has gone beyond the primary ambient pollutants (e.g., SO_2 or O_3) and governments require air toxics management plans. One component of this process is the characterization of the air quality via sampling.

Most of the airborne chemicals classified as "air toxics" are organic compounds with physical and chemical properties ranging from those of formaldehyde found in the gas phase to polynuclear aromatic hydrocarbons (PAHs) which may be absorbed on particle surfaces. This range of volatility and reactivity represented by air toxics requires a variety of sampling techniques—from grab sampling to filter techniques followed by extraction and detailed derivatization techniques. When these compounds are present in the atmosphere, the concentration level can be quite low, in the parts per billion (ppb) to sub-ppb range for gases and the picogram/m^3 range for particulate components. This generally requires extended sampling times and very sensitive analytical techniques for laboratory analysis.

Two examples of this type of ambient sampling are described. The U.S. Environmental Protection Agency established a pilot Toxics Air Monitoring System network for sampling ambient volatile organic compounds (VOCs) at ppb levels in Boston, Chicago, and Houston for a two-year period (8). Evacuated stainless steel canisters were used to collect air at 3 cm^3/min for 24 hours. The canisters were returned to a central laboratory and analyzed by cryogenic concentration of the VOCs, separation by gas chromatography, and mass-selective detection. This system provided information on 13 VOCs in three classes: chlorofluorocarbons, aromatics, and chlorinated alkanes.

A second sampling program in Southern California sampled for polychlorinated dioxins and polychlorinated dibenzofurans at seven locations (9). Because of the semivolatile nature of these compounds, a tandem sampler was used with a glass fiber filter to collect the particulate-associated components followed by a polyurethene foam sorbent trap to collect the vapor-phase portion. These samples were returned to the laboratory, where they were extracted and analyzed with high-resolution gas chromatography and high-resolution mass spectrometry. The observed concentrations were in the picogram/m^3 range.

Each of these examples suggest that air toxics sampling is complex and expensive and requires careful attention to quality assurance.

REFERENCES

1. Pagnotto, L. D., and Keenan, R. G., Sampling and analysis of gases and vapors, *in* "The Industrial Environment—Its Evaluation and Control," pp. 167–179. U.S. Department of Health, Education, and Welfare, U.S. Government Printing Office, Washington, DC, 1973.
2. Tanaka, T., *J. Chromatogr.* **153,** 7–13 (1978).
3. Slanina, J., De Wild, P. J., and Wyers, G. P., *Adv. Environ. Sci. Technol.* **24,** 129–154 (1992).
4. Cadle, R. D., "The Measurement of Airborne Particles." Academic Press, New York, 1976.
5. Liu, B. Y. H. (ed.), "Fine Particles." Academic Press, New York, 1976.
6. Palmes, E. D., Gunnison, A. F., DiMatto, J., and Tomczyk, C., *Am. Ind. Hyg. Assoc. J.* **37,** 570–577 (1976).
7. West, P. W., *Am. Lab.* **12,** 35–39 (1980).
8. Evans, G. F., Lumpkin, T. A., Smith, D. L., and Somerville, M. C., *J. Air Waste Manage. Assoc.* **42,** 1319–1323, (1992).
9. Hunt, G. T., and Maisel, B. E., *J. Air Waste Manage. Assoc.* **42,** 672–680 (1992).

SUGGESTED READING

Friedlander, S. K., "Smoke, Dust and Haze." Wiley, New York, 1977.

Hering, S. V., "Air Sampling Instruments for Evaluation of Air Contaminants." ACGIH, Cincinnati, OH, 1989

Noll, K. E., and Miller, T. L., "Air Monitoring Survey Design." Ann Arbor Science Publishers, Ann Arbor, MI, 1977.

Stanley-Wood, N. G., and Lines, R. W. (eds.), "Particle Size Analysis." Royal Society of Chemistry, Cambridge, UK, 1992.

Willeke, K., and Baron, P. A., "Aerosol Measurement—Principles, Techniques, and Applications." Van Nostrand Reinhold, New York, 1993

QUESTIONS

1. Describe the four components of a sampling system.
2. List three examples of the four components, e.g., a metal bellows pump.
3. A solid sorbent Tenax cartridge has a capacity of 100 μg of toluene. If samples were collected at a rate of 5 liters/min, calculate the maximum ambient concentration which can be determined by an hourly sample and a 15-min sample.

4. Describe the sampling approaches used for air pollutants by your state or local government.
5. List the possible sources of loss or error in sampling for particulate matter.
6. Why is sampling velocity not an important parameter when sampling for gases?
7. List the advantages of static sampling systems.
8. Describe the precautions which should be considered when determining the location of the sampling manifold inlet for an ambient monitoring system.

14

Ambient Air Pollutants: Analysis and Measurement

I. ANALYSIS AND MEASUREMENT OF GASEOUS POLLUTANTS

The two major goals of testing for air pollutants are identification and quantification of a sample of ambient air. Air pollution measurement techniques generally pass through evolutionary stages. The first is the qualitative identification stage. This is followed by separate collection and quantification stages. The last stage is the concurrent collection and quantification of a given pollutant.

Gaseous SO_2 is an example. Very early procedures detected the presence of SO_2 in ambient air by exposing a lead peroxide candle for a period of time and then measuring the amount of lead sulfate formed. Because the volume of air in contact with the candle was not measured, the technique could not quantify the amount of SO_2 per unit volume of air.

The next stage involved passing a known volume of ambient air through an absorbing solution in a container in the field and then returning this container to the laboratory for a quantitative determination of the amount of absorbed SO_2. The United Nations Environmental Program–World Health Organization's worldwide air sampling and analysis network used this method for SO_2, the only gaseous pollutant measured by the network. The final evolutionary step has been the concurrent collection and quantification of SO_2. An example of this is the flame photometric SO_2 analyzer, in which SO_2-laden air is fed into an H_2 flame, and light emissions from electronically excited combustion products are detected by a photomultiplier tube. Prior calibration of the analyzer permits the rapid determination of SO_2. This is but one of the many methods available for the measurement of SO_2.

Hundreds of chemical species are present in urban atmospheres. The gaseous air pollutants most commonly monitored are CO, O_3, NO_2, SO_2, and nonmethane volatile organic compounds (NMVOCs). Measurement of specific hydrocarbon compounds is becoming routine in the United States for two reasons: (1) their potential role as air toxics and (2) the need for detailed hydrocarbon data for control of urban ozone concentrations. Hydrochloric acid (HCl), ammonia (NH_3), and hydrogen fluoride (HF) are occasionally measured. Calibration standards and procedures are available for all of these analytic techniques, ensuring the quality of the analytical results.

A. Carbon Monoxide

The primary reference method used for measuring carbon monoxide in the United States is based on nondispersive infrared (NDIR) photometry (1, 2). The principle involved is the preferential absorption of infrared radiation by carbon monoxide. Figure 14-1 is a schematic representation of an NDIR analyzer. The analyzer has a hot filament source of infrared radiation, a chopper, a sample cell, reference cell, and a detector. The reference cell is filled with a non-infrared-absorbing gas, and the sample cell is continuously flushed with ambient air containing an unknown amount of CO. The detector cell is divided into two compartments by a flexible membrane, with each compartment filled with CO. Movement of the membrane causes a change in electrical capacitance in a control circuit whose signal is processed and fed to a recorder.

The chopper intermittently exposes the two cells to infrared radiation. The reference cell is exposed to a constant amount of infrared energy which is transmitted to one compartment of the detector cell. The sample cell, which contains varying amounts of infrared-absorbing CO, transmits to the detector cell a reduced amount of infrared energy that is inversely proportional to the CO concentration in the air sample. The unequal amounts of energy received by the two compartments in the detector cell cause the membrane to move, producing an alternating current (AC) electrical signal whose frequency is established by the chopper spacing and the speed of chopper rotation.

Water vapor is a serious interfering substance in this technique. A moisture trap such as a drying agent or a water vapor condenser is required to remove water vapor from the air to be analyzed.

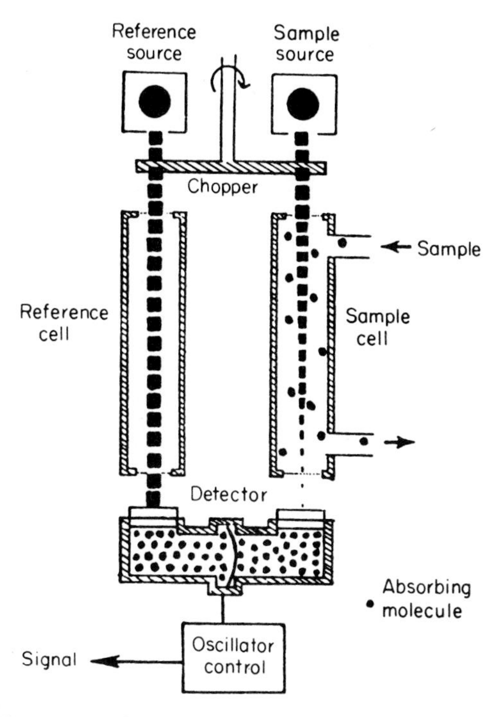

Fig. 14-1. Nondispersive infrared (NDIR) analyzer. Source: Bryan, R. J., Ambient air quality surveillance, *in* "Air Pollution," 3rd ed., Vol. III (A. C. Stern, ed.). Academic Press, New York, 1976, p. 375.

Instruments based on other techniques are available which meet the performance specifications outlined in Table 14-1.

B. Ozone

The principal method used for measuring ozone is based on chemiluminescence (3). When ozone and ethylene react chemically, products are formed which are in an excited electronic state. These products fluoresce, releasing light. The principal components are a constant source of ethylene, an inlet sample line for ambient air, a reaction chamber, a photomultiplier tube, and signal-processing circuitry. The rate at which light is received by the photomultiplier tube is dependent on the concentrations of O_3 and ethylene. If the concentration of ethylene is made much higher than the ozone concentration to be measured, the light emitted is proportional only to the ozone concentration.

Instruments based on this principle may be calibrated by a two-step process shown in Fig. 14-2 (4). A test atmosphere with a known source of

TABLE 14-1

Performance Specifications for Automated Analytical Methods for Measuring Carbon Monoxide

Range	0–57 mg/m^3 (0–50 ppm)
Noise	0.6 mg/m^3 (0.50 ppm)
Lower detectable limit	1.2 mg/m^3 (1.0 ppm)
Interference equivalent	
Each interfering substance	±1.2 mg/m^3 (±1.0 ppm)
Total interfering substances	1.7 mg/m^3 (1.5 ppm)
Zero drift	
12 hr	±1.2 mg/m^3 (±1.0 ppm)
24 hr	±1.2 mg/m^3 (±1.0 ppm)
Span drift, 24 hr	
20% of upper range limit	±10.0%
80% of upper range limit	±2.5%
Lag time	10 min
Rise time	5 min
Fall time	5 min
Precision	
20% of upper range limit	0.6 mg/m^3 (0.5 ppm)
80% of upper range limit	0.6 mg/m^3 (0.5 ppm)

Definitions:
Range: Nominal minimum and maximum concentrations that a method is capable of measuring.
Noise: The standard deviation about the mean of short-duration deviations in output that are not caused by input concentration changes.
Lower detectable limit: The minimum pollutant concentration that produces a signal of twice the noise level.
Interference equivalent: Positive or negative response caused by a substance other than the one being measured.
Zero drift: The change in response to a zero pollutant concentration during continuous unadjusted operation.
Span drift: The percentage change in response to an upscale pollutant concentration during continuous unadjusted operation.
Lag time: The time interval between a step change in input concentration and the first observable corresponding change in response.
Rise time: The time interval between the initial response and 95% of the final response.
Fall time: The time interval between the initial response to a step decrease in concentration and 95% of the final response.
Precision: Variation about the mean of repeated measurements of the same pollutant concentration expressed as one standard deviation about the mean.
Source: *Fed. Regist.* **40,** 7042–7070 (1975).

ozone is produced by an ozone generator, a device capable of generating stable levels of O_3. Step 1 involves establishing the concentration of ozone in the test atmosphere by ultraviolet photometry. This is followed by step 2, calibration of the instrument's response to the known concentration of ozone in the test atmosphere.

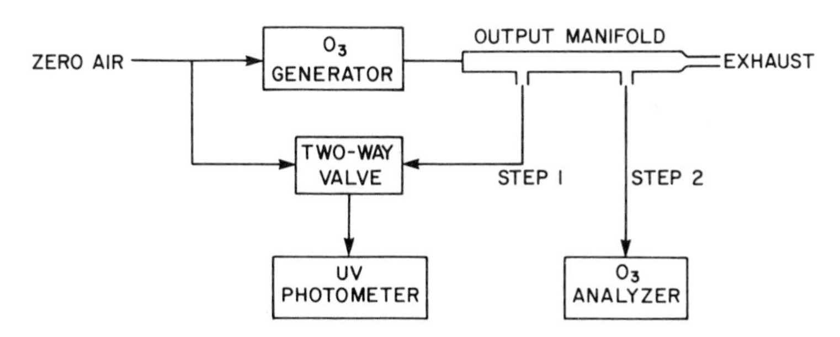

Fig. 14-2. Calibration apparatus for ozone analyzer (UV, ultraviolet).

C. Nitrogen Dioxide

The principal method used for measuring NO_2 is also based on chemiluminescence (Fig. 14-3) (5). NO_2 concentrations are determined indirectly from the difference between the NO and NO_x (NO + NO_2) concentrations in the atmosphere. These concentrations are determined by measuring the light emitted from the chemiluminescent reaction of NO with O_3 (similar to the reaction of O_3 with ethylene noted for the measurement of O_3), except that O_3 is supplied at a high constant concentration, and the light output is proportional to the concentration of NO present in the ambient air stream.

Figure 14-4 illustrates the analytical technique based on this principle. To determine the NO_2 concentration, the NO and NO_x (NO + NO_2) concentrations are measured. The block diagram shows a dual pathway through the instrument, one to measure NO and the other to measure NO_x. The NO pathway has an ambient air stream containing NO (as well as NO_2), an ozone stream from the ozone generator, a reaction chamber, a photomultiplier tube, and signal-processing circuitry. The NO_x pathway has the same components, plus a converter for quantitatively reducing NO_2 to NO. The instrument can also electronically subtract the NO from NO_x and yield as output the resultant NO_2.

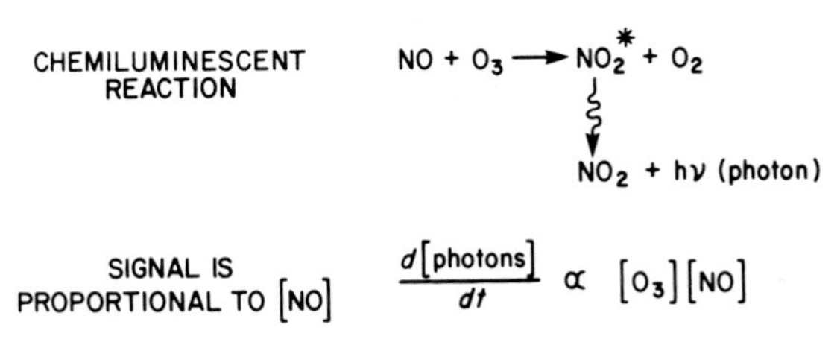

Fig. 14-3. NO_2 chemiluminescent detection principle based on the reaction of NO with O_3.

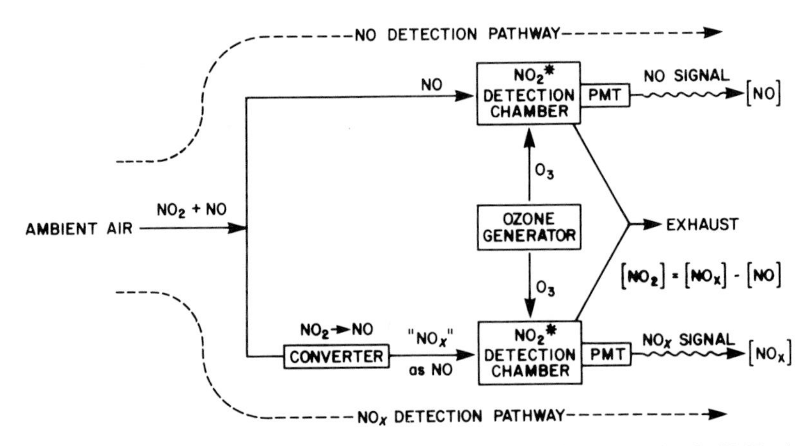

Fig. 14-4. Schematic diagram of chemiluminescent detector for NO_2 and NO. PMT, photomultiplier tube.

Air passing through the NO pathway enters the reaction chamber, where the NO present reacts with the ozone. The light produced is measured by the photomultiplier tube and converted to an NO concentration. The NO_2 in the air stream in this pathway is unchanged. In the NO_x pathway, the NO- and NO_2-laden air enters the converter, where the NO_2 is reduced to form NO; all of the NO_x exits the converter as NO and enters the reaction chamber. The NO reacts with O_3 and the output signal is the total NO_x concentration. The NO_2 concentration in the original air stream is the difference between NO_x and NO. Calibration techniques use gas-phase titration of an NO standard with O_3 or an NO_2 permeation device.

D. Sulfur Dioxide

Several manual and continuous analytical techniques are used to measure SO_2 in the atmosphere. The manual techniques involve two-stage sample collection and measurement. Samples are collected by bubbling a known volume of gas through a liquid collection medium. Collection efficiency is dependent on the gas–liquid contact time, bubble size, SO_2 concentration, and SO_2 solubility in the collection medium. The liquid medium contains chemicals which stabilize SO_2 in solution by either complexation or oxidation to a more stable form. Field samples must be handled carefully to prevent losses from exposure to high temperatures. Samples are analyzed at a central laboratory by an appropriate method.

The West-Gaeke manual method is the basis for the U.S. Environmental Protection Agency reference method for measurement of SO_2 (6). The method uses the colorimetric principle; i.e., the amount of SO_2 collected is proportional to the amount of light absorbed by a solution. The collection medium is an aqueous solution of sodium or potassium tetrachloromercurate (TCM). Absorbed SO_2 forms a stable complex with TCM. This enhanced stability permits the collection, transport, and short-term storage of samples at a central laboratory. The analysis proceeds by adding bleached pararosaniline dye and formaldehyde to form red–purple pararosaniline methylsulfonic acid. Optical absorption at 548 nm is linearly proportional to the SO_2 concentration. Procedures are followed to minimize interference by O_3, oxides of nitrogen, and heavy metals.

The continuous methods combine sample collection and the measurement technique in one automated process. The measurement methods used for continuous analyzers include conductometric, colorimetric, coulometric, and amperometric techniques for the determination of SO_2 collected in a liquid medium (7). Other continuous methods utilize physicochemical techniques for detection of SO_2 in a gas stream. These include flame photometric detection (described earlier) and fluorescence spectroscopy (8). Instruments based on all of these principles are available which meet standard performance specifications.

E. Nonmethane Volatile Organic Compounds

The large number of individual hydrocarbons in the atmosphere and the many different hydrocarbon classes make ambient air monitoring a very difficult task. The ambient atmosphere contains an ubiquitous concentration of methane (CH_4) at approximately 1.6 ppm worldwide (9). The concentration of all other hydrocarbons in ambient air can range from 100 times less to 10 times greater than the methane concentration for a rural versus an urban location. The terminology of the concentration of hydrocarbon compounds is potentially confusing. Hydrocarbon concentrations are referred to by two units—parts per million by volume (ppmV) and parts per million by carbon (ppmC). Thus, 1 μl of gas in 1 liter of air is 1 ppmV, so the following is true:

Mixing ratio	ppmV	ppmC
$\frac{1\ \mu\text{l of } O_3}{1 \text{ liter of air}}$ =	1 ppm ozone	—
$\frac{1\ \mu\text{l of } SO_2}{1 \text{ liter of air}}$ =	1 ppmV SO_2	—
$\frac{1\ \mu\text{l of } CH_4}{1 \text{ liter of air}}$ =	1 ppmV CH_4	1 ppmC CH_4
$\frac{1\ \mu\text{l of } C_2H_6}{1 \text{ leter of air}}$ =	1 ppmV C_2H_6	2 ppmC C_2H_6

The unit parts per million by carbon takes into account the number of carbon atoms contained in a specific hydrocarbon and is the generally accepted way to report ambient hydrocarbons. This unit is used for three reasons: (1) the number of carbons atoms is a very crude indicator of the total reactivity of a group of hydrocarbon compounds, (2) historically, analytical techniques have expressed results in this unit, and (3) considerable information has been developed on the role of hydrocarbons in the atmosphere in terms of concentrations determined as parts per million by carbon.

Historically, measurements have classified ambient hydrocarbons in two classes: methane (CH_4) and all other nonmethane volatile organic compounds (NMVOCs). Analyzing hydrocarbons in the atmosphere involves a three-step process: collection, separation, and quantification. Collection involves obtaining an aliquot of air, e.g., with an evacuated canister. The principal separation process is gas chromatography (GC), and the principal quantification technique is with a calibrated flame ionization detector (FID). Mass spectroscopy (MS) is used along with GC to identify individual hydrocarbon compounds.

A simple schematic diagram of the GC/FID principle is shown in Fig. 14-5. Air containing CH_4 and other hydrocarbons classified as NMVOCs pass through a GC column and the air, CH_4, and NMVOC molecules are clustered into groups because of different absorption/desorption rates. As CH_4 and NMVOC groups exit the column, they are "counted" by the flame ionization detector. The signal output of the detector is proportional to the two groups and may be quantified when compared with standard concentrations of gases. This simplified procedure has been used extensively to collect hydrocarbon concentration data for the ambient atmosphere. A major disadvantage of this technique is the grouping of all hydrocarbons other than CH_4 into one class. Hydrocarbon compounds with similar structures are detected by an FID in a proportional manner, but for compounds with significantly different structures the response may be different. This difference in sensitivity results in errors in measurements of NMVOC mixtures.

More sophisticated GC columns and techniques perform more detailed separations of mixtures of hydrocarbons into discrete groups. Table 12-3

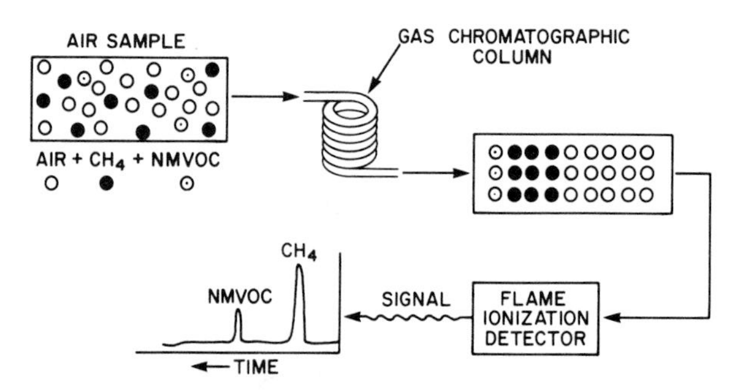

Fig. 14-5. Schematic diagram of hydrocarbon detection by gas chromatography. NMVOC, nonmethane volatile organic carbon.

lists individual hydrocarbons measured in ambient air by advanced GC techniques.

Other types of detectors include the flame photometric detector (FPD) and the electron capture detector (ECD). The FID is composed of an H_2 flame through which the hydrocarbon gases are burned, forming charged carbon atoms, and an electrometer grid which generates a signal current proportional to the number of carbon atoms in the flame. The example of 1 ppmV methane, CH_4, and 1 ppmV (but 2 ppmC) ethane, C_2H_6, is related to this detection principle. One ppmV of CH_4 and 1 ppmV of C_2H_6 in air have the same number of molecules of hydrocarbon in a given volume of air, but if an aliquot of each mixture were run through an FID, the signal for ethane would be nearly twice the methane signal: 2 ppmC ethane compared to 1 ppmC methane.

The FPD is also used to measure sulfur-containing compounds and therefore is useful for measurement of sulfur-containing hydrocarbons such as dimethylsulfide or furan. The FPD has an H_2 flame in which sulfur-containing gases are burned. In the combustion process, electronically excited S_2^* is formed. A photomultiplier tube detects light emitted from the excited sulfur at ~395 nm. The ECD is preferred for measuring nitrogen-containing compounds such as PAN and other peroxyacyl nitrate compounds. The ECD contains a radioactive source which establishes a stable ion field. Nitrogen-containing compounds capture electrons in passing through the field. Alterations in the electronic signal are related to the concentration of the nitrogen species.

F. General

The methods that have been discussed require specially designed instruments. Laboratories without such instruments can measure these gases using general-purpose chemical analytical equipment. A compendium of methods for these laboratories is the "Manual on Methods of Air Sampling and Analysis"published by the American Public Health Association. (10).

II. ANALYSIS AND MEASUREMENT OF PARTICULATE POLLUTANTS

The three major characteristics of particulate pollutants in the ambient atmosphere are total mass concentration, size distribution, and chemical composition. In the United States, the PM_{10} concentration, particulate matter with an aerodynamic diameter <10 μm, is the quantity measured for an air quality standard to protect human health from effects caused by inhalation of suspended particulate matter. As shown in Chapter 7, the size distribution of particulate pollutants is very important in understanding the transport and removal of particles in the atmosphere and their deposition behavior in the human respiratory system. Their chemical composition may determine the type of effects caused by particulate matter on humans, vegetation, and materials.

Mass concentration units for ambient measurements are mass (μg) per unit volume (m^3). Size classification involves the use of specially designed inlet configurations, e.g., PM_{10} sampling. To determine mass concentration, all the particles are removed from a known volume of air and their total mass is measured. This removal is accomplished by two techniques, filtration and impaction, described in Chapter 13. Mass measurements are made by pre- and postweighing of filters or impaction surfaces. To account for the absorption of water vapor, the filters are generally equilibrated at standard conditions (T = 20°C and 50% relative humidity).

Size distributions are determined by classifying airborne particles by aerodynamic diameter, electrical mobility, or light-scattering properties. The most common technique is the use of multistage impactors, each stage of which removes particles of progressively smaller diameter. Figure 14-6 shows a four-stage impactor. The particulate matter collected on each stage is weighed to yield a mass size distribution or is subjected to chemical analysis to obtain data on its chemical size distribution. Impactors are used to determine size distributions for particle diameters of ~0.1 μm and larger.

Electrical mobility is utilized to obtain size distribution information in the 0.01–1.0 μm diameter range. This measurement method requires unipolar

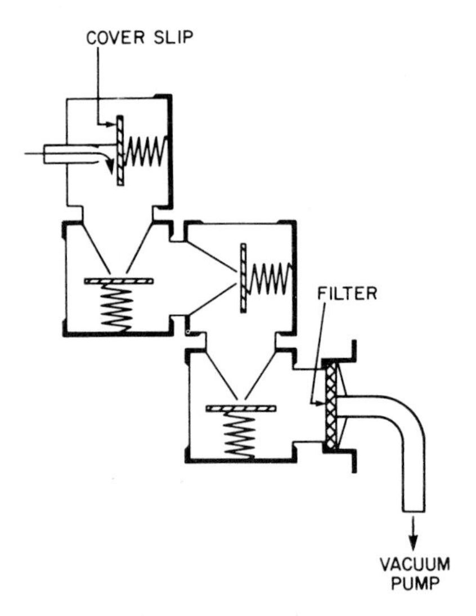

Fig. 14-6. Schematic diagram of a four-stage cascade impactor. Source: Giever, P. M., Particulate matter sampling and sizing, *in* "Air Pollution," 3rd ed., Vol. III (A. C. Stern, ed.). Academic Press, New York, 1976, p. 41.

charging of particles and their separation by passage through an electrical field (11). By incrementing the electrical field strength, progressively larger charged particles may be removed from a flowing air stream. The change in the amount of charge collected by a electrometer grid is then related to the number of particles present in a particular size increment. Instruments based on this principle yield a number size distribution.

Light-scattering properties of particles are also utilized to determine a number size distribution (12). Individual particles interact with a light beam and scatter light at an angle to the original direction of the beam. The intensity of the scattered light is a function of the diameter and the refractive index of the particle. Inlet systems are designed to dilute a particle-laden air stream sufficiently to permit only one particle in the beam at a time. The intensity of the scattered light, as measured by a photomultiplier tube, is proportional to particle size. The number of electrical pulses of each magnitude is accumulated in a multichannel analyzer. By sampling at a known flow rate, the number of particles of different diameters are counted with this type of instrument.

The chemical composition of particulate pollutants is determined in two forms: specific elements, or specific compounds or ions. Knowledge of their chemical composition is useful in determining the sources of airborne particles and in understanding the fate of particles in the atmosphere. Elemental analysis yields results in terms of the individual elements present in a sample such as a given quantity of sulfur, S. From elemental analysis techniques we do not obtain direct information about the chemical form of S in a sample such as sulfate ($S0_4^{2-}$) or sulfide. Two nondestructive techniques used for direct elemental analysis of particulate samples are X-ray fluorescence spectroscopy (XRF) and neutron activation analysis (NAA).

XRF is a technique in which a sample is bombarded by x-rays (13). Inner shell electrons are excited to higher energy levels. As these excited electrons return to their original state, energy with wavelengths characteristic of each element present in the sample is emitted. These high-energy photons are detected and analyzed to give the type and quantity of the elements present in the sample. The technique is applicable to all elements with an atomic number of 11 (sodium) or higher. In principle, complex mixtures may be analyzed with this technique. Difficulties arise from a matrix effect, so that care must be taken to use appropriate standards containing a similar matrix of elements. This technique requires relatively expensive equipment and highly trained personnel.

NAA involves the bombardment of the sample with neutrons, which interact with the sample to form different isotopes of the elements in the sample (14). Many of these isotopes are radioactive and may be identified by comparing their radioactivity with standards. This technique is not quite as versatile as XRF and requires a neutron source.

Pretreatment of the collected particulate matter may be required for chemical analysis. Pretreatment generally involves extraction of the particulate matter into a liquid. The solution may be further treated to transform the material into a form suitable for analysis. Trace metals may be determined by atomic absorption spectroscopy (AA), emission spectroscopy, polarography, and anodic stripping voltammetry. Analysis of anions is possible by colorimetric techniques and ion chromatography. Sulfate ($S0_4^{2-}$), sulfite (SO_3^{2-}), nitrate (NO_3^-), chloride Cl^-), and fluoride (F^-) may be determined by ion chromatography (15).

Analytical methods available to laboratories with only general-purpose analytical equipment may be found in the "Methods of Air Sampling and Analysis" cited at the end of the previous section.

III. ANALYSIS AND MEASUREMENT OF ODORS

Odorants are chemical compounds such as H_2S, which smells like rotten eggs, and may be measured by chemical or organoleptic methods. Organoleptic methods are those which rely on the response to odor of the human nose. Although chemical methods may be useful in identifying and quantifying specific odorants, human response is the only way to assess the degree of acceptability of odorants in the atmosphere. This is due to several factors: the nonlinear relationship between odorant concentration and human response, the variability of individual responses to a given odorant concentration, and the sensory attributes of odor.

Four characteristics of odor are subject to measurement by sensory techniques: intensity, detectability, character (quality), and hedonic tone (pleasantness–unpleasantness) (16). Odor intensity is the magnitude of the perceived sensation and is classified by a descriptive scale, e.g., faint–moderate–strong, or a 1–10 numerical scale. The detectability of an odor or threshold limit is not an absolute level but depends on how the odorant is present, e.g., alone or in a mixture. Odor character or quality is the characteristic which permits its description or classification by comparison to other odors, i.e., sweet or sour, or like that of a skunk. The last characteristic is the hedonic type, which refers to the acceptability of an odorant. For the infrequent visitor, the smell of a large commercial bread bakery may be of high intensity but pleasant. For the nearby resident, the smell may be less acceptable.

The sensory technique used for assessing human perception of odors is called *olfactometry*. The basic technique is to present odorants at different concentrations to a panel of subjects and assess their response. The process favored by the U.S. National Academy of Sciences is dynamic olfactometry (16). This technique involves a sample dilution method in which a flow of clean, nonodorous air is mixed with the odorant under dynamic or constant flow conditions. With this type of apparatus and standard operating conditions, it is possible to determine the detection threshold and the recognition threshold. At high dilution, the panel will be able to tell only whether an odorant is present or absent. Only at concentrations higher, typically by a factor of 2–10, will the subjects be able to identify the odorant.

The olfactometric procedure contains the following elements:

1. Dynamic dilution.
2. Delivery of diluted odorant for smelling through a mask or port.
3. Schedule of presentation of various dilutions and blanks.
4. Obtaining responses from the panelists.
5. Calculation of a panel threshold from experimental data.
6. Panelist selection criteria.

The first element, dynamic dilution, provides a reproducible sample for each panelist. The system must minimize the loss of the odorant to the walls of the delivery apparatus, provide clean dilution air of odor-free quality, maintain a constant dilution ratio for the duration of a given test, and have no memory effect when going from high to low concentrations or switching between odorants of different character. The type of mask or port and the delivery flow rate have been found to influence the response of panelists in determining odor threshold and intensity.

The schedule of presentation may influence the results. The sensory effects are judgment criterion, anticipation, and adaptation. The judgment criterion determines how the panelist will respond when asked whether or not an odor is sensed. Individuals differ in their readiness to be positive or negative. The anticipation effect is a tendency to expect an odor over a given series of trials. Subjects show some positive response when no odorant is present. The adaptation effect is the temporary desensitization after smelling an odorant. This is also called olfactory fatigue and often occurs in occupational settings. Because of olfactory fatigue, investigators evaluating odor concentration in the field must breathe air deodorized by passage through an activated carbon canister before and after sniffing the ambient air being evaluated.

Individuals differ in their sensitivity to odor. Figure 14-7 shows a typical distribution of sensitivities to ethylsulfide vapor (17). There are currently no guidelines on inclusion or exclusion of individuals with abnormally high or low sensitivity. This variability of response complicates the data treatment procedure. In many instances, the goal is to determine some mean value for the threshold representative of the panel as a whole. The small size of panels (generally fewer than 10 people) and the distribution of individual sensitivities require sophisticated statistical procedures to find the threshold from the responses.

Thresholds may also be determined by extrapolation of dose–response plots. In this approach, the perceived odor intensity is measured at several

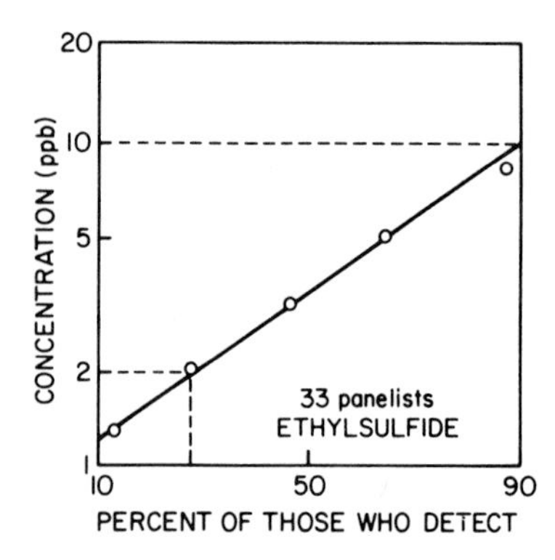

Fig. 14-7. Distribution of sensitivity to ethylene sulfide odor in 33 individuals. The abscissa is the percentage of the individuals who detected the presence of ethylene sulfide at various levels. Source: Dravnicks, A., and Jarke, F., *J. Air Pollut. Control Assoc.* **30,** 1284–1289 (1980).

dilutions using some intensity rating method (Fig. 14-7). The threshold value may be selected at some value (e.g., zero intensity) and the concentration determined with the dilution ratio.

IV. ANALYSIS AND MEASUREMENT OF VISIBILITY

Impairment of visibility is a degradation of our ability to perceive objects through the atmosphere. As discussed in Chapter 10, several components influence our concept of visibility: the characteristics of the source, the human observer, the object, and the degree of pollution in the atmosphere. Our attempts to measure visibility at a given location can take two approaches: human observations and optical measurements. In pristine locations such as national parks, use of human observers has permitted us to gain an understanding of the public's concept of visibility impairment. Although it is difficult to quantify the elements of human observations, this type of research, when coupled with optical measurements, provides a better measure of visibility at a given location (18).

Optical measurements permit the quantification of visibility degradation under different conditions. Several instruments are capable of measuring visual air quality, e.g., cameras, photometers, telephotometers, transmissometers, and scattering instruments.

Photography can provide a permanent record of visibility conditions at a particular place and time. This type of record can preserve a scene in a photograph in a form similar to the way it is seen. Photometers measure light intensity by converting brightness to representative electric signals with a photodetector. Different lenses and filters may be used to determine color and other optical properties. When used in combination with long-range lenses, photometers become telephotometers. This type of instrument may view distant objects with a much smaller viewing angle. The output of the photodetector is closely related to the perceived optical properties of distant targets. Telephotometers are often used to measure the contrast between a distant object and its surroundings, a measurement much closer to the human observer's perception of objects.

A transmissometer is similar to a telephotometer except that the target is a known light source. If we know the characteristics of the source, the average extinction coefficient over the path of the beam may be calculated. Transmissometers are not very portable in terms of looking at a scene from several directions. They are also very sensitive to atmospheric turbulence, which limits the length of the light beam.

Scattering instruments are also used to measure visibility degradation. The most common instrument is the integrating nephelometer, which measures the light scattered over a range of angles. The physical design of the instrument, as shown in Fig. 14-8, permits a point determination of the scattering coefficient of extinction, b_{ext} (19). In clean areas, b_{ext} is dominated by scattering, so that the integrating nephelometer yields a measure of the extinction coefficient. As noted in Chapter 10, b_{ext} can be related to visual range through the Koschmieder relationship.

Other measurements important to visual air quality are pollutant related, i.e., the size distribution, mass concentration, and number concentration of airborne particles and their chemical composition. From the size distribution, the Mie theory of light scattering can be used to calculate the scattering coefficient (20). Table 14-2 summarizes the different types of visual monitoring methods (21).

V. ANALYSIS AND MEASUREMENT OF ACIDIC DEPOSITION

The two components of acidic deposition described in Chapter 10 are wet deposition and dry deposition. The collection and subsequent analysis

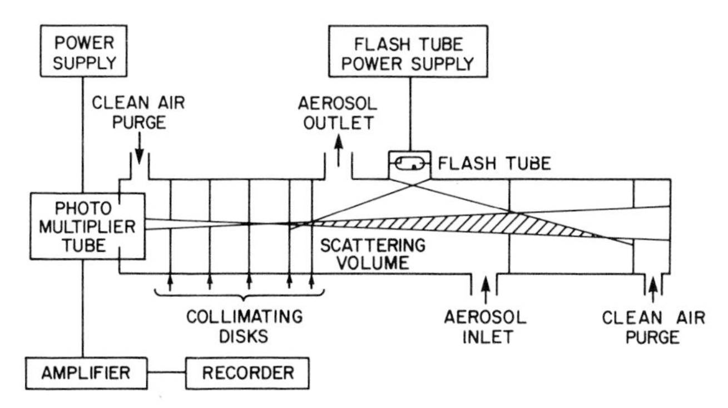

Fig. 14-8. Schematic diagram of the integrating nephelometer. Source: Ahliquist, N. C., and Charlson, R. J., *J. Air Pollut. Control Assoc.* **17**, 467 (1967).

TABLE 14-2

Visibility Monitoring Methods

Method	Parameters measured	Advantages	Limitations	Preferred use
Human observer	Perceived visual quality, atmospheric color, plume blight, visual range	Flexibility, judgment; large existing data base (airport visual range)	Labor intensive; variability in observer perception; suitable targets for visual range not generally available	Complement to instrumental observations; areas with frequent plume blight, discoloration; visual ranges with available target distances
Integrating nephelometer	Scattering coefficient (b_{scat}) at site	Continuous readings; unaffected by clouds, night; b_{scat} directly relatable to fine aerosol concentration at a point; semiportable; used in a number of previous studies; sensitive models available; automated	Point measurement, requires assumption of homogeneous distribution of particles; neglects extinction from absorption, coarse particles (>3–10 μm; must consider humidity effects at high relative humidity	Areas experiencing periodic, well-mixed general haze; medium to short viewing distances; small absorption coefficient (b_{abs}); relating to point composition measurements
Multiwavelength telephotometer	Sky and/or target radiance, contrast at various wavelengths	Measurement over long view path (up to 100 km) with suitable illumination and target, contrast transmittance, total extinction, and chromaticity over sight path can be determined; includes scattering and absorption from all sources; can detect plume blight; automated	Sensitive to illumination conditions; useful only in daylight; relationship to extinction, aerosol relationship possible only under cloudless skys; requires large, uniform targets	Areas experiencing mixed or inhomogeneous haze, significant fugitive dust; medium to long viewing distances (one-fourth of visual range); areas with frequent discoloration; horizontal sight path

Transmissometer	Long path extinction coefficient (b_{ext})	Measurement over medium view path (10–25 km); measures total extinction, scattering and absorption; unaffected by clouds, night	Calibration problems; single wavelength; equivalent to point measurement in areas with long view paths (50–100 km); limited applications to date still under development	Areas experiencing periodic mixed general haze, medium to short viewing distance areas with significant absorption (b_{abs})
Photography	Visual quality, plume blight, color, contrast (limited)	Related to perception of visual quality; documentation of vista conditions	Sensitive to lighting conditions; degradation in storage; contrast measurement from film subject to significant errors	Complement to human observation, instrumental methods; areas with frequent plume blight, discoloration
Particle samplers	Particles	Permit evaluation of causes of impairment	Not always relatable to visual air quality; point measurement	Complement to visibility measurements
Hi vol.	TSP	Large data base, amenable to chemical analysis; coarse particle analysis	Does not separate sizes; sampling artifacts for nitrate, sulfate; not automated	Not useful for visibility sites
Cascade impactor	Size-segregated particles (more than two stages)	Detailed chemical, size evaluation	Particle bounce, wall losses; labor intensive	Detailed studies of scattering by particles $<2\ \mu m$
Dichotomous and fine particle samplers (several fundamentally different types)	Fine particles ($<2.5\ \mu m$) coarse particles (2.5–15 μm) inhalable particles (0–15 μm)	Size cut enhances resolution, optically important aerosol analysis, low artifact potential, particle bounce; amenable to automated compositional analysis; automated versions available; large networks under development	Some large-particle penetration; 24 hr or longer sample required in clean areas for mass measurement; automated version relatively untested in remote locations	Complement to visibility measurement, source assessment for general haze, ground-level plumes

Source: U.S. Environmental Protection Agency. "Protecting Visibility," EPA 450/5-79-008. Office of Air Quality Planning and Standards, Research Triangle Park, NC, 1979.

of wet deposition are intuitively straightforward. A sample collector opens to collect rainwater at the beginning of a rainstorm and closes when the rain stops. The water is then analyzed for pH, anions (negative ions), and cations (positive ions). The situation for dry deposition is much more difficult (22). Collection of particles settling from the air is very dependent on the surface material and configuration. The surfaces of trees, plants, and grasses are considerably different from that of the round, open-top canister often used to collect dry deposited particles. After collection, the material must be suspended or dissolved in pure water for subsequent analysis.

An overview of acid rain monitoring activities in North America shows several national and regional programs in operation in the United States,

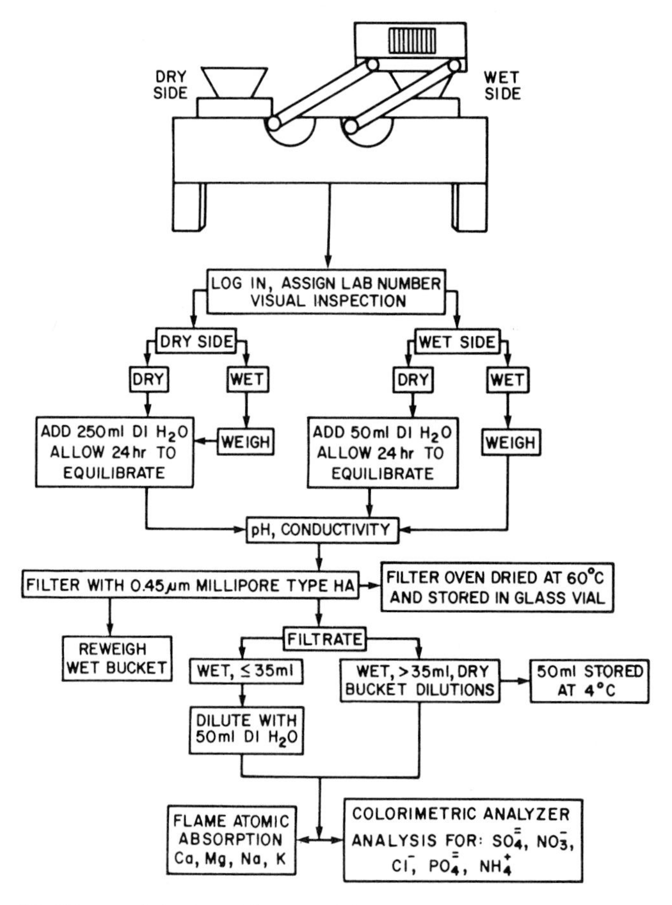

Fig. 14-9. Wet/dry precipitation collector and flow chart for analysis of samples. (DI H_2O: distilled water). Source: "NADP Quality Assurance Report," Central Analytical Laboratory, Illinois Institute of Natural Resources, Champaign, Ill., March 1980.

Canada, and Mexico (23). The National Atmospheric Deposition Program has established the nationwide sampling network of ~100 stations in the United States. The sampler is shown in Fig. 14-9 with a wet collection container. The wet collection bucket is covered with a lid when it is not raining. A sensor for rain moves the lid to open the wet collector bucket and cover the dry bucket at the beginning of a rainstorm. This process is reversed when the rain stops.

The primary constituents to be measured are the pH of precipitation, sulfates, nitrates, ammonia, chloride ions, metal ions, phosphates, and specific conductivity. The pH measurements help to establish reliable long-term trends in patterns of acidic precipitation. The sulfate and nitrate information is related to anthropogenic sources where possible. The measurements of chloride ions, metal ions, and phosphates are related to sea spray and wind-blown dust sources. Specific conductivity is related to the level of dissolved salts in precipitation.

Figure 14-9 also shows a flowchart for analysis of wet and dry precipitation. The process involves weight determinations, followed by pH and conductivity measurements, and finally chemical analysis for anions and cations. The pH measurements are made with a well-calibrated pH meter, with extreme care taken to avoid contaminating the sample. The metal ions Ca^{2+}, Mg^{2+}, Na^{+}, and K^{+} are determined by flame photometry, which involves absorption of radiation by metal ions in a hot flame. Ammonia and the anions Cl^{-}, SO_4^{2-}, NO_3^{-}, and PO_4^{3-} are measured by automated colorimetric techniques.

REFERENCES

1. Dailey, W. V., and Fertig, G. H., *Anal. Instrum.* **77,** 79–82 (1978).
2. U.S. Environmental Protection Agency, **40** CFR, Part 50, App. C, July 1992.
3. Stevens, R. K., and Hodgeson, J. A., *Anal. Chem.* **45,** 443A–447A (1973).
4. U.S. Environmental Protection Agency, "Transfer Standards for Calibration of Air Monitoring Analyzers for Ozone," EPA-600/4-79-056. Office of Air Quality Planning and Standards, Research Triangle Park, NC, 1979.
5. U.S. Environmental Protection Agency, *Fed. Regist.* **41,** 52686–52695 (1976).
6. U.S. Environmental Protection Agency, **40** CFR, Part 50, App. A, July 1992.
7. Hollowell, C. D., Gee, G. Y., and McLaughlin, R. D., *Anal. Chem.* **45,** 63A–72A (1973).
8. Okake, H., Splitstone, P. L., and Ball, J. J., *J. Air Pollut. Control Assoc.* **23,** 514–516 (1973)
9. National Oceanic and Atmospheric Administration, "United States Standard Atmosphere." U.S. Government Printing Office, Washington, DC, 1976.
10. Lodge, J. P. (ed.), "Methods of Air Sampling and Analysis," 3rd ed. American Public Health Association, Washington, DC, 1989.
11. Liu, B. Y. H., Pui, D. Y. H., and Kapadia, A., Electrical aerosol analyzer, *in* "Aerosol Measurement" (Lundgren, D. A., Harris, F. S., Jr., Marlow, W. H., Lippmann, M., Clark, W. E., and Durham, U. D., eds.), pp. 341–384. University Presses of Florida, Gainesville, FL, 1979.

12. Whitby, K. T., and Willeke, K., Single particle optical particle counters, *in* "Aerosol Measurement" (Lundgren, D. A., Harris, F. S., Jr., Marlow, W. H., Lippmann, M., Clark, W. E., and Durham, U. D., eds.), pp. 241–284. University Presses of Florida, Gainesville, FL, 1979.
13. Dzubay, T. G., "X-Ray Fluorescence Analysis of Environmental Samples." Ann Arbor Science Publishers, Ann Arbor, MI, 1977.
14. Heindryckx, R., and Dams, R., *Proc. Nucl. Energy* **3,** 219–252 (1979).
15. Mulik, J. D., and Sawicki, E., "Ion Chromatographic Analysis of Environmental Pollutants," Vol. 2. Ann Arbor Science Publishers, Ann Arbor, MI, 1979.
16. National Research Council, "Odors from Stationary and Mobile Sources." National Academy of Sciences, Washington, DC, 1979.
17. Dravnicks, A., and Jarke, F., *J. Air Pollut. Control Assoc.* **30,** 1284–1289 (1980).
18. Malm, W., Kelley, K., Molenar, J., and Daniel, T., *Atmos. Environ.* **15,** 1875–1890 (1981).
19. Ahliquist, N. C., and Charlson, R. J., *J. Air Pollut. Control Assoc.* **17,** 467 (1967).
20. Twomey, S., "Atmospheric Aerosols," pp. 200–216. Elsevier North-Holland, New York, 1977.
21. U.S. Environmental Protection Agency, "Protecting Visibility," EPA-450/5-79-008. Office of Air Quality Planning and Standards, Research Triangle Park, NC, 1979.
22. Hicks, B. B., Wesely, M. L., and Durham, J. L., "Critique of Methods to Measure Dry Deposition—Workshop Summary," EPA-600/9-80-050. Environmental Sciences Research Laboratory, Research Triangle Park, NC, 1980.
23. Wisniewski, J., and Kinsman, J. D., *Bull. Am. Meteorol. Soc.* **63,** 598–618 (1982).

SUGGESTED READING

Harrison, R. M., and Young, R. J. (eds.), "Handbook of Air Pollution Analysis," 2nd Ed. Chapman & Hall, London, 1986.

Lundgren, D. A., Harris, F. S., Jr., Marlow, W. H., Lippmann, M., Clark, W. E., and Durham, U. D. (eds.), "Aerosol Measurement." University Presses of Florida, Gainesville, 1979.

Newman, L. (ed.), "Measurement Challenges in Atmospheric Chemistry." American Chemical Society, Washington, DC, 1993.

Sickles, J.E., II, *Adv. Environ. Sci. Technol.* **24,** 51–128 (1992).

Winegar, E. D., and Keith, L.H., "Sampling and Analysis of Airborne Pollutants." Lewis Publishers, Boca Raton, FL, 1993.

QUESTIONS

1. Describe the rationale for the U.S. Environmental Protection Agency's establishment of a standard reference method for measurement of National Ambient Air Quality Standard air pollutants.
2. Under what conditions can another method be substituted for a standard reference method?
3. Describe the potential interferences (a) in the nondispersive infrared (NDIR) method for measuring CO and (b) in the chemiluminescent method for measuring NO_2.
4. The electrical aerosol analyzer and the optical counter are used to measure particle size distributions. Describe the size range and resolution characteristics of each of these instruments.

5. How can human observers, optical measurements along a line of sight, and point measurements by nephelometry provide conflicting information about visual air quality in the same location?
6. Using the Code of Federal Regulations, list the current reference methods for measuring NO_2, O_3, SO_2, CO, total suspended particulate matter, and lead.
7. List two types of calibration sources for gas analyzers.
8. Review the air pollution literature and describe the difficulties in establishing a standard reference method for measuring NO_2.
9. Describe the deficiencies of a total suspended particulate measurement for relating ambient concentrations to potential human health effects.

15

Air Pollution Monitoring and Surveillance

I. STATIONARY MONITORING NETWORKS

The U.S. Environmental Protection Agency has established National Ambient Air Quality Standards (NAAQS) for protection of human health and welfare. These standards are defined in terms of concentration and time span for a specific pollutant; for example, the NAAQS for carbon monoxide is 9 ppmV for 8 hr, not to be exceeded more than once per year. For a state or local government to establish compliance with a National Ambient Air Quality Standard, measurements of the actual air quality must be made. To obtain these measurements, state and local governments have established stationary monitoring networks with instrumentation complying with federal specifications, as discussed in Chapter 14. The results of these measurements determine whether a given location is violating the air quality standard.

Stationary monitoring networks are also operated to determine the impact of new sources of emissions. As part of the environmental impact statement and Prevention of Significant Deterioration processes, the projected impact of a new source on existing air quality must be assessed. Air quality monitoring is one means of making this type of assessment. A monitoring network is established at least 12 months before construction to determine prior air quality. Once the facility is completed and in operation, the network data determine the actual impact of the new source.

Long-term trends are measured by stationary air quality monitoring networks. Figure 4-4 shows the long-term decrease of SO_2 in the atmosphere resulting from the implementation of air pollution control technology. The trends in other atmospheric trace gases such as methane, NO, NO_2, and CO are similarly measured in rural as well as urban locations. Atmospheric budgets of various gases are developed to allow estimation of whether sources are anthropogenic or natural.

A stationary monitoring network should yield the following information: (1) background concentration levels, (2) highest concentration levels, (3) representative concentration levels in high-density areas, (4) the impact of local sources, (5) the impact of remote sources, and (6) the relative impact of natural and anthropogenic sources.

The spatial scale of a stationary network is determined by monitoring objectives. Spatial scales include microscale (1–100 m), middle scale (100 m–0.5 km), neighborhood scale (0.5–4.0 km), urban scale (4–50 km), and regional scale (tens to hundreds of km). Table 15-1 shows the relationship between spatial scale and monitoring objectives (1).

Sampler siting within a network must meet the limitations of any individual sampling site and the relationship of sampling sites with each other (2). The overall approach for selection of sampling sites is to (1) define the purpose of the collected data, (2) assemble site selection aids, (3) define the general areas for samplers based on chemical and meteorological constraints, and (4) determine the final sites based on sampling requirements and surrounding objects (3). Several purposes of air quality monitoring were mentioned earlier, such as air quality standard compliance, long-term trends, and new facility siting.

The tools available for site selection include climatological data, topography, population data, emission inventory data, and diffusion modeling. Climatological data are useful in relating meteorology to emission patterns. For example, elevated levels of photochemical oxidant are generally related

TABLE 15-1

Relationship of the Scale of Representativeness and Monitoring Objectives

Siting scales	Monitoring objectives
Micro, middle, neighborhood, (sometimes urban)	Highest concentration affecting people
Neighborhood, urban	High-density population exposure
Micro, middle, neighborhood	Source impact
Neighborhood, region	General/background concentration

to stagnant meteorological conditions and warm temperatures. Seasonal climatological patterns of prevailing winds and frequency of inversions will influence the location of sampling stations. Various types of maps are useful for establishing topography, population density, and location of sources of various pollutants. Wind roses overlaid with emission sources and population densities help to locate the general areas for location of samplers.

Various types of diffusion models are available which can use as input emission patterns, climatological data, and population data to rank sampling locations by concentration threshold, resolution of peak concentrations, and frequency of exposure (4) or to rank sampling locations for maximum sensitivity to source emission changes, to provide coverage of as many sources or to cover as large a geographic area as possible (5).

The last step in selecting specific sites is based on the following: availability of land and electrical power, security from vandalism, absence of nearby structures such as large buildings, probe height (inlet >3 m), and cost.

An example of matching scale and objective is the determination of CO exposure of pedestrians on sidewalks in urban street canyons. The location of a station to meet this objective would be an elevation of ~ 3 m on a street with heavy vehicular traffic and large numbers of pedestrians.

Figure 15-1 shows the Los Angeles, California, basin stationary air monitoring network, one of the most extensive in the United States (6). At most of these locations, automated instruments collect air quality data continuously. Five pollutant gases are monitored, and particulate matter filter samples are collected periodically.

II. MOBILE MONITORING AND SURVEILLANCE

Mobile monitoring is accomplished from a movable platform, i.e., an aircraft or vehicle. Emissions are measured by source monitoring techniques (see Chapter 31). Atmospheric transport and chemical transformation processes occur in the region between the source and the receptor. By using mobile platforms containing air pollution instrumentation, one can obtain data to help understand the formation and transport of photochemical smog, acidic deposition, and the dispersion of air pollutants from sources. Mobile monitoring platforms may also be moved to *hot spots,* areas suspected of having high concentrations of specific air pollutants. These areas may be nearby locations downwind of a large source or a particular location that is an unfavorable receptor due to meteorological conditions. Vehicular and aircraft monitoring systems can also be moved to locations where hazardous chemical spills, nuclear and chemical plant accidents, or volcanoes or earthquakes have occurred.

The major advantage of a mobile monitoring system is its ability to obtain air quality information in the intermediate region between source monitors

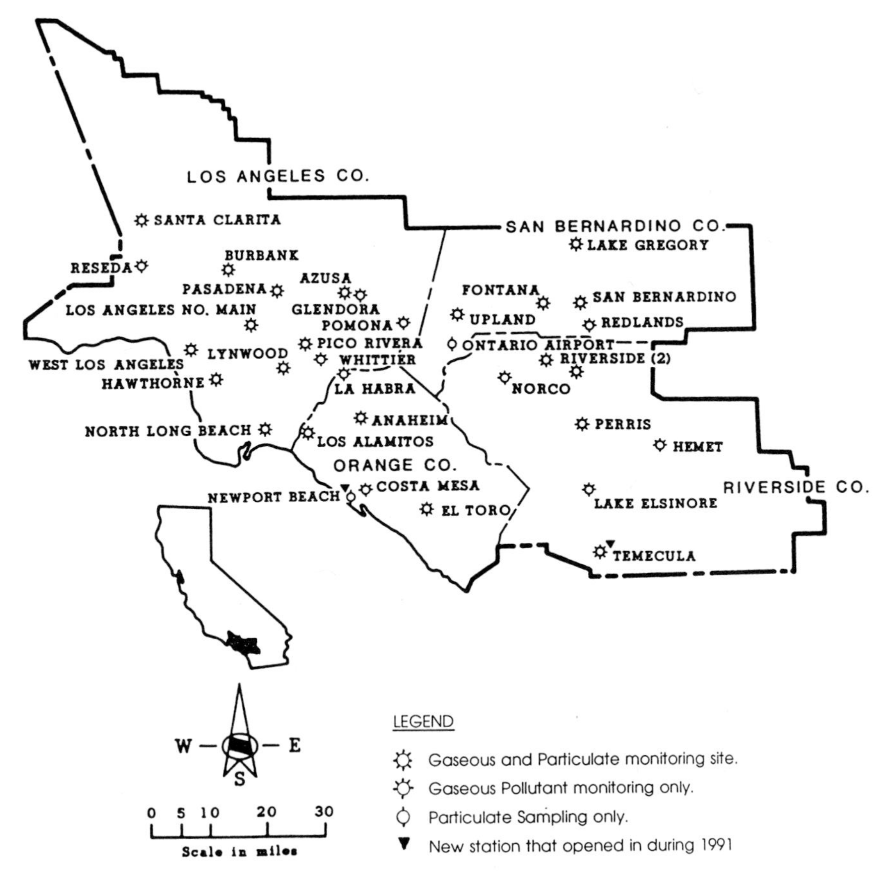

Fig. 15-1. California South Coast Air Basin stationary monitoring locations operating during 1991. (L.A., Los Angeles). Source: California Air Resources Board, "Summary of 1991 Air Quality Data, Gaseous and Particulate Pollutants," Vol. 23, 1991.

and stationary fixed monitors. The major disadvantage is the sparsity of suitable instrumentation that operates properly in the mobile platform environment. Limitations of existing instrumentation for use on movable platforms are inadequate temperature and pressure compensation; incompatible power, size, and weight requirements; and excessive response time. Most movable platforms are helicopters, airplanes, trucks, or vans. These platforms do not provide the relatively constant-temperature environment required by most air quality instrumentation. Equipment mounted in aircraft is subject to large pressure variations with changing altitude. Most instrumentation is designed to operate with alternating current (AC) electrical power, whereas relatively low amounts of direct current (DC) power are available in aircraft or vans. Space is at a premium, and response times are often too slow to permit observation of rapid changes in concentration as aircraft move in and out of a plume.

Despite these limitations, mobile monitoring systems have been used to obtain useful information, such as the verification and tracking of the St. Louis, Missouri, urban plume. The measurement of a well-defined urban plume spreading northeastward from St. Louis is shown in Fig. 15-2 (7). These data were collected by a combination of instrumented aircraft and mobile vans. Cross-sectional paths were flown by the aircraft at increasing distances downwind. Meteorological conditions of low wind speed in the same direction helped to maintain this urban plume in a well-defined

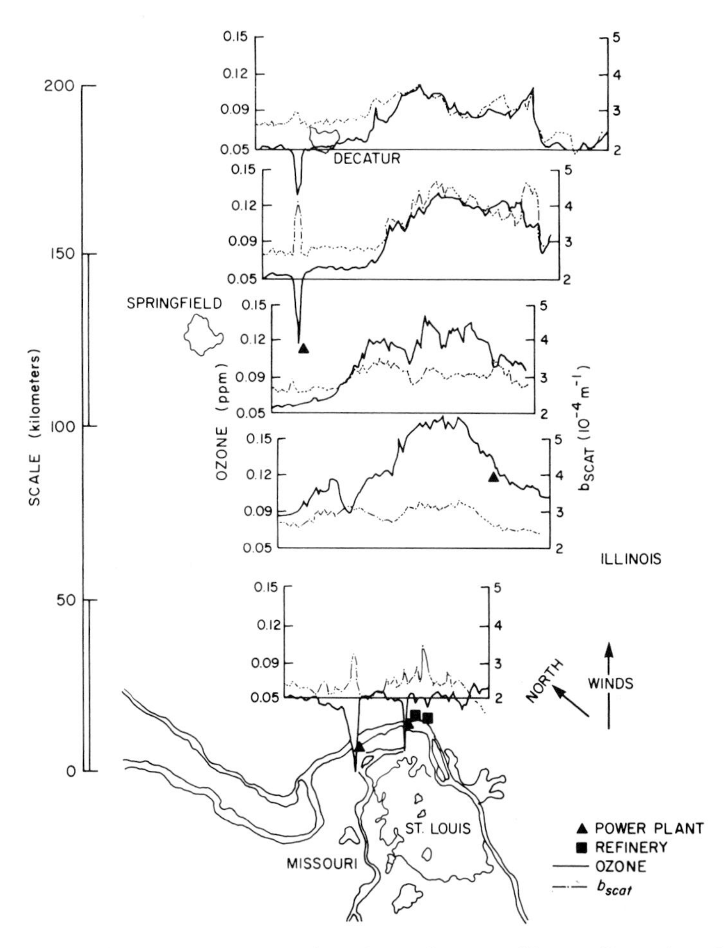

Fig. 15-2. The St. Louis, Missouri, urban plume. Ozone and b_{scat} profiles at four distances downwind of St. Louis track a detectable urban plume for 150 km. Source: Wilson, W. E., Jr., *Atmos. Environ.* **12,** 537–547 (1978).

condition for several hundred kilometers downwind. The presence of large point sources of NO and SO_2 was observed by the changes in the O_3 and b_{scat} profiles. Sharp decreases in ozone concentration occurred when large amounts of NO were present, and rapid increases in b_{scat} were caused by the primary and secondary particulate matter from power plant plumes embedded in the larger urban plume. The overall increasing level of ozone and b_{scat} at greater downwind distances was caused by the photochemical reactions as the urban plume was transported farther away from St. Louis. This type of plume mapping can be accomplished only by mobile monitoring systems.

III. REMOTE SENSING

Remote sensing involves monitoring in which the analyzer is physically removed from the volume of air being analyzed. Satellites have been used to monitor light-scattering aerosol over large areas (8). Large point sources such as volcanic activity and forest fires can be tracked by satellite. The development and movement of hazy air masses have been observed by satellite imagery (see Fig. 10-9). Remote sensing methods are available for determining the physical structure of the atmosphere with respect to turbulence and temperature profiles (see Chapter 17).

Differential absorption lidar (DIAL) is used for remote sensing of gases and particles in the atmosphere. *Lidar* is an acronym for *light detection and ranging,* and the technique is used for measuring physical characteristics of the atmosphere. DIAL measurements consist of probing the atmosphere with pulsed laser radiation at two wavelengths. One wavelength is efficiently absorbed by the trace gas, and the other wavelength is less efficiently absorbed. The radiation source projects packets of energy through the atmosphere, which interact with the trace gas. The optical receiver collects radiation backscattered from the target. By controlling the timing of source pulses and processing of the optical receiver signal, one can determine the concentration of the trace gas over various distances from the analyzer. This capability permits three-dimensional mapping of pollutant concentrations. Applications are plume dispersion patterns and three-dimensional gaseous pollutant profiles in urban areas.

SO_2 and O_3 are detected by an ultraviolet DIAL system operating at wavelengths near 300 nm (9). Tunable infrared CO_2 lasers are used in applications of IR-DIAL systems which are capable of measuring SO_2, CO, HCl, CH_4, CO_2, H_2O, N_2O, NH_3, and H_2S (10). The components of this type of system are shown in Fig. 15-3 (11). The laser source is switched between the low-absorption and high-absorption frequencies for the trace gas to be detected. The system is pointed toward a target, and focusing lenses are used to collect the returning signal. The beam splitter diverts

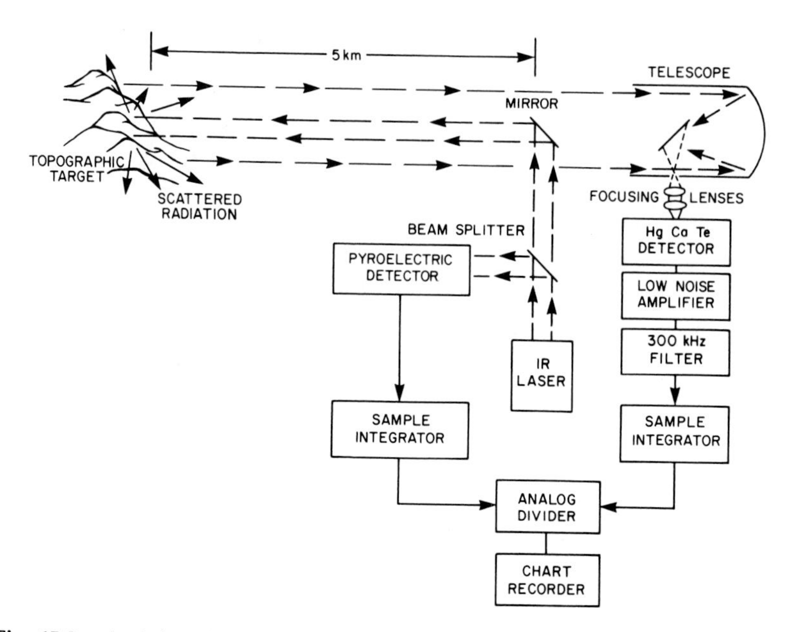

Fig. 15-3. An infrared DIAL system. Source: Murray, E. A., and Van der Laan, J. E., *Appl. Opt.* **17,** 814–817 (1978).

a portion of the transmitted beam to a detector. The backscattered and transmitted pulses are integrated to yield DC signals. Figure 15-4 shows ethylene concentrations measured on a 5-km path in Menlo Park, California, with this system.

Satellites are being used to detect global distributions of large areas of CO and O_3 (12).

IV. QUALITY ASSURANCE

Air quality monitoring for standards compliance, new facility siting, and long-term trend measurement has been going on for many years. Historically, a large number of federal, state, and local organizations, both governmental and nongovernmental, have been using a variety of technologies and approaches to obtain air quality data. This has resulted in multiple data sets of variable accuracy and precision. Questionable or conflicting air quality data are of little value in ascertaining compliance with air quality standards, determining whether air quality is improving or worsening in a given region over an extended period, or understanding the chemistry and physics of the atmosphere.

In order to minimize the collection of questionable air quality data, the U.S. Environmental Protection Agency has established and implemented

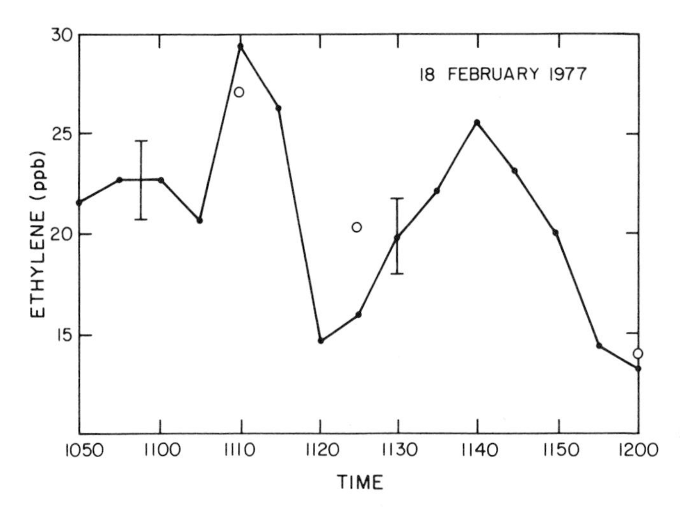

Fig. 15-4. Ethylene concentrations measured by the DIAL system along a 5-km path length near Menlo Park, California. The open circles are ethylene concentrations of samples taken at three ground-level locations near the line of sight and analyzed by gas chromatography. Source: Murray, E. A., and Van der Laan, J. E., *Appl. Opt.* **17,** 814–817 (1978).

stringent regulations requiring well-documented quality assurance programs for air quality monitoring activities (13).

Quality assurance programs are designed to serve two functions: (1) assessment of collected air quality data and (2) improvement of the data collection process. These two functions form a loop; as air quality data are collected, procedures are implemented to determine whether the data are of acceptable precision and accuracy. If they are not, increased quality control procedures are implemented to improve the data collection process.

The components of a quality assurance program are designed to serve the two functions just mentioned—control and assessment. Quality control operations are defined by operational procedures, specifications, calibration procedures, and standards and contain the following components:

1. Description of the methods used for sampling and analysis
2. Sampling manifold and instrument configuration
3. Appropriate multipoint calibration procedures
4. Zero/span checks and record of adjustments
5. Control specification checks and their frequency
6. Control limits for zero, span, and other control limits
7. The corrective actions to be taken when control limits are exceeded
8. Preventative maintenance
9. Recording and validation of data
10. Documentation of quality assurance activities

Table 15-2 contains a specific example of these components for ambient monitoring for ozone.

In addition to fulfilling the in-house requirements for quality control, state and local air monitoring networks which are collecting data for compliance purposes are required to have an external performance audit on an annual basis. Under this program, an independent organization supplies externally calibrated sources of air pollutant gases to be measured by the instrumentation undergoing audit. An audit report summarizes the performance of the instruments. If necessary, further action must be taken to eliminate any major discrepancies between the internal and external calibration results.

Data quality assessment requirements are related to precision and accuracy. Precision control limits are established, i.e., +10% of span value, as calculated from Eq. (15-1). The actual results of the may be used to calculate an average deviation (Eq. 15-3):

$$d_i = (y_i - x_i)/x_i \times 100 \tag{15-1}$$

where d_i is the percentage difference, y_i the analyzer's indicated concentration of the test gas for the ith precision check, and x_i the known concentration of the test gas for the ith precision check.

$$d_{\text{av}} = \frac{1}{n}\sum_{i=1}^{n} d_i \tag{15-2}$$

TABLE 15-2

Quality Control Components for Ambient Ozone Monitoring

Component	Description
Method	Chemiluminescent O_3 monitor Calibration method by certified ozone UV transfer method
Manifold/instrument configuration	Instrument connected to sampling manifold which draws ambient air at 3 m into instrument shelter
Calibration	Multipoint calibration on 0.5-ppm scale at 0.0, 0.1, 0.2, and 0.4 ppm weekly
Zero/span check	Zero check ± 0.005 ppm Span check 0.08–0.10 on a 1.0-ppm full scale daily
Control specification checks	Ethylene flow Sample flow, daily
Corrective limits	± 0.005 ppm zero and span
Corrective action	Do multipoint calibration; invalidate data collection since last zero/span check within control limits
Preventive maintenance	Manufacturer's procedures to be followed
Recording and validating data	Data reported weekly to quality assurance coordinator, with invalid data flagged
Documentation	Data volume includes all quality control forms, e.g., zero/span control charts and multipoint calibration results

TABLE 15-3

Explanation of Gaseous Pollutant Summary

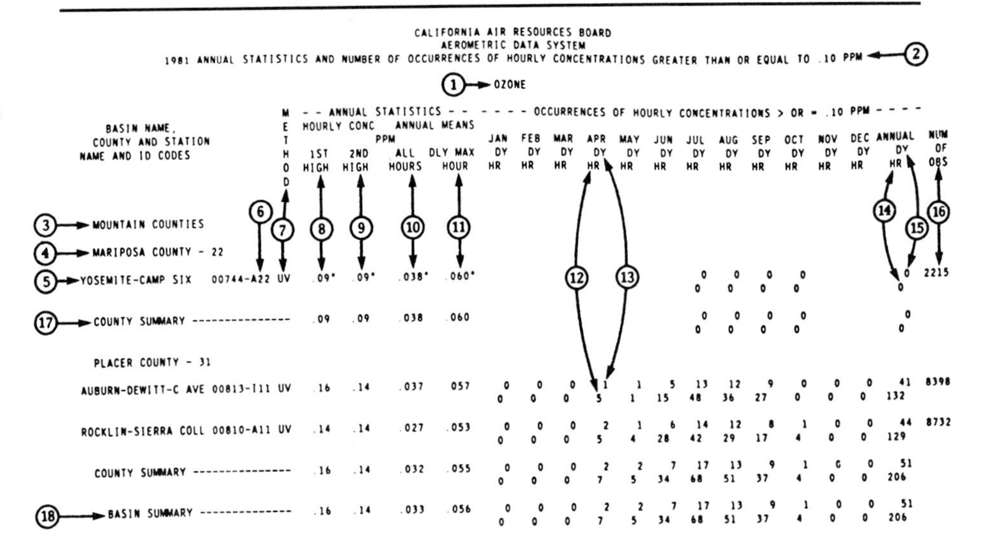

CALIFORNIA AIR RESOURCES BOARD
AEROMETRIC DATA SYSTEM
1981 ANNUAL STATISTICS AND NUMBER OF OCCURRENCES OF HOURLY CONCENTRATIONS GREATER THAN OR EQUAL TO .10 PPM ← (2)

(1) → OZONE

BASIN NAME, COUNTY AND STATION NAME AND ID CODES	METHOD	ANNUAL STATISTICS: HOURLY CONC PPM 1ST HIGH	2ND HIGH	ANNUAL MEANS: ALL HOURS	DLY MAX HOUR	OCCURRENCES OF HOURLY CONCENTRATIONS > OR = .10 PPM: JAN DY/HR	FEB DY/HR	MAR DY/HR	APR DY/HR	MAY DY/HR	JUN DY/HR	JUL DY/HR	AUG DY/HR	SEP DY/HR	OCT DY/HR	NOV DY/HR	DEC DY/HR	ANNUAL DY/HR	NUM OF OBS
(3) → MOUNTAIN COUNTIES																			
(4) → MARIPOSA COUNTY - 22																			
(5) → YOSEMITE-CAMP SIX 00744-A22	UV	.09*	.09*	.038*	.060*							0 / 0	0 / 0	0 / 0	0 / 0			0 / 0	2215
(17) → COUNTY SUMMARY		.09	.09	.038	.060							0 / 0	0 / 0	0 / 0	0 / 0			0 / 0	
PLACER COUNTY - 31																			
AUBURN-DEWITT-C AVE 00813-I11	UV	.16	.14	.037	057	0 / 0	0 / 0	0 / 0	1 / 5	1 / 1	5 / 15	13 / 48	12 / 36	9 / 27	0 / 0	0 / 0	0 / 0	41 / 132	8398
ROCKLIN-SIERRA COLL 00810-A11	UV	.14	.14	.027	.053	0 / 0	0 / 0	0 / 0	2 / 5	1 / 4	6 / 28	14 / 42	12 / 29	8 / 17	1 / 4	0 / 0	0 / 0	44 / 129	8732
COUNTY SUMMARY		.16	.14	.032	.055	0 / 0	0 / 0	0 / 0	2 / 7	2 / 5	7 / 34	17 / 68	13 / 51	9 / 37	1 / 4	C / 0	0 / 0	51 / 206	
(18) → BASIN SUMMARY		.16	.14	.033	.056	0 / 0	0 / 0	0 / 0	2 / 7	2 / 5	7 / 34	17 / 68	13 / 51	9 / 37	1 / 4	0 / 0	0 / 0	51 / 206	

1. Ozone
   Pollutant summarized in table
2. Concentration ≥ 0.10 ppm (parts per million)
   California Ambient Ozone Air Quality Standard
3. Mountain counties
   Air basin in which station is located
4. Mariposa County—22
   County name—County code
5. Yosemite–Camp Six 00744
   Station Name and ID code
6. A 22
   Agency and project code
7. UV (ultraviolet photometric)
   Measurement method
8. Highest annual hourly concentration: 0.09 ppm
9. Second highest annual hourly concentration: 0.09 ppm
10. Mean of all hours monitored: 0.038 ppm
11. Mean of all daily maximum values: 0.060 ppm
12. Number of hourly exceedances in 1 month: There are no samples at Yosemite—Camp Six which exceed the specified level. There are 5 hourly exceedances recorded at Auburn-Dewitt-C Avenue in April.
13. Number of daily exceedances in 1 month: There is only one daily exceedance recorded at Auburn-Dewitt-C Avenue in April.
14. Total number of hourly exceedances for year: There are no samples which exceed the specified level during the 4 months of observations.
15. Total number of daily exceedances for year: There are no samples which exceed the specified level during the 4 months of operation.
16. Total number of observations for year: 2215
17. Same statistics as 7–15 above, except for entire county, e.g., highest hourly concentration in county. Note: sum of occurrences at individual stations must equal or exceed county totals.
18. Same as 17 above, except for entire air basin.

$$S_j = \sqrt{\frac{1}{n-1}\left[\sum_{i=1}^{n} d_i^2 - \frac{1}{n}\left(\sum_{i=1}^{n} d_i\right)^2\right]} \tag{15-3}$$

The external audit results are used to determine the accuracy of the measurements. Accuracy is calculated from percentage differences, d_i, for the audit concentrations and the instrument response.

V. DATA ANALYSIS AND DISPLAY

In general, air quality data are classified as a function of time, location, and magnitude. Several statistical parameters may be used to characterize a group of air pollution concentrations, including the arithmetic mean, the median, and the geometric mean. These parameters may be determined over averaging times of up to 1 year. In addition to these three parameters, a measure of the variability of a data set, such as the standard deviation

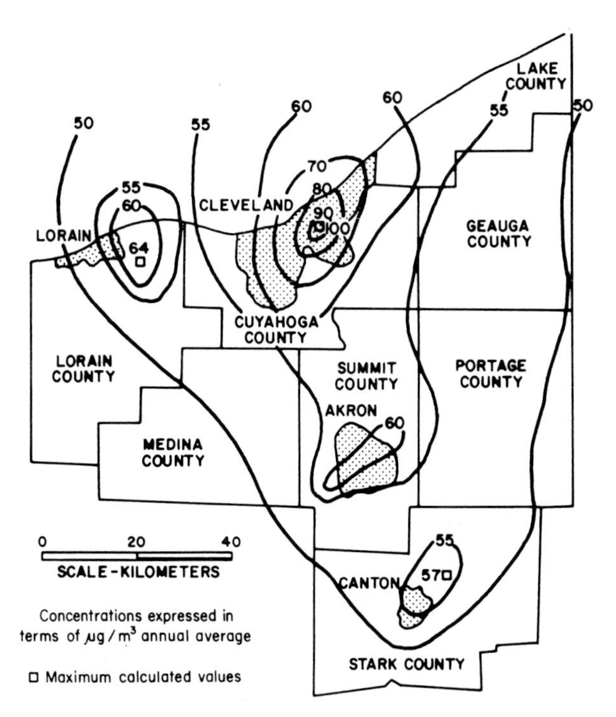

Fig. 15-5. Concentration isopleth diagram of ambient particulate matter calculated by a computer model. Source: Zimmer, C. E., "Air Pollution," 3rd ed., Vol. III (A. C. Stern, ed.). Academic Press, New York, 1976.

or the geometric standard deviation, indicates the range of data around the value selected to represent the data set.

Raw data must be analyzed and transformed into a format useful for specific purposes. Summary tables, graphs, and geographic distributions are some of the formats used for data display. Air quality information often consists of a large body of data collected at a variety of locations and over different seasons. Table 15-3 shows the tabular format used by the California Air Resources Board to reduce ozone hourly measurements to a format which shows information about compliance with air quality standards (6). The format has location, maximum values, annual means, and number of occurrences of hourly values above a given concentration as a function of the month of the year. One can quickly determine which areas are violating a standard, at what time of the year elevated concentrations are occurring, and the number of good data points collected.

Pollutant concentration maps may be constructed as shown in Fig. 15-5 (14). In this example, elevated levels of ambient particulate matter are associated with population centers. For a given geographic area, isopleths, lines showing equal concentrations of a pollutant, are drawn on a map. Regions of high concentration are quickly identified. Further action may be taken to determine the cause, such as review of emission inventories of additional sampling.

REFERENCES

1. Code of Federal Regulations, Title 40, Part 58, Ambient Air Quality Surveillance, Appendix D—Network Design for State and Local Air Monitoring Stations (SLAMS). U.S. Government Printing Office, Washington, DC, July 1992, pp. 158–172.
2. Harrison, R. M., and Young, R. J. (eds.), "Handbook of Air Pollution Analysis," 2nd ed. Chapman & Hall, London, 1986.
3. U.S. Environmental Protection Agency, "Optimum Site Exposure Criteria for SO_2 Monitoring," EPA 450/3-77-13. Office for Quality Planning and Standards, Research Triangle Park, NC, 1977.
4. Smith, D. G., and Egan, B. A., Design of monitoring networks to meet multiple criteria, *in* "Quality Assurance in Air Pollution Measurement" (E. D. Frederick, ed.). Air Pollution Control Association, Pittsburgh, 1979, pp. 139–150.
5. Houghland, E. S., Air quality monitoring network design by analytical techniques III, *in* "Quality Assurance in Air Pollution Measurement" (E. D. Frederick, ed.). Air Pollution Control Association, Pittsburgh, 1979, pp. 181–187.
6. California Air Resources Board, "Summary of 1991 Air Quality Data, Gaseous and Particulate Pollutants." California Air Resources Board, Sacramento, 1991.
7. Wilson, W. E., Jr., *Atmos. Environ.* **12**, 537–547 (1978).
8. Alfodi, T. T., Satellite remote sensing for smoke plume definition, *in* Proceedings of the 4th Joint Conference on Sensing of Environmental Pollutants. American Chemical Society, Washington, DC, 1978, pp. 258–261.
9. Browell, E. V., Lidar remote sensing of tropospheric pollutants and trace gases, *in* Proceed-

ings of the 4th Joint Conference on Sensing of Environmental Pollutants. American Chemical Society, Washington, DC, 1978, pp. 395–402.
10. Grant, W. B., *Appl. Opt.* **21**, 2390–2394 (1982).
11. Murray, E. A., and Van der Laan, J. E., *Appl. Opt.* **17**, 814–817 (1978).
12. Fishman, J., *Environ. Sci. Technol.* **25:** 613–21 (1991).
13. Code of Federal Regulations, Title 40, Part 58, Ambient Air Quality Surveillance, Appendix A—Quality Assurance Requirements for State and Local Air Monitoring Stations (SLAMS). U.S. Government Printing Office, Washington, DC, July 1992, pp. 137–150.
14. Zimmer, C. E., "Air Pollution," 3rd ed., Vol. 111 (A. C. Stern, ed.). Academic Press, New York, 1976, p. 476.

SUGGESTED READING

Barrett, E. C., and Curtis, L. F., "Introduction to Environmental Remote Sensing," 3rd ed. Chapman & Hall, London, 1992.

Beer, R., "Remote Sensing by Fourier Transform Spectroscopy." Wiley, New York, 1992.

Cracknell, A. P., "Introduction to Remote Sensing." Taylor & Francis, New York, 1991.

Keith, L. H., "Environmental Sampling and Analysis." Lewis Publishers, Chelsea, MI, 1991.

QUESTIONS

1. List the two major functions of a quality assurance program and describe how they are interrelated.
2. List the advantages and disadvantages of remote sensing techniques by optical methods.
3. Determine which month and location have the greatest number of hours with ozone concentrations ≥.01 ppm, using Table 15-3.
4. Determine how many monitoring stations per million people are located in the four counties in southern California shown in Fig. 15-1.
5. What are the physical constraints in placing instrumentation in aircraft or motor vehicles?
6. What are appropriate uses of mobile platforms for monitoring?
7. Describe the chemical behavior of the b_{scat} and ozone concentration profiles of the St. Louis urban plume in Fig. 15-2. What is the reason for the sharp increase of b_{scat} and the sharp decrease of ozone in the vicinity of power plants?
8. List the reasons for establishing a stationary air monitoring network.

16

Air Pathways from Hazardous Waste Sites

I. INTRODUCTION

This chapter addresses the potential for hazardous air emissions from environmental remediation sites. These emissions can occur at hazardous spill locations, at undisturbed remediation sites, and during cleanup of remediation sites under the Comprehensive Environmental Response, Compensation, and Liability Act (CERCLA) or the Superfund Amendments and Reauthorization Act (SARA). Air emissions may pose a potential health risk at these sites.

The U. S. Environmental Protection Agency (EPA) developed the Hazard Ranking System (HRS) (1) to determine priorities among releases, or threatened releases, from remediation sites. The HRS applies the appropriate consideration of each of the following site-specific characteristics of such facilities:

- The quantity, toxicity, and concentrations of hazardous constituents that are present in such waste and a comparison with other wastes.
- The extent of, and potential for, release of such hazardous constituents to the environment.

• The degree of risk to human health and the environment posed by such constituents.

II. MULTIMEDIA TRANSPORT

Air contaminant releases from hazardous waste sites can occur from wastes placed aboveground or belowground. The following are categories of air contaminant releases:

- Fugitive dust resulting from:
  —Wind erosion of contaminated soils
  —Vehicle travel over contaminated roadways
- Volatilization release from:
  —Covered landfills (with and without gas generation)
  —Spills, leaks, and landforming
  —Lagoons

The Environmental Protection Agency has detailed procedures for conducting air pathway analysis for Superfund applications (2). Decision network charts are given for all expected situations.

Figure 16-1 and 16-2 present the decision networks that guide contaminant release screening analysis. Figure 16-1 deals with contaminants in or under the soil and Fig. 16-2 addresses aboveground wastes. Any release mechanisms evident at the site will require a further screening evaluation to determine the likely environmental fate of the contaminants involved.

III. CONTAMINANT FATE ANALYSIS

Simplified environmental fate estimation procedures are based on the predominant mechanisms of transport within each medium, and they generally disregard intermedia transfer or transformation processes. In general, they produce conservative estimates (i.e., reasonable upper bounds) for final ambient concentrations and the extent of hazardous substance migration. However, caution should be taken to avoid using inappropriate analytical methods that underestimate or overlook significant pathways that affect human health.

When more in-depth analysis of environmental fate is required, the analyst must select the modeling procedure that is most appropriate to the circumstances. In general, the more sophisticated models are more data, time, and resource intensive.

Figures 16-3 through 16-5 present the decision network for screening contaminant fate in air, surface water, ground water, and biota. Pathways must be further evaluated to determine the likelihood of population exposure.

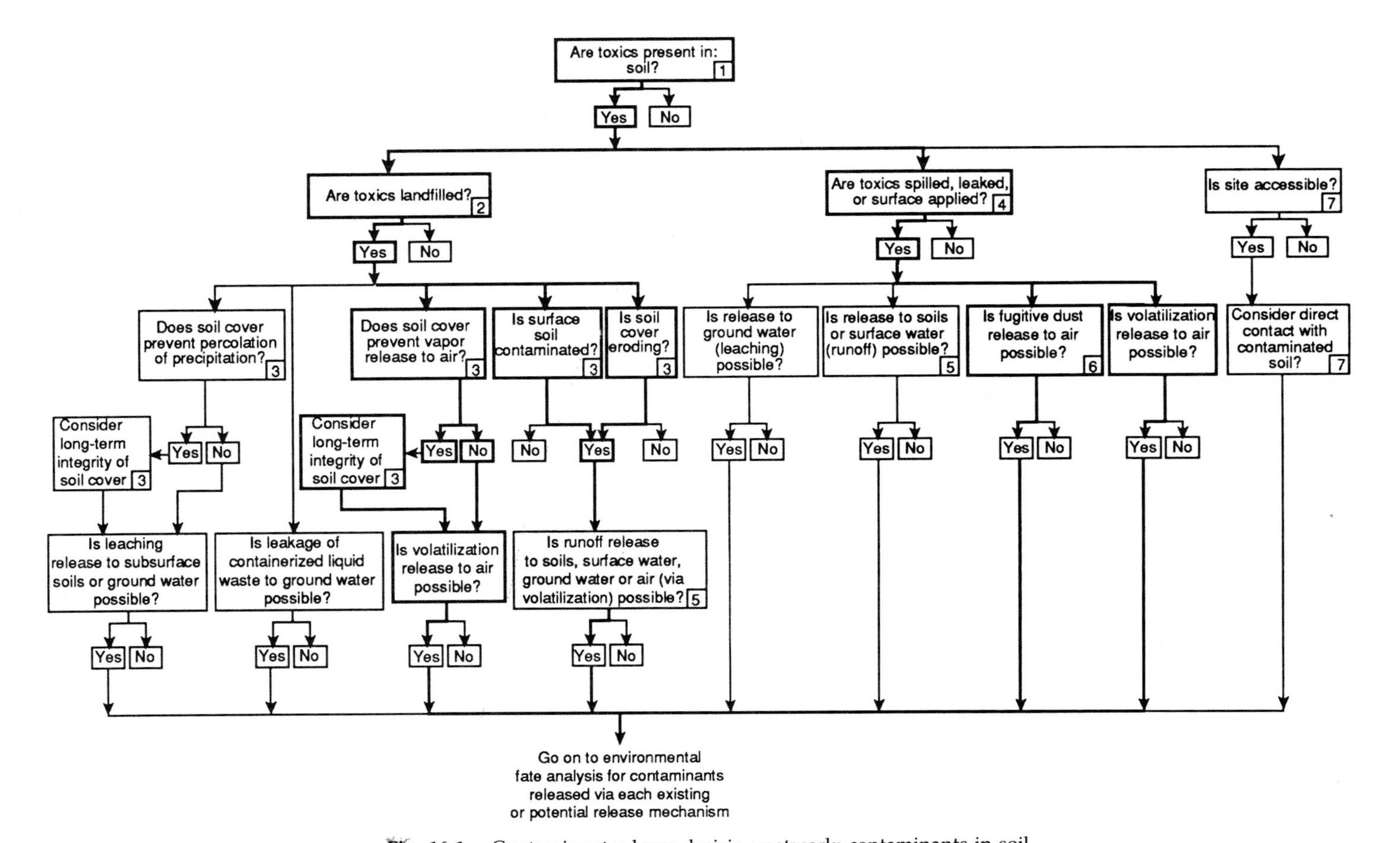

Fig. 16-1. Contaminant release decision network: contaminants in soil.

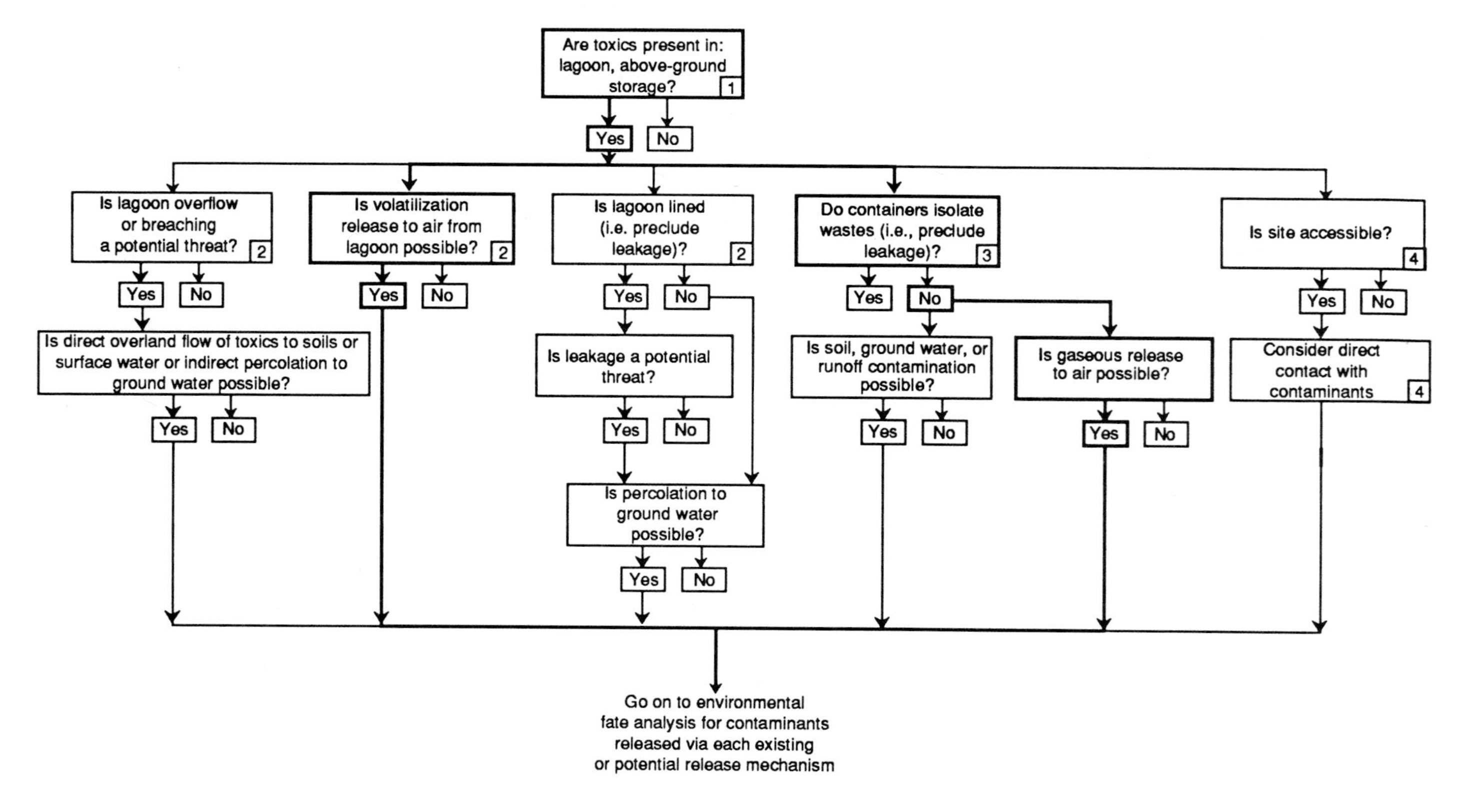

Fig. 16-2. Contaminant release decision network: contaminants above ground.

A. Atmospheric Fate

The following numbered paragraphs refer to particular numbered boxes in Fig. 16-3.

1. The atmospheric fate of contaminants must be assessed whenever it is determined that significant gaseous or airborne particulate contaminants are released from the site. The atmospheric fate of contaminants released originally to other media, but eventually partitioned to the atmosphere beyond site boundaries, must also be assessed whenever this intermedia transfer is likely to be significant.

2. The predominant directions of contaminant movement will be determined by relative directional frequencies of wind over the site (as reflected in area-specific wind rose data). Atmospheric stability and wind speeds determine off-site areas affected by ambient concentrations of gaseous contaminants. Usually, high stability and low wind speed conditions result in higher atmospheric concentrations of gaseous contaminants close to the site. High stability and moderate wind speeds result in moderate concentrations over a larger downwind area. Low stability or high wind speed conditions cause greater dispersion and dilution of contaminants, resulting in lower concentrations over larger areas.

For particulate contaminants (including those adsorbed to dust or soil particles), ambient concentrations in the atmosphere and areas affected by airborne contaminants are determined by wind speed and stability and also by particle size distribution. High winds result in greater dispersion and cause particles to remain airborne longer (which may also increase release rates). Low winds and high stability result in rapid settling of particles and in a more concentrated contaminant plume closer to the site. Larger particles settle rapidly, decreasing the atmospheric concentrations with distance from the site. Finer particles remain airborne longer, and their behavior more closely approximates that of gaseous contaminants, as described.

3. Settling and rainout are important mechanisms of contaminant transfer from the atmospheric media to both surface soils and surface waters. Rates of contaminant transfer caused by these mechanisms are difficult to assess qualitatively; however, they increase with increasing soil adsorption coefficients, solubility (for particulate contaminants or those adsorbed to particles), particle size, and precipitation frequency.

Areas affected by significant atmospheric concentrations of contaminants exhibiting the foregoing physical and chemical properties should also be considered as potentially affected by contaminant rainout and settling to surface media. Contaminants dissolved in rainwater may percolate to ground water, run off or fall directly into surface waters, and adsorb to

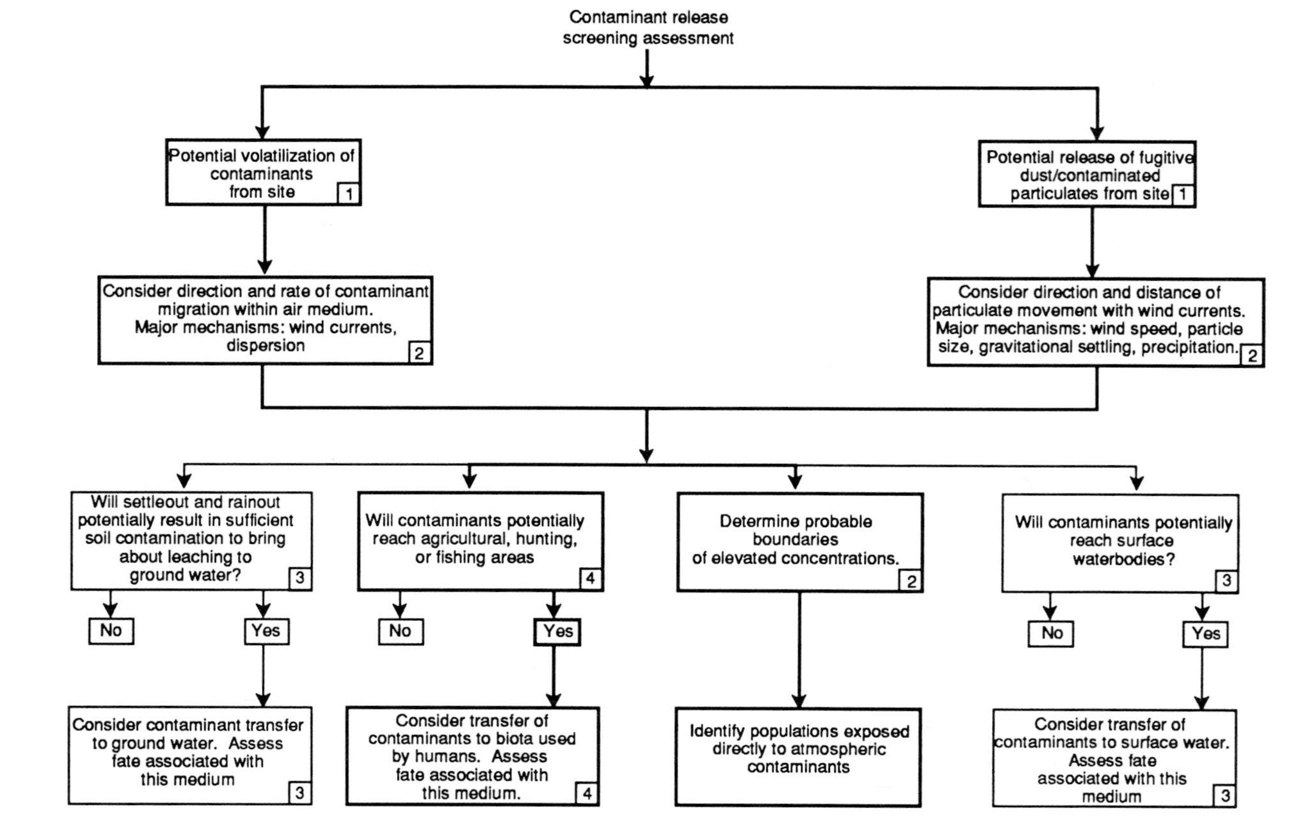

Fig. 16-3. Environmental fate screening assessment decision network: atmosphere.

unsaturated soils. Contaminants settling to the surface through dry deposition may dissolve in or become suspended in surface waters or may be leached into unsaturated soils and ground water by subsequent rainfall. Dry deposition may also result in formation of a layer of relatively high contamination at the soil surface. When such intermedia transfers are likely, one should assess the fate of contaminants in the receiving media.

4. If areas identified as likely to receive significant atmospheric contaminant concentrations include areas supporting edible biota, the biouptake of contaminants must be considered as a possible environmental fate pathway. Direct biouptake from the atmosphere is a potential fate mechanism for lipophilic contaminants. Biouptake from soil or water following transfer of contaminants to these media must also be considered as part of the screening assessments of these media.

B. Surface Water Fate

The following numbered paragraphs refer to particular numbered boxes in Fig. 16-4.

1. The aquatic fate of contaminants released from the CERCLA site as well as those transferred to surface water from other media beyond site boundaries must be considered.
2. Direction of contaminant movement is usually clear only for contaminants introduced into rivers and streams. Currents, thermal stratification or eddies, tidal pumping, and flushing in impoundments and estuaries render qualitative screening assessment of contaminant directional transport highly conjectural for these types of water bodies. In most cases, entire water bodies receiving contaminants must be considered potentially significant human exposure points. More in-depth analyses or survey data may subsequently identify contaminated and unaffected regions of these water bodies.
3. Similarly, contaminant concentrations in rivers or streams can be roughly assessed based on rate of contaminant introduction and dilution volumes. Estuary or impoundment concentration regimes are highly dependent on the transport mechanisms enumerated. Contaminants may be localized and remain concentrated or may disperse rapidly and become diluted to insignificant levels. The conservative approach is to conduct a more in-depth assessment and use model results or survey data as a basis for determining contaminant concentration levels.
4. Important intermedia transfer mechanisms that must be considered where significant surface water contamination is expected include transfers to ground water where hydrogeology of the area indicates significant surface water–ground water exchange, transfers to biota where waters contaminated with lipophilic substances support edible biotic species, and transfer

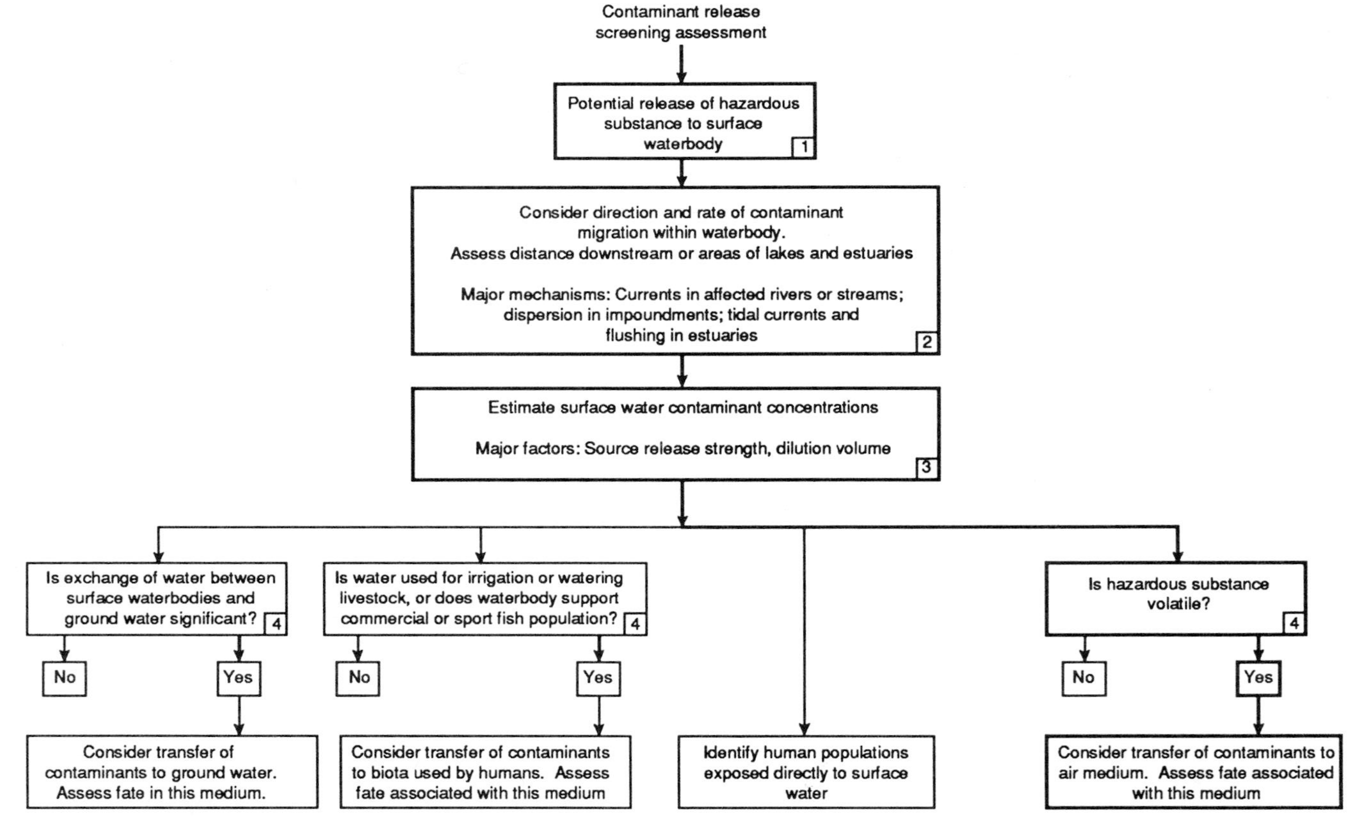

Fig. 16-4. Environmental fate screening assessment decision network: surface water.

to the atmosphere where surface water is contaminated by volatile substances. High temperatures, high surface area/volume ratios, high wind conditions, and turbulent stream flow also enhance volatilization rates.

Contaminant transfer to bed sediments represents another significant transfer mechanism, especially in cases where contaminants are in the form of suspended solids or are dissolved hydrophobic substances that can become adsorbed by organic matter in bed sediments. For the purposes of this chapter, sediments and water are considered part of a single system because of their complex interassociation. Surface water–bed sediment transfer is reversible; bed sediments often act as temporary repositories for contaminants and gradually rerelease contaminants to surface waters. Sorbed or settled contaminants are frequently transported with bed sediment migration or flow. Transfer of sorbed contaminants to bottom-dwelling, edible biota represents a fate pathway potentially resulting in human exposure. Where this transfer mechanism appears likely, the biotic fate of contaminants should be assessed.

C. Soil and Ground Water Fate

The following numbered paragraphs refer to particular numbered boxes in Fig. 16-5.

1. The fate of contaminants in the soil medium is assessed whenever the contaminant release atmospheric, or fate screening, assessment results show that significant contamination of soils is likely.
2. The most significant contaminant movement in soils is a function of liquid movement. Dry, soluble contaminants dissolved in precipitation, run-on, or human applied water will migrate through percolation into the soil. Migration rates are a function of net water recharge rates and contaminant solubility.
3. Important intermedia transfer mechanisms affecting soil contaminants include volatilization or resuspension to the atmosphere and biouptake by plants and soil organisms. These, in turn, introduce contaminants into the food chain.

IV. MODELING

An extremely difficult task is the estimation of emissions from hazardous waste sites. Frequently, both the amounts of materials existing within the site and the compounds and mixtures that are represented are not known. Even if both of these pieces of information are reasonably well known, the conditions of the containers holding these chemicals are not initially known.

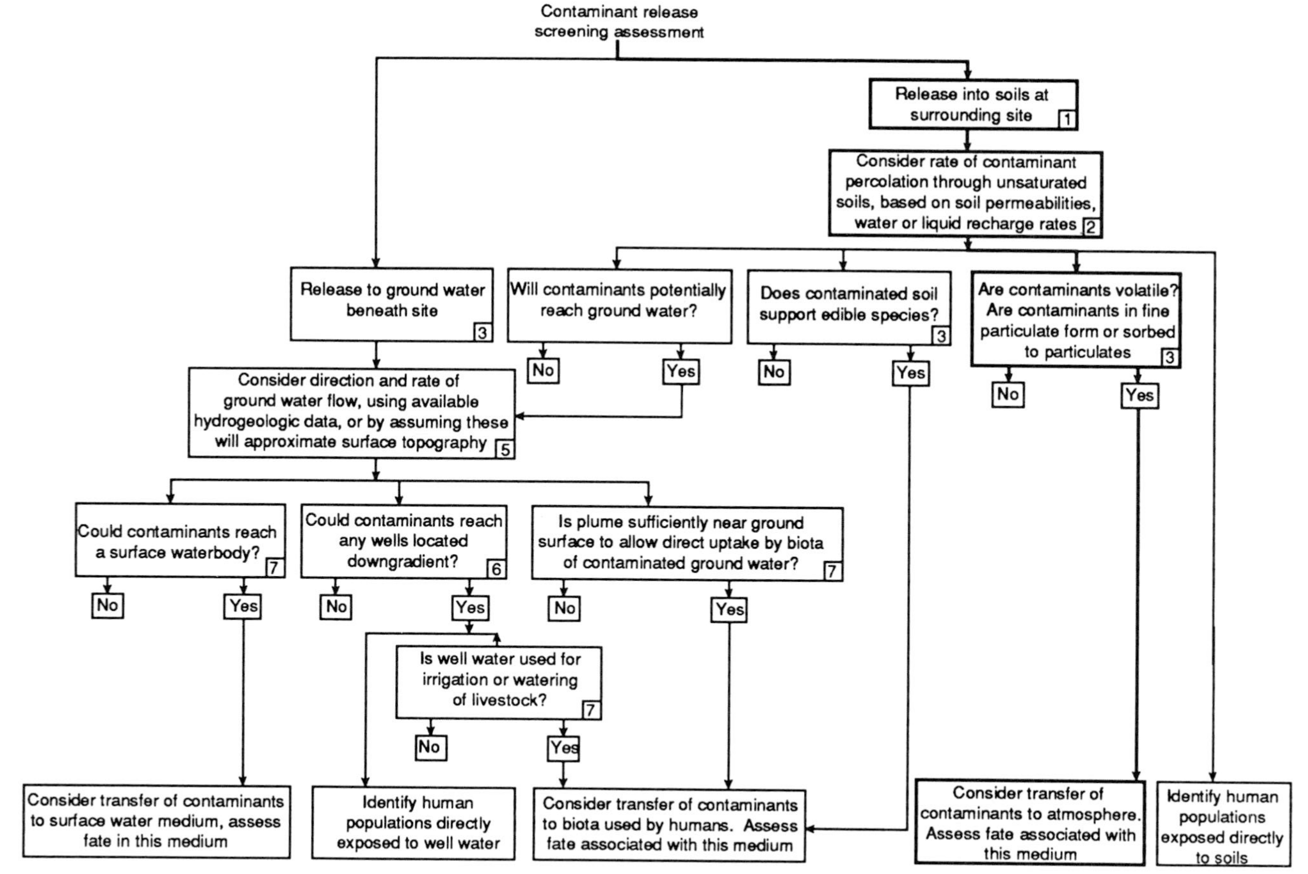

Fig. 16-5. Environmental fate screening assessment decision network: soils and ground water.

Hazardous materials may enter the air pathway by evaporation from leaking containers and release of these gases through fissures and spaces between soil particles. Another pathway may release hazardous substances to the air if they are water soluble. Then ground water passing leaking containers may carry substances to or near the surface, where they may be released to the air near the original source or at locations at significant distances.

A. Estimates of Long-Term Impact

If the foregoing problems of emissions estimation can be overcome, or if it is possible to make estimates of maximum possible and minimum possible emissions, then it is quite easy to make estimates of resulting long-term impact on the surrounding area. The representation of the emissions may be through consideration of an area source or area sources; or if vent pipes are releasing material or flaring the gases, point sources should be used.

A single finite line source method is used to simulate area sources in the long-term (seasons to years) model ISCLT (3). Although this method has been criticized as frequently underestimating concentrations for receptors that are quite close to the area source (within two or three side lengths away), this model is usually used for these estimates. In addition to the long-term estimate of emission rate for each constituent to be modeled, the ISCLT model requires meteorological data in the form of a joint frequency distribution of three parameters: wind direction (in 16 classes), wind speed (in 6 classes), and Pasquill stability class (in 6 classes). As long as the emissions can be considered relatively constant over the period of simulation, the long-term estimates will represent mean concentrations over the period represented by the meteorological data.

B. Estimates of Short-Term Impact during Remediation

If it is necessary to consider short-term (hours or days) impact, the model PAL (4) will do a superior simulation of the area sources and a similar simulation of any point sources as done by the ISCLT model.

In addition to short-term emission estimates, normally for hourly periods, the meteorological data include hourly wind direction, wind speed, and Pasquill stability class. Although of secondary importance, the hourly data also include temperature (only important if buoyant plume rise needs to be calculated from any sources) and mixing height.

The short-term model can then be used to estimate resulting concentrations during specific periods or to estimate concentrations for suspected adverse meteorological conditions, so that changes can be incorporated in the remediation process if concentrations are expected to be higher than desirable.

REFERENCES

1. Federal Register, Part II, Environmental Protection Agency, 40 CFR Part 300, Hazard Ranking System: Final Rule, Vol. 55, No. 241, December 14, 1990.
2. Procedures for Conducting Air Pathway Analysis for Superfund Applications, Vol. I, Application of Air Pathway Analyses for Superfund Activities, EPA-450/1-89-001, July 1989.
3. U.S. EPA, User's Guide for the Industrial Source Complex (ISC2) Dispersion Models. EPA-450/4-92-008a (Vol. I—User Instructions), EPA-450/4-92-008b (Vol. II—Description of Model Algorithms), EPA-450/4-92-008c (Vol. III—Guide for Programmers). U.S. Environmental Protection Agency, Research Triangle Park, NC, 1992.
4. Petersen, W. B., and Rumsey, E. D., "User's Guide for PAL 2.0—A Gaussian Plume Algorithm for Point, Area, and Line Sources," EPA/600/8-87/009. U.S. Environmental Protection Agency, Research Triangle Park, NC, 1987 (NTIS Accession No. PB87-168 787).

SUGGESTED READING

"Air/Superfund National Technical Guidance Study Series," Volume II: "Estimation of Baseline Air Emission at Superfund Sites," EPA-450/1-89-002a, August 1990. Volume III: "Estimation of Air Emissions from Clean-up Activities at Superfund Sites," EPA-450/1-89-003, January 1989. Volume IV: "Procedures for Conducting Air Pathway Analyses for Superfund Applications," EPA-450/1-89-004, July 1989.

An Act to Amend the Clean Air Act, Public Law 101-549, U.S. Congress, November 15, 1990.

Summerhays, B. E., Procedures for estimating emissions from the cleanup of Superfund sites, *J. Air Waste Management Assoc.* **40**(1), January 1990.

"Superfund Exposure Assessment Manual," U.S. Environmental Protection Agency, EPA/540/1-88/001, OSWER Directive 9285.5-1, April 1988.

QUESTIONS

1. How would the release of a volatile gas from contaminated soil be affected by the soil temperature?
2. The EPA Hazardous Ranking System computes a numerical score for hazardous waste. If the score exceeds a predetermined value, the waste site is placed on the National Priority List (NPL) for Superfund cleanup. Discuss the pros and cons of such a ranking system.
3. Describe a possible situation in which an air contaminant is controlled but the control system used transfers the contaminant problem to another medium, such as water or soil.

Part IV

The Meteorology of Air Pollution

17

The Physics of the Atmosphere

As indicated in previous chapters, the atmosphere serves as the medium through which air pollutants are transported and dispersed. While being transported, the pollutants may undergo chemical reactions and, in addition to removal by chemical transformations, may be removed by physical processes such as gravitational settling, impaction, and wet removal.

This chapter provides an introduction to basic concepts of meteorology necessary to an understanding of air pollution meteorology without specific regard to air pollution problems. The relationship of meteorology to air pollution is discussed in the following four chapters.

I. SUN, ATMOSPHERE SYSTEM, AND HEAT BALANCE

All of the energy that drives the atmosphere is derived from a minor star in the universe—our sun. The planet that we inhabit, earth, is 150 million km from the sun. The energy received from the sun is radiant energy—electromagnetic radiation. The electromagnetic spectrum is shown in Fig. 17-1. Although this energy is, in part, furnished to the atmosphere, it is primarily received at the earth's surface and redistributed by several

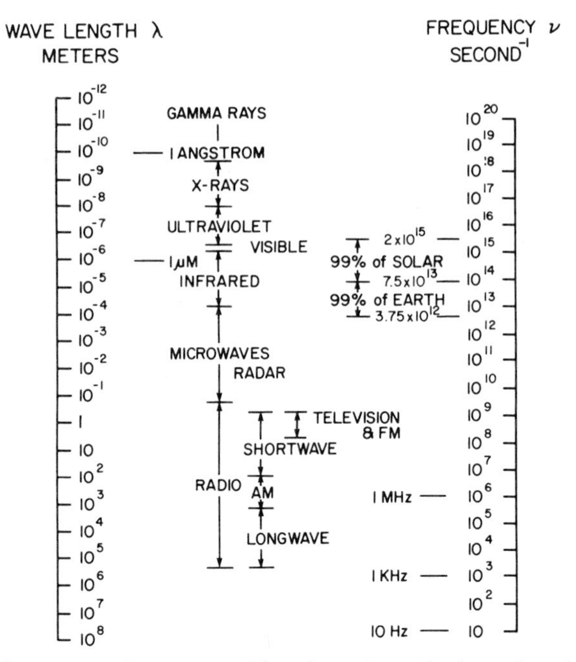

Fig. 17-1. Electromagnetic spectrum. Note the regions of solar and earth radiation.

processes. The earth's gravity keeps the thin layer of gases that constitute the atmosphere from escaping. The combination of solar heating and the spin of the earth causes internal pressure forces in the atmosphere, resulting in numerous atmospheric motions. The strength of the sun's radiation, the distance of the earth from the sun, the mass and diameter of the earth, and the existence and composition of the atmosphere combine to make the earth habitable. This particular combination of conditions would not be expected to occur frequently throughout the universe.

As noted in Chapter 2, the atmosphere is approximately 76% nitrogen, 20% oxygen, 3% water, 0.9% argon, and 0.03% carbon dioxide; the rest consists of relatively inert gases such as neon, helium, methane, krypton, nitrous oxide, hydrogen, and xenon. Compared with the average radius of the earth, 6370 km, the atmosphere is an incredibly thin veil; 90% is below 12 km and 99% below 30 km. In spite of its thinness, however, the total mass of the atmosphere is about 5×10^{18} kg. Therefore, its heat content and energy potential are very large.

A. Radiation from a Blackbody

Blackbody is the term used in physics for an object that is a perfect emitter and absorber of radiation at all wavelengths. Although no such object exists in nature, the properties describable by theory are useful for comparison with materials found in the real world. The amount of radiation, or radiant flux over all wavelengths (F), from a unit area of a blackbody is dependent on the temperature of that body and is given by the Stefan–Boltzmann law:

$$F = \sigma T^4 \tag{17-1}$$

where σ is the Stefan–Boltzmann constant and equals 8.17×10^{-11} cal $\text{cm}^{-2}\ \text{min}^{-1}\ \text{deg}^{-4}$ and T is the temperature in degrees K. Radiation from a blackbody ceases at a temperature of absolute zero, 0 K.

In comparing the radiative properties of materials to those of a blackbody, the terms *absorptivity* and *emissivity* are used. Absorptivity is the amount of radiant energy absorbed as a fraction of the total amount that falls on the object. Absorptivity depends on both frequency and temperature; for a blackbody it is 1. Emissivity is the ratio of the energy emitted by an object to that of a blackbody at the same temperature. It depends on both the properties of the substance and the frequency. Kirchhoff's law states that for any substance, its emissivity at a given wavelength and temperature equals its absorptivity. Note that the absorptivity and emissivity of a given substance may be quite variable for different frequencies.

As seen in Eq. (17-1), the total radiation from a blackbody is dependent on the fourth power of its absolute temperature. The frequency of the maximum intensity of this radiation is also related to temperature through Wien's displacement law (derived from Planck's law):

$$\nu_{\text{max}} = 1.04 \times 10^{11}\, T \tag{17-2}$$

where frequency ν is in s^{-1} and the constant is in $\text{s}^{-1}\ \text{K}^{-1}$.

The radiant flux can be determined as a function of frequency from Planck's distribution law for emission:

$$E_\nu\, d\nu = c_1 \nu^3 [\exp(c_2\, \nu/T) - 1]^{-1}\, d\nu \tag{17-3}$$

where

$$\begin{aligned} c_1 &= 2\pi h/c^2 & & \\ h &= 6.55 \times 10^{-27}\ \text{erg s} & & \text{(Planck's constant)} \\ c &= 3 \times 10^8\ \text{m s}^{-1} & & \text{(speed of light)} \\ c_2 &= h/k & & \end{aligned}$$

and

$$k = 1.37 \times 10^{-16}\ \text{erg K}^{-1} \qquad \text{(Boltzmann's constant)}$$

The radiation from a blackbody is continuous over the electromagnetic spectrum. The use of the term black in blackbody, which implies a particular color, is quite misleading, as a number of nonblack materials approach blackbodies in behavior. The sun behaves almost like a blackbody; snow radiates in the infrared nearly as a blackbody. At some wavelengths, water vapor radiates very efficiently. Unlike solids and liquids, many gases absorb (and reradiate) selectively in discrete wavelength bands, rather than smoothly over a continuous spectrum.

B. Incoming Solar Radiation

The sun radiates approximately as a blackbody, with an effective temperature of about 6000 K. The total solar flux is 3.9×10^{26} W. Using Wien's law, it has been found that the frequency of maximum solar radiation intensity is $6.3 \times 10^{14}\ s^{-1}$ ($\lambda = 0.48\ \mu m$), which is in the visible part of the spectrum; 99% of solar radiation occurs between the frequencies of $7.5 \times 10^{13}\ s^{-1}$ ($\lambda = 4\ \mu m$) and $2 \times 10^{15}\ s^{-1}$ ($\lambda = 0.15\ \mu m$) and about 50% in the visible region between $4.3 \times 10^{14}\ s^{-1}$ ($\lambda = 0.7\ \mu m$) and $7.5 \times 10^{14}\ s^{-1}$ ($\lambda = 0.4\ \mu m$). The intensity of this energy flux at the distance of the earth is about 1400 W m^{-2} on an area normal to a beam of solar radiation. This value is called the *solar constant*. Due to the eccentricity of the earth's orbit as it revolves around the sun once a year, the earth is closer to the sun in January (perihelion) than in July (aphelion). This results in about a 7% difference in radiant flux at the outer limits of the atmosphere between these two times.

Since the area of the solar beam intercepted by the earth is πE^2, where E is the radius of the earth, and the energy falling within this circle is spread over the area of the earth's sphere, $4\pi E^2$, in 24 hr, the average energy reaching the top of the atmosphere is 338 W m^{-2}. This average radiant energy reaching the outer limits of the atmosphere is depleted as it attempts to reach the the earth's surface. Ultraviolet radiation with a wavelength less than 0.18 μm is strongly absorbed by molecular oxygen in the ionosphere 100 km above the earth; shorter x-rays are absorbed at even higher altitudes above the earth's surface. At 60–80 km above the earth, the absorption of 0.2–0.24 μm wavelength radiation leads to the formation of ozone; below 60 km there is so much ozone that much of the 0.2–0.3 μm wavelength radiation is absorbed. This ozone layer in the lower mesosphere and the top of the stratosphere shields life from much of the harmful ultraviolet radiation. The various layers warmed by the the absorbed radiation reradiate in wavelengths dependent on their temperature and spectral emissivity. Approximately 5% of the total incoming solar radiation is absorbed above 40 km. Under clear sky conditions, another 10–15% is absorbed by the lower atmosphere or scattered back to space by the atmospheric aerosols and molecules; as a result, only 80–85% of the incoming radiation reaches the earth's surface. With average cloudiness, only about 50% of the incoming radiation reaches the earth's surface, because of the additional interference of the clouds.

C. Albedo and Angle of Incidence

The portion of the incoming radiation reflected and scattered back to space is the *albedo*. The albedo of clouds, snow, and ice-covered surfaces is around 0.5–0.8, that of fields and forests is 0.03–0.3, and that of water is 0.02–0.05 except when the angle of incidence becomes nearly parallel to the water surface. Table 17-1 shows the albedo of a water surface as a function of the angle of incidence. The albedo averaged over the earth's surface is about 0.35.

Although events taking place on the sun, such as sun spots and solar flares, alter the amount of radiation, the alteration is almost entirely in the x-ray and ultraviolet regions and does not affect the amount in the wavelengths reaching the earth's surface. Therefore, the amount of radiation from the sun that can penetrate to the earth's surface is remarkably constant.

In addition to the effect of albedo on the amount of radiation that reaches the earth's surface, the angle of incidence of the radiation compared to the perpendicular to the surface affects the amount of radiation flux on an area. The flux on a horizontal surface S_h is as follows:

$$S_h = S \cos Z \tag{17-4}$$

where S is the flux through an area normal to the solar beam and Z is the zenith angle (between the local vertical, the zenith, and the solar beam).

Because of the tilt of the earth's axis by 23.5° with respect to the plane of the earth's revolution around the sun, the north pole is tilted toward the sun on June 22 and away from the sun on December 21 (Fig. 17- 2). This tilt causes the solar beam to have perpendicular incidence at different latitudes depending on the date. The zenith angle Z is determined from:

$$\cos Z = \sin \phi \sin \delta + \cos \phi \cos \delta \cos \eta \tag{17-5}$$

where ϕ is latitude (positive for Northern Hemisphere, negative for South-

TABLE 17-1

Percent of Incident Radiation Reflected by a Water Surface (Albedo of Water)[a]

Angle of incidence	Percent reflected	Percent absorbed
90	2.0	98.0
70	2.1	97.9
50	2.5	97.5
40	3.4	96.6
30	6.	94.
20	13.	87.
10	35.	65.
5	58.	42.

[a] Adapted from Fig. 3-13 of Battan (1).

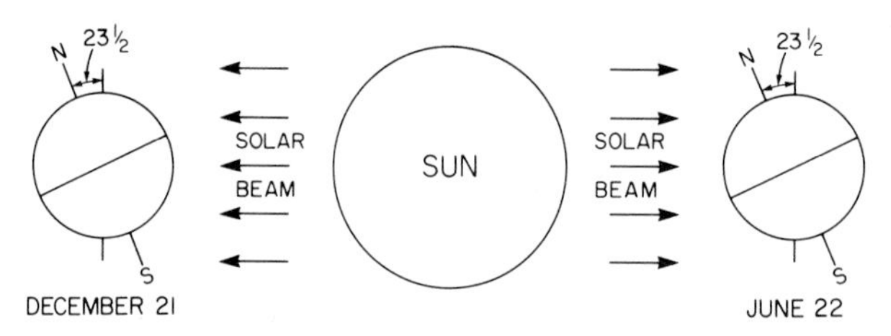

Fig. 17-2. Orientation of the earth to the solar beam at the extremes of its revolution around the sun.

ern Hemisphere), δ is solar declination (see Table 17-2), and η is hour angle, $15° \times$ the number of hours before or after local noon.

The solar azimuth ω is the angle between south and the direction toward the sun in a horizontal plane:

$$\sin \omega = (\cos \delta \sin \eta)/\sin Z \tag{17-6}$$

Since many surfaces receiving sunlight are not horizontal, a slope at an angle i from the horizontal facing an azimuth ω' degrees from south experiences an intensity of sunlight (neglecting the effects of the atmosphere) of

$$S_s = S[\cos Z \cos i + \sin Z \sin i \cos(\omega - \omega')] \tag{17-7}$$

Here ω and ω' are negative to the east of south and positive to the west.

At angles away from the zenith, solar radiation must penetrate a greater thickness of the atmosphere. Consequently, it can encounter more scattering due to the presence of particles and greater absorption due to this greater thickness.

TABLE 17-2

Solar Declination[a]

Date	Declination degree	Date	Declination degree
Jan. 21	−20.90	Jul.21	20.50
Feb. 21	−10.83	Aug. 21	12.38
Mar. 21	0.	Sep. 21	1.02
Apr. 21	11.58	Oct. 21	−10.42
May 21	20.03	Nov. 21	−19.75
June 21	23.45	Dec. 21	−23.43

[a] Adapted from Table 2-1 of Byers (2).

D. Outgoing Longwave Radiation

Because most ultraviolet radiation is absorbed from the solar spectrum and does not reach the earth's surface, the peak of the solar radiation which reaches the earth's surface is in the visible part of the spectrum. The earth reradiates nearly as a blackbody at a mean temperature of 290 K. The resulting infrared radiation extends over wavelengths of 3–80 μm, with a peak at around 11 μm. The atmosphere absorbs and reemits this longwave radiation primarily because of water vapor but also because of carbon dioxide in the atmosphere. Because of the absorption spectrum of these gases, the atmosphere is mostly opaque to wavelengths less than 7 μm and greater than 14 μm and partly opaque between 7 and 8.5 μm and between 11 and 14 μm. The atmosphere loses heat to space directly through the nearly transparent window between 8.5 and 11 μm and also through the absorption and successive reradiation by layers of the atmosphere containing these absorbing gases.

Different areas of the earth's surface react quite differently to heating by the sun. For example, although a sandy surface reaches fairly high temperatures on a sunny day, the heat capacity and conductivity of sand are relatively low; the heat does not penetrate more than about 0.2–0.3 m and little heat is stored. In contrast, in a body of water, the sun's rays penetrate several meters and slowly heat a fairly deep layer. In addition, the water can move readily and convection can spread the heat through a deeper layer. The heat capacity of water is considerably greater than that of sand. All these factors combine to allow considerable storage of heat in water bodies.

E. Heat Balance

Because of the solar beam's more direct angle of incidence in equatorial regions, considerably more radiation penetrates and is stored by water near the equator than water nearer the poles. This excess is not compensated for by the outgoing longwave radiation, yet there is no continual buildup of heat in equitorial regions. Figure 17-3 shows the annual mean incoming and outgoing radiation averaged over latitude bands. There is a transfer of heat poleward from the equatorial regions to make up for a net outward transfer of heat near the poles. This heat is transferred by air and ocean currents as warm currents move poleward and cool currents move equatorward. Considerable heat transfer occurs by the evaporation of water in the tropics and its condensation into droplets farther poleward, with the release of the heat of condensation. Enough heat is transferred to result in no net heating of the equatorial regions or cooling of the poles. The poleward flux of heat across various latitudes is shown in Table 17-3.

Taking the earth as a whole over a year or longer, because there is no appreciable heating or cooling, there is a heat balance between the incoming

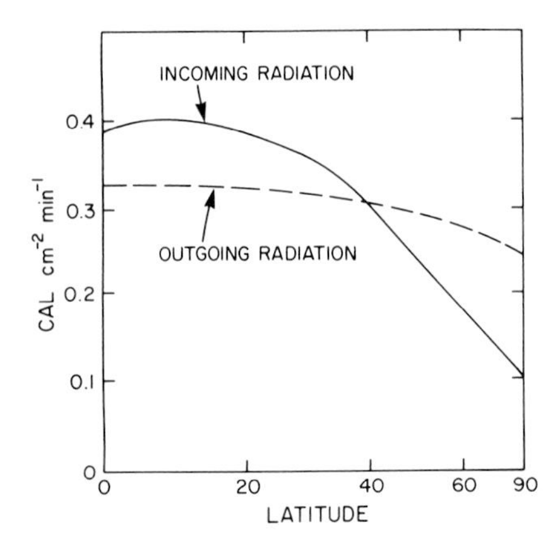

Fig. 17-3. Annual mean radiation by latitude. Note that the latitude scale simulates the amount of the earth's surface area between the latitude bands. Incoming radiation is that absorbed by earth and atmosphere. Outgoing radiation is that leaving the atmosphere. Source: After Byers (2) (1 cal cm^{-2} min^{-1} = 697.58 W m^{-2}).

solar radiation and the radiation escaping to space. This balance is depicted in Fig. 17-4.

II. STABILITY AND INSTABILITY

Vertical air motions affect both weather and the mixing processes of importance to air pollution. Upward vertical motions can be caused by lifting over terrain, lifting over weather fronts, and convergence toward low-pressure centers. Downward vertical motions can be caused by sinking to make up for divergence near high-pressure centers. One must know whether the atmosphere enhances or suppresses these vertical motions to

TABLE 17-3

Poleward Flux of Heat across Latitudes (10^{19} kcal per year)[a]

Latitude	Flux	Latitude	Flux
10	1.21	50	3.40
20	2.54	60	2.40
30	3.56	70	1.25
40	3.91	80	0.35

[a] Adapted from Table 12 of Sellers (3).

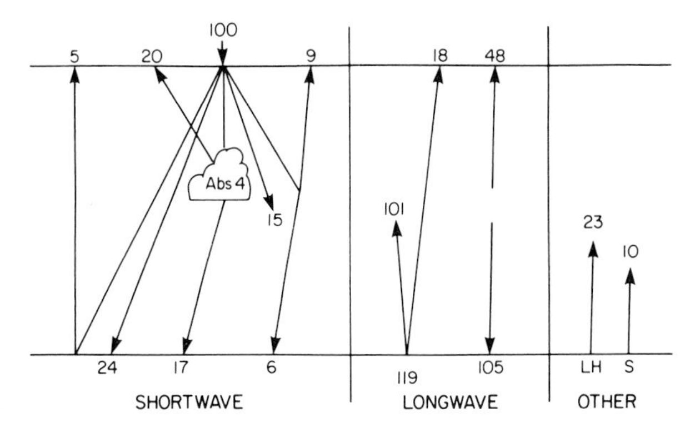

Fig. 17-4. Radiation heat balance. The 100 units of incoming shortwave radiation are distributed: reflected from earth's surface to space, 5; reflected from cloud surfaces to space, 20; direct reaching earth, 24; absorbed in clouds, 4; diffuse reaching earth through clouds, 17; absorbed in atmosphere, 15; scattered to space, 9; scattered to earth, 6. The longwave radiation comes from (1) the earth radiating 119 units: 101 to the atmosphere and 18 directly to space, and (2) the atmosphere radiating 105 units back to earth and 48 to space. Additional transfers from the earth's surface to the atmosphere consist of latent heat, 23; and sensible heat, 10. Source: After Lowry (4).

assess their effects. When the atmosphere resists vertical motions, it is called *stable;* when the atmosphere enhances vertical motions, it is called *unstable* or in a state of *instability.*

In incompressible fluids, such as water, the vertical structure of temperature very simply reveals the stability of the fluid. When the lower layer is warmer and thus less dense than the upper layer, the fluid is unstable and convective currents will cause it to overturn. When the lower layer is cooler than the upper layer, the fluid is stable and vertical exchange is minimal. However, because air is compressible, the determination of stability is somewhat more complicated. The temperature and density of the atmosphere normally decrease with elevation; density is also affected by moisture in the air.

The relationship between pressure p, volume V, mass m, and temperature T is given by the equation of state:

$$pV = RmT \tag{17-8}$$

where R is a specific gas constant equal to the universal gas constant divided by the gram molecular weight of the gas. Since the density ρ is m/V, the equation can be rewritten as

$$p = R\rho T \tag{17-9}$$

or considering specific volume $\alpha = 1/\rho$ as

$$\alpha p = RT \tag{17-10}$$

These equations combine Boyle's law, which states that when temperature is held constant the volume varies inversely with the pressure, and the law of Guy-Lussac, which states that when pressure is held constant the volume varies in proportion to the absolute temperature.

A. First Law of Thermodynamics

If a volume of air is held constant and a small amount of heat Δh is added, the temperature of the air will increase by a small amount ΔT. This can be expressed as

$$\Delta h = c_v \, \Delta T \tag{17-11}$$

where c_v is the specific heat at constant volume. In this case, all the heat added is used to increase the internal energy of the volume affected by the temperature. From the equation of state (Eq. 17-8), it can be seen that the pressure will increase.

If, instead of being restricted, the volume of air considered is allowed to remain at an equilibrium constant pressure and expand in volume, as well as change temperature in response to the addition of heat, this can be expressed as

$$\Delta h = c_v \, \Delta T + p \, \Delta v \tag{17-12}$$

By using the equation of state, the volume change can be replaced by a corresponding pressure change:

$$\Delta h = c_p \, \Delta T + v \, \Delta p \tag{17-13}$$

where c_p is the specific heat at constant pressure and equals $c_v + R_d$, where R_d is the gas constant for dry air.

B. Adiabatic Processes

An adiabatic process is one with no loss or gain of heat to a volume of air. If heat is supplied or withdrawn, the process is *diabatic* or *nonadiabatic*. Near the earth's surface, where heat is exchanged between the earth and the air, the processes are diabatic.

However, away from the surface, processes frequently are adiabatic. For example, if a volume (parcel) of air is forced upward over a ridge, the upward-moving air will encounter decreased atmospheric pressure and will expand and cool. If the air is not saturated with water vapor, the process is called *dry adiabatic*. Since no heat is added or subtracted, Δh in Eq. (17-13) can be set equal to zero, and introducing the hydrostatic equation

$$-\Delta p = \rho g \, \Delta z \tag{17-14}$$

and combining equations results in

$$-\Delta T / \Delta z = g / c_p \tag{17-15}$$

Thus air cools as it rises and warms as it descends. Since we have assumed an adiabatic process, $-\Delta T/\Delta z$ defines γ_d, the dry adiabatic process lapse rate, a constant equal to 0.0098 K/m, is nearly 1 K/100 m or 5.4°F/1000 ft.

If an ascending air parcel reaches saturation, the addition of latent heat from condensing moisture will partially overcome the cooling due to expansion. Therefore, the saturated adiabatic lapse rate (of cooling) γ_w is smaller than γ_d.

C. Determining Stability

By comparing the density changes undergone by a rising or descending parcel of air with the density of the surrounding environment, the enhancement or suppression of the vertical motion can be determined. Since pressure decreases with height, there is an upward- directed pressure gradient force. The force of gravity is downward. The difference between these two forces is the buoyancy force. Using Newton's second law of motion, which indicates that a net force equals an acceleration, the acceration a of an air parcel at a particular position is given by

$$a = g(T_p - T_e)/T_p \tag{17-16}$$

where g is the acceleration due to gravity (9.8 m s^{-2}), T_p the temperature of an air parcel that has undergone a temperature change according to the process lapse rate, and T_e the temperature of the surrounding environment at the same height. (Temperatures are expressed in degrees Kelvin.)

Figure 17-5 shows the temperature change undergone by a parcel of air forced to rise 200 m in ascending a ridge. Assuming that the air is dry, and therefore that no condensation occurred, this figure also represents the warming of the air parcel if the flow is reversed so that the parcel moves downslope from B to A.

Comparing the temperature of this parcel to that of the surrounding environment (Fig. 17-6), it is seen that in rising from 100 to 300 m, the parcel undergoes the temperature change of the dry adiabatic process lapse rate. The dashed line is a dry adiabatic line or dry adiabat. Suppose that

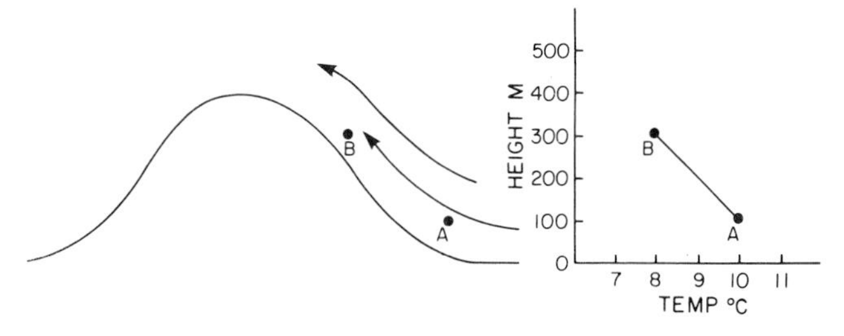

Fig. 17-5. Cooling of ascending air. Dry air forced to rise 200 m over a ridge cools adiabatically by 2°C.

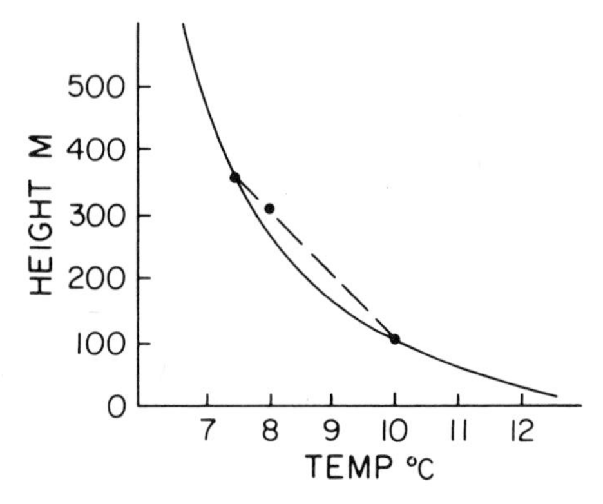

Fig. 17-6. Temperature of a parcel of air forced to rise 200 m compared to the superadiabatic environmental lapse rate. Since the parcel is still warmer than the environment, it will continue to rise.

the environmental temperature structure is shown by the solid curve. Since the lapse rate of the surrounding environment in the lowest 150–200 m is steeper than the adiabatic lapse rate (superadiabatic)—that is, since the temperature drops more rapidly with height—this part of the environment is thermally unstable. At 300 m the parcel is 0.2°C warmer than the environment, the resulting acceleration is upward, and the atmosphere is enhancing the vertical motion and is unstable. The parcel of air continues to rise until it reaches 350 m, where its temperature is the same as that of the environment and its acceration drops to zero. However, above 350 m the lapse rate of the surrounding environment is not as steep as the adiabatic lapse rate (subadiabatic), and this part of the environment is thermally stable (it resists upward or downward motion).

If the temperature structure, instead of being that of Fig. 17-6, differs primarily in the lower layers, it resembles Fig. 17-7, where a temperature inversion (an increase rather than a decrease of temperature with height) exists. In the forced ascent of the air parcel up the slope, dry adiabatic cooling produces parcel temperatures that are everywhere cooler than the environment; acceration is downward, resisting displacement; and the atmosphere is stable.

Thermodynamic diagrams which show the relationships between atmospheric pressure (rather than altitude), temperature, dry adiabatic lapse rates, and moist adiabatic lapse rates are useful for numerous atmospheric thermodynamic estimations. The student is referred to a standard text on meteorology (see Suggested Reading) for details. In air pollution meteorology, the thermodynamic diagram may be used to determine the current mixing height (the top of the neutral or unstable layer). The mixing height

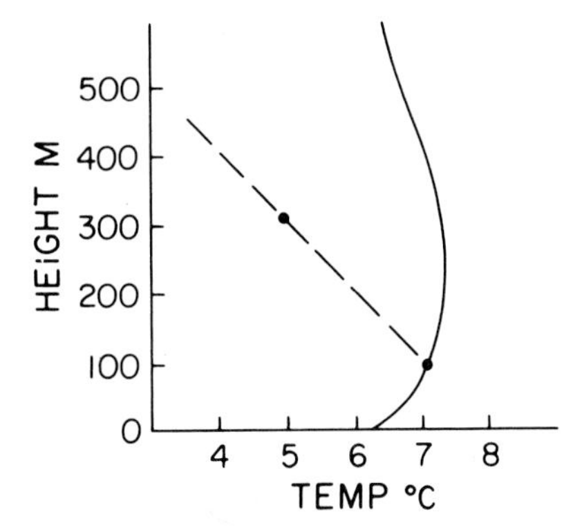

Fig. 17-7. Temperature of a parcel of air forced to rise 200 m compared to an inversion environmental lapse rate. Since the parcel is cooler than the environment, it will sink back to its original level.

at a given time may be estimated by use of the morning radiosonde ascent plotted on a thermodynamic chart. The surface temperature at the given time is plotted on the diagram. If a dry adiabat is drawn through this temperature, the height aboveground at the point where this dry adiabat intersects the morning sounding is the mixing height for that time. The mixing height for the time of maximum temperature is the maximum mixing height. Use of this sounding procedure provides an approximation because it assumes that there has been no significant advection since the time of the sounding.

D. Potential Temperature

A useful concept in determining stability in the atmosphere is *potential temperature*. This is a means of identifying the dry adiabat to which a particular atmospheric combination of temperature and pressure is related. The potential temperature θ is found from

$$\theta = T(1000/p)^{0.288} \tag{17-17}$$

where T is temperature and p is pressure (in millibars, mb). This value is the same as the temperature that a parcel of dry air would have if brought dry adiabatically to a pressure of 1000 mb.

If the potential temperature decreases with height, the atmosphere is unstable. If the potential temperature increases with height, the atmosphere is stable. The average lapse rate of the atmosphere is about 6.5°C/km; that is, the potential temperature increases with height and the average state of the atmosphere is stable.

E. Effect of Mixing

The mixing of air in a vertical layer produces constant potential temperature throughout the layer. Such mixing is usually mechanical, such as air movement over a rough surface. In Fig. 17-8 the initial temperature structure is subadiabatic (solid line). The effect of mixing is to achieve a mean potential temperature throughout the layer (dashed line), which in the lower part is dry adiabatic. The bottom part of the layer is warmed; the top is cooled. Note that above the vertical extent of the mixing, an inversion is formed connecting the new cooled portion with the old temperature structure above the zone of mixing. If the initial layer has considerable moisture, although not saturated, cooling in the top portion of the layer may decrease the temperature to the point where some of the moisture condenses, forming clouds at the top. An example of this is the formation of an inversion and a layer of stratus clouds along the California coast.

F. Radiation or Nocturnal Inversions

An inversion caused by mixing in a surface layer was just discussed above. Inversions at the surface are caused frequently at night by radiational cooling of the ground, which in turn cools the air near it.

G. Subsidence Inversions

There is usually some descent (subsidence) of air above surface high-pressure systems. This air warms dry adiabatically as it descends, decreasing the relative humidity and dissipating any clouds in the layer. A subsidence inversion forms as a result of this sinking. Since the descending air compresses as it encounters the increased pressures lower in the atmo-

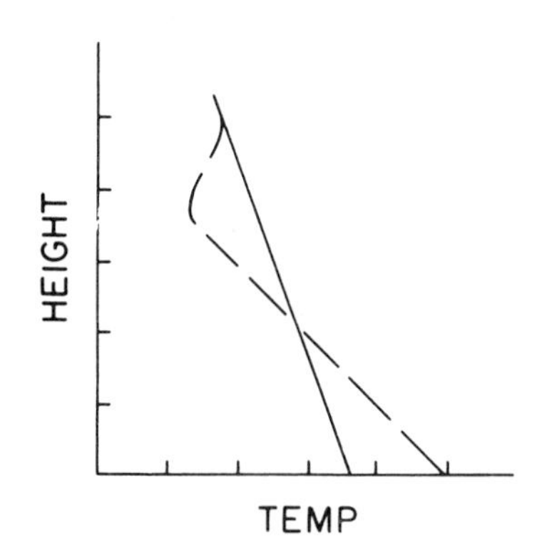

Fig. 17-8. Effect of forced mixing (dashed) on the environmental subadiabatic lapse rate (solid). Note the formation of an inversion at the top of the mixed layer.

sphere, the top portion of the descending layer will be further warmed due to its greater descent than will the bottom portion of the layer (Fig. 17-9). Occasionally a subsidence inversion descends all the way to the surface, but usually its base is well above the ground.

Inversions are of considerable interest in relation to air pollution because of their stabilizing influence on the atmosphere, which suppresses the vertical motion that causes the vertical spreading of pollutants.

III. LAWS OF MOTION

The atmosphere is nearly always in motion. The scales and magnitude of these motions extend over a wide range. Although vertical motions certainly occur in the atmosphere and are important to both weather processes and the movement of pollutants, it is convenient to consider wind as only the horizontal component of velocity.

On the regional scale (hundreds to thousands of kilometers), the winds are most easily understood by considering the balance of various forces in the atmosphere. The applicable physical law is Newton's second law of motion, $F = ma$; if a force F is exerted on a mass m, the resulting acceleration a equals the force divided by the mass. This can also be stated as the rate of change of momentum of a body, which is equal to the sum of the forces that act on the body. It should be noted that all the forces to be discussed are vectors; that is, they have both magnitude and direction. Although Newton's second law applies to absolute motion, it is most convenient to consider wind relative to the earth's surface. These create some slight difficulties, but they can be rather easily managed.

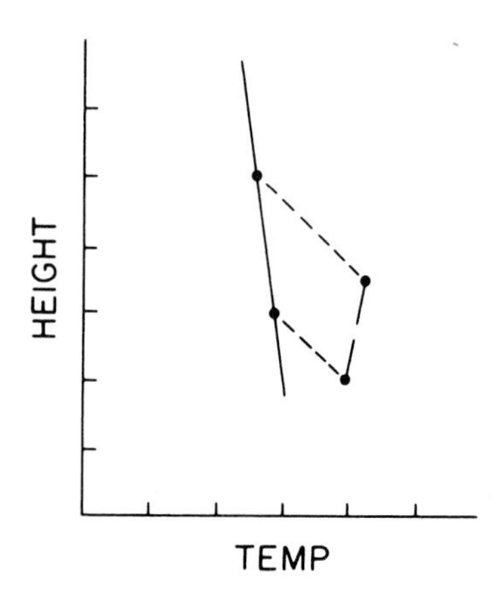

Fig. 17-9. Formation of a subsidence inversion in subsiding (sinking) air. Note the vertical compression of the sinking layer which is usually accompanied by horizontal divergence.

A. Pressure Gradient Force

Three forces of importance to horizontal motion are the pressure gradient force, gravity, and friction. Atmospheric pressure equals mass times the acceleration of gravity. Considering a unit volume, $p = \rho g$; the gravitational force on the unit volume is directed downward. Primarily because of horizontal temperature gradients, there are horizontal density gradients and consequently horizontal pressure gradients. The horizontal pressure gradient force $p_h = \Delta p/\rho \, \Delta x$ where Δp is the horizontal pressure difference over the distance Δx. The direction of this force and of the pressure difference measurement is locally perpendicular to the lines of equal pressure (isobars) and is directed from high to low pressure.

B. Coriolis Force

If the earth were not rotating, the wind would blow from high to low pressure. Close to the earth, it would be slowed by friction between the atmosphere and the earth's surface but would maintain the same direction with height. However, since the earth undergoes rotation, there is an apparent force acting on horizontal atmospheric motions when examined from a point of reference on the earth's surface. For example, consider a wind of velocity 10 m s^{-1} blowing at time 1 in the direction of the 0° longitude meridian across the north pole (Fig. 17-10). The wind in an absolute sense continues to blow in this direction for 1 hr, and a parcel of air starting at the pole at time 1 travels 36 km in this period. However, since the earth turns 360° every 24 hr, or 15° per hr, it has rotated 15° in the hour and we find that at time 2 (60 min after time 1) the 15° meridian is now beneath the wind vector. As viewed from space (the absolute frame of reference), the flow has continued in a straight line. However, as viewed from the earth, the flow has undergone an apparent deflection to the right. The force required to produce this apparent deflection is the coriolis force and is equal to $D = vf$ where f, the coriolis parameter, equals $2\Omega \sin \phi$.

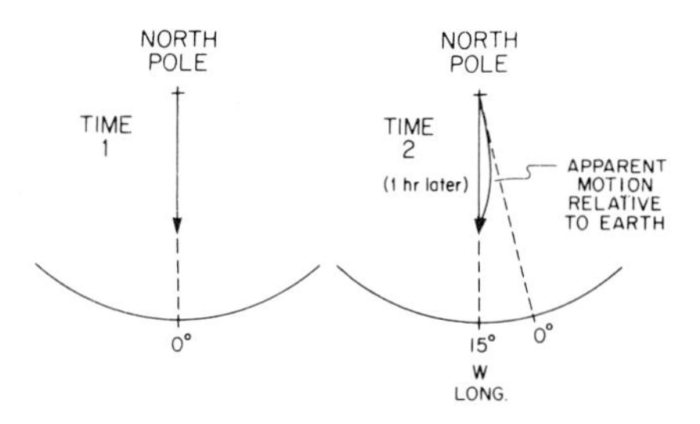

Fig. 17-10. Effect of the coriolis force. The path of air moving from the north pole to the south as viewed from space is straight; as viewed from the earth's surface it is curved.

Here Ω is the angular speed of the earth's rotation, $2\pi/(24 \times 60 \times 60) = 7.27 \times 10^{-5}\ s^{-1}$, and ϕ is the latitude. It is seen that f is maximal at the poles and zero at the equator. The deflecting force is to the right of the wind vector in the Northern Hemisphere and to the left in the Southern Hemisphere. For the present example, the deflecting force is $1.45 \times 10^{-3}\ m\ s^{-2}$, and the amount of deflection after the 36-km movement in 1 hr is 9.43 km.

C. Geostrophic Wind

Friction between the atmosphere and the earth's surface may generally be neglected at altitudes of about 700 m and higher. Therefore, large-scale air currents closely represent a balance between the pressure gradient force and the coriolis force. Since the coriolis force is at a right angle to the wind vector, when the coriolis force is equal in magnitude and opposite in direction to the pressure gradient force, a wind vector perpendicular to both of these forces occurs, with its direction along the lines of constant pressure (Fig. 17-11). In the Northern Hemisphere, the low pressure is to the left of the wind vector (Buys Ballot's law); in the Southern Hemisphere, low pressure is to the right. The geostrophic velocity is

$$v_g = -\Delta p/\rho f\ \Delta d \tag{17-18}$$

When the isobars are essentially straight, the balance between the pressure gradient force and the coriolis force results in a geostrophic wind parallel to the isobars.

D. Gradient Wind

When the isobars are curved, an additional force, a centrifugal force outward from the center of curvature, enters into the balance of forces. In the case of curvature around low pressure, a balance of forces occurs when the pressure gradient force equals the sum of the coriolis and centrifugal forces (Fig. 17-12) and the wind continues parallel to the isobars. In the

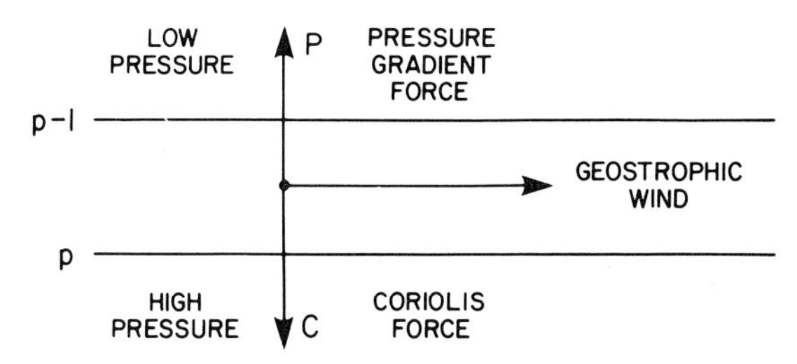

Fig. 17-11. Balance of forces resulting in geostrophic wind.

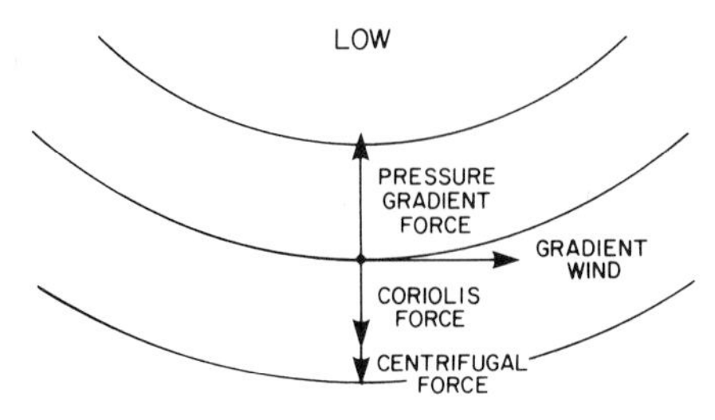

Fig. 17-12. Balance of forces resulting in gradient wind around low pressure.

case of curvature around high pressure, a balance of forces occurs when the sum of the pressure gradient and centrifugal forces equals the coriolis force (Fig. 17- 13). To maintain a given gradient wind speed, a greater pressure gradient force (tighter spacing of the isobars) is required in the flow around low-pressure systems than in the flow around high-pressure systems.

E. The Effect of Friction

The frictional effect of the earth's surface on the atmosphere increases as the earth's surface is approached from aloft. Assuming that we start with geostrophic balance aloft, consider what happens to the wind as we move downward toward the earth. The effect of friction is to slow the wind velocity, which in turn decreases the coriolis force. The wind then turns toward low pressure until the resultant vector of the frictional force and the coriolis force balances the pressure gradient force (Fig. 17-14). The greater the friction, the slower the wind and the greater the amount of turning toward low pressure. The turning of the wind from the surface through the friction layer is called the *Ekman spiral*. A radial plot, or hodograph, of the winds through the friction layer is shown diagrammatically in Fig. 17-15.

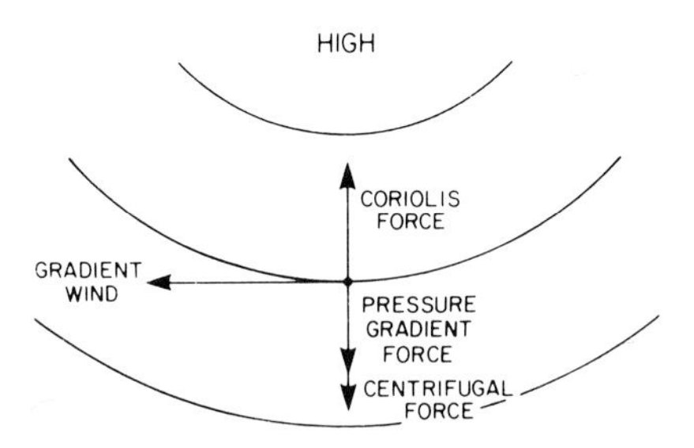

Fig. 17-13. Balance of forces resulting in gradient wind around high pressure. Note that the wind speed is greater for a given pressure gradient force than that around low pressure.

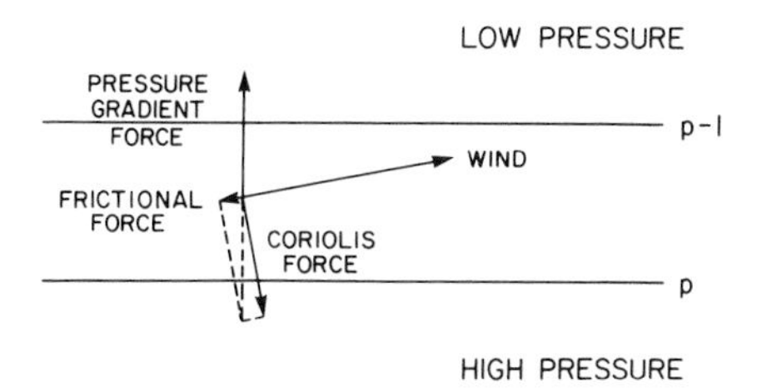

Fig. 17-14. Effect of friction on the balance of forces, causing wind to blow toward low pressure.

Note that this frictional effect will cause pollutants released at two different heights to tend to move in different directions.

In the friction layer where the isobars are curved, the effect of frictional drag is added to the forces discussed under gradient wind. The balance of the pressure gradient force, the coriolis deviating force, the centrifugal force, and the frictional drag in the vicinity of the curved isobars results in wind flow around low pressure and high pressure in the Northern Hemisphere, as shown in Fig. 17-16.

F. Vertical Motion—Divergence

So far in discussing motion in the atmosphere, we have been emphasizing only horizontal motions. Although of much smaller magnitude than horizontal motions, vertical motions are important both to daily weather formation and to the transport and dispersion of pollutants.

Persistent vertical motions are linked to the horizontal motions. If there is divergence (spreading) of the horizontal flow, there is sinking (downward vertical motion) of air from above to compensate. Similarly, converging (negative divergence) horizontal air streams cause upward vertical motions,

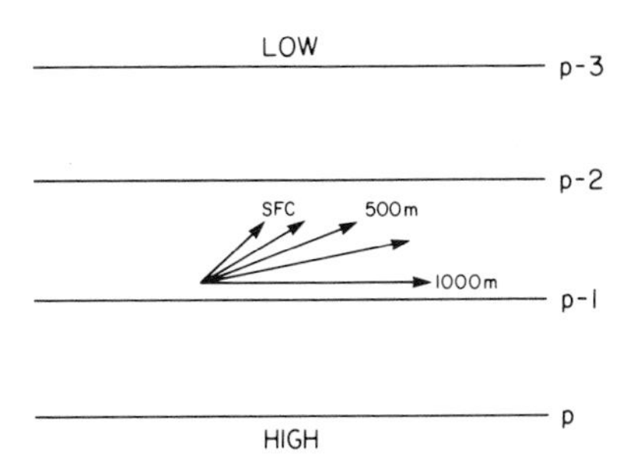

Fig. 17-15. Hodograph showing variation of wind speed and direction with height above ground. SFC = surface wind.

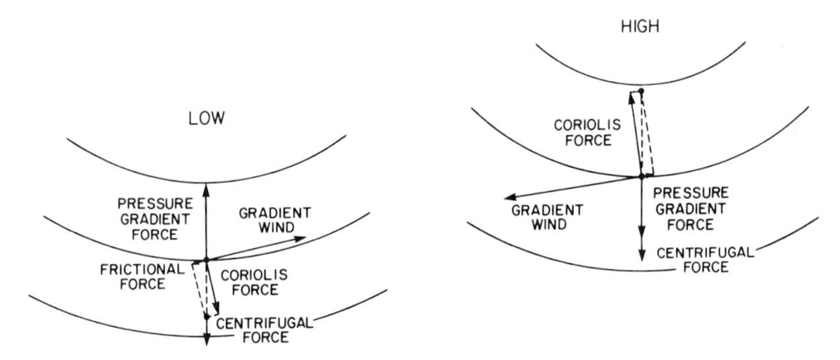

Fig. 17-16. Effect of friction upon gradient wind around low (left) and high (right) pressures.

producing condensation and perhaps precipitation in moist air masses, as well as transport of air and its pollutants from near the surface to higher altitudes.

IV. LOCAL WIND SYSTEMS

Frequently, local wind systems are superimposed on the larger-scale wind systems just discussed. These local flows are especially important and may dominate when the larger-scale flow becomes light and indefinite. Local wind systems are usually quite significant in terms of the transport and dispersion of air pollutants.

A. Sea and Land Breezes

The sea breeze is a result of the differential heating of land and water surfaces by incoming solar radiation. Since solar radiation penetrates several meters of a body of water, it warms very slowly. In contrast, only the upper few centimeters of land are heated, and warming occurs rapidly in response to solar heating. Therefore, especially on clear summer days, the land surface heats rapidly, warming the air near the surface and decreasing its density. This causes the air to rise over the land, decreasing the atmospheric pressure near the surface relative to the pressure at the same altitude over the water surface. The rising air increases the pressure over the land relative to that above the water at altitudes of approximately 100–200 m. The air that rises over the land surface is replaced by cooler air from over the water surface. This air, in turn, is replaced by subsiding air from somewhat higher layers of the atmosphere over the water. Air from the higher-pressure zone several hundred meters above the surface then flows from over the land

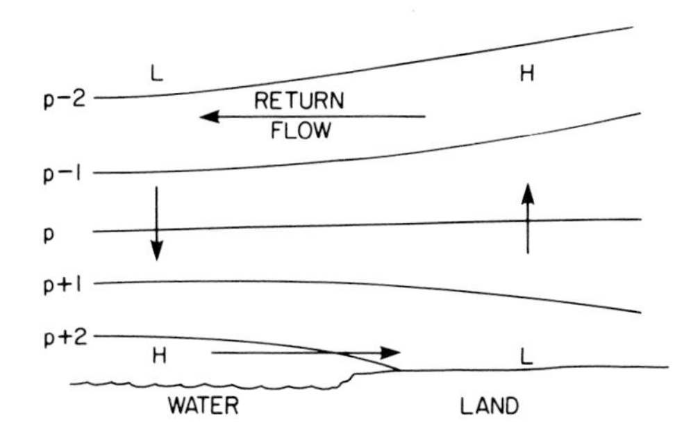

Fig. 17-17. Sea breeze due to surface heating over land, resulting in thermals, and subsidence over water.

surface out over the water, completing a circular or cellular flow (Fig. 17-17). Any general flow due to large-scale pressure systems will be superimposed on the sea breeze and may either reinforce or inhibit it. Ignoring the larger-scale influences, the strength of the sea breeze will generally be a function of the temperature excess of the air above the land surface over that above the water surface.

Just as heating in the daytime occurs more quickly over land than over water, at night radiational cooling occurs more quickly over land. The pressure pattern tends to be the reverse of that in the daytime. The warmer air tends to rise over the water, which is replaced by the land breeze from land to water, with the reverse flow (water to land) completing the circular flow at altitudes somewhat aloft. Frequently at night, the temperature differences between between land and water are smaller than those during the daytime, and therefore the land breeze has a lower speed.

B. Mountain and Valley Winds

Solar heating and radiational cooling influence local flows in terrain situations. Consider midday heating of a south-facing moutainside. As the slope heats, the air adjacent to the slope warms, its density is decreased, and the air attempts to ascend (Fig. 17-18). Near the top of the slope, the air tends to rise vertically. Along each portion of the slope farther down the mountain, it is easier for each rising parcel of air to move upslope, replacing the parcel ahead of it rather than rising vertically. This upslope flow is the valley wind.

At night when radiational cooling occurs on slopes, the cool dense air near the surface descends along the slope (Fig. 17-19). This is the downslope wind. To compensate for this descending air, air farther from the slope that is cooled very little is warmer relative to the descending air and rises, frequently resulting in a closed circular path. Where the downslope winds

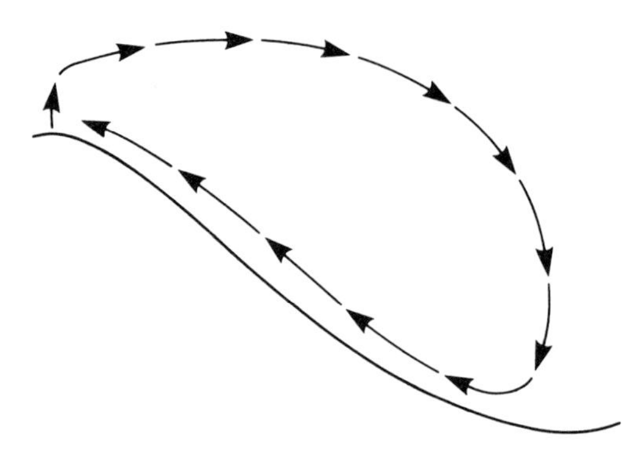

Fig. 17-18. Upslope wind (daytime) due to greater solar heating on the valley's side than in its center.

occur on opposite slopes of a valley, the cold air can accumulate on the valley floor. If there is any slope to the valley floor, this pool of cold air can move down the valley, resulting in a drainage or canyon wind.

Different combinations of valley and mountain slope, especially with some slopes nearly perpendicular to the incoming radiation and others in deep shadow, lead to many combinations of wind patterns, many nearly unique. Also, each local flow can be modified by the regional wind at the time which results from the current pressure patterns. Table 17-4 gives characteristics of eight different situations depending on the orientation of the ridgeline and valley with respect to the sun, wind direction perpendicular or parallel to the ridgeline, and time of day. Figure 17-20 shows examples of some of the mountain and valley winds listed in Table 17-4. These are rather idealized circulations compared to observed flows at any one time.

The effect of solar radiation is different with valley orientation. An east–west valley has only one slope that is significantly heated—the south-

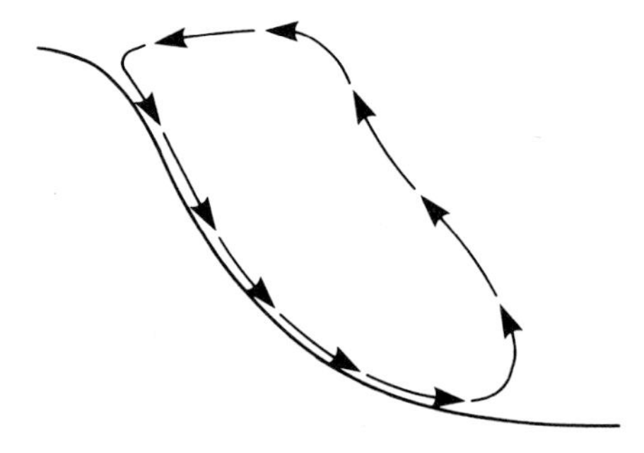

Fig. 17-19. Downslope wind (night) due to more rapid radiational cooling on the valley's slope than in its center.

TABLE 17-4

Generalized Mesoscale Windflow Patterns Associated with Different Combinations of Wind Direction and Ridgeline Orientation

Wind direction relative to ridgeline	Time of day	Ridgeline orientation: East–west	Ridgeline orientation: North–south
Parallel	Day	1[a] South-facing slope is heated—single helix	2 Upslope flow on both heated slopes—double helix
	Night	3 Downslope flow on both slopes–double helix	4 Downslope flow on both slopes—double helix
Perpendicular	Day	South-facing slope is heated. 5a North wind—stationary eddy fills valley 5b South wind—eddy suppressed, flow without separation	6 Upslope flow on both heated slopes—stationary eddy on one side of valley
	Night	7 Indefinite flow—extreme stagnation in valley bottom	8 Indefinite flow—extreme stagnation in valley bottom

[a] Numbers refer to Fig. 17-20.

facing slope may be near normal with midday sunshine. A north–south valley will have both slopes heated at midday. The effect of flow in relation to valley orientation is such that flows perpendicular to valleys tend to form circular eddies and encourage local flows; flows parallel to valleys tend to discourage local flows and to sweep clean the valley, especially with stronger wind speeds.

Keep in mind that the flows occurring result from the combination of the general and local flows; the lighter the general flow, the greater the opportunity for generation of local flows.

Complicated terrain such as a major canyon with numerous side canyons will produce complicated and unique flows, especially when side canyon drainage flows reinforce the drainage flow in the main valley.

C. Urban–Rural Circulations

Urban areas have roughness and thermal characteristics different from those of their rural surroundings. Although the increased roughness affects both the vertical wind profile and the vertical temperature profile, the effects due to the thermal features are dominant. The asphalt, concrete,

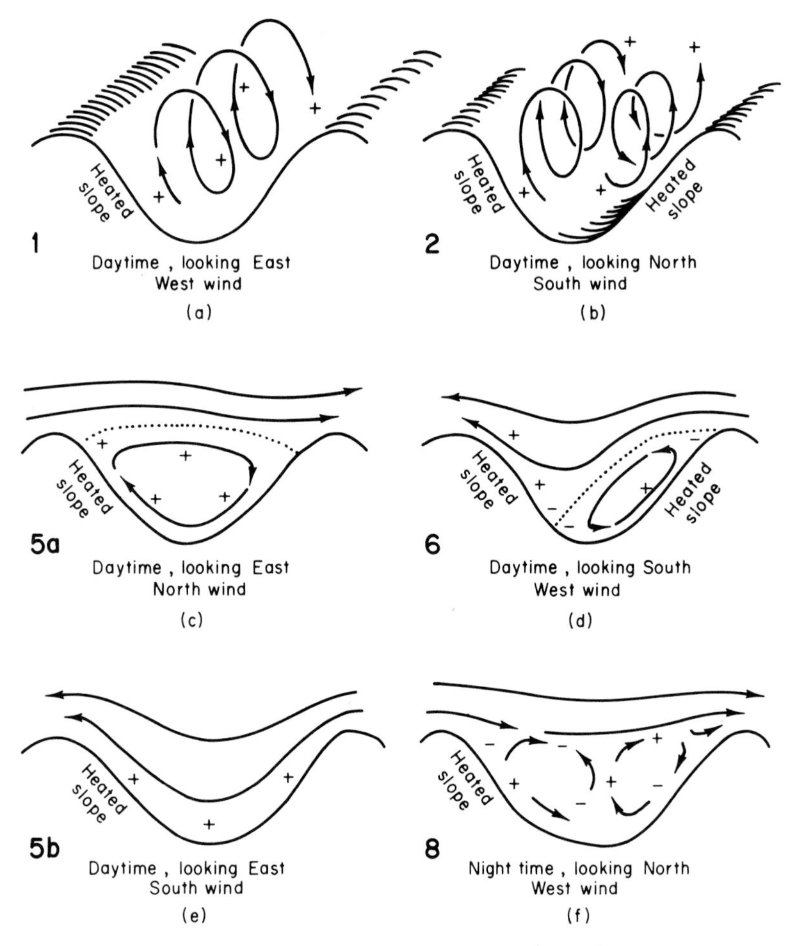

Fig. 17-20. Local valley–ridge flow patterns. (Numbers refer to Table 17-4.)

and steel of urban areas heat quickly and have a high heat-storing capability compared to the soil and vegetation of rural areas. Also, some surfaces of buildings are normal to the sun's rays just after sunrise and also before sunset, allowing warming throughout the day. The result is that the urban area becomes warmer than its surroundings during the day and stores sufficient heat that reradiation of the stored heat during the night keeps the urban atmosphere considerably warmer than its rural surroundings throughout most nights with light winds.

Under the lightest winds, the air rises over the warmest part of the urban core, drawing cooler air from all directions from the surroundings (Fig. 17-21). Subsidence replaces this air in rural areas, and a closed torus (doughnut)-shaped circulation occurs with an outflow above the urban

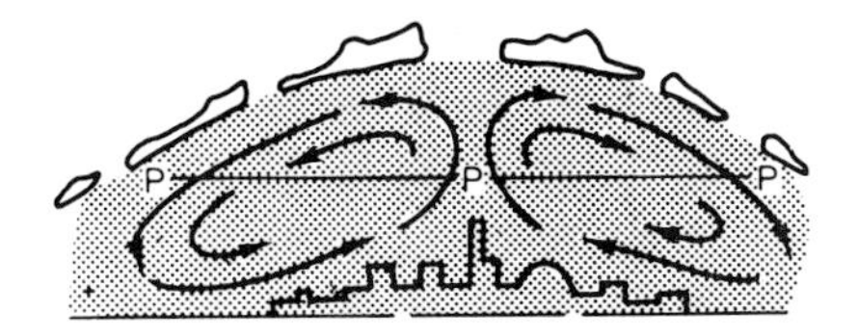

Fig. 17-21. Urban heat island (light regional wind).

area. This circulation is referred to as the *urban heat island*. The strength of the resulting flow is dependent on the difference in temperature between the urban center and its surroundings.

When the regional wind allows the outflow to take place in primarily one direction and the rising warm urban air moves off with this regional flow, the circulation is termed the *urban plume* (Fig. 17-22). Under this circumstance, the inflow to the urban center near the surface may also be asymmetric, although it is more likely to be symmetric than the outflow at higher altitudes.

The urban area also gives off heat through the release of gases from combustion and industrial processes. Compared to the heat received through solar radiation and subsquently released, the combustion and process heat is usually quite small, although it may be 10% or more in major urban areas. It can be of significance in the vicinity of a specific local source, such as a steam power plant (where the release of heat is large over a small area) and during light-wind winter conditions.

D. Flow around Structures

When the wind encounters objects in its path such as an isolated structure, the flow usually is strongly perturbed and a turbulent wake is formed in the vicinity of the structure, especially downwind of it. If the structure is semistreamlined in shape, the flow may move around it with little disturbance. Since most structures have edges and corners, generation of a turbulent wake is quite common. Figure 17-23 shows schematically the flow in the vicinity of a cubic structure. The disturbed flow consists of a cavity

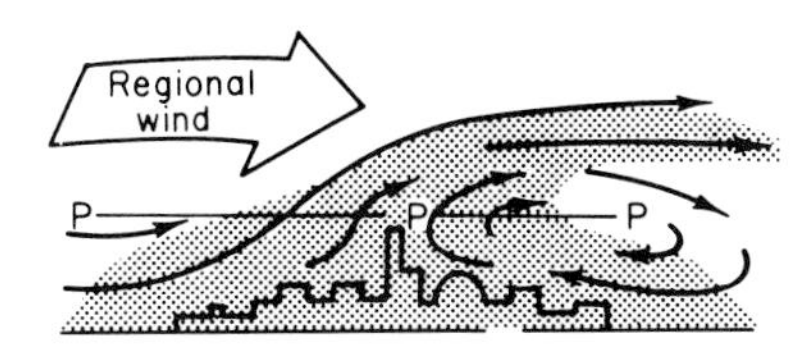

Fig. 17-22. Urban plume (moderate regional wind).

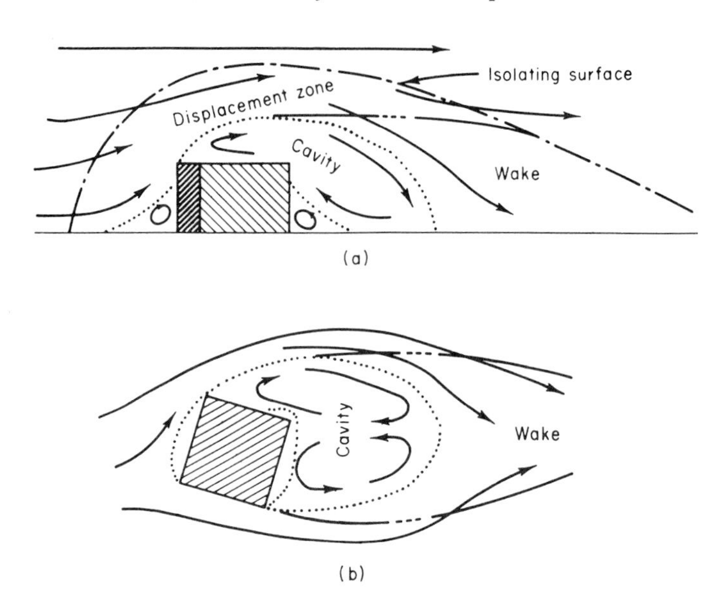

Fig. 17-23. Aerodynamic flow around a cube: (a) side view, (b) plan view. Source: After Halitsky (5).

with strong turbulence and mixing, a wake extending downwind from the cavity a distance equivalent to a number of structure side lengths, a displacement zone where flow is initially displaced before entering the wake, and a region of flow that is displaced away from the structure but does not get caught in the wake. Wind tunnels, water channels, and/or towing tanks are extremely useful in studying building wake effects.

V. GENERAL CIRCULATION

Atmospheric motions are driven by the heat from incoming solar radiation and the redistribution and dissipation of this heat to maintain constant temperatures on the average. The atmosphere is inefficient, because only about 2% of the received incoming solar radiation is converted to kinetic energy, that is, air motion; even this amount of energy is tremendous compared to that which humans are able to produce. As was shown in Section I, a surplus of radiant energy is received in the equatorial regions and a net outflux of energy occurs in the polar regions. Many large-scale motions serve to transport heat poleward or cooler air toward the equator.

If the earth did not rotate or if it rotated much more slowly than it does, a meridional (along meridians) circulation would take place in the troposphere (Fig. 17-24). Air would rise over the tropics, move poleward, sink over the poles forming a subsidence inversion, and then stream equa-

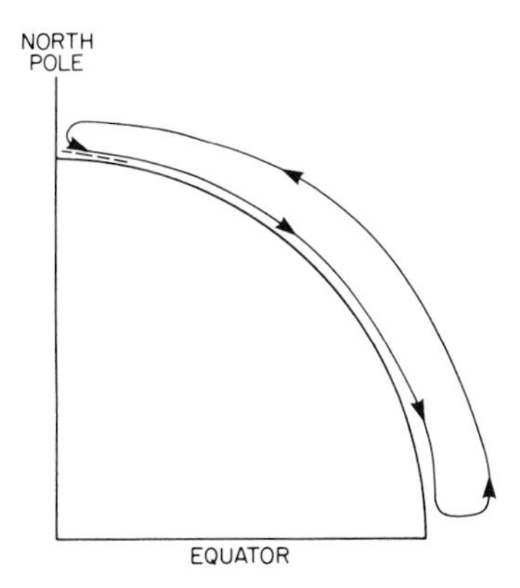

Fig. 17-24. Meridional single-cell circulation (on the sunny side of a nonrotating earth).

torward near the earth's surface. However, since the earth's rotation causes the apparent deflection due to the coriolis force, meridional motions are deflected to become zonal (along latitude bands) before moving more than 30°. Therefore, instead of the single cell consisting of dominantly meridional motion (Fig. 17-24), meridional transport is accomplished by three cells between the equator and the pole (Fig. 17-25). This circulation results in subsidence inversions and high pressure where there is sinking toward the earth's surface and low pressure where there is upward motion.

A. Tropics

Associated with the cell nearest the equator are surface winds moving toward the equator which are deflected toward the west. In the standard

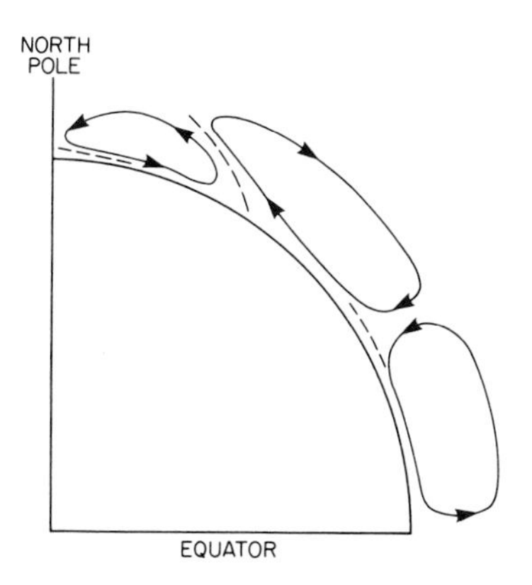

Fig. 17-25. Meridional three-cell circulation (rotating earth).

terminology of winds, which uses the direction from which they come, these near-surface winds are referred to as *easterlies* (Fig. 17-26), also called *trade winds*. Since most of the earth's surface near the equator is ocean, these winds absorb heat and moisture on their way to the equator.

Where the trade winds from each hemisphere meet is a low-pressure zone, the *intertropical convergence zone*. This zone of light winds or doldrums shifts position with the season, moving slightly poleward into the summer hemisphere. The rising air with high humidity in the convective motions of the convergence zone causes heavy rainfall in the tropics. This giant convective cell, or Hadley cell, absorbs heat and the latent heat of evaporation at low levels, releasing the latent heat as the moisture condenses in the ascending air. Some of this heat is lost through infrared radiation from cloud tops. The subsiding air, which warms adiabatically as it descends in the vicinity of 30° latitude (horse latitudes), feeds warm air into the mid-latitudes. Although the position of the convergence zone shifts somewhat seasonally, the Hadley cell circulation is quite persistent, resulting in a fairly steady circulation.

B. Mid-Latitudes

Because at higher latitudes the coriolis force deflects wind to a greater extent than in the tropics, winds become much more zonal (flow parallel to lines of latitude). Also in contrast to the persistent circulation of the tropics, the mid-latitude circulations are quite transient. There are large temperature contrasts, and temperature may vary abruptly over relatively short distances (frontal zones). In these regions of large temperature contrast, potential energy is frequently released and converted into kinetic energy as wind. Near the surface there are many closed pressure sys-

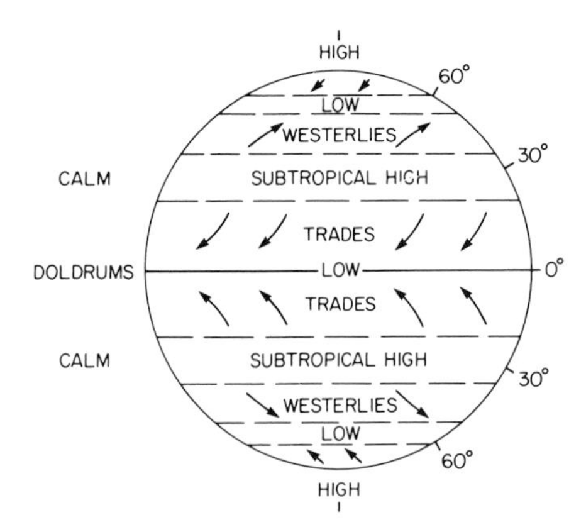

Fig. 17-26. Near-surface winds for various latitude belts.

tems—cyclones and anticyclones, which are quite mobile, causing frequent changes in weather at any given location. In contrast to the systems near the earth's surface, the motions aloft (above about 3 km) have few closed centers and are mostly waves moving from west to east. The core where speeds are highest in this zonal flow is the jet stream at about 11–14 km aboveground (Fig. 17-27). Where the jet stream undergoes acceleration, divergence occurs at the altitude of the jet stream. This, in turn, promotes convergence near the surface and encourages cyclogenesis (formation of cyclonic motion). Deceleration of the jet stream conversely causes convergence aloft and subsidence near the surface, intensifying high-pressure systems. The strength of the zonal flow is determined by the zonal index, which is the difference in average pressure of two latitude circles such as

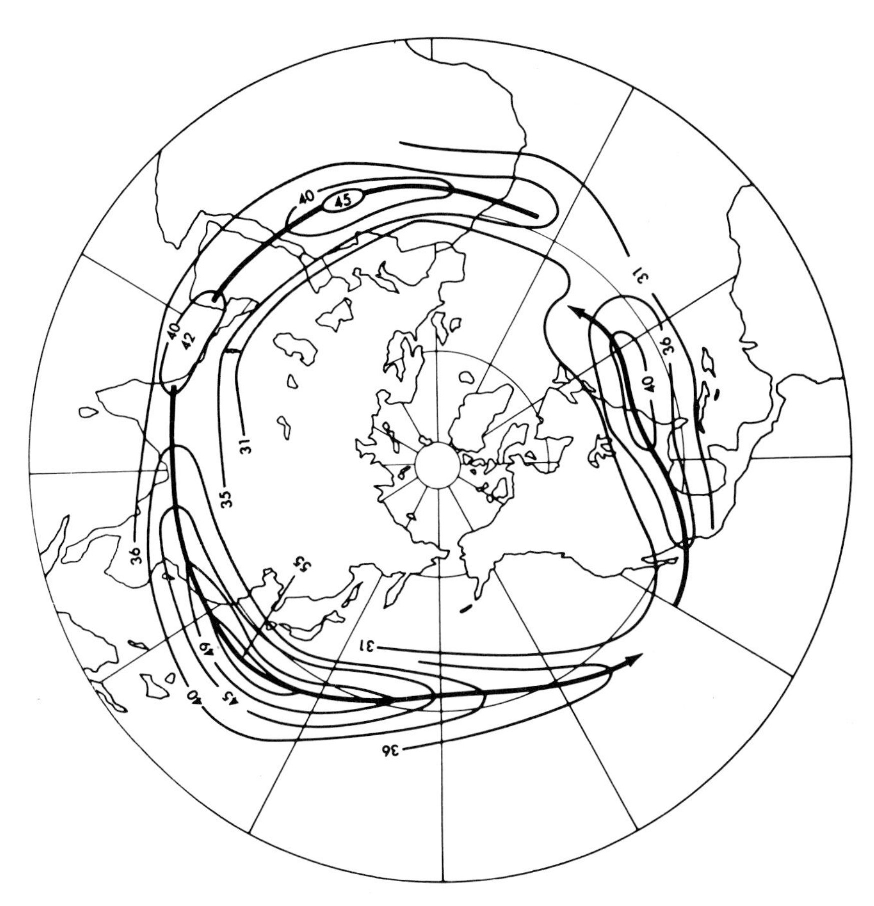

Fig. 17-27. Average position and strength of the jet stream in January between 11 and 14 km above the earth's surface (speeds are in m s^{-1}). Source: After Battan (1).

35° and 55°. A high index thus represents strong zonal flow; a low index indicates weak zonal flow. A low index is frequently accompanied by closed circulations which provide a greater degree of meridional flow. In keeping with the transient behavior of the mid-latitude circulation, the zonal index varies irregularly, cycling from low to high in periods ranging from 20 to 60 days.

The jet stream is caused by strong temperature gradients, so it is not surprising that it is frequently above the polar front, which lies in the convergence zone between the mid-latitude loop of the general circulation and the loop nearest the poles (Fig. 17-25). The positions of both the polar front and the jet stream are quite variable, shifting poleward with surface surges of warm air and moving toward the equator with outbreaks of cold air.

C. Polar Region

The circulation cells nearest the poles include rising air along the polar front, movement toward the poles aloft, sinking in the polar regions causing subsidence inversions, and flow toward the equator near the earth's surface. These motions contribute to the heat balance as the moisture in the air rising over the polar front condenses, releasing the heat that was used to evaporate the water nearer the equator. Also, the equatorward-moving air is cold and will be warmed as it is moved toward the tropics.

D. Other Factors

Of considerable usefulness in transporting heat toward the poles are the ocean currents. They are particularly effective because of the high heat content of water. Significant poleward-moving currents are the Brazil, Kuroshio, and Gulf Stream currents. Currents returning cold water toward the equator are the Peru and California currents.

The pressure pattern changes from winter to summer in response to temperature changes. Because most of the Southern Hemisphere consists of ocean, the summer-to- winter temperature differences are moderated. However, the increased land mass in the Northern Hemisphere allows high continental temperatures in summer, causing small equator-to-pole temperature differences; cooling over the continents in winter produces more significant equator-to-pole temperature differences, increasing the westerly winds in general and the jet stream in particular.

REFERENCES

1. Battan, L. J. "Fundamentals of Meteorology." Prentice-Hall, Englewood Cliffs, NJ, 1979.
2. Byers, H. R., "General Meteorology," 4th ed. McGraw-Hill, New York, 1974.

3. Sellers, W. D., "Physical Climatology." University of Chicago Press, Chicago, 1965.
4. Lowry, W., "Weather and Life: An Introduction to Biometeorology." Academic Press, New York, 1970.
5. Halitsky, J., Gas diffusion near buildings, *in* "Meteorology and Atomic Energy—1968" (D. Slade, ed.), TID-24190. U.S. Atomic Energy Commission, Oak Ridge, TN, 1968, pp. 221–255.

SUGGESTED READING

Arya, S. P., "Introduction to Micrometeorology," International Geophysics Series, Vol. 42. Academic Press, Troy, MO, 1988.

Critchfield, H. J., "General Climatology," 4th ed. Prentice-Hall, Englewood Cliffs, NJ, 1983.

Landsberg, H. E., "The Urban Climate." Academic Press, New York, 1981.

Neiburger, M., Edinger, J. G., and Bonner, W. D., "Understanding Our Atmospheric Environment." Freeman, San Francisco, 1973.

Petterssen, S., "Introduction to Meteorology," 3rd ed. McGraw-Hill, New York, 1969.

Stull, R. B., "An Introduction to Boundary Layer Meteorology." Kluwer Academic Press. Hingham, MA, 1989.

Wallace, J. M., and Hobbs, P. V., "Atmospheric Science—An Introductory Survey." Academic Press. Orlando, FL, 1977.

Wanta, R. C., and Lowry, W. P., The meteorological setting for dispersal of air pollutants, *in* "Air Pollution," 3rd ed., Vol. I, "Air Pollutants, Their Transformation and Transport" (A. C. Stern, ed.). Academic Press, New York, 1976.

QUESTIONS

1. Verify the intensity of the energy flux from the sun in cal cm^{-2} min^{-1} reaching the outer atmosphere of the earth from the total solar flux of 5.6×10^{27} cal min^{-1} and the fact that the earth is 1.5×10^8 km from the sun. (The surface area of a sphere of radius r is $4\pi r^2$.)
2. Compare the difference in incoming radiation on a horizontal surface at noon on June 22 with that at noon on December 21 at a point at 23.5°N latitude.
3. What is the zenith angle at 1000 local time on May 21 at a latitude of 36°N?
4. At what local time is sunset on August 21 at 40°S latitude?
5. Show the net heating of the atmosphere, on an annual basis, by determining the difference between heat entering the atmosphere and heat radiating to the earth's surface and to space. (See Fig. 17-4.)
6. If the universal gas constant is 8.31×10^{-2} mb m^{-3} $(\text{g-mole})^{-1}$ K^{-1} and the gram molecular weight of dry air is 28.9, what is the mass of a cubic meter of air at a temperature of 293 K and an atmospheric pressure of 996 mb?
7. On a particular day, temperature can be considered to vary linearly with height between 28°C at 100 m aboveground and 26°C at 500 m aboveground. Do you consider the layer between 100 and 500 m aboveground to be stable or unstable?
8. What is the potential temperature of air having a temperature of 288 K at a pressure of 890 mb?
9. Using Wien's displacement law, determine the mean effective temperature of the earth–atmosphere system if the resulting longwave radiation peaks at 11 μm. Contrast the magnitude of the radiant flux at 11 μm with that at 50 μm.

10. What accompanies horizontal divergence near the earth's surface? What effect is this likely to have on the thermal stability of this layer?
11. At what time of day and under what meteorological conditions is maximum ground-level pollution likely to occur at locations several kilometers inland from a shoreline industrial complex whose pollutants are released primarily from stacks of moderate height (about 40–130 m)?
12. When the regional winds are light, at what time of day and what location might high ground-level concentrations of pollutants occur from the low-level sources (less than 20 m) of a town in a north–south-oriented valley whose floor slopes down to the north? Can you answer this question for sources releasing in the range 50–70 m aboveground?

18

The Meteorological Bases of Atmospheric Pollution

Air pollutants reach receptors by being transported and perhaps transformed in the atmosphere (Fig. 18-1). The location of receptors relative to sources and atmospheric influences affect pollutant concentrations, and the sensitivity of receptors to these concentrations determines the effects. The location, height, and duration of release, as well as the amount of pollutant released, are also of importance. Some of the influences of the atmosphere on the behavior of pollutants, primarily the large-scale effects, are discussed here, as well as several effects of pollutants on the atmosphere.

I. VENTILATION

If air movement past a continuous pollutant source is slow, pollutant concentrations in the plume moving downwind will be much higher than they would be if the air were moving rapidly past the source. If polluted air continues to have pollution added to it, the concentration will increase. Generally, a source emits into different volumes of air over time. However,

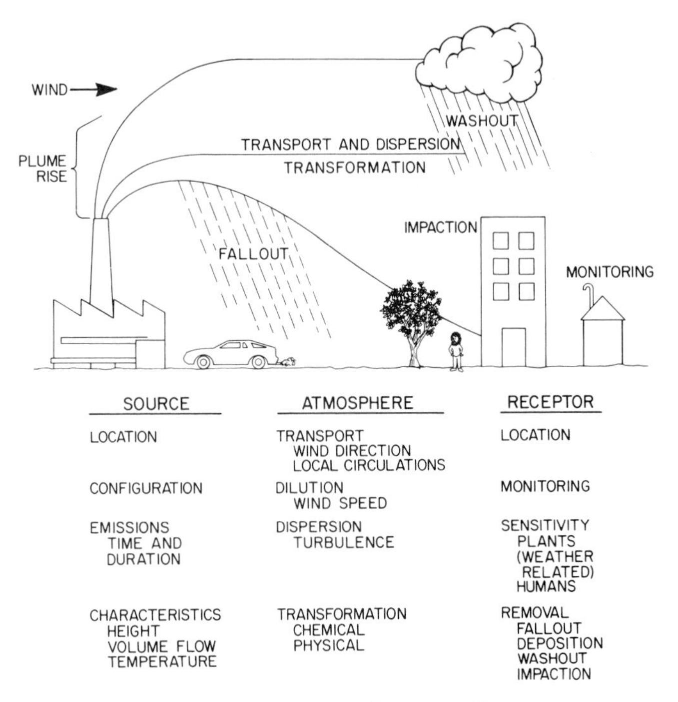

Fig. 18-1. The atmosphere's role in air pollution.

there can be a buildup of concentration over time even with significant air motion if there are many sources.

Low- and high-pressure systems have considerably different ventilation characteristics. Air generally moves toward the center of a low (Fig. 18-2) in the lower atmosphere, due in part to the frictional turning of the wind toward low pressure. This convergence causes upward vertical motion near the center of the low. Although the winds very near the center of the low are generally light, those away from the center are moderate, resulting in increased ventilation rates. Note the increased wind in the area to the west of the low in Fig. 18-2. Low-pressure systems generally cover relatively small areas (although the low-pressure system shown in Fig. 18-2 covers an extensive area) and are quite transient seldom remaining the same at a given area for a significant period of time. Lows are frequently accompanied by cloudy skies, which may cause precipitation. The cloudy skies minimize the variation in atmospheric stability from day to night. Primarily because of moderate horizontal wind speeds and upward vertical motion, ventilation (i. e., total air volume moving past a location) in the vicinity of low-pressure sytems is quite good.

High-pressure systems are characteristicaly the opposite of lows. Since the winds flow outward from the high-pressure center, subsiding air from higher in the atmosphere compensates for the horizontal transport of mass.

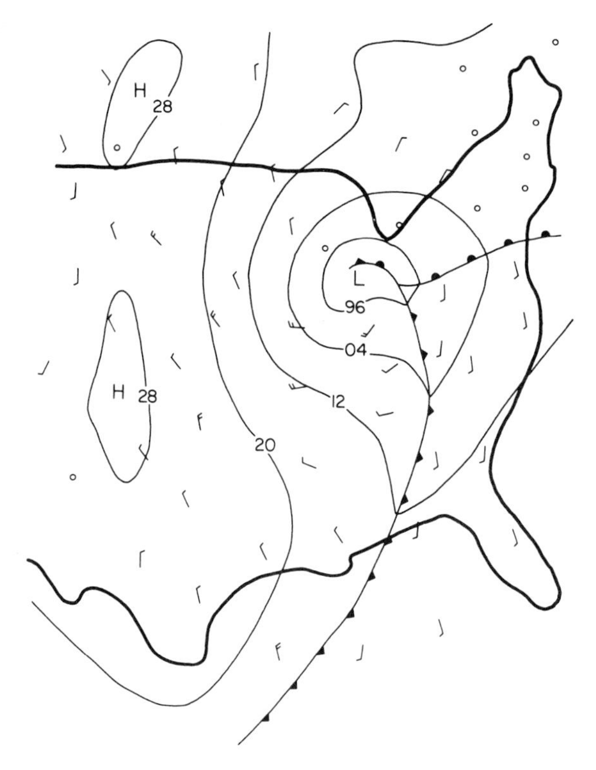

Fig. 18-2. Surface chart for 06Z Friday, November 20, 1981. Contours are isobars of atmospheric pressure; 12 is 1012 mb. Line with triangles, cold front; line with semicircles, warm front; line with both triangles and semicircles, an occluded front (a cold front that has caught up with a warm front). Wind direction is with the arrow; wind speed is 10 knots for 1 barb, 5 knots for one-half barb. Small station circles indicate calm. H, center of high pressure; L, center of low pressure.

This sinking air causes a subsidence inversion. Partially because of the subsiding vertical motion, the skies are usually clear, allowing maximum radiation—incoming during the day and outgoing at night—causing extremes of stability; there is instability during the day and stability at night, with frequent radiation inversions. Highs generally occupy large areas, and although they are transient, they are usually slow-moving. Winds over large areas are generally light; note the winds to the south of the high in the lower left corner of Fig. 18-2. Thus, the ventilation in the vicinity of high-pressure systems is generally much less than that of lows.

II. STAGNATION

At times the ventilation rate becomes very low. Such a lack of air motion usually occurs in the weak pressure gradient near the center of an anticyclone (i.e., of a high). If the high has a warm core, there is likely to be very little air movement near the center, i.e., stagnation. Under such circumstances, winds are very light. Skies are usually cloudless, contributing to the formation of surface-based radiation inversions at night. Although the clear skies contribute to instability in the daytime, the depth of the unstable layer (i.e., mixing height) may be severely limited due to the subsidence inversion over the high.

The mixing height at a given time may be estimated by use of the morning radiosonde ascent plotted on a thermodynamic chart. The surface temperature at the given time is plotted on the diagram. If a dry adiabat is drawn through this temperature, the height aboveground at the point where this dry adiabat intersects the morning sounding is the mixing height for that time. The mixing height for the time of maximum temperature is the maximum mixing height. Use of this sounding procedure provides an approximation because it assumes that there has been no significant advection since the time of the sounding.

III. METEOROLOGICAL CONDITIONS DURING HISTORIC POLLUTION EPISODES

A. Meuse Valley, Belgium

During the period December 1–5, 1930, an intense fog occupied the heavily industrialized Meuse Valley between Liege and Huy (about 24 km) in eastern Belgium (1). Several hundred persons had respiratory attacks primarily beginning on the 4th and 63 persons died on the 4th and 5th after a few hours of sickness. On December 6 the fog dissipated; the respiratory difficulties improved and, in general, rapidly ceased.

The fog began on December 1 under anticyclonic conditions. What little air motion occurred was from the east, causing air to drift upvalley, moving smoke from the city of Liege and the large factories southwest of it into the narrow valley. The valley sides extend to about 100 m, and the width of the valley is about 1 km. A temperature inversion extended from the ground to a height of about 90 m, transforming the valley essentially into a tunnel deeper than the height of the stacks in the valley, which were generally around 60 m. Much of the particulate matter was in the 2–6 μm range. The fog was cooled by radiation from the top and warmed by contact with the ground. This caused a gentle convection in the "tunnel," mixing the pollutants uniformly and resulting in nearly uniform temperature with height.

The symptoms of the first patients began on the afternoon of December 3 and seemed to occur simultaneously along the entire valley. Deaths took place only on December 4 and 5, with the majority at the Liege end of the valley. Those affected were primarily elderly persons who had lung or heart problems. However, some previously healthy persons were among the seriously ill. There were no measurements of pollutants during the episode, but the five Liege University professors who participated in the subsequent inquiry indicated that part of the sulfur dioxide was probably oxidized to sulfuric acid.

Roholm (2), in discussing the episode, noted that 15 of the 27 factories in the area were capable of releasing gaseous fluorine compounds and suggested that the release of these compounds was of significance.

During the 30 years prior to the episode, fogs lasting for more than 3 days had occurred only five times, always in the winter, in 1901, 1911, 1917, 1919, and 1930. Some respiratory problems were also noted in 1911. Industrial activity was at a low level in 1917 and 1919.

It is prophetic that Firket (1), in speaking about public anxiety about potential catastrophes, said, "This apprehension was quite justified, when we think that proportionately, the public services of London, for example, might be faced with the responsibility of 3200 sudden deaths if such phenomenon occurred there" (p. 1192). In 1952, such a catastrophe occurred (see Section III,C).

B. Donora, Pennsylvania

A severe episode of atmospheric pollution occurred in Donora, Pennsylvania, during the period Ocotober 25–31, 1948 (3). Twenty persons died, 17 of them within 14 hours on October 30.

During this period, a polar high-pressure area remained nearly stationary, with its center in the vicinity of northeastern Pennsylvania. This caused the regional winds, both at the ground and through the lowest layers, to be extremely light. Donora is southeast of Pittsburgh and is in the Monongahela River valley. Cold air accumulated in the bottom of the river valley and fog formed, which persisted past midday for 4 consecutive days. The top of the fog layer has a high albedo and reflects solar radiation, so that only part of the incoming radiation is available to heat the fog layer and eliminate it (Fig. 18-3, left). During the night, longwave radiation leaves the top of the fog layer, further cooling and stabilizing the layer (Fig. 18-3, right). Wind speeds at Donora were less than 3.1 m s^{-1} (7 mi h^{-1}) from the surface up to 1524 m (5000 ft) for 3 consecutive days, so that pollutants emitted into the air within the valley were not transported far from their point of emission. Maximum temperatures at Donora at an elevation of 232 m (760 ft) mean sea level were considerably lower than those at the Pittsburgh airport, elevation 381 m (1250 ft), indicating the extreme vertical stability of the atmosphere. In the vicinity of Donora there were sources of sulfur dioxide, particulate matter, and carbon monoxide. Previous recorded periods of stagnation had occurred in Donora in October 5–13, 1923 and October 7–18, 1938.

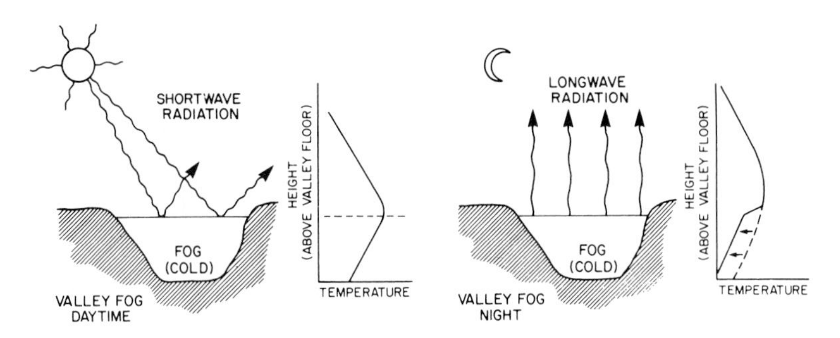

Fig. 18-3. Dense fog maintaining stability in a valley. It reflects shortwave radiation during the day and radiates heat from the top of the fog at night. Source: Adapted from Schrenk *et al.* (3).

C. London, England

A dense 4-day fog occurred in London and its surroundings during December 5–9, 1952 (4, 5). The fog began as the area came under the influence of an anticyclone approaching from the northwest early on December 5. This system became stationary, so that there was almost no wind until milder weather spread into the area from the west on December 9. Temperatures remained near freezing during the fog. The visibility was unusually restricted, with a 4-day average of less than 20 m over an area approximately 20 by 40 km and of less than 400 m over an area 100 by 60 km. The density of the fog was enhanced by the many small particles in the air available for condensation of fog droplets. The result was a very large number of very small fog droplets, more opaque and persistent than fog formed in cleaner air. The depth of the fog layer was somewhat variable, but was generally 100 m or less.

Measurements of particulate matter less than approximately 20 μm in diameter and of sulfur dioxide were made at 12 sites in the greater London area. The measurements were made by pumping air through a filter paper and then through a hydrogen peroxide solution. The smoke deposit on the filter was analyzed by reflectometer; the sulfur dioxide was determined by titrating the hydrogen peroxide with standard alkali, eliminating interference by carbon dioxide, Using thesampling procedure, sulfur dioxide existing as a gas and dissolved in fine fog droplets was measured. Any sulfur dioxide associated with larger fog droplets or adsorbed on particles collected on the filter would not be measured.

Smoke concentrations ranged from 0.3 to more than 4 mg m^{-3}. Daily means of the sampling stations are shown in Fig. 18-4. Sulfur dioxide measurements ranged from less than 0.1 ppm (260 $\mu g\ m^{-3}$) to 1.34 ppm (3484 $\mu g\ m^{-3}$). Also, 4 of the 11 stations had at least one daily value in excess of 1 ppm, and 9 of the 11 stations had at least one daily value in excess

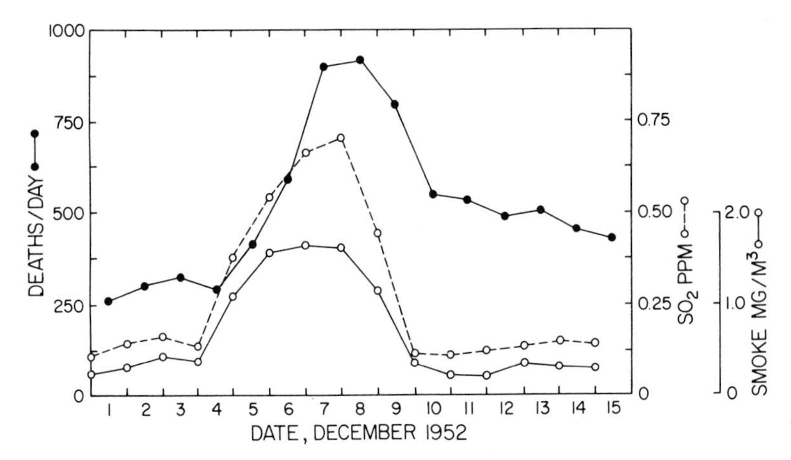

Fig. 18-4. Daily air pollution (SO_2 and smoke) and deaths during the 1952 London episode. Source: Adapted from Wilkins (4).

of 0.5 ppm. The U.S. primary standard for sulfur dioxide is a maximum 24-hr concentration of 365 $\mu g\ m^{-3}$ (0.14 ppm), not to be exceeded more than once per year. Daily means are shown in Fig. 18-4. The smoke and SO_2 means rose and later decreased in parallel. Daily concentrations of smoke averaged over all stations rose to about five times normal and sulfur dioxide to about six times normal, peaking on December 7 and 8, respectively. In addition to the daily measurements at 12 sites, monthly measurements at 117 sites were made using lead peroxide candles. This allowed determination of the spatial pattern. The December 1952 concentrations were about 50% higher than those of December 1951.

From the commencement of the fog and low visibility, many people experienced difficulty breathing, the effects occurring more or less simultaneously over a large area of hundreds of square kilometers. The rise in the number of deaths (Fig. 18-4) paralleled the mean daily smoke and sulfur dioxide concentrations; daily deaths reached a peak on December 8 and 9, with many of them related to respiratory troubles. Although the deaths decreased when the concentrations decreased, the deaths per day remained considerably above the pre-episode level for some days. Would most of the persons who died have died soon afterward anyway? If this were the case, a below-normal death rate would have occurred following the episode. This situation did not seem to exist, but detailed analysis was complicated by increased deaths in January and February 1953 which were attributed primarily to an influenza outbreak.

Those who analyzed these excess deaths (the number of deaths above the normal number for each calendar day) believed that the level of sulfur dioxide was not near the toxic limit of 10 ppm necessary to affect healthy persons. They attributed the deaths to the synergistic effect of fine particles and sulfur dioxide combined. They believed that considerable sulfuric acid mist was formed from the oxidation of sulfur dioxide, but since no measurements were made, its amount was speculative.

D. Similarities of the Three Episodes

In the Meuse Valley, Donora, and London episodes, the areas were influenced by high pressure with nearly nonexistent surface air motion. Surface inversions caused the condensation of fog, which, once formed, persisted throughout the day, even during midafternoon. In each case the fog layer was relatively shallow, extending only about 100 m. The persistence of the fog past the third day and the lack of any air transport out of the region, as well as the existence of considerable emissions of pollutants, seem to separate these episodes from more common meteorological occurrences. Both the Meuse Valley and Donora had topography constraining the volume in which the pollutants were confined. This constraint apparently resulted from the lack of any transport wind in the London 1952 episode. Measurements of pollutant concentrations were made only in London.

E. Other Episodes

A number of somewhat less severe episodes are discussed in Goldsmith and Friberg (6). Mention of the more important ones follows.

An air pollution episode responsible for approximately 300 excess deaths occurred in London between November 26 and December 1, 1948. Concentrations of smoke and sulfur dioxide were 50–70% of the values during the 1952 episode.

An accident complicated by fog, weak winds, and a surface inversion occurred in Poza Rica, Mexico, in the early morning of November 24, 1950, when hydrogen sulfide was released from a plant for the recovery of sulfur from natural gas. There were 22 deaths, and 320 persons were hospitalized.

In November and December 1962, a number of air pollution episodes occurred in the Northern Hemisphere. In London a fog occurred during the period December 3–7, with sulfur dioxide as high as during the 1952 episode, but with particulate concentrations considerably lower due to the partial implementation of the 1956 British Clean Air Act. Excess deaths numbered 340. High pollution levels were measured in the eastern United States between November 27 and December 5, 1962, notably in Washington, DC, Philadelphia, New York, and Cincinnati. Between December 2 and 7 elevated pollution levels were found in Rotterdam; Hamburg, Frankfurt, and the Ruhr area; Paris; and Prague. Pollution levels were high in Osaka between December 7 and 10, and mortality studies, which were under way, indicated 60 excess deaths.

F. Air Pollution Emergencies

Government authorities increasingly are facing emergencies that may require lifesaving decisions to be made rapidly by those on the scene. Of increasing frequency are transportation accidents involving the movement of volatile hazardous materials. A railroad derailment accident of a tank car of liquefied chlorine on February 26, 1978, at Youngtown, Florida, in which seven people died, and an accident in Houston, Texas, involving a truck carrying anhydrous ammonia on May 11, 1976, which also claimed seven lives, are examples. Two potentially dangerous situations involved barges with tanks of chlorine: one which sank in the lower Mississippi River and another which came adrift and came to rest on the Ohio River dam at Louisville, Kentucky; neither resulted in release of material.

Releases of radioactive materials from nuclear power plants have occurred, as at Three-Mile Island, Pennsylvania. In such situations, releases may be sufficient to require evacuation of residents.

Bhopal, India—On December 2, 1984 the contents of a methyl isocyanate (MIC) storage tank at the Union Carbide India plant in Bhopal became hot. Pressure in the tank became high. Nearly everything that could go wrong did. The refrigerator unit for the tank, which would have slowed the reactions, was turned off. After midnight, when the release valve blew, the vent gas scrubber that was to neutralize the gas with caustic soda failed to work. The flare tower, which would have burned the gas to harmless byproducts, was down for repairs. As a result many tons of MIC were released from the tank. The gas spread as a foglike cloud over a large, highly populated area to the south and east of the plant (7). The number of fatalitites was in excess of 2000 with thousands of others injured. Although little is available in the way of meteorological measurements, it is assumed that winds were quite light and that the atmosphere at this time of day was relatively stable.

Chernobyl, USSR—On April 26, 1986, shortly after midnight local time, a serious accident occurred at a nuclear power plant in Chernobyl in the Ukraine. It is estimated that 4% of the core inventory was released between April 26 and May 6. Quantities of Cs-137 (cesium) and I-131 (iodine) were released and transported, resulting in contamination, primarily by wet deposition of cesium, in Finland, northern Sweden and Norway, the Alps, and the northern parts of Greece. Because of temperatures of several thousand K during the explosionlike release, the resulting pollutant cloud is assumed to have reached heights of 2000 m or more. The estimated southeast winds at plume level initially moved the plume toward Finland, northern Sweden, and northern Norway. As winds at plume level gradually turned more easterly and finally north and northwesterly, contaminated air affected the region of the Alps and northern Greece. A number of investigators, including Hass et al. (8), modeled the long-range transport including wet and dry removal processes. These attempts were considered quite successful, as radioactivity measurements provided some confirmation of the regions affected. Elevated levels of radioactivity were measured throughout the Northern Hemisphere. Because of the half-life of about 30 years for Cs-137, the contamination will endure.

In such emergencies, it is most important to know the local wind direction at the accident site, so that the area that should be immediately evacuated can be determined. The next important factor is the wind speed, so that the travel time to various areas can be determined, again primarily for evacuation purposes. Both of these can be estimated on-site by simple means such as watching the drift of cigarette smoke. It would be well to keep in mind that wind speeds are higher above ground and that wind direction is usually different.

As evacuation is taking place, it is important to determine whether meteorological events will cause a wind direction shift later on, requiring a change in the evacuation scenario. Particularly in coastal areas, or areas of significant terrain, authorities should be alert to a possible change in wind direction in going from night to day or vice versa. Useful advice may be obtained from the nearest weather forecaster, although accurate forecasting of wind direction for specific locations is not easy.

If the situation is one of potential rather than current release, specific concentrations at various distances and localities may be estimated for various conditions.

IV. EFFECTS OF POLLUTION ON THE ATMOSPHERE

Pollutant effects on the atmosphere include increased particulate matter, which decreases visibility and inhibits incoming solar radiation, and increased gaseous pollutant concentrations, which absorb longwave radiation and increase surface temperatures. For a detailed discussion of visibility effects, see Chapter 10.

A. Turbidity

The attenuation of solar radiation has been studied by McCormick and his associates (9, 10) utilizing the Voltz sun photometer, which uses measurements at a wavelength of 0.5 μm. The ratio of ground-level solar intensity at 0.5 μm to extraterrestrial solar intensity can be as high as 0.5 in clean atmospheres but can drop to 0.2–0.3 in polluted areas, indicating that ground-level solar intensity can be decreased as much as 50% by pollution in the air. By making measurements using aircraft at various heights, the vertical extent of the polluted air can be determined. The turbidity coefficient can also be derived from the measurements and used to estimate the aerosol loading of the atmosphere. By assuming a particle size distribution in the size range 0.1–10 μm and a particle density, the total number of particles can be estimated. The mass loading per cubic meter can also be estimated. Because of the reasonable cost and simplicity of the sun photometer, it is useful for making comparative measurements around the world.

B. Precipitation

Depending on its concentration, pollution can have opposite effects on the precipitation process. Addition of a few particles that act as ice nuclei can cause ice particles to grow at the expense of supercooled water droplets, resulting in particles large enough to fall as precipitation. An example of this is commercial cloud seeding, with silver iodide particles released from aircraft to induce rain. If too many such particles are added, none of them will grow sufficiently to cause precipitation. Therefore, the effects of pollution on precipitation are not at all straightforward.

There have been some indications, although controversial, of increased precipitation downwind of major metropolitan areas. Urban addition of nuclei and moisture and urban enhancement of vertical motion due to increased roughness and the urban heat island effect have been suggested as possible causes.

C. Fogs

As mentioned in the previous section, the increased number of nuclei in polluted urban atmospheres can cause dense persistent fogs due to the many small droplets formed. Fog formation is very dependent on humidity and, in some situations, humidity is increased by release of moisture from industrial processes. Low atmospheric moisture content can also occur, especially in urban areas; two causes are lack of vegetation and rapid runoff of rainwater through storm sewers. Also, slightly higher temperatures in urban areas lower the relative humidity.

D. Solar Radiation

In the early part of this century, the loss of ultraviolet light in some metropolitan areas due to heavy coal smoke was of concern because of the resulting decrease in the production of natural vitamin D which causes the disease rickets. Recently, measurements in Los Angeles smog have revealed much greater decreases in ultraviolet than visible light. This is due to both absorption by ozone of wavelengths less than 0.32 μm and absorption by nitrogen dioxide in the 0.36–0.4 μm range. Heavy smog has decreased ultraviolet radiation by as much as 90%.

V. REMOVAL MECHANISMS

Except for fine particulate matter (0.2 μm or less), which may remain airborne for long periods of time, and gases such as carbon monoxide, which do not react readily, most airborne pollutants are eventually removed from the atmosphere by sedimentation, reaction, or dry or wet deposition.

A. Sedimentation (Settling by Gravity)

Particles less than about 20 μm are treated as dispersing as gases, and effects due to their fall velocity are generally ignored. Particles greater than about 20 μm have appreciable settling velocities. The fall velocity of smooth spheres as a function of particle size has been plotted (Fig. 18-5) by Hanna *et al.* (11). Particles in the range 20–100 μm are assumed to disperse approximately as gases, but with their centroid moving downward in the atmosphere according to the fall velocity. This can be accounted for by subtracting $v_g t$ from the effective height of release, where v_g is the gravitational fall velocity of the particles and t, in seconds, is x/u, where x is downwind distance from the source in m and u is wind speed. This is called the *tilted plume model.* The model may be modified to decrease the strength of the

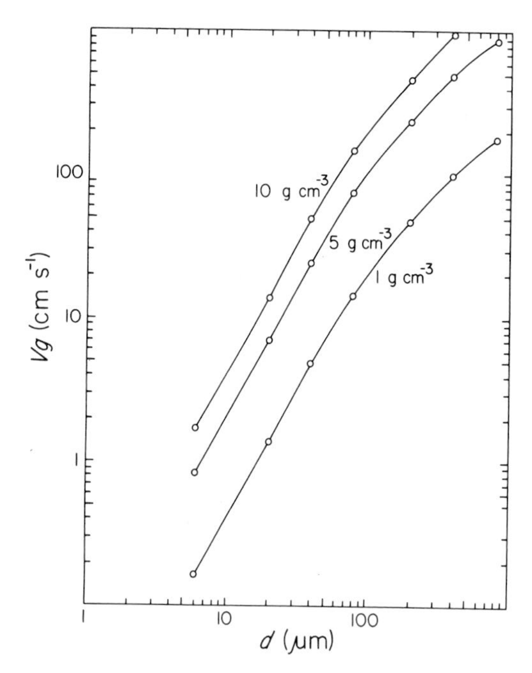

Fig. 18-5. Fall velocity of spherical particles as a function of particle diameter and density. Source: Adapted from Hanna et al. (11).

source with distance from the source to account for the particles removed by deposition.

For 20–100 μm particles, the deposition w on the ground is

$$w = v_g \chi(x, y, z) \tag{18-1}$$

where the air concentration χ is evaluated for a height above ground z of about 1 m.

Particles larger than 100 μm fall through the atmosphere so rapidly that turbulence has less chance to act upon and disperse them. The trajectories of such particles are treated by a ballistic approach.

B. Reaction (Transformation)

Transformations due to chemical reactions throughout the plume are frequently treated as exponential losses with time. The concentration $\chi(t)$ at travel time t when pollutant loss is considered compared to the concentration χ at the same position with no loss is

$$\chi(t)/\chi = \exp - (0.693t/L) \tag{18-2}$$

where L is the half-life of the pollutant in seconds. The half-life is the time required to lose 50% of the pollutant.

C. Dry Deposition

Although it does not physically explain the nature of the removal process, deposition velocity has been used to account for removal due to impaction with vegetation near the surface or for chemical reactions with the surface. McMahon and Denison (12) gave many deposition velocities in their review paper. Examples (in cm s^{-1}) are sulfur dioxide, 0.5–1.2; ozone, 0.1–2.0; iodine, 0.7–2.8; and carbon dioxide, negligible.

D. Wet Deposition

Scavenging of particles or gases may take place in clouds (rainout) by cloud droplets or below clouds(washout) by precipitation. A scavenging ratio or washout ratio W can be defined as

$$W = k\rho/\chi \tag{18-3}$$

where k is concentration of the contaminant in precipitation in $\mu g\ g^{-1}$; ρ is the density of air, approximately 1200 g m^{-3}; and χ is the concentration, $\mu g\ m^{-3}$, of the pollutant in the air prior to scavenging. McMahon and Denison (12) gave a table of field observations of washout ratios. The values for various pollutants range from less than 100 to more than 4000. These values are a function of particle size and rainfall intensity, generally decreasing with the latter and increasing with the former.

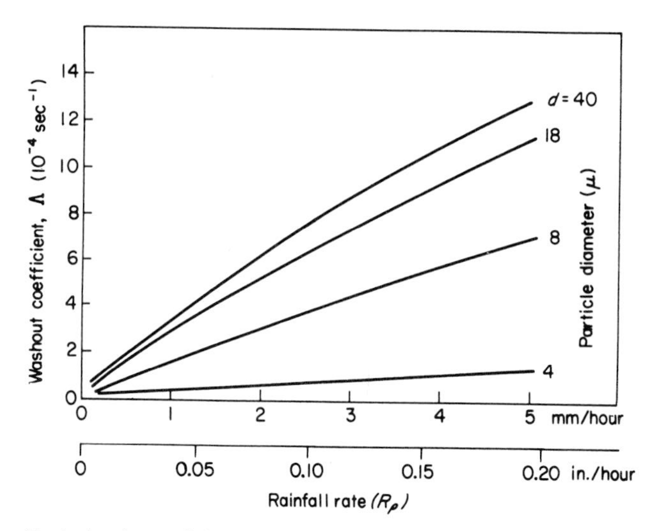

Fig. 18-6. Typical values of the washout coefficient as a function of rainfall rate and particle diameter. Source: After Engelmann (13).

Scavenging may also be considered as an exponential decay process:

$$\chi(t) = \chi(0)e^{-\Lambda t} \tag{18-4}$$

where $\chi(t)$ is the concentration in $\mu g\ m^{-3}$ at time t in seconds, $\chi(0)$ is the concentration at time 0, and Λ is the scavenging or washout coefficient, s^{-1}. Figure 18-6, after Engelmann (13), gives the washout coefficient as a function of particle diameter and rainfall rate. McMahon and Denison (12) give a table of field measurements of scavenging coefficients. This same concept can be applied to gaseous pollutants. Fewer data are available for gases. Values ranging from 0.4×10^{-5} to 6×10^{-5} for SO_2 are given by McMahon and Denison (12) and compare reasonably well with an equation for SO_2 by Chamberlain (14):

$$\Lambda = 10 \times 10^{-5}\ J^{0.53} \tag{18-5}$$

where J is rainfall intensity in mm h^{-1}.

REFERENCES

1. Firket, J., *Trans. Faraday Soc.* **32,** 1192–1197 (1936).
2. Roholm, K., *J. Ind. Hyg.* **19,** 126–137 (1937).
3. Schrenk, H. H., Heimann, H., Clayton, G. D., Gafafer, W. M., and Wexler, H., "Air Pollution in Donora, Pa," Public Health Bulletin 306. U.S. Public Health Service, Washington, DC, 1949.

4. Wilkins, E. T., *J. R. Sanit. Inst.* **74,** 1–15 (1954).
5. Wilkins, E. T., *Q. J. R. Meteorol. Soc.* **80,** 267–271 (1954).
6. Goldsmith, J. R., and Friberg, L. T., Effects of air pollution on human health, *in* "Air Pollution," 3rd ed., Vol. II, "The Effects of Air Pollution" (A. C. Stern, ed.). Academic Press, New York, 1976, pp. 457–610.
7. Heylin, M., *Chem. Eng. News* (Feb. 11, 1985).
8. Hass, H., Memmesheimer, M., Jakobs, H. J., Laube, M. and Ebel, A., *Atmos. Environ.* **24A,** 673–692 (1990).
9. McCormick, R. A., and Baulch, D. M., *J. Air Pollut. Control Assoc.* **12,** 492–496 (1962).
10. McCormick, R. A., and Kurfis, K. R., *Q. J. R. Meteorol. Soc.* **92,** 392–396 (1966).
11. Hanna, S. R., Briggs, G. A., and Hosker, R. P., Jr., "Handbook on Atmospheric Diffusion," DOE/TIC-11223. Technical Information Center, U.S. Department of Energy, Oak Ridge, TN, 1982.
12. McMahon, T. A., and Denison, P. J., *Atmos. Environ.* **13,** 571–585 (1979).
13. Engelmann, R. J., The calculation of precipitation scavenging, *in* "Meteorology and Atomic Energy—1968" (D. Slade, ed.), TID-24190. U.S. Atomic Energy Commission, Oak Ridge, TN, 1968, pp. 208–220.
14. Chamberlain, A. C., "Aspects of Travel and Deposition of Aerosol and Vapour Clouds," Atomic Energy Research Establishment HP/R-1261. Her Majesty's Stationery Office, London, 1953.

SUGGESTED READING

ApSimon, H. M., and Wilson, J. J. N., Modeling atmospheric dispersal of the Chernobyl release across Europe. *Boundary-Layer Meteorol.* **41,** 123–133 (1987).

Godish, T., "Air Quality," 2nd ed. Lewis Publishers, Boca Raton, FL, 1991.

Hanna, S. R., and Drivas, P. J., "Guidelines for Use of Vapor Cloud Dispersion Models." American Institute of Chemical Engineers. New York, 1987.

Knap, A. H., (ed.), "The Long-Range Atmospheric Transport of Natural and Contaminant Substances." Kluwer Academic Press, Hingham, MA, 1989.

Kramer, M. L., and Porch, W. M., "Meteorological Aspects of Emergency Response." American Meteorological Society, Boston, 1990.

Puttock, J. S. (ed.), "Stably Stratified Flow and Dense Gas Dispersion." Oxford University Press, New York, 1988.

Sandroni, S. (ed.), "Regional and Long-Range Transport of Air Pollution." Elsevier Science Publishers, New York, 1987.

Scorer, R. S., "Air Pollution." Pergamon, Oxford, 1968.

Seinfeld, J. H., "Air Pollution—Physical and Chemical Fundamentals." McGraw-Hill, New York, 1975.

QUESTIONS

1. Characterize the conditions typical of low-pressure systems, particularly as they relate to ventilation.
2. Characterize the conditions typical of high- pressure systems, particularly as they relate to ventilation.

3. What atmospheric characteristics are usually associated with stagnating high-pressure systems?
4. What factors contribute to a high mixing height?
5. Discuss the simularities of the three major episodes of pollution (Meuse Valley, Donora, and London).
6. A railroad tank car has derailed and overturned, and some material is leaking out and apparently evaporating. The car is labeled "Toxic." In order to take appropriate emergency action, which meteorological factors would you consider and how would you assess them?
7. In addition to air pollutants, what meteorological factor has a profound effect on decreasing visibility, and what is the approximate threshold of its influence?
8. What pollution factors may affect precipitation?
9. What is the approximate lowering of the centroid of a dispersing cloud of particles at 2 km from the source whose mass medium diameter is 30 μm and whose particle density is 1 g cm^{-3} in a 5 m s^{-1} wind?
10. Prior to the onset of rain at the rate of 2.5 mm h^{-1}, the average concentration of 10-μm particles in a pollutant plume is 80 μg m^{-3}. What is the average concentration after 30 min of rain at this rate?

19

Transport and Dispersion of Air Pollutants

I. WIND VELOCITY

A. Wind Direction

The initial direction of transport of pollutants from their source is determined by the wind direction at the source. Air pollutant concentrations from point sources are probably more sensitive to wind direction than any other parameter. If the wind is blowing directly toward a receptor (a location receiving transported pollutants), a shift in direction of as little as 5° (the approximate accuracy of a wind direction measurement) causes concentrations at the receptor to drop about 10% under unstabie conditions, about 50% under neutral conditions, and about 90% under stable conditions. The direction of plume transport is very important in source impact assessment where there are sensitive receptors or two or more sources and in trying to assess the performance of a model through comparison of measured air quality with model estimates.

There is normally considerable wind direction shear (change of direction) with height, especially near the ground. Although surface friction causes the wind to shift clockwise (veer) with height near the ground, the horizontal thermal structure of the atmosphere may exert a dominating influence at higher altitudes, such that the wind will shift counterclockwise (back) with additional height. Cold air advection in an air layer will cause the wind to back with height through that layer. Warm air advection will cause veering with height.

B. Wind Speed

Wind speed generally increases with height. A number of expressions describe the variation of wind speed in the surface boundary layer. A power law profile has frequently been used in air pollution work:

$$u(z) = u(z_a)\ (z/z_a)^p \tag{19-1}$$

where $u(z)$ is the wind speed at height z, $u(z_a)$ the wind speed at the anemometer measurement height z_a, and p an exponent varying from about 0.1 to 0.4. Figure 19-1 gives the measured wind speed variation with height for specific instances for five locations. The result of using the power law profile (Eq. 19-1) is also shown (open circles and dashed lines) using a value of p of 1/7. It should be noted that the power law wind profiles do not necessarily represent the data well. The exponent actually varies with atmospheric stability, surface roughness, and depth of the layer (1).

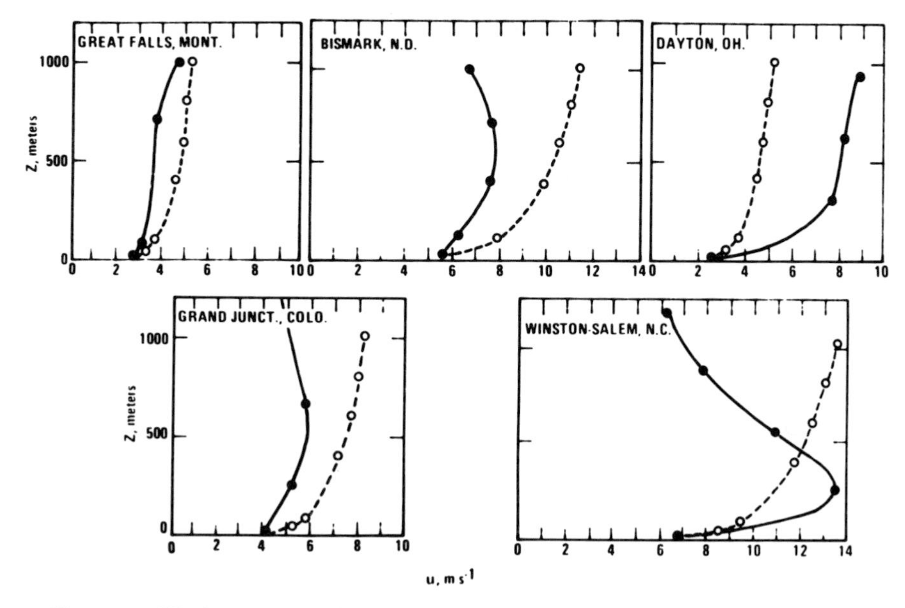

Fig. 19-1. Wind variation with height—measured (solid lines) and one-seventh power law (dashed lines).

One of the effects of wind speed is to dilute continuously released pollutants at the point of emission. Whether a source is at the surface or elevated, this dilution takes place in the direction of plume transport. Figure 19-2 shows this effect of wind speed for an elevated source with an emission of 6 mass units per second. For a wind speed of 6 m s^{-1}, there is 1 unit between the vertical parallel planes 1 m apart. When the wind is slowed to 2 m s^{-1}, there are 3 units between those same vertical parallel planes 1 m apart. Note that this dilution by the wind takes place at the point of emission. Because of this, wind speeds used in estimating plume dispersion are generally estimated at stack top.

Wind speed also affects the travel time from source to receptor; halving of the wind speed will double the travel time. For buoyant sources, plume rise is affected by wind speed; the stronger the wind, the lower the plume. Specific equations for estimating plume rise are presented in Chapter 20.

II. TURBULENCE

Turbulence is highly irregular motion of the wind. The atmosphere does not flow smoothly but has seemingly random, rapidly varying erratic motions. This uneven flow superimposed on the mean flow has swirls or

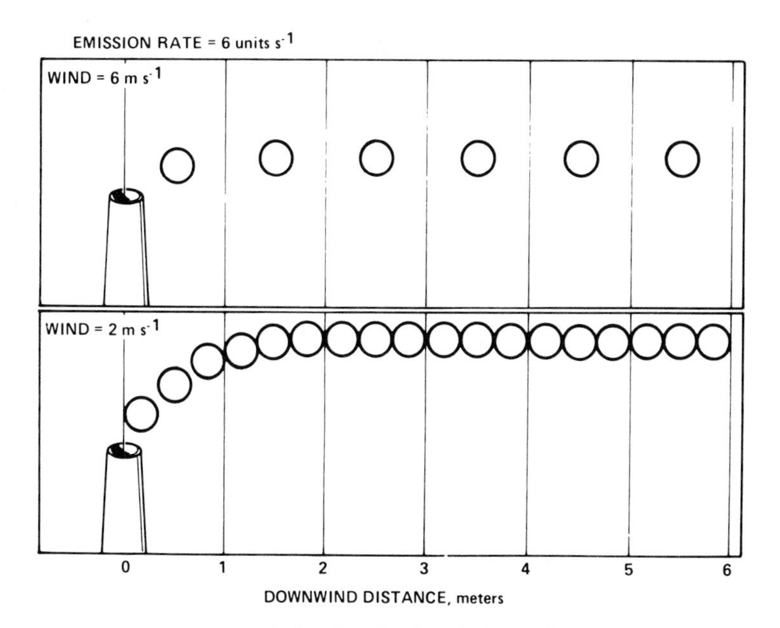

Fig. 19-2. Dilution by wind speed.

eddies in a wide range of sizes. The energy cascades through the eddy sizes, which are described by L. F. Richardson in verse:

Big whirls have little whirls that feed on their velocity
And little whirls have lesser whirls and so on to viscosity.

There are basically two different causes of turbulent eddies. Eddies set in motion by air moving past objects are the result of mechanical turbulence. Parcels of superheated air rising from the heated earth's surface, and the slower descent of a larger portion of the atmosphere surrounding these more rapidly rising parcels, result in thermal turbulence. The size and, hence, the scale of the eddies caused by thermal turbulence are larger than those of the eddies caused by mechanical turbulence.

The manifestation of turbulent eddies is gustiness and is displayed in the fluctuations seen on a continuous record of wind or temperature. Figure 19-3 displays wind direction traces during (a) mechanical and (b) thermal turbulence. Fluctuations due to mechanical turbulence tend to be quite regular; that is, eddies of nearly constant size are generated. The eddies generated by thermal turbulence are both larger and more variable in size than those due to mechanical turbulence.

The most important mixing process in the atmosphere which causes the dispersion of air pollutants is called *eddy diffusion*. The atmospheric eddies cause a breaking apart of atmospheric parcels which mixes polluted air with relatively unpolluted air, causing polluted air at lower and lower concentrations to occupy successively larger volumes of air. Eddy or turbulent dispersion is most efficient when the scale of the eddy is similar to that of the pollutant puff or plume being diluted. Smaller eddies are effective only at tearing at the edges of the pollutant mass. On the other hand, larger eddies will usually only transport the mass of polluted air as a whole.

The size and influence of eddies on the vertical expansion of continuous plumes have been related to vertical temperature structure (3). Three ap-

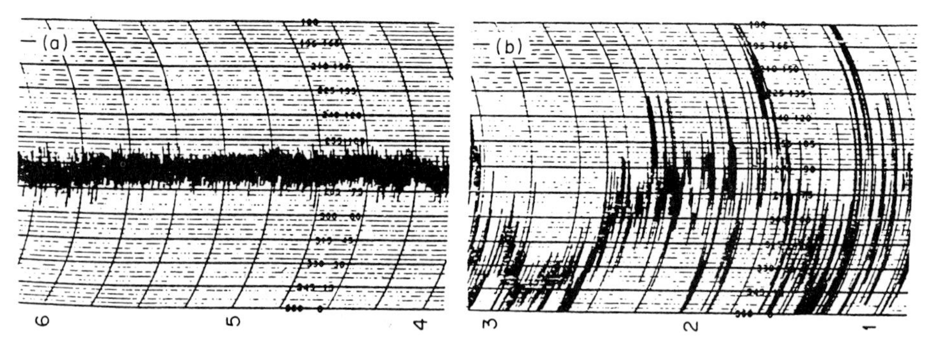

Fig. 19-3. Examples of turbulence on wind direction records: (a) mechanical, (b) thermal. Source: From Smith (2).

pearances of instantaneous plumes related to specific lapse rates and three appearances of instantaneous plumes related to combinations of lapse rates are shown in Fig. 19-4. Strong lapse is decrease of temperature with height in excess of the adiabatic lapse rate. Weak lapse is decrease of temperature with height at a rate between the dry adiabatic rate and the isothermal condition (no change of temperature with height).

A number of methods have been used to measure or estimate the level of turbulence in the atmosphere and, in turn, its dispersive ability. These

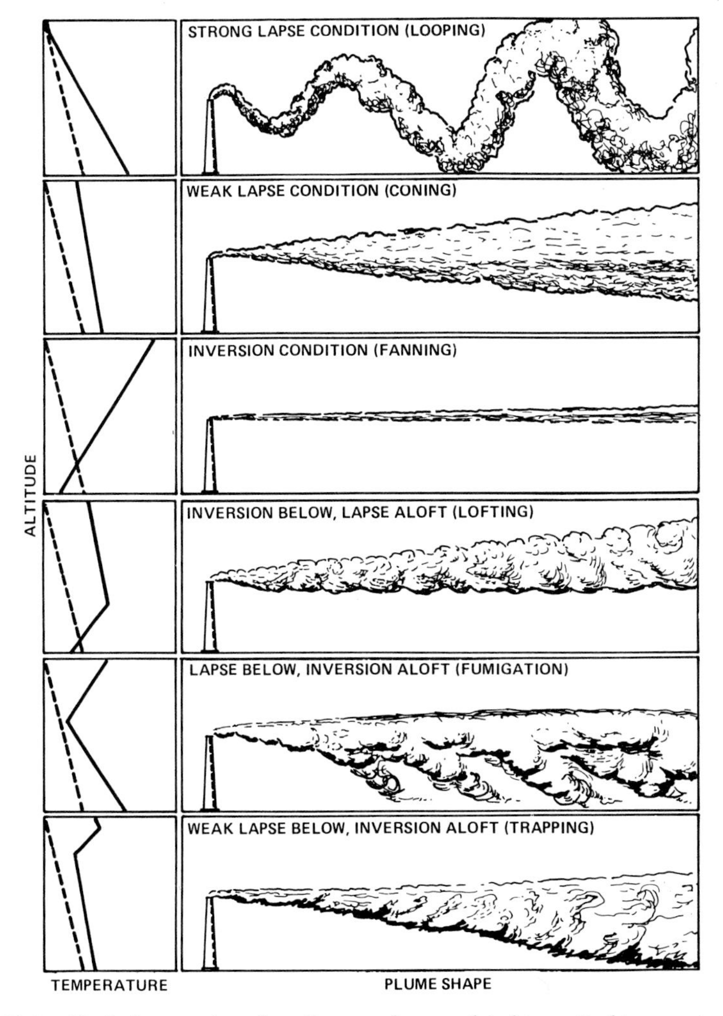

Fig. 19-4. Vertical expansion of continuous plumes related to vertical temperature structure. The dashed lines correspond to the dry adiabatic lapse rate for reference.

methods vary from direct measurement of wind fluctuations by sensitive wind measurement systems; to classification based on the appearance of the chart record of the wind direction trace; to classification of atmospheric stability indirectly by wind speed and estimates of insolation (incoming solar radiation) or outgoing longwave radiation. Details of these methods are given in the next section.

III. ESTIMATING CONCENTRATIONS FROM POINT SOURCES

The principal framework of empirical equations which form a basis for estimating concentrations from point sources is commonly referred to as the *Gaussian plume model.* Employing a three-dimensional axis system of downwind, crosswind, and vertical with the origin at the ground, it assumes that concentrations from a continuously emitting plume are proportional to the emission rate, that these concentrations are diluted by the wind at the point of emission at a rate inversely proportional to the wind speed, and that the time- averaged (about 1 h) pollutant concentrations crosswind and vertically near the source are well described by Gaussian or normal (bell-shaped) distributions. The standard deviations of plume concentration in these two directions are empirically related to the levels of turbulence in the atmosphere and increase with distance from the source.

In its simplest form, the Gaussian model assumes that the pollutant does not undergo chemical reactions or other removal processes in traveling away from the source and that pollutant material reaching the ground or the top of the mixing height as the plume grows is eddy-reflected back toward the plume centerline.

A. The Gaussian Equations

All three of the Gaussian equations (19-2 through 19-4) are based on a coordinate scheme with the origin at the ground, x downwind from the source, y crosswind, and z vertical. The normal vertical distribution near the source is modified at greater downwind distances by eddy reflection at the ground and, when the mixing height is low, by eddy reflection at the mixing height. *Eddy reflection* refers to the movement away ("reflection') of circular eddies of air from the earth's surface, since they cannot penetrate that surface. Cross sections in the horizontal and vertical at two downwind distances through a plume from a 20-m-high source with an additional 20 m of plume rise (to result in a 40-m effective height) are shown in Fig. 19-5. The following symbols are used:

χ, concentration, g m^{-3}
Q, emission rate, g s^{-1}

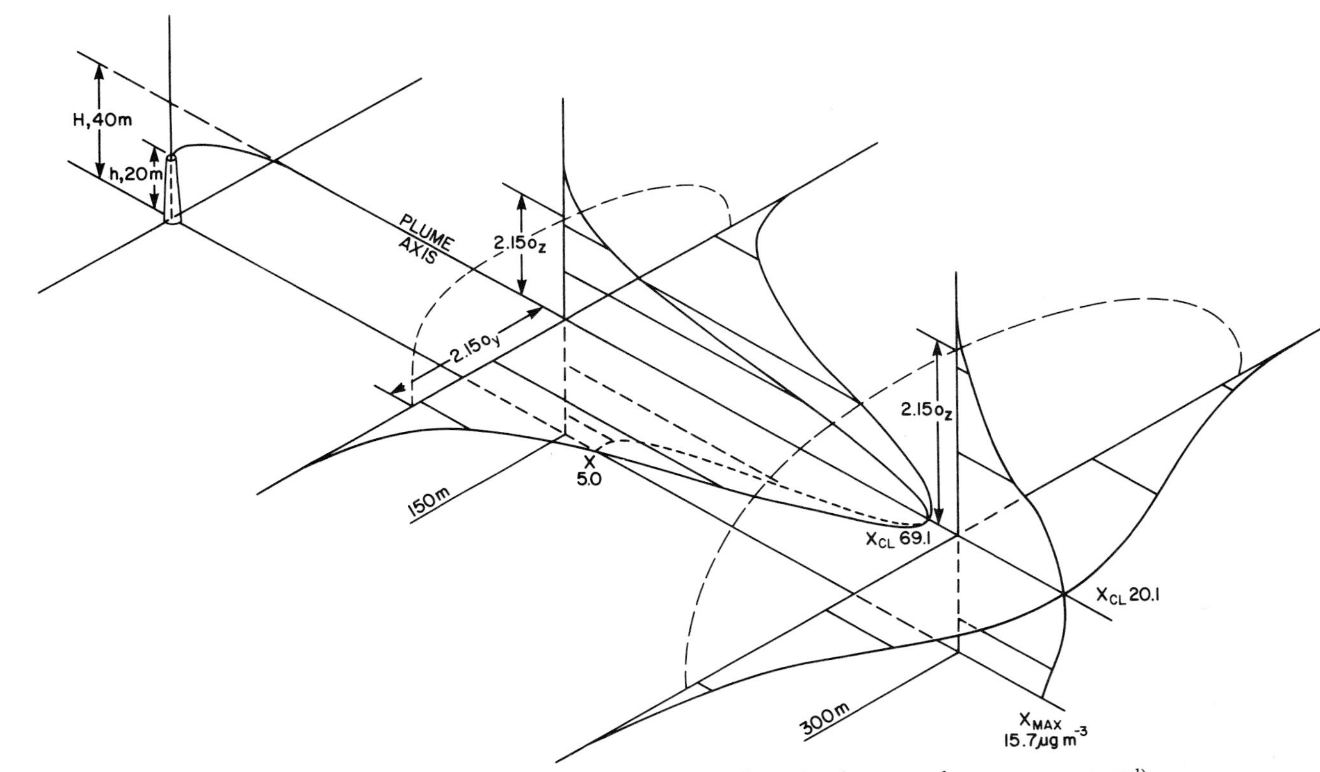

Fig. 19-5. Two cross sections through a Gaussian plume (total mass under curves conserved).

u, wind speed, m s^{-1}
σ_y, standard deviation of horizontal distribution of plume concentration (evaluated at the downwind distance x and for the appropriate stability), m
σ_z, standard deviation of vertical distribution of plume concentration (evaluated at the downwind distance x and for the appropriate stability), m
L, mixing height, m
h, physical stack height, m
H, effective height of emission, m
x, downwind distance, m
y, crosswind distance, m
z, receptor height above ground, m

The concentration χ resulting at a receptor at (x, y, z) from a point source located at $(0, 0, H)$ is given by one of the three following equations. (Methods for obtaining values for the dispersion parameters σ_y and σ_z in the following equations are discussed later in this chapter.)

For stable conditions or unlimited vertical mixing (a very high mixing height), use

$$\chi = Q(1/u)\{g_1/[(2\pi)^{0.5}\,\sigma_y]\}\{g_2/[(2\pi)^{0.5}\,\sigma_z]\} \tag{19-2}$$

where

$$g_1 = \exp(-0.5y^2/\sigma_y^2)$$
$$g_2 = \exp[-0.5(H - z)^2/\sigma_z^2] + \exp[-0.5(H + z)^2/\sigma_z^2]$$

Note that if $y = 0$, or $z = 0$, or both z and H are 0, this equation is greatly simplified. For locations in the vertical plane containing the plume centerline, $y = 0$ and $g_1 = 1$.

For unstable or neutral conditions, where σ_z is greater than $1.6L$, use

$$\chi = Q(1/u)\{g_1/[(2\pi)^{0.5}\,\sigma_y]\}(1/L) \tag{19-3}$$

For these large σ_z values, eddy reflection has occurred repeatedly both at the ground and at the mixing height, so that the vertical expanse of the plume has been uniformly mixed through the mixing height, i.e., $1/L$.

For unstable or neutral conditions, where σ_z is less than $1.6L$, use the following equation provided that both H and z are less than L:

$$\chi = Q(1/u)\{g_1/[(2\pi)^{0.5}\sigma_y]\}\{g_3/[(2\pi)^{0.5}\,\sigma_z]\} \tag{19-4}$$

where

$$g_3 = \sum_{N=-\infty}^{\infty} \{\exp[-0.5(H - z + 2NL)^2/\sigma_z^2] + \exp[-0.5(H + z + 2NL)^2/\sigma_z^2]\}$$

This infinite series converges rapidly, and evaluation with N varying from -4 to $+4$ is usually sufficient . These equations are used when evaluating by computer, as the series g_3 can easily be evaluated.

When estimates are being made by hand calculations, Eq. (19-2) is frequently applied until $\sigma_z = 0.8L$. This will cause an inflection point in a plot of concentrations with distance.

By adding Eq. (19-4), which includes multiple eddy reflections, and changing the criteria for the use of Eq. (19-3) to situations in which σ_z is evaluated as being greater than $1.6L$, a smooth transition to uniform mixing, Eq. (19-3), is achieved regardless of source or receptor height. By differentiating Eq. (19-2) and setting it equal to zero, an equation for maximum concentration can be derived:

$$\chi_{\max} = \frac{2Q}{\pi u e H^2} \frac{\sigma_z}{\sigma_y} \tag{19-5}$$

and the distance to maximum concentration is at the distance where $\sigma_z = H/(2)^{0.5}$. This equation is strictly correct only if the σ_z/σ_y ratio is constant with distance.

B. Alternate Coordinate Systems for the Gaussian Equations

For estimating concentrations from more than one source, it is convenient to use map coordinates for locations. Gifford (4) has pointed out that the resulting calculated concentration is the same whether the preceding axis system is used or whether an origin is placed at the ground beneath the receptor, with the x-axis oriented upwind, the z-axis remaining vertical, and the y-axis crosswind.

This latter axis system is convenient in assessing the total concentration at a receptor from more than one source provided that the wind direction can be assumed to be the same over the area containing the receptor and the sources of interest.

Given an east–north coordinate system (R, S) the upward distance x and the crosswind distance y of a point source from a receptor are given by

$$x = (S_p - S_r) \cos \theta + (R_p - R_r) \sin \theta \tag{19-6}$$

$$y = (S_p - S_r) \sin \theta - (R_p - R_r) \cos \theta \tag{19-7}$$

where R_p, S_p are the coordinates of the point source; R_r, S_r are the coordinates of the receptor; and θ is the wind direction (the direction from which the wind blows). The units of x and y will be the same as those of the coordinate system R, S. In order to determine plume dispersion parameters, distances must be in kilometers or meters. A conversion may be required to convert x and y above to the appropriate units.

C. Determination of Dispersion Parameters

1. *By Direct Measurements of Wind Fluctuations*

Hay and Pasquill (5) and Cramer (6, 7) have suggested the use of fluctuation statistics from fixed wind systems to estimate the dispersion taking place within pollutant plumes over finite release times. The equation used for calculating the variance of the bearings (azimuth) from the point of release of the particles, σ_p^2, at a particular downwind location is

$$\sigma_p^2 = \sigma_a^2(\tau, s) \tag{19-8}$$

where σ_a^2 is the variance of the azimuth angles of a wind vane over the sampling period τ calculated from average wind directions averaged over averaging periods of duration s; s equals T/β, where T is the travel time to the downwind location; T is equivalent to x/u, where x is the downwind distance from the source and u is the transport wind speed. Here β is the ratio of the time scale of the turbulence moving with the air stream (Lagrangian) to the time scale of the turbulence at a fixed point (Eulerian). Although β has considerable variation (from about 1 to 9), a reasonable fit to field data has been found using a value of 4 for β.

A similar equation can be written for vertical spread from an elevated source. The standard deviation of the vertical distribution of pollutants at the downwind distance x is given by

$$\sigma_z = \sigma_e(\tau, s)x \tag{19-9}$$

where σ_z is in meters and σ_e is the standard deviation of the elevation angle, in radians, over the sampling period τ calculated from averaged elevation angles over averaging periods s. Here, as before, s equals T/β where T is travel time, and β can be approximated as equal to 4; x in Eq. (19-9) is in meters. In application, σ values can be calculated over several set averaging periods s. The distances to which each σ applies are then given by $x = \beta us$.

To calculate plume dispersion directly from fluctuation measurements, Draxler (8) used equations in the form

$$\sigma_y = x\sigma_a f_y \tag{19-10}$$

$$\sigma_z = x\sigma_e f_z \tag{19-11}$$

He analyzed dispersion data from 11 field experiments in order to determine the form of the functions f_y and f_z, including release height effects. Irwin (9) has used simplified expressions for these functions where both f_y and f_z have the form

$$f = 1/[1 + 0.9(T/T_0)^{0.5}] \tag{19-12}$$

where travel time T is x/u; T_0 is 1000 for f_y; T_0 is 500 for f_z for unstable (including daytime neutral) conditions; and T_0 is 50 for f_z for stable (including nighttime neutral) conditions.

2. By Classification of Wind Direction Traces

Where specialized fluctuation data are not available, estimates of horizontal spreading can be approximated from convential wind direction traces. A method suggested by Smith (2) and Singer and Smith (10) uses classification of the wind direction trace to determine the turbulence characteristics of the atmosphere, which are then used to infer the dispersion. Five turbulence classes are determined from inspection of the analog record of wind direction over a period of 1 h. These classes are defined in Table 19-1. The atmosphere is classified as A, B_2, B_1, C, or D. At Brookhaven National Laboratory, where the system was devised, the most unstable category, A, occurs infrequently enough that insufficient information is available to estimate its dispersion parameters. For the other four classes, the equations, coefficients, and exponents for the dispersion parameters are given in Table 19-2, where the source to receptor distance x is in meters.

3. By Classification of Atmospheric Stability

Pasquill (11) advocated the use of fluctuation measurements for dispersion estimates but provided a scheme "for use in the likely absence of special measurements of wind structure, there was clearly a need for broad estimates" of dispersion "in terms of routine meteorological data" (p. 367). The first element is a scheme which includes the important effects of thermal stratification to yield broad categories of stability. The necessary parameters for the scheme consist of wind speed, insolation, and cloudiness, which are basically obtainable from routine observations (Table 19-3).

Pasquill's dispersion parameters were restated in terms of σ_y and σ_z by Gifford (14, 15) to allow their use in the Gaussian plume equations. The

TABLE 19-1

Brookhaven Gustiness Classes (Based on Variations of Horizontal Wind Direction over 1 Hr at the Height of Release)

A:	Fluctuations of wind direction exceeding 90°
B_2:	Fluctuations ranging from 45° to 90°
B_1:	Similar to A and B_2, with fluctuations confined to a range of 15–45°
C:	Distinguished by the unbroken solid core of the trace, through which a straight line can be drawn for the entire hour without touching "open space." The fluctuations must be 15°, but no upper limit is imposed
D:	The trace approximates a line. Short-term fluctuations do not exceed 15°

Source: From Singer and Smith (10).

TABLE 19-2

Coefficients and Exponents for Brookhaven Gustiness Classes

Type	*a*	*b*	*c*	*d*
B_2	0.40	0.91	0.41	0.91
B_1	0.36	0.86	0.33	0.86
C	0.32	0.78	0.22	0.78
D	0.31	0.71	0.06	0.71

Note: $\sigma_y = ax^b$; $\sigma_z = cx^d$ (x is in meters).
Source: Adapted from Table 1 of Gifford (12).

parameters σ_y and σ_z are found by estimation from the graphs (Fig. 19-6), as a function of the distance between source and receptor, from the appropriate curve, one for each stability class (12). Alternatively, σ_y and σ_z can be calculated using the equations given in Tables 19-4 and 19-5, which are used in the point source computer techniques PTDIS and PTMTP (16). These parameter values are most applicable for releases near the ground (within about 50 m).

Other estimations of σ_y and σ_z by Briggs for two different situations, urban and rural, for each Pasquill stability class, as a function of distance between source and receptor, are given in Tables 19-6 and 19-7 (12).

TABLE 19-3

Pasquill Stability Categories

Surface wind speed (m/s)	Isolation			Night	
	Strong	Moderate	Slight	Thinly overcast or ≥ 4/8 low cloud	≤ 3/8 cloud
<2	A	A–B	B	—	—
2–3	A–B	B	C	E	F
3–5	B	B–C	C	D	E
5–6	C	C–D	D	D	D
>6	C	D	D	D	D
(for A–B, take the average of values for A and B, etc.)					

Notes:
1. Strong insolation corresponds to sunny midday in midsummer in England; slight insolation to similar conditions in midwinter.
2. Night refers to the period from 1 hr before sunset to 1 hr after sunrise.
3. The neutral category D should also be used, regardless of wind speed, for overcast conditions during day or night and for any sky conditions during the hour preceding or following night as defined above.

Source: From Pasquill (13).

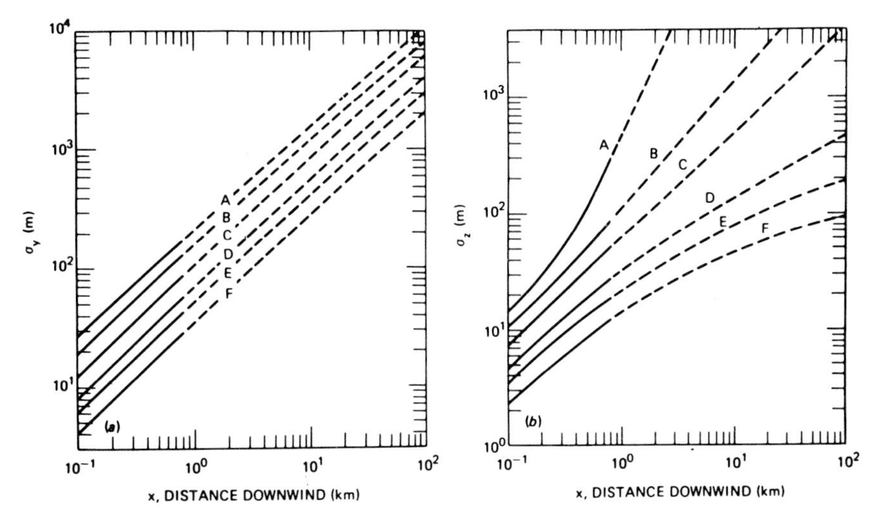

Fig. 19-6. Pasquill–Gifford σ_y (left) and σ_z (right). Source: From Gifford (12).

D. Example of a Dispersion Calculation

As an example of the use of the Gaussian plume equations using the Pasquill–Gifford dispersion parameters, assume that a source releases 0.37 g s^{-1} of a pollutant at an effective height of 40 m into the atmosphere with the wind blowing at 2 m s^{-1}. What is the approximate distance of the maximum concentration, and what is the concentration at this point if the atmosphere is appropriately represented by Pasquill stability class B?

Solution: The maximum occurs approximately when $\sigma_z = H/(2)^{1/2} = 28.3$ m. Under B stability, this occurs at $x = 0.28$ km. At this point

TABLE 19-4

Pasquill–Gifford Horizontal Dispersion Parameter

Stability	Parameter
A	$T = 24.167 - 2.5334 \ln x$
B	$T = 18.333 - 1.8096 \ln x$
C	$T = 12.5 - 1.0857 \ln x$
D	$T = 8.3333 - 0.72382 \ln x$
E	$T = 6.25 - 0.54287 \ln x$
F	$T = 4.1667 - 0.36191 \ln x$

Note: σ_y (meters) $= 465.116\ x \tan T$; x is downwind distance in km; T is one-half Pasquill's θ (degrees).

TABLE 19-5

Pasquill–Gifford Vertical Dispersion Parameter

Stability	Distance (km)	a	b	σ_z[a]
A	>3.11			5000 m
	0.5–3.11	453.85	2.1166	
	0.4–0.5	346.75	1.7283	104.7
	0.3–0.4	258.89	1.4094	71.2
	0.25–0.3	217.41	1.2644	47.4
	0.2–0.25	179.52	1.1262	37.7
	0.15–0.2	170.22	1.0932	29.3
	0.1–0.15	158.08	1.0542	21.4
	<0.1	122.8	0.9447	14.0
B	>35			5000 m
	0.4–35	109.30	1.0971	
	0.2–0.4	98.483	0.98332	40.0
	<0.2	90.673	0.93198	20.2
C	all x	61.141	0.91465	
D	>30	44.053	0.51179	
	10–30	36.650	0.56589	251.2
	3–10	33.504	0.60486	134.9
	1–3	32.093	0.64403	65.1
	0.3–1	32.093	0.81066	32.1
	<0.3	34.459	0.86974	12.1
E	>40	47.618	0.29592	
	20–40	35.420	0.37615	141.9
	10–20	26.970	0.46713	109.3
	4–10	24.703	0.50527	79.1
	2–4	22.534	0.57154	49.8
	1–2	21.628	0.63077	33.5
	0.3–1	21.628	0.75660	21.6
	0.1–0.3	23.331	0.81956	8.7
	<0.1	24.260	0.83660	3.5
F	>60	34.219	0.21716	
	30–60	27.074	0.27436	83.3
	15–30	22.651	0.32681	68.8
	7–15	17.836	0.4150	54.9
	3–7	16.187	0.4649	40.0
	2–3	14.823	0.54503	27.0
	1–2	13.953	0.63227	21.6
	0.7–1.0	13.953	0.68465	14.0
	0.2–0.7	14.457	0.78407	10.9
	<0.2	15.209	0.81558	4.1

[a] σ_z at boundary of distance range for all values except 5000 m.
Note: σ_z (meters) $= ax^b$; x is downwind distance in kilometers.

TABLE 19-6

Urban Dispersion Parameters by Briggs (for Distances between 100 and 10000 m)

Pasquill type	σ_y, m	σ_z, m
A–B	$0.32x(1 + 0.0004x)^{-0.5}$	$0.24x(1 + 0.001x)^{0.5}$
C	$0.22x(1 + 0.0004x)^{-0.5}$	$0.20x$
D	$0.16x(1 + 0.0004x)^{-0.5}$	$0.14x(1 + 0.0003x)^{-0.5}$
E–F	$0.11x(1 + 0.0004x)^{-0.5}$	$0.08x(1 + 0.0015x)^{-0.5}$

Source: From Gifford (12).

σ_y = 49.0 m (from Table 19- 4). First, the maximum can be estimated by Eq. (19-5):

$$\chi_{\max} = \frac{2Q}{\pi u e H^2}\frac{\sigma_z}{\sigma_y} = \frac{2(0.37)}{\pi 2e40^2}\frac{28.3}{49.0} = 1.56 \times 10^{-5}\,\mathrm{g\,m^{-3}}$$

To see if this is approximately the distance of the maximum, the equation

$$\chi = [Q/(\pi u \sigma_y \sigma_z)]\exp[-0.5(H/\sigma_z)^2)] \tag{19-13}$$

which results from Eq. (19-2) with y and z equal to 0, is evaluated at three distances: 0.26, 0.28, and 0.30 km. The parameter values and the resulting concentrations are given in the following table:

x,km	σ_z,m	σ_y,m	χ,g m^{-3}
0.26	26.2	45.9	1.53×10^{-5}
0.28	28.2	49.0	1.56×10^{-5}
0.30	30.1	52.2	1.55×10^{-5}

σ_z is obtained from equations in Table 19-5.
σ_y is obtained from equations in Table 19-4.

TABLE 19-7

Rural Dispersion Parameters by Briggs (for Distances between 100 and 10000 m)

Pasquill type	σ_y, m	σ_z, m
A	$0.22x(1 + 0.0001x)^{-0.5}$	$0.20x$
B	$0.16x(1 + 0.0001x)^{-0.5}$	$0.12x$
C	$0.11x(1 + 0.0001x)^{-0.5}$	$0.08x(1 + 0.0002x)^{-0.5}$
D	$0.08x(1 + 0.0001x)^{-0.5}$	$0.06x(1 + 0.0015x)^{-0.5}$
E	$0.06x(1 + 0.0001x)^{-0.5}$	$0.03x(1 + 0.0003x)^{-1}$
F	$0.04x(1 + 0.0001x)^{-0.5}$	$0.016x(1 + 0.0003x)^{-1}$

Source: From Gifford (12).

which verifies that, to the nearest 20 m, the maximum is at 0.28 km. Note that the concentration obtained from this equation is the same as that obtained from the approximation equation for the maximum.

Buoyancy-induced dispersion, which is caused near the source due to the rapid expansion of the plume during the rapid rise of the thermally buoyant plume after its release from the point of discharge, should also be included for buoyant releases (15). The effective vertical dispersion σ_{ze} is found from

$$\sigma_{ze}^2 = (\Delta H/3.5)^2 + \sigma_z^2 \tag{19-14}$$

where ΔH, the plume rise, and σ_z are evaluated at the distance x from the source. Beyond the distance to the final rise, ΔH is a constant. At shorter distances, it is evaluated for the gradually rising plume (see Chapter 20).

Since in the initial growth phases of a buoyant plume the plume is nearly symmetrical about its centerline, the buoyancy-induced dispersion in the crosswind (horizontal) direction is assumed to be equal to that in the vertical. Thus, the effective horizontal dispersion σ_{ye} is found from

$$\sigma_{ye}^2 = (\Delta H/3.5)^2 + \sigma_y^2 \tag{19-15}$$

The Gaussian plume equations are then used by substituting the value of σ_{ye} for σ_y and σ_{ze} for σ_z.

IV. DISPERSION INSTRUMENTATION

A. Measurements near the Surface

Near-surface (within 10 m of the ground) meteorological instrumentation always includes wind measurements and should include turbulence measurements as well. Such measurements can be made at 10 m above ground by using a guyed tower. A cup anemometer and wind vane (Fig. 19- 7), or a vane with a propeller speed sensor mounted in front (Fig. 19-8), can be the basic wind system. The wind sensor should have a threshold starting speed of less than 0.5 m s^{-1}, an accuracy of 0.2 m s^{-1} or 5%, and a distance constant of less than 5 m for proper response. The primary quantity needed is the hourly average wind speed. A representative value may be obtained from values taken each minute, although values taken at intervals of 1–5 sec are better.

The vane can be used for both average wind direction and the fluctuation statistic σ_a, both determined over hourly intervals. The vane should have a distance constant of less than 5 m and a damping ratio greater than or equal to 0.4 to have a proper response. Relative accuracy should be 1° and absolute accuracy should be 5°. In order to estimate σ_a accurately, the direction should be sampled at intervals of 1–5 sec. This can best be accom-

Fig. 19-7. Microvane and three-cup anemometer. Source: Photo courtesy of R. M. Young Co.

plished by microcircuitry (minicomputer) designed to sample properly the output from the vane and perform the calculations for both mean wind and σ_a, taking into account crossover shifts of the wind past the 360° and 0° point.

The elevation angle, and through appropriate data processing σ_e, can be measured with a bivane (a vane pivoted so as to move in the vertical as well as the horizontal). Bivanes require frequent maintenance and calibration and are affected by precipitation and formation of dew. A bivane is therefore more a research instrument than an operational one. Vertical fluctuations may be measured by sensing vertical velocity w and calculating σ_w from the output of a propeller anemometer mounted on a vertical shaft.

Fig. 19-8. Propeller vane wind system. Source: Photo courtesy of R. M. Young Co.

The instrument should be placed away from other instrumentation and the propeller axis carefully aligned to be vertical. The specifications of this sensor are the same as those of the wind sensor. Because this instrument will frequently be operating near its lower threshold and because the elevation angle of the wind vector is small, such that the propeller will be operating at yaw angles where it has least accuracy, this method of measuring vertical velocity is not likely to be as accurate as the measurement of horizontal fluctuation.

Rather than using separate systems for horizontal and vertical wind measurements, a *u-v-w* anemometer system (Fig. 19-9) sensing wind along three orthogonal axes, with proper processing to give average wind direction and σ_a from the combination of the u and v components and w and σ_w from the w component may be used.

Fig. 19-9. *U, V, W* wind system. Source: Photo courtesy of R. M. Young Co.

Additional near-surface measurements may also be required to support calculated quantities such as the bulk Richardson number (a stability parameter):

$$\mathrm{Ri_B} = \frac{gh}{T}\frac{\theta_h - \theta_z}{u_h^2}$$

which requires a temperature gradient, a temperature, and wind speed at the height of the boundary layer h. For this purpose, in addition to the wind speed at 10 m from the instrumentation, a vertical temperature difference measurement is needed. This can be obtained for the interval of 2–10 m aboveground using two relatively slow response sensors wired to give the temperature difference directly. Again, hourly averages are of greatest interest. The specifications are response time of 1 min, accuracy of 0.1°C, and resolution of 0.02°C. Both sensors should use good-quality aspirated radiation shields to give representative values. Sensor sampling about every 30 sec yields good hourly averages.

Radiation instruments are useful in determining stability such as F. B. Smith's (17) stability parameter *P*. Although somewhat similar to the Pasquill stability class (Table 19-3), *P* is continuous (rather than a discrete class) and is derived from wind speed and measurement of upward heat flux or, lacking this, incoming solar radiation (in daytime) and cloud amount at night. Pyranometers measure total sun and sky radiation. Net radiometers measure both incoming (mostly shortwave) radiation and outgoing (mostly longwave) radiation. Data from both are useful in turbulence characterization, and the values should be integrated over hourly periods. Care should be taken to avoid shadows on the sensors. The net radiometer is very sensitive to the condition of the ground surface over which it is exposed.

B. Measurements above the Surface

Measurements above the surface are also important to support pollutant impact evaluation. The radiosonde program of the National Weather Service (Fig. 19-10), established to support forecast and aviation weather activities, is a useful source of temperatures and data on winds aloft, although it has the disadvantage that measurements are made at 12-hr intervals and the surface layer is inadequately sampled because of the fast rate of rise of the balloon. Mixing height, the height aboveground of the neutral or unstable layer, is calculated from the radiosonde information (see Chapter 17, Section II).

Measurements of wind, turbulence, and temperature aloft may also be made at various heights on meteorological towers taller than 10 m. Where possible, the sensors should be exposed on a boom at a distance from the tower equal to two times the diameter of the tower at that height.

Aircraft can take vertical temperature soundings and can measure air pollutant and tracer concentrations and turbulence intensity. Airborne lidar can measure plume heights, and integrating nephelometers can determine particle size distributions.

Since operating aircraft, building towers, and establishing instruments on towers are extremely expensive, considerable attention has focused on indirect upper-air sounding from the ground. Mixing height within the range of measurement (approximately 500–600 m aboveground) can be determined by either the Doppler or the monostatic version of sodar (sound direction and ranging) with a spatial resolution of about 30 m. Data on wind and turbulence can be determined by Doppler sodar, FM-CW radar, and lidar. Doppler sodar measurements of wind components are within approximately 0.5 m s^{-1} of tower measurements. Measurements represent 30-m volume averages in the vertical. A height of 500 m aboveground, and sometimes over 1000 m, can be reached routinely.

Fig. 19-10. Radiosonde launch; sensor-transmitter (inset). Source: Photos courtesy of National Oceanic and Atmospheric Administration.

Some measurements that are completely impractical for routine measurement programs are useful during periods of intensive field programs. Winds and temperatures can be measured through frequent releases of balloon-carried sensors. Lidar is useful for determining plume dimensions. The particle lidar measures backscatter of laser radiation from particles in the plume and particles in the free air. The differential absorption lidar uses two wavelengths, one with strong absorption by sulfur dioxide and the other for weak absorption. The difference determines the amount of sulfur dioxide in the plume. Positioning of the lidar and its scanning mode determines whether vertical or horizontal dimensions of the plume are measured.

C. Data Reduction and Quality Assurance

A meteorological measurement program includes data reduction, calculation of quantities not directly measured, data logging, and archiving. Special-purpose minicomputers are used for sampling sensor output at frequent intervals (down to fractions of a second), calculating averages, and determining standard deviations. The output from the minicomputer should go to a data logger so that the appropriate information can be recorded on magnetic tape or disk or paper tape. If only hourly values must be archived, a considerable period of record for all data from a site can be contained on a single tape, disk, or cassette. Hard copy from a printer is usually also obtained. Immediate availability of this copy can aid in detecting system or sensor malfunctions. Sometimes analog charts are maintained for each sensor to provide backup data recovery (in case of reduction error or data logger malfunction) and to detect sensor malfunction.

An extremely important part of a measurement program is an adequate quality assurance program. Cost cutting in this part of the program can result in useless measurements. A good-quality assurance program includes calibration of individual components and of the entire system in the laboratory; calibration of the system upon installation in the field; scheduled maintenance and servicing; recalibration (perhaps quarterly); and daily examination of data output for unusual or unlikely values. More frequent servicing than that recommended by manufacturers may be required when sensors are placed in polluted atmospheres which may cause relatively rapid corrosion of instrument parts.

V. ATMOSPHERIC TRACERS

A. Technique

Tracer studies are extremely important in furthering our knowledge of atmospheric dispersion. These studies consist of release of a known quantity of a unique substance (the tracer), with measurements of that substance at one or more downwind sampling locations. Early experiments released uranine dye as a liquid spray; the water evaporated, leaving fine fluorescent particles to be sampled. Later, dry fluorescent particles (e.g., zinc–cadmium sulfide) having a relatively narrow range of particle sizes were used. Since the early 1970s, the gas sulfur hexafluoride has been used for most tracer studies, with collection in bags at sampling locations for later laboratory analysis using electron-capture gas chromatography.

Tracer studies are generally conducted by going into the field for a 2-week to 1-month intensive study period. The tracer is released, generally from a constant height, continuously at a constant rate for a set period (perhaps 2–3 h) on a day selected for its meteorological conditions with the wind forecast to blow toward the sampling network. Sampling equipment is arranged at ground level on constant-distance arcs usually at three or four distances. The samplers begin at a set time as switched on by the field crew or by radio control. More sophisticated samplers allow the unattended collection sequentially of several samples. Sampling time varies from around 20 min to several hours. This procedure measures horizontal dispersion at the height of the samplers.

Although it is highly desirable to determine vertical dispersion as well by direct measurement, it is seldom practical. Sampling in the vertical can be done by sampling on fixed towers or arranging samplers along the cables of captive balloons. Both of these methods are extremely expensive in terms of both equipment and personnel. Although it is possible to sample the tracer with aircraft, the pass through the pollutant plume occurs at such high speed that it is difficult to relate this instantaneous sample to what would occur over a longer sampling time of from 20 min to 1 h.

B. Computations

If the tracer concentration is χ_i measured at each sampling position that has its position at y_i on a scale along the arc (either in degrees or in meters), estimates of the mean position of the plume at ground level and the variance of the groundlevel concentration distribution are given by:

$$y = \frac{\sum \chi_i y_i}{\sum \chi_i} \tag{19-16}$$

$$\sigma_y^2 = \frac{\sum \chi_i \sum \chi_i y_i^2 - (\sum \chi_i y_i)^2}{(\sum \chi_i)^2} \tag{19-17}$$

In the example shown in Fig. 19-11, measurements were made every 2° on an arc 5 km from the source. The mean position of the plume is at an azimuth of 97.65° and the standard deviation is 4.806°.

$$\sigma_y \text{ (meters)} = \sigma_y \text{ (degrees)} \frac{\pi}{180} x \text{ (meters)} \tag{19-18}$$

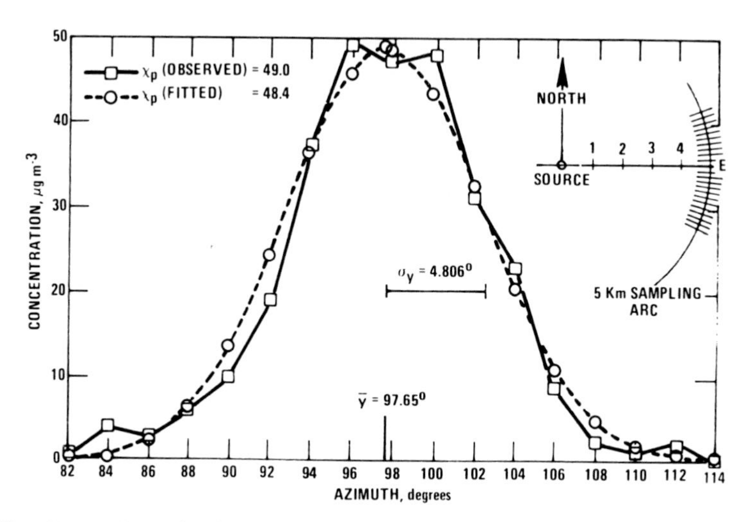

Fig. 19-11. Example of tracer concentration measurements along a sampling arc.

In this case σ_y is 419 m. The peak concentration can be found from the measurements, or from the Gaussian distribution fitted to the data and the peak concentration obtained from the fitted distribution. Provided that the emission rate Q, the height of release H, and the mean wind speed u are known, the standard deviation of the vertical distribution of the pollutant can be approximated from either the peak concentration (actual or fitted) or the crosswind integrated (CWI) concentration from one of the following equations:

$$\sigma_z \exp 0.5(H/\sigma_z)^2 = 2Q/[(2\pi)^{0.5}u\chi_{\mathrm{CWI}}] \tag{19-19}$$

$$\sigma_z \exp 0.5(H/\sigma_z)^2 = Q/(\pi u \sigma_y \chi_{\mathrm{peak}}) \tag{19-20}$$

The CWI concentration in g m^{-2} may be approximated from the tracer measurements from

$$\chi_{\mathrm{CWI}} = \text{sampler spacing (meters)} \sum \chi_i \tag{19-21}$$

Using the data from Fig. 19-11, the calculated σ_z from the CWI concentration is 239 m; from the observed peak concentration it is 232 m; and from the fitted peak concentration it is 235 m. Note that errors in any ofthe parameters H, Q, or u, will cause errors in the estimated σ_z.

Although extremely useful, tracer experiments require considerable capital expenditures and personnel. In addition to the difficulties and uncertainty in making estimates of various parameters, especially σ_z, one of the difficulties in interpreting tracer studies is relating the atmospheric conditions under which the study was conducted to the entire spectrum of atmospheric conditions. For example, trying to interpret a series of tracer experiments, even if conducted over a relatively large number of hours, in relation to the conditions that cause the second highest concentration once a year is extremely difficult, if not impossible.

VI. CONCENTRATION VARIATION WITH AVERAGING TIME

If emission and meteorological conditions remained unchanged hour after hour, concentrations at various locations downwind would remain the same. However, since such conditions are ever-changing, concentrations vary with time. Even under fairly steady meteorological conditions, with the mean wind direction remaining nearly the same over a period of some hours, as the averaging time increases, greater departures in wind direction from the mean are experienced, thus spreading the time-averaging plume more and reducing the longer averaging time concentration compared with that experienced for shorter averaging times at the location of the highest concentrations. This effect is more pronounced for receptors influenced by single point sources than for those influenced by a number of point sources or by a combination of point and area sources, because there will be many hours when the wind is not blowing from the source to the receptor.

Figure 19-12 shows the maximum sulfur dioxide concentrations for eight averaging times over a 1-year period (1976) for two air monitoring stations in the Regional Air Monitoring (RAM) network in St. Louis. These two monitoring stations, 104 and 113, have the highest and lowest maximum

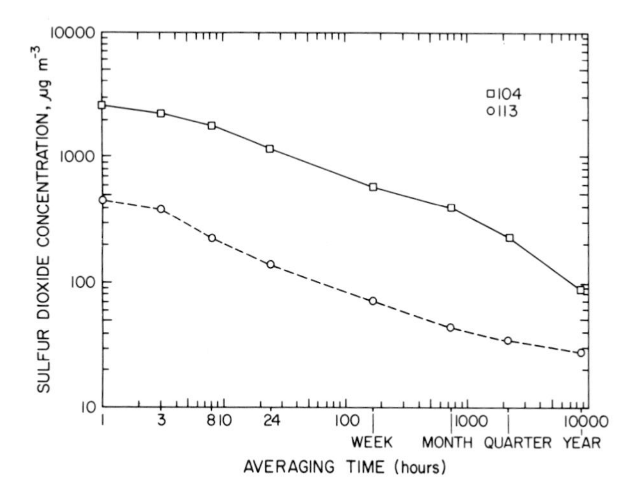

Fig. 19-12. Variation of St. Louis SO_2 maximum concentrations with sampling time for locations with highest (station 104) and lowest (station 113) maximum 1-hr concentrations.

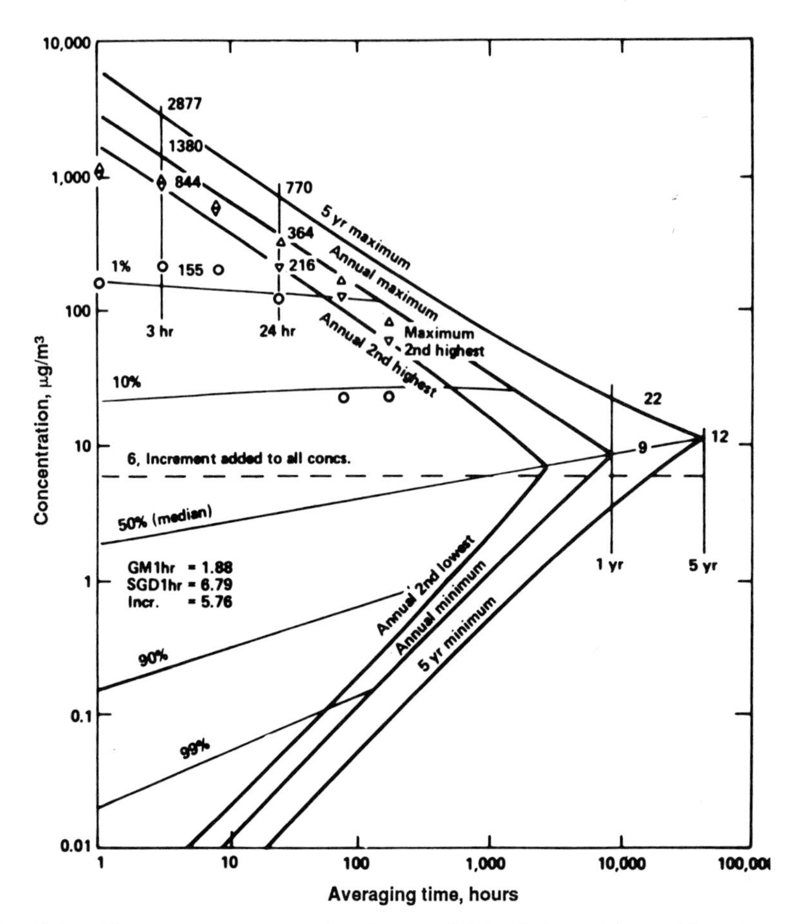

Fig. 19-13. Three-parameter averaging-time model fitted through the arithmetic mean and the second highest 3-hr and 24-hr SO_2 concentrations measured in 1972 a few miles from a coal-burning power plant. Source: From Larsen (21).

1-h concentrations of the 13 stations with sulfur dioxide measurements. These maximum concentrations deviate only slightly from a power law relation:

$$\chi_p = at_p^b \tag{19-22}$$

where χ_p is the maximum concentration for the period p, t_p is the averaging time in hours, and a and b are appropriate constants. The power b is -0.28 for station 104 and -0.33 for station 113.

Larsen (18–21) has developed averaging time models for use in analysis and interpretation of air quality data. For urban areas where concentrations for a given averaging time tend to be lognormally distributed, that is, where a plot of the log of concentration versus the cumulative frequency of occurrence on a normal frequency distribution scale is nearly linear, the two-parameter averaging time model (Fig. 4-6) is adequate. The two parameters are the geometric mean and the standard geometric deviation. If these two parameters for a pollutant at a site can be determined for an averaging time, the model can calculate them and the annual maximum concentration expected for any other averaging time. For receptors in the vicinity of point sources, where for a given averaging time many concentrations will be zero, a three-parameter averaging time model is required. The third parameter is an increment (positive or negative) that is added to every observed concentration. In Fig. 19-13, showing the three-parameter model applied to data from the vicinity of a power plant, 6 $\mu g\ m^{-3}$ have been added to each observed concentration.

REFERENCES

1. Irwin, J. S., *Atmos. Environ.* **13,** 191–194 (1979).
2. Smith, M. E., *Meteorol. Monogr.* No. 4, 50–55 (1951).
3. Church, P. E., *Ind. Eng. Chem.* **41,** 2753–2756 (1949).
4. Gifford, F. A., *Int. J. Air Pollut.* **2,** 109–110 (1959).
5. Hay, J. S., and Pasquill, F., Diffusion from a continuous source in relation to the spectrum and scale of turbulence, *in* "Advances in Geophysics," Vol. 6, "Atmospheric Diffusion and Air Pollution" (F. N. Frenkiel and P. A. Sheppard, eds.). Academic Press, New York, 1959, pp. 345–365.
6. Cramer, H. E., *Am. Ind. Hyg. Assoc. J.* **20**(3), 183–189 (1959).
7. Cramer, H. E., Improved techniques for modeling the dispersion of tall stack plumes. "Proceedings of the Seventh International Technical Meeting on Air Pollution Modeling and Its Application." North Atlantic Treaty Organization Committee on Challenges of Modern Society. Pub. No. 51. Brussels, 1976. (National Technical Information Service PB-270 799.)
8. Draxler, R. R., *Atmos. Environ.* **10,** 99–105 (1976).
9. Irwin, J. S., *J. Clim. Appl. Meteorol.* **22**(1), 92–114 (1983).
10. Singer, I. A., and Smith, M. E., *J. Meteorol.*, **10**(2), 121–126 (1953)
11. Pasquill, F., "Atmospheric Diffusion," 2nd ed. Halstead Press, New York, 1974.
12. Gifford, F. A., *Nucl. Safety* **17**(1), 68–86 (1976).
13. Pasquill, F., *Meteorol. Mag.* **90**(1063), 33–49 (1961).
14. Gifford, F. A., *Nucl. Safety* **2,** 47–55 (1961).
15. Pasquill, F., "Atmospheric Dispersion Parameters in Gaussian Plume Modeling, Part II. Possible Requirements for Change in the Turner Workbook Values," EPA-600/4-76-030b. U.S. Environmental Protection Agency, Research Triangle Park, NC, 1976.
16. Turner, D. B., and Busse, A. D., "User's Guide to the Interactive Versions of Three Point Source Dispersion Programs: PTMAX, PTDIS, and PTMTP." Meteorology Laboratory, U.S. Environmental Protection Agency, Research Triangle Park, NC, 1973.
17. Smith, F. B., A scheme for estimating the vertical dispersion of a plume from a source near ground level. "Proceedings of the Third Meeting of the Expert Panel on Air Pollution Modeling," Oct. 2–3, 1972, Paris. North Atlantic Treaty Organization Committee on the Challenges of Modern Society. Pub. No. 14. Brussels, 1972. (National Technical Information Service, PB-240 574.)

18. Larsen, R. I., *J. Air Pollut. Control Assoc.* **23,** 933–940 (1973).
19. Larsen, R. I., *J. Air Pollut. Control Assoc.* **24,** 551–558 (1974).
20. Larsen, R. I., and Heck, W. W., *J. Air Pollut. Control Assoc.* **26,** 325–333 (1976).
21. Larsen, R. I., *J. Air Pollut. Control Assoc.* **27,** 454–459 (1977).

SUGGESTED READING

Draxler, R. R., "A Summary of Recent Atmospheric Diffusion Experiments." National Oceanic and Atmospheric Administration Technical Memorandum ERL ARL-78. Silver Spring, MD, 1979.

Electric Power Research Institute, "Preliminary Results from the EPRI Plume Model Validation Project—Plains Site." EPRI EA-1788, RP 1616. Palo Alto, CA. Interim Report. Prepared by TRC Environmental Consultants, 1981.

Gryning, S. E., and Lyck, E., "Comparison between Dispersion Calculation Methods Based on In-Situ Meteorological Measurements and Results from Elevated- Source Tracer Experiments in an Urban Area." National Agency of Environmental Protection, Air Pollution Laboratory, MST Luft - A40. Riso National Laboratory, Denmark, 1980.

Haugen, D. A. (ed)., "Lectures on Air Pollution and Environmental Impact Analyses." American Meteorological Society, Boston, 1975.

Hewson, E. W., Meteorological measurements, *in* "Air Pollution," 3rd ed., Vol. I, "Air Pollutants, Their Transformation and Transport" A. C. Stern (ed.) Academic Press, New York, 1976, pp. 563–642.

Lenschow, D. H. (ed.), "Probing the Atmospheric Boundary Layer." American Meteorological Society, Boston, 1986.

Lockhart, T. J., "Quality Assurance Handbook for Air Pollution Measurement Systems," Vol. IV, "Meteorological Measurements." U.S. Environmental Protection Agency, Research Triangle Park, NC, 1989.

Munn, R. E., "Boundary Layer Studies and Applications." Kluwer Academic Press. Hingham, MA, 1989.

Strimaitis, D., Hoffnagle, G., and Bass, A., "On-Site Meteorological Instrumentation Requirements to Characterize Diffusion from Point Sources—Workshop Report," EPA-600/9-81-020. U.S. Environmental Protection Agency, Research Triangle Park, NC, 1981.

Turner, D. B., The transport of Pollutants, pp. 95–144 *in* "Air Pollution," 3rd ed., Vol. VI, "Supplement to Air Pollutants, Their Transformation, Transport, and Effects" (A. C. Stern, ed.) Academic Press, Orlando, FL, 1986.

Vaughan, R. A., "Remote Sensing Applications in Meteorology and Climatology." Reidel, Norwell, MA, 1987.

QUESTIONS

1. In a situation under stable conditions with a wind speed of 4 m s^{-1} and $\sigma_a = 0.12$ radians, the wind is blowing directly toward a receptor 1 km from the source. How much must the wind shift , in degrees, to reduce the concentration to 10% of its previous value? (At 2.15σ from the peak, the Gaussian distribution is 0.1 of the value at the peak.)
2. If the variation in wind speed with height is well approximated with a power law wind profile having an exponent equal to 0.15, how much stronger is the wind at 100 m above ground than at 10 m?

3. At a particular downwind distance, a dispersing plume is approximately 40 m wide. Which of the following three turbulent eddy diameters—5 m, 30 m, or 100 m—do you believe would be more effective in further dispersing this plume?
4. A pollutant is released from an effective height of 50 m and has a ground-level concentration of 300 $\mu g\ m^{-3}$ at a position directly downwind where the σ_z is 65 m. How does the concentration at 50 m above this point, that is, the plume centerline, compare with this ground-level concentration?
5. At a downwind distance of 800 m from a 75-m source having an 180-m plume rise, σ_y is estimated as 84 m and σ_z is estimated as 50 m. If one considers buoyancy-induced dispersion as suggested by Pasquill, by how much is the plume centerline concentration reduced at this distance?
6. Consider the parameters that need to be measured and the instrumentation needed to make the measurements for monitoring dispersion of effluent from a 30-m stack.
7. A tracer experiment includes sampling on an arc at 1000 m from the source. If the horizontal spread is expected to result in a σ_y between 120 and 150 m at this distance, and if the wind direction is within $\pm 15°$ azimuth of that forecast, how many samplers should be deployed and what should be that spacing? It is desirable to have above seven measurements within $\pm 2\sigma_y$ of the plume centerline and at least one sample on each side of the plume.
8. For a tracer release that can be considered to be at ground level, approximate the vertical dispersion σ_z at the downwind distance where measurements indicate that the concentration peak is 7×10^{-8} g m^{-3}, the horizontal σ_y is 190 m, and the crosswind integrated concentration is 3.16×10^{-5} g m^{-3}. The tracer release rate is 0.01 g s^{-1} and the wind speed is 3.7 m s^{-1}.
9. The maximum 1-h concentration at an urban monitoring station is 800 $\mu g\ m^{-3}$. If the concentration varies with averaging time with a power law relation when the power is -0.3, what is the expected maximum concentration for a 1-week averaging time?

20

Air Pollution Modeling and Prediction

In order to build new facilities or expand existing ones without harming the environment, it is desirable to assess the air pollution impact of a facility prior to its construction, rather than construct and monitor to determine the impact and whether it is necessary to retrofit additional controls. Potential air pollution impact is usually estimated through the use of air quality simulation models. A wide variety of models is available. They are usually distinguished by type of source, pollutant, transformations and removal, distance of transport, and averaging time. No attempt will be made here to list all the models in existence at the time of this writing.

In its simplest form, a model requires two types of data inputs: information on the source or sources including pollutant emission rate, and meteorological data such as wind velocity and turbulence. The model then simulates mathematically the pollutant's transport and dispersion, and perhaps its chemical and physical transformations and removal processes. The model output is air pollutant concentration for a particular time period, usually at specific receptor locations.

Impact estimates by specific models are required to meet some regulatory requirements.

I. PLUME RISE

Gases leaving the tops of stacks rise higher than the stack top when they are either of lower density than the surrounding air (buoyancy rise) or ejected at a velocity high enough to give the exit gases upward kinetic energy (momentum rise). Buoyancy rise is sometimes called *thermal rise* because the most common cause of lower density is higher temperature. Exceptions are emissions of gases of higher density than the surrounding air and stack downwash, discussed next. To estimate effective plume height, the equations of Briggs (1–5) are used. The wind speed u in the following equations is the measured or estimated wind speed at the physical stack top.

A. Stack Downwash

The lowering below the stack top of pieces of the plume by the vortices shed downwind of the stack is simulated by using a value h' in place of the physical stack height h. This is somewhat less than the physical height when the stack gas exit velocity v_s is less than 1.5 times the wind speed u, m s^{-1}.

$$h' = h \quad \text{for } v_s \geq 1.5u \quad (20\text{-}1)$$

$$h' = h + 2d[(v_s/u) - 1.5] \quad \text{for } v_s < 1.5u \quad (20\text{-}2)$$

where d is the inside stack-top diameter, m. This h' value is used with the buoyancy or momentum plume rise equations that follow. If stack downwash is not considered, h is substituted for h' in the equations.

B. Buoyancy Flux Parameter

For most plume rise estimates, the value of the buoyancy flux parameter F in m^4 s^{-3} is needed.

$$F = gv_sd^2(T_s - T)/(4T_s)$$

$$F = 2.45v_sd^2(T_s - T)/T_s \quad (20\text{-}3)$$

where g is the acceleration due to gravity, about 9.806 m s^{-2}, T_s is the stack gas temperature in K, T is ambient air temperature in K, and the other parameters are as previously defined.

C. Unstable–Neutral Buoyancy Plume Rise

The final effective plume height H, in m, is stack height plus plume rise. Where buoyancy dominates, the horizontal distance x_f from the stack to where the final plume rise occurs is assumed to be at $3.5x^*$, where x^* is the horizontal distance, in km, at which atmospheric turbulence begins to dominate entrainment.

For unstable and neutral stability situations, and for F less than 55, H, in m, and x_f, in km, are

$$H = h' + 21.425F^{3/4}/u \qquad x_f = 0.049F^{5/8} \qquad \text{(20-4a,b)}$$

For F equal to or greater than 55, H and x_f are

$$H = h' + 38.71F^{3/5}/u \qquad x_f = 0.119F^{2/5} \qquad \text{(20-5a,b)}$$

D. Stability Parameter

For stable situations, the stability parameter s is calculated by

$$s = g(\Delta\theta/\Delta z)/T$$

where $\Delta\theta/\Delta z$ is the change in potential temperature with height.

E. Stable Buoyancy Plume Rise

For stable conditions when there is wind, H and x_f are

$$H = h' + 2.6[F/(us)]^{1/3} \qquad x_f = 0.00207us^{-1/2} \qquad \text{(20-6a,b)}$$

For calm conditions (i.e., no wind) the stable buoyancy rise is

$$H = h' + 4F^{1/4}s^{-3/8} \qquad \text{(20-7)}$$

Under stable conditions, the lowest value of Eq. (20- 6a) or (20-7) is usually taken as the effective stack height.

The wind speed that yields the same rise from Eq. (20- 6a) as that from Eq. (20-7) for calm conditions is

$$u = 0.2746F^{1/4}s^{1/8} \qquad \text{(20-8)}$$

F. Gradual Rise—Buoyancy Conditions

Plume rise for distances closer to the source than the distance to the final rise can be estimated from

$$H = h' + 160F^{1/3}x^{2/3}/u \qquad \text{(20-9)}$$

where x is the source-to-receptor distance, km. If this height exceeds the final effective plume height, that height should be substituted.

G. Unstable–Neutral Momentum Plume Rise

If the stack gas temperature is below or only slightly above the ambient temperature, the plume rise due to momentum will be greater than that due to buoyancy. For unstable and neutral situations:

$$H = h' + 3dv_s/u \tag{20-10}$$

This equation is most applicable when v_s/u exceeds 4. Since momentum plume rise occurs quite close to the source, the horizontal distance to the final plume rise is considered to be zero.

H. Stable Momentum Plume Rise

For low-buoyancy plumes in stable conditions, plume height due to momentum is given by

$$H = h' + 1.5[(v_s^2 d^2 T)/(4T_s u)]^{1/3} s^{-1/6} \tag{20-11}$$

Equation (20-10) should also be evaluated and the lower value used.

I. Momentum–Buoyancy Crossover

There is a specific difference between stack gas temperature and ambient air temperature that gives the same result for buoyancy rise as for momentum rise. For unstable or neutral conditions this is as follows: For F less than 55,

$$(T_s - T)_c = 0.0297 T_s v_s^{1/3}/d^{2/3} \tag{20-12}$$

For F equal to or greater than 55,

$$(T_s - T)_c = 0.00575 T_s v_s^{2/3}/d^{1/3} \tag{20-13}$$

For stable conditions,

$$(T_s - T)_c = 0.01958 T v_s s^{1/2} \tag{20-14}$$

J. Maximum Concentrations as a Function of Wind Speed and Stability

Using the source in the example in Chapter 19 ($Q = 0.37$, $h = 20$, $d = 0.537$, $v_s = 20$, and $T_s = 350$), with plume rise calculated using the above equations, maximum ground-level concentrations are shown (Fig. 20-1) as functions of stability class and wind speed calculated using the Gaussian model with Pasquill–Gifford dispersion parameters. Maximum concentrations are nearly the same for stabilities A, B, and C and occur at wind speeds of 1.5–2.0 m s^{-1}. The maximum for D stability occurs at around $u = 2.5$ m s^{-1}. Because of the competing effects of dilution by wind and lower effective stack heights with higher wind speeds, concentrations

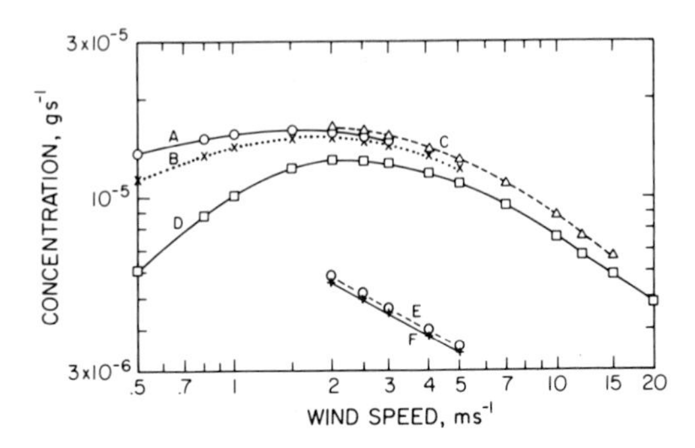

Fig. 20-1. Concentration of an air pollutant at the point of maximum ground-level concentration as a function of wind speed and Pasquill stability category (A–F).

do not change rapidly with wind speed. For E and F stabilities the concentrations are nearly the same (assuming that $\Delta\theta/\Delta z$ is 0.02 K m^{-1} for E stability and 0.035 for F stability), but are considerably less than for the unstable and neutral cases.

II. MODELING TECHNIQUES

Gaussian techniques, discussed in Chapter 19, are reasonable for estimates of concentrations of nonreative pollutants within 20 km of point sources. It is preferable to utilize on-site wind fluctuation measurements to estimate the horizontal and vertical spreading of a pollutant plume released from a point source.

In addition to the Gaussian modeling techniques already discussed, four other methods will be considered.

A. Box Model

Models which assume uniform mixing throughout the volume of a three-dimensional box are useful for estimating concentrations, especially for first approximations. For steady-state emission and atmospheric conditions, with no upwind background concentrations, the concentration is given by

$$\chi = \Delta x \, q_a/(z_i u) \tag{20-15}$$

where χ is the steady-state concentration, Δx is the distance over which the emissions take place, q_a is the area emission rate, z_i is the mixing height, and u is the mean wind speed through the vertical extent of the box.

When there is an upwind background concentration χ_b and the mixing height is rising with time into a layer aloft having an average concentration of χ_a, the equation of continuity is

$$\delta\chi/\delta t = [\Delta x\, q_a + u z_i(\chi_b - \chi) + \Delta x(\Delta z_i/\Delta t)\,(\chi_a - \chi)]/\Delta x\, z_i \qquad (20\text{-}16)$$

This forms a basis for an urban photochemical box model (6) discussed later in this chapter.

B. Narrow Plume Hypothesis

By assuming that the principal contributors to the concentration at a receptor are the sources directly upwind, especially those nearby, the concentration due to area sources can be calculated using the vertical growth rate rather than uniform vertical mixing and considering the specific area emission rate of each area upwind of the receptor. Area emission rate changes in the crosswind direction are neglected as being relatively unimportant. The expansion in the vertical is usually considered using the Gaussian vertical growth (7, 8).

C. Gradient Transport Models

The mean turbulent flux of concentration in the vertical direction z is $w'\chi'$. Assuming that this turbulent flux is proportional to the gradient of concentration, and in the direction from higher to lower concentrations, an overall diffusivity K can be defined:

$$\overline{w'\chi'} = -K(\delta\chi/\delta z) \qquad (20\text{-}17)$$

The change in concentration with respect to time can be written as

$$\frac{\delta\chi}{\delta t} + \left(\bar{u}\frac{\delta\chi}{\delta x} + \bar{v}\frac{\delta\chi}{\delta y} + \bar{w}\frac{\delta\chi}{\delta z}\right)$$

$$= \frac{\delta}{\delta x}K_x\frac{\delta\chi}{\delta x} + \frac{\delta}{\delta y}K_y\frac{\delta\chi}{\delta y} + \frac{\delta}{\delta z}K_z\frac{\delta\chi}{\delta z} + S \qquad (20\text{-}18)$$

where the term in parentheses on the left accounts for advection. The terms on the right account for diffusivities in three directions, K_x, K_y, and K_z (where K_x is in the direction of the wind, K_y is horizontally crosswind, and K_z is vertically crosswind), and S represents emissions. This equation is the basis for the gradient transport model, which can handle varying wind and diffusivity fields. The vector speeds $\bar{u}$, $\bar{v}$, and $\bar{w}$ (where $\bar{u}$ is in the direction of the wind, $\bar{v}$ is horizontally crosswind, and $\bar{w}$ is vertically crosswind) and concentrations imply both time and space scales. Fluctuations over times and distances less than these scales are considered as turbulence and are included in the diffusivities.

The gradient transport model is most appropriate when the turbulence is confined to scales that are small relative to the pollutant volume. It is therefore most applicable to continuous line and area sources at ground level, such as automobile pollutants in urban areas, and to continuous or instantaneous ground-level area sources. It is not appropriate for elevated point source diffusion until the plume has grown larger than the space scale. Numerical rather than analytical solutions of Eq. (20-18) are used.

Errors in advection may completely overshadow diffusion. The amplification of random errors with each succeeding step causes numerical instability (or distortion). Higher-order differencing techniques are used to avoid this instability, but they may result in sharp gradients, which may cause negative concentrations to appear in the computations. Many of the numerical instability (distortion) problems can be overcome with a second-moment scheme (9) which advects the moments of the distributions instead of the pollutants alone. Six numerical techniques were investigated (10), including the second-moment scheme; three were found that limited numerical distortion: the second-moment, the cubic spline, and the chapeau function.

In the application of gradient transfer methods, horizontal diffusion is frequently ignored, but the variation in vertical diffusivity must be approximated (11–14).

D. Trajectory Models

In its most common form, a trajectory model moves a vertical column, with a square cross section intersecting the ground, at the mean wind speed, with pollutants added to the bottom of the column as they are generated by each location over which the column passes. Treatment of vertical dispersion varies among models, from those which assume immediate vertical mixing throughout the column to those which assume vertical dispersion using a vertical coefficient K_z with a suitable profile (15).

Modeling a single parcel of air as it is being moved along allows the chemical reactions in the parcel to be modeled. A further advantage of trajectory models is that only one trajectory is required to estimate the concentration at a given endpoint. This minimizes calculation because concentrations at only a limited number of points are required, such as at stations where air quality is routinely monitored. Since wind speed and direction at the top and the bottom of the column are different, the column is skewed from the vertical. However, for computational purposes, the column is usually assumed to remain vertical and to be moved at the wind speed and direction near the surface. This is acceptable for urban application in the daytime, when winds are relatively uniform throughout the lower atmosphere.

Trajectory models of a different sort are used for long-range transport, because it is necessary to simulate transport throughout a diurnal cycle in which the considerable wind shear at night transports pollutants in different directions. Expanding Gaussian puffs can be used, with the expanded puff breaking, at the time of maximum vertical mixing, into a series of puffs initially arranged vertically but subsequently moving with the appropriate wind speed and direction for each height.

III. MODELING NONREACTIVE POLLUTANTS

A. Seasonal or Annual Concentrations

In estimating seasonal or annual concentrations from point or area sources, shortcuts can generally be taken rather than attempting to integrate over short intervals, such as hour-by-hour simulation. A frequent shortcut consists of arranging the meteorological data by joint frequency of wind direction, wind speed, and atmospheric stability class, referred to as a *STability ARray*, STAR. The ISCLT (16) is a model of this type and is frequently used to satisfy regulatory requirements where concentrations averaged over 1 year (but not shorter averaging times) or longer are required. Further simplification may be achieved by determining a single effective wind speed for each stability–wind direction sector combination by weighting $1/u$ by the frequency of each wind speed class for each such wind direction–stability combination. Calculations for each sector are made, assuming that the frequency of wind direction is uniform across the sector.

B. Single Sources—Short-Term Impact

Gaussian plume techniques have been quite useful for determining the maximum impact of single sources, which over flat terrain, occurs within 10–20 km of the source. The ISCST model (16) is usually used to satisfy regulatory requirements. Because the combination of conditions that produces multihour high concentrations cannot be readily identified over the large range of source sizes, it has been common practice to calculate the impact of a source for each hour of the year for a large number of receptors at specific radial distances from the source for 36 directions from the source, e.g., every 10°. Averaging and analysis can proceed as the calculations are made to yield, upon completion of a year's simulation, the highest and second-highest concentrations over suitable averaging times, such as 3 and 24 h. Frequently, airport surface wind data have been utilized as input for such modeling, extrapolating the surface wind speed to stack top using a power law profile, with the exponent dependent on stability class, which is also determined from the surface data. Although the average hourly wind direction at stack top and plume level is likely to be different from that at the surface, this has been ignored because hourly variations in wind direction at plume level closely parallel surface directional variations. Although the true maximum concentration may occur in a somewhat different direction from that calculated, its magnitude will be closely approximated.

Several point source algorithms, HPDM (17, 18) and TUPOS (19, 20), incorporate the use of fluctuation statistics (the standard deviations of horizontal and vertical wind directions) and non-Gaussian algorithms for strongly convective conditions. Because during strong convective conditions, thermals with updrafts occupy about 30–35% of the area and slower descending downdrafts occupy 65–70% of the area, the resulting distribution of vertical motions are not Gaussian but have a smaller number of upward motions, but with higher velocity, and a larger number of downward motions, but with lower velocity. These skewed vertical motion distributions then cause non-Gaussian vertical distributions of pollutant concentrations.

C. Multiple Sources and Area Sources

The problem, already noted, of not having the appropriate plume transport direction takes on added importance when one is trying to determine the effects of two or more sources some distance apart, since an accurate estimate of plume transport direction is necessary to identify critical periods when plumes are superimposed, increasing concentrations.

In estimating concentrations from area sources, it is important to know whether there is one source surrounded by areas of no emissions or whether the source is just one element in an area of continuous but varying emissions.

To get an accurate estimate of the concentrations at all receptor positions from an isolated area source, an integration should be done over both the alongwind and crosswind dimensions of the source. This double integration is accomplished in the PAL model (21) by approximating the area source using a number of finite crosswind line sources. The concentration due to the area source is determined using the calculated concentration from each line source and integrating numerically in the alongwind direction.

If the receptor is within an area source, or if emission rates do not vary markedly from one area source to another over most of the simulation area, the narrow-plume hypothesis can be used to consider only the variation in emission rates from each area source in the alongwind direction. Calculations are made as if from a series of infinite crosswind line sources whose emission rate is assigned from the area source emission rate directly upwind of the receptor at the distance of the line source. The ATDL model (22) accomplishes this for ground-level area sources. The RAM model (8) does this for ground-level or elevated area sources.

Rather than examine the variation of emissions with distance upwind from the receptor as already described, one can simplify further by using the area emission rate of only the emission square in which the receptor resides (23). The concentration c is then given by

$$\chi = Cq_a/u \qquad (20\text{-}19)$$

where qa is the area emission rate, u is the mean wind speed over the simulation period, and the constant C is dependent on the stability, the effective height of emission of the sources, and the characteristics of the pollutant. For estimation of annual concentrations with this method, $C = 50$, 200, and 600 for SO_2, particulate matter, and CO, respectively (24).

D. Pollutants That Deposit

The Fugitive Dust Model, FDM (25), was formulated to estimate air concentrations as well as deposition from releases of airborne dust. It has a greatly improved deposition mechanism compared with that in previous models. It considers the mass removed from the plume through deposition as the plume is moved downwind. Up to 20 particle size fractions are available. The particle emissions caused as material is raised from the surface by stronger winds is built internally into this model. It has the capacity of making calculations for point, line, and area sources. The area source algorithm has two options, simulation of the area source by five finite line sources perpendicular to wind flow, or a converging algorithm which provides greater accuracy. Currently, the model should be used for releases at or below 20 m above ground level.

E. Dispersion from Sources over Water

The Offshore and Coastal Dispersion (OCD) model (26) was developed to simulate plume dispersion and transport from offshore point sources to receptors on land or water. The model estimates the overwater dispersion by use of wind fluctuation statistics in the horizontal and the vertical measured at the overwater point of release. Lacking these measurements the model can make overwater estimates of dispersion using the temperature difference between water and air. Changes taking place in the dispersion are considered at the shoreline and at any points where elevated terrain is encountered.

F. Dispersion over Complex Terrain

Development efforts in complex terrain of the EPA using physical modeling, both wind tunnel and towing tank, and field studies at three locations have resulted in the CTDMPLUS, Complex Terrain Dispersion Model plus the calculation of concentrations for unstable conditions (27). Complex terrain is the situation in which there are receptor locations above stack top. Using the meteorological conditions and the description of the nearby terrain feature, the model calculates the height of a dividing streamline. Releases that take place below the height of this streamline tend to seek a path around the terrain feature; releases above the streamline tend to rise over the terrain feature. Because the dispersion is calculated from fluctuation statistics, the meteorological measurements to provide data input for the model are quite stringent, requiring the use of tall instrumented towers. Evaluations both of the model (28) and of a screening technique derived from the model (29) indicate that the model does a better job of estimating concentrations than previous complex terrain models.

IV. MODELING POLLUTANT TRANSFORMATIONS

A. Individual Plumes

An understanding of the transformation of SO_2 and NO_x into other constituents no longer measurable as SO_2 and NO_x is needed to explain mass balance changes from one plume cross section to another. This loss of the primary pollutant SO_2 has been described as being exponential, and rates up to 1% per hour have been measured (30). The secondary pollutants generated by transformation are primarily sulfates and nitrates.

The horizontal dispersion of a plume has been modeled by the use of expanding cells well mixed vertically, with the chemistry calculated for each cell (31). The resulting simulation of transformation of NO to NO_2 in a power plant plume by infusion of atmospheric ozone is a peaked distribution of NO_2 that resembles a plume of the primary pollutants, SO_2 and NO. The ozone distribution shows depletion across the plume, with maximum depletion in the center at 20 min travel time from the source, but relatively uniform ozone concentrations back to initial levels at travel distances 1 h from the source.

B. Urban-Scale Transformations

Approaches used to model ozone formation include box, gradient transfer, and trajectory methods. Another method, the particle-in-cell method, advects centers of mass (that have a specific mass assigned) with an effective velocity that includes both transport and dispersion over each time step. Chemistry is calculated using the total mass within each grid cell at the end of each time step. This method has the advantage of avoiding both the numerical diffusion of some gradient transfer methods and the distortion due to wind shear of some trajectory methods.

It is not feasible to model the reaction of each hydrocarbon species with oxides of nitrogen. Therefore, hydrocarbon species with similar reactivities are lumped together, e.g., into four groups of reactive hydrocarbons: olefins, paraffins, aldehydes, and aromatics (32).

In addition to possible errors due to the steps in the kinetic mechanisms, there may be errors in the rate constants due to the smog chamber data bases from which they were derived. A major shortcoming is the limited amount of quality smog chamber data available.

The emission inventory and the initial and boundary conditions of pollutant concentrations have a large impact on the ozone concentrations calculated by photochemical models.

To model a decrease in visibility, the chemical formation of aerosols from sulfur dioxide and oxidants must be simulated.

In a review of ozone air quality models, Seinfeld (33) indicates that the most uncertain part of the emission inventories is the hydrocarbons. The models are especially sensitive to the reactive organic gas levels, speciation, and the concentrations aloft of the various species. He points out the need for improvement in the three-dimensional wind fields and the need for hybrid models that can simulate sub-grid-scale reaction processes to incorporate properly effects of concentrated plumes. Schere (34) points out that we need to improve the way vertical exchange processes are included in the model. Also, although the current models estimate ozone quite well, the atmospheric chemistry needs improvement to better estimate the concentrations of other photochemical components such as peroxyacyl nitrate (PAN), the hydroxyl radical (OH), and volatile organic compounds (VOCs). In addition to the improvement of data bases , including emissions, boundary concentrations, and meteorology, incorporation of the urban ozone with the levels at larger scales is needed.

C. Regional-Scale Transformations

In order to formulate appropriate control strategies for oxidants in urban areas, it is necessary to know the amount of oxidant already formed in the air reaching the upwind side of the urban area under various atmospheric conditions. Numerous physical and chemical processes are involved in modeling transformations (35, 36) on the regional scale (several days, 1000 km): (1) horizontal transport; (2) photochemistry, including very slow reactions; (3) nighttime chemistry of the products and precursors of photochemical reactions; (4) nighttime wind shear, stability stratification, and turbulence episodes associated with the nocturnal jet; (5) cumulus cloud effects—venting pollutants from the mixed layer, perturbing photochemical reaction rates in their shadows, providing sites for liquid-phase reactions, influencing changes in the mixed-layer depth, perturbing horizontal flow; (6) mesoscale vertical motion induced by terrain and horizontal divergence of the large-scale flow; (7) mesoscale eddy effects on urban plume trajectories and growth rates; (8) terrain effects on horizontal flows, removal, and diffusion; (9) sub-grid-scale chemistry processes resulting from emissions from sources smaller than the model's grid can resolve; (10) natural sources of hydrocarbons, NO_x, and stratospheric ozone; and (11) wet and dry removal processes, washout, and deposition.

Approaches to long-term (monthly, seasonal, annual) regional exchanges are EURMAP (37) for Europe and ENAMAP (38) for eastern North America. These two models can calculate SO_2 and sulfate air concentrations as well as dry and wet deposition rates for these constituents. The geographic region of interest (for Europe, about 2100 km N–S by 2250 km E–W) is divided into an emissions grid having approximately 50 by 50 km resolution. Calculations are performed by releasing a 12-h average emission increment or "puff" from each cell of the grid and tracking the trajectories of each puff by 3-h time steps according to the 850-mb winds interpolated objectively for the puff position from upper-air data.

Uniform mixing in the vertical to 1000 m and uniform concentrations across each puff as it expands with the square root of travel time are assumed. A 0.01 h^{-1} transformation rate from SO_2 to sulfate and 0.029 and 0.007 h^{-1} dry deposition rates for SO_2 and sulfate, respectively, are used. Wet deposition is dependent on the rainfall rate determined from the surface observation network every 6 h, with the rate assumed to be uniform over each 6-h period. Concentrations for each cell are determined by averaging the concentrations of each time step for the cell, and deposition is determined by totaling all depositions over the period.

The EURMAP model has been useful in estimating the contribution to the concentrations and deposition on every European nation from every other European nation. Contributions of a nation to itself range as follows: SO_2 wet deposition, 25–91%; SO_2 dry deposition, 31–91%; sulfate wet deposition, 2–46%; sulfate dry deposition, 4–57%.

In the application of the model to eastern North America, the mixing height is varied seasonally, and hourly precipitation data are used.

V. MODEL PERFORMANCE, ACCURACY, AND UTILIZATION

A. Methodology of Assessing Model Performance

A number of statistics have been suggested (39, 40) as measures of model performance. Different types of models and the use of models for different purposes may require different statistics to measure performance.

In time series of measurements of air quality and estimates of atmospheric concentration made by a model, residuals d can be computed for each location. The residual d is the difference between values paired timewise.

$$d = M - E \tag{20-20}$$

where M is the measured value and E is the value estimated by the model.

A measure of bias $\bar{d}$ is the first moment of the distribution of these differences, or the average difference:

$$\bar{d} = \frac{1}{N} \Sigma\, d \tag{20-21}$$

A measure of the variability of the differences is the variance S^2, which is the second moment of the distribution of these differences:

$$S^2 = \Sigma(d - \bar{d})^2/N = \frac{\Sigma d^2}{N} - \left(\frac{\Sigma d}{N}\right)^2 \tag{20-22}$$

The square root of the variance is the standard deviation.

The mean absolute error MAE is

$$\text{MAE} = \Sigma |d|/N \tag{20-23}$$

The mean square error MSE is

$$\text{MSE} = \Sigma d^2/N \tag{20-24}$$

The root-mean-square error is the square root of the mean square error. Note that since the root-mean-square error involves the square of the differences, outliers have more influence on this statistic than on the mean absolute error.

The fractional error FE (41) is

$$\text{FE} = (M - E)/0.5(M + E) \tag{20-25}$$

The mean fractional error MFE is

$$\text{MFE} = \Sigma\ \text{FE}/N \tag{20-26}$$

The fractional error is logarithmically unbiased; that is, an M which is k times E produces the same magnitude fractional error (but of opposite sign) as an M which is $1/k$ times E.

In addition to analyzing the residuals, it may be desirable to determine the degree of agreement between sets of paired measurements and estimates. The linear correlation coefficient r_{EM} is

$$r_{EM} = \frac{N \Sigma E \cdot M - \Sigma E \Sigma M}{\{[N \Sigma E^2 - (\Sigma E)^2][N \Sigma M^2 - (\Sigma M)^2]\}^{0.5}} \tag{20-27}$$

The slope b and intercept a of the least-squares line of best fit of the relation $M = a + bE$ are

$$\text{Intercept:} \quad a = \frac{\Sigma E^2 \Sigma M - \Sigma E \Sigma E \cdot M}{N \Sigma E^2 - (\Sigma E)^2} \tag{20-28}$$

$$\text{Slope:} \quad b = \frac{N \Sigma E \cdot M - \Sigma E \Sigma M}{N \Sigma E^2 - (\Sigma E)^2} \tag{20-29}$$

The temporal correlation coefficient at each monitoring location can be calculated by analysis of the paired values over a time period of record. The spatial correlation coefficient at a given time can be calculated by analysis of the paired values from each station. For the spatial correlation coefficient to have much significance, there should be 20 or more monitoring locations.

Techniques to use for evaluations have been discussed by Cox and Tikvart (42), Hanna (43) and Weil *et al.* (44). Hanna (45) shows how resampling of evaluation data will allow use of the bootstrap and jackknife techniques so that error bounds can be placed about estimates.

The use of various statistical techniques has been discussed (46) for two situations. For standard air quality networks with an extensive period of record, analysis of residuals, visual inspection of scatter diagrams, and comparison of cumulative frequency distributions are quite useful techniques for assessing model performance. For tracer studies the spatial coverage is better, so that identification of maximum measured concentrations during each test is more feasible. However, temporal coverage is more limited with a specific number of tests not continuous in time.

The evaluations cited in the following sections are examples of the use of various measures of performance.

B. Performance of Single-Source Models

Since wind direction changes with height above ground and plume rise will cause the height of the plume to vary in time, it is difficult to model accurately plume transport direction. Because of this transport wind direction error, analyses of residuals and correlations are not frequently utilized for assessment of single-source model performance. Instead, cumulative frequency distributions of estimated and measured concentrations for specific averaging times at each location are examined, as well as comparison of the highest concentrations such as the five highest for each averaging time of interest, e.g., 3 and 24 h.

The performance of one specific model, CRSTER (47), for the second-highest once-a-year 24-h concentrations, is summarized in Fig. 20-2 in terms of ratios of estimates to measurements. Overestimates by more than a factor of 2 occur for all receptors whose elevations are near or exceed stack top. The value ΔE is the elevation of the receptor minus the elevation of the stack base; h is the physical stack height. Over seven sites, having stacks varying from 81 to 335 m, for receptors having elevations above the stack base less than 0.7 of the stack height, the second-highest, once-a-year 24-h estimates were within a factor of 2 of the measurements for 25 of 35 monitoring stations.

Comparisons (49) of measured concentrations of SF_6 tracer released from a 36-m stack, and those estimated by the PTMPT model for 133 data pairs over Pasquill stabilities varying from B through F, had a linear correlation coefficient of 0.81. Here 89% of the estimated values were within a factor of 3 of the measured concentrations. The calculations were most sensitive to the selection of stability class. Changing the stability classification by one varies the concentration by a factor of 2 to 4.

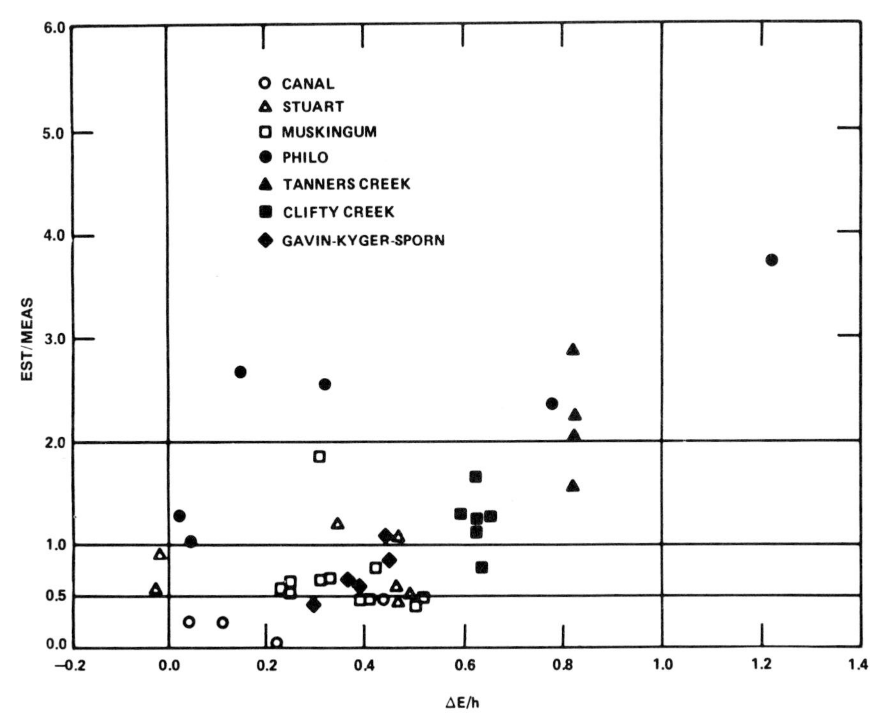

Fig. 20-2. Ratio of second highest 24-hr estimated concentrations from the CRSTER model (47) to measured concentrations as a function of the excess of receptor elevation over stack base evaluation ΔE relative to the stack height h. Names with each symbol are power plants. Source: From Turner and Irwin (48).

C. Performance of Urban Models for Nonreactive Pollutants

Because there are multiple sources of most pollutants in urban areas, and because the meteorology of urban areas is modified so that the extremes of stability are avoided, concentrations tend to vary much less in urban than in rural areas.

For sources having a large component of emissions from low-level sources, the simple Gifford–Hanna model given previously as Eq. (20-19), $\chi = Cq_a/u$, works well, especially for long-term concentrations, such as annual ones. Using the derived coefficients of 225 for particulate matter and 50 for SO_2, an analysis of residuals (measured minus estimated) of the dependent data sets (those used to determine the values of the coefficient C) of 29 cities for particulate matter and 20 cities for SO_2 and an independent data set of 15 cities for particulate matter is summarized in Table 20-1. For the dependent data sets, overestimates result. The standard deviations of the residuals and the mean absolute errors are about equal for particulates and sulfur dioxide. For the independent data set the mean residual shows

TABLE 20-1

Residual Analysis for the Gifford-Hanna Model (Eq. 18-19)

Pollutant	N	$\overline{M}$	S_M	$\overline{E}$	S_E	$\overline{d}$	S_d	$\lvert \overline{d} \rvert$
Particulate	29	110.7	27.8	171.9	111.5	−61.2	103.7	81.3
SO_2	20	89.8	83.1	116.25	72.0	−26.5	101.2	82.5
Independent data set particulate	15	91.6	38.7	93.2	51.7	−1.6	24.9	21.3

Note: N is the number of cities, M is measured, E is estimated from the model, d is residual (measured − estimated) and S values are standard deviations.
Source: Data are from Gifford and Hanna (23).

only slight overestimation, and the standard deviation of residuals and mean absolute error are considerably smaller.

A version of the Gifford–Hanna model was evaluated (50) using 1969 data for 113 monitoring stations for particulate matter and 75 stations for SO_2 in the New York metropolitan area. This version differed from Eq. (20-19) in considering major point source contributions and the stack height of emission release. This model produced results (Table 20-2) comparable to those of the much more complicated CDM model (51).

The urban RAM model was evaluated (52) using 1976 sulfur dioxide data from 13 monitoring locations in St. Louis on the basis of their second-highest once-a-year concentrations. The ratio of estimated to measured 3-h average concentrations was from 0.28 to 2.07, with a median of 0.74. Half of the values were between 0.61 and 1.11. For the 24-h average concentrations the ratios ranged from 0.18 to 2.31, with a median of 0.70. Half of the values were between 0.66 and 1.21. Thus, the urban RAM model generally underestimates concentrations by about 25%.

TABLE 20-2

Residual Analysis Using 1969 New York Data[a]

Model	Pollutant	N	$\overline{M}$	S_m	$\overline{E}$	S_d	$\lvert \overline{d} \rvert$
CDM	Particulate	113	82	23	74	22	16
6B	Particulate	113	82	23	67	25	19
CDM	SO_2	75	135	72	138	52	37
6B	SO_2	75	135	72	127	56	38

[a] Adapted from Turner *et al.* (50).
Note: N is the number of monitoring stations, M is measured, E is estimated from the model, d is residual (measured − estimated), and S values are standard deviations.

D. Performance of Photochemical Models

The performance of photochemical models is evaluated by their ability to estimate the magnitude, time, and location of occurrence of the secondary pollutant oxidant (ozone). Several photochemical models were evaluated using 10 days' data from St. Louis (Table 20-3). Among those evaluated were the Photochemical Box Model PBM (6), which features a single variable-volume reacting cell. This model is most appropriate for use with light winds, so that most of the emissions will remain in the cell and react, rather than being rapidly transported through the downwind side of the cell. The maximum measured concentrations for the 10 days ranged from 0.08 to 0.18 ppm; the maximum model estimates ranged from 0.07 to 0.22 ppm.

The second model evaluated was the Lagrangian Photochemical Model LPM (54), a trajectory model. Backward trajectories were first determined so that starting positions could be used which would allow trajectories to reach station locations at the times of measurement. Measured concentrations ranged from 0.20 to 0.26 ppm and estimated concentrations from 0.05 to 0.53 ppm.

The third model evaluated was the Urban Airshed Model UAM (55), a three-dimensional grid-type model. The area modeled was 60 by 80 km, consisting of cells 4 km on a side. There are four layers of cells in the vertical, the bottom simulating the mixing layer. The model provides both spatially and temporally resolved estimates. Hour-averaged estimates of each pollutant species at each monitoring site within the model domain are given. Measured concentrations ranged from 0.17 to 0.24 ppm and estimated concentrations ranged from 0.10 to 0.17 ppm. The model underestimates the concentration at the point of the observed maximum each day.

Seinfeld (33) indicates that photochemical models estimating peak ozone concentrations in urban areas are generally within 30% of measured peaks.

TABLE 20-3

Residual Analyses of Three Photochemical Models

		Ozone (ppm)		
Model	N	$\bar{d}$	S_d	$\lvert \bar{d} \rvert$
Photochemical box model	10	−0.033	0.041	0.039
Lagrangian photochemical model	10	−0.004	0.110	0.080
Urban airshed model	10	0.074	0.033	0.074

Note: N is the number of days, d is residual (measured − estimated), and S is the standard deviation.
Source: Adapted from Shreffler and Schere (53).

E. Utilization of Models

For the air quality manager to place model estimates in the proper perspective to aid in making decisions, it is becoming increasingly important to place error bounds about model estimates. In order to do this effectively, a history of model performance under circumstances similar to those of common model use must be established for the various models. It is anticipated that performance standards will eventually be set for models.

F. Modeling to Meet Regulatory Requirements

In the United States if the anticipated air pollution impact is sufficiently large, modeling has been a requirement for new sources in order to obtain a permit to construct. The modeling is conducted following guidance issued by the U.S. Environmental Protection Agency (56, 57). The meeting of all requirements is examined on a pollutant-by-pollutant basis. Using the assumptions of a design that will meet all emission requirements, the impact of the new source, which includes all new sources and changes to existing sources at this facility, is modeled to determine pollutant impact. This is usually done using a screening-type model such as SCREEN (58). The impacts are compared to the modeling significance levels for this pollutant for various averaging times. These levels are generally about 1/50 of the National Air Quality Standards. If the impact is less than the significance level, the permit can usually be obtained without additional modeling. If the impact is larger than the significance level, a radius is defined which is the greatest distance to the point at which the impact falls to the significance level. Using this radius, a circle is defined which is the area of significance for this new facility. All sources (not only this facility, but all others) emitting this pollutant are modeled to compare anticipated impact with the National Ambient Air Quality Standards and with the Prevention of Significant Deterioration increments.

The implementation of the Clean Air Act Amendments of 1990 (CAAA90) will require not only permitting for new sources but also permits for existing facilities. There is also a requirement for reexamination of these permits at intervals not longer than 5 years. The permit programs are set up and administered by the states under the review and guidance of EPA. These programs include the collection of permit fees sufficient to support the program. Implementation of CAAA90 will require the application of Maximum Achievable Control Technology (MACT) on an industry-by-industry basis as specified by EPA. Following the use of MACT, a calculation will be made of the residual risk due to the remaining pollutant emissions. This will be accomplished using air quality dispersion modeling that is interpreted using risk factors derived from health information with the inclusion of appropriate safety factors.

Because of their inclusion of the effects of building downwash, the ISC2 (Industrial Source Complex) models (16), ISCST and ISCLT (short-term performing calculations by hourly periods and long-term using the joint frequencies of wind direction, wind speed, and Pasquill stability, commonly referred to as STAR, STability ARray data) are recommended. The ISC2 models differ only slightly from the earlier ISC models (several small errors that make little difference in the resulting calculations were corrected). The recoding of the model to result in ISC2 was requested by EPA to overcome a difficult-to-understand code that had resulted from a number of changes that had been incorporated. The ISC models will make calculations for three types of sources: point, area, and volume. There are recognized deficiencies in the area source algorithm incorporated in the ISC models. EPA is supporting work to create an improved algorithm for incorporation in the models.

G. Availability of Models

Until 1989 models were made available from the National Technical Information Service of the Department of Commerce on 9-track magnetic tape. EPA's Office of Research and Development explored the use of electronic bulletin boards in the mid-1980s and began 24-hour operation of the UNAMAP bulletin board in the fall of 1987. This has continued as the AMRB (Applied Modeling Research Branch) Bulletin Board Service (BBS) furnishing modeling information from the Office of Research and Development. To use a BBS, all that is needed is a personal computer with a modem (to connect to a telephone line) and some file-transfer protocol such as Kermit or Ymodem. The AMRB BBS in Research Triangle Park, NC is reached on (919) 541-1325 and the system operator on (919) 541-1376.

Seeing the success of the UNAMAP BBS, EPA's Office of Air Quality Planning and Standards started a BBS for information on regulatory models in June 1989. This has expanded to a BBS called TTN, Technology Transfer Network. This BBS, in Durham, NC, is reached on (919) 541-5742 and the system operator on (919) 541-5384. A part of this BBS called SCRAM , Support Center for Regulatory Air Models, contains model FORTRAN codes, model executable codes for use on personal computers, meteorological data, and in some cases model user's guides. Much of the information is downloaded in "packed" form, and software to unpack the files must also be downloaded from the bulletin board.

Private vendors are another source of models and information. Specially designed packages are available for many of the regulatory approved models. These packages usually provide easier data entry through use of data-entry screens and capability of output graphics. The data-entry screens usually save professional time, especially for occasional model users. Additional advantages are technical support from the vendor and, if a maintenance agreement has been purchased, obtaining the latest version of the model without having to check the sources of model information periodically.

REFERENCES

1. Briggs, G. A., "Plume Rise." United States Atomic Energy Commission Critical Review Series, TID-25075. National Technical Information Service, Springfield, VA, 1969.
2. Briggs, G. A., Some recent analyses of plume rise observation, *in* "Proceedings of the Second International Clean Air Congress" (H. M. Englund and W. T. Beery, eds.). Academic Press, New York, 1971, pp. 1029–1032.
3. Briggs, G. A., *Atmos. Environ.* **6**, 507–510 (1972).
4. Briggs, G. A., "Diffusion Estimation for Small Emissions." Atmospheric Turbulence and Diffusion Laboratory, Contribution File No. 79. (draft). Oak Ridge, TN, 1973.
5. Briggs, G. A., Plume rise predictions, *in* "Lectures on Air Pollution and Environmental Impact Analysis" (D. A. Haugen, ed.). American Meteorological Society, Boston, 1975, Chapter 3 (59–111).
6. Schere, K. L., and Demerjian, K. L., A photochemical box model for urban air quality simulation, *in* "Proceedings of the Fourth Joint Conference on Sensing of Environmental Pollutants." American Chemical Society, Washington, DC, 1978, pp. 427–433.
7. Hanna, S. R., *J. Air Pollut. Control Assoc.* **21**, 774–777 (1971).
8. Novak, J. H., and Turner, D. B., *J. Air Pollut. Control Assoc.* **26**, 570–575 (1976).
9. Egan, B. A., and Mahoney, J., *J. Appl. Meteorol.* **11**, 312–322 (1972).
10. Long, P. E., and Pepper, D. W., A comparison of six numerical schemes for calculating the advection of atmospheric pollution, *in* "Proceedings of the Third Symposium on Atmospheric Turbulence, Diffusion and Air Quality." American Meteorological Society, Boston, 1976, pp. 181–186.
11. Businger, J. A., Wyngaard, J. C., Izumi, Y., and Bradley, E. F., *J. Atmos. Sci.* **28**, 181–189 (1971).
12. Smith, F. B., A scheme for estimating the vertical dispersion of a plume from a source near ground level, *in* "Proceedings of the Third Meeting of the Expert Panel on Air Pollution Modeling." North Atlantic Treaty Organization Committee on the Challenges of Modern Society Pub. No. 14. Brussels, 1972. (National Technical Information Service PB 240-574.)
13. Shir, C. C., *J. Atmos. Sci.* **30**, 1327–1339 (1973).
14. Hanna, S. R., Briggs, G. A., and Hosker, R. P., Jr., "Handbook on Atmospheric Diffusion," DOE/TIC-11223. Technical Information Center, U.S. Department of Energy, Oak Ridge, TN, 1982.
15. Eschenroeder, A. Q., and Martinez, J. R., "Mathematical Modeling of Photochemical Smog," No. IMR- 1210. General Research Corp., Santa Barbara, CA, 1969.
16. U.S. Environmental Protection Agency, User's Guide for the Industrial Source Complex (ISC2) Dispersion Models. Vol. I—User Instructions, Vol. II—Description of Model Algorithms, Vol. III—Guide to Programmers. Technical Support Division, Office of Air Quality Planning and Standards, Research Triangle Park, NC, 1992.
17. Hanna, S. R., and Paine, R. J., *J. Appl. Meteorol.* **28**, 3, 206–224 (1989).
18. Hanna, S. R., and Chang, J. C., Modification of the Hybrid Plume Dispersion Model (HPDM) for urban conditions and its evaluation using the Indianapolis data set. Report

Number A089-1200.I, EPRI Project No. RP-02736-1. Prepared for the Electric Power Research Institute by Sigma Research Corporation, Westford, MA, 1990.

19. Turner, D. B., Chico, T., and Catalano, J. A., TUPOS—A multiple source Gaussian dispersion algorithm using on-site turbulence data, EPA/600/8-86/010. U.S. Environmental Protection Agency, Research Triangle Park, NC, 1986.
20. Turner, D. B., Bender, L. W., Paumier, J. O., and Boone, P. F., *Atmos. Environ.* **25A,** 2187–2201 (1991).
21. Petersen, W. B., "User's Guide for PAL—A Gaussian-Plume Algorithm for Point, Area, and Line Sources," Pub. No. EPA-600/4-78-013. U.S. Environmental Protection Agency, Research Triangle Park, NC, 1978.
22. Gifford, F. A., and Hanna, S. R., Urban air pollution modeling, *in* "Proceedings of the Second International Clean Air Congress (H. M. Englund and W. T. Beery, eds.). Academic Press, New York, 1971, pp. 1146–1151.
23. Gifford, F. A., and Hanna, S. R., *Atmos. Environ.* **7,** 131–136 (1973).
24. Hanna, S. R., *J. Air Pollut. Control Assoc.* **28,** 147–150 (1978).
25. Winges, K. D., User's Guide for the Fugitive Dust Model (FDM) (revised), EPA-910/9-88-202R. U.S. Environmental Protection Agency, Region 10, Seattle, WA, 1990.
26. Hanna, S. R., Schulman, L. L., Paine, R. J., and Pleim, J. E., "User's Guide to the Offshore and Coastal Dispersion (OCD) Model," DOI/SW/MT-88/007a. Environmental Research & Technology, Concord, MA for Minerals Management Service, Reston, VA, 1988. (NTIS Accession Number PB88-182 019.)
27. Perry, S. G., Burns, D. J., Adams, L. A., Paine, R. J., Dennis, M. G., Mills, M. T., Strimaitis, D. G., Yamartino, R. J., and Insley, E. M., "User's Guide to the Complex Terrain Dispersion Model plus Algorithms for Unstable Conditions (CTDMPLUS)," Vol. I: "Model Description and User Instructions," EPA/600/8-89/041. U.S. Environmental Protection Agency, Research Triangle Park, NC, 1989.
28. Perry, S. G., Paumier, J. O., and Burns, D. J., Evaluation of the EPA Complex Terrain Dispersion Model (CTDMPLUS) with the Lovett Power Plant Data Base, pp 189–192 *in* "Preprints of Seventh Joint Conference on Application of Air Pollution Meteorology with AWMA," Jan. 14–18, 1991, New Orleans, American Meteorological Society, Boston, 1991.
29. Burns, D. J., Perry, S. G., and Cimorelli, A. J., An advanced screening model for complex terrain applications, pp. 97–100 *in* "Preprints of Seventh Joint Conference on Application of Air Pollution Meteorology with AWMA," Jan. 14–18, 1991, New Orleans. American Meteorological Society, Boston, 1991.
30. Gartrell, F. E., Thomas, F. W., and Carpenter, S. B., *Am. Ind. Hyg. J.* **24,** 113–120 (1963).
31. Liu, M. K., Stewart, D. A., and Roth, P. M., An improved version of the reactive plume model (RPM-II), *in* "Ninth International Technical Meeting on Air Pollution Modeling." North Atlantic Treaty Organization Committee on the Challenges of Modern Society Pub. No. 103, Umweltbundesamt, Berlin, 1978.
32. Demerjian, K. L., and Schere, K. L., Application of a photochemical box model for O_3 air quality in Houston, TX, *in* "Proceedings of Ozone/Oxidants: Interactions with the Total Environment II." Air Pollution Control Association, Pittsburgh, 1979, pp. 329–352.
33. Seinfeld, J. H., *J. Air Pollut. Control Assoc.* **38,** 616–645 (1988).
34. Schere, K. L., *J. Air Pollut. Control Assoc.* **38,** 1114–1119 (1988).
35. Viebrock, H. J., (ed.), "Fiscal Year 1980 Summary Report of NOAA Meteorology Laboratory Support to the Environmental Protection Agency." National Oceanic and Atmospheric Administration Tech. Memo. ERL ARL-107. Air Resources Laboratories, Silver Spring, MD, 1981.
36. Lamb, R. G., "A Regional Scale (1000 km) Model of Photochemical Air Pollution. Part

I: Theoretical Formulation," EPA-600/3-83-035. U.S. Environmental Protection Agency, Research Triangle Park, NC, 1983.

37. Johnson, W. B., Wolf, D. E., and Mancuso, R. L., *Atmos. Environ.* **12,** 511–527 (1978).
38. Bhumralkar, C. M., Johnson, W. B., Mancuso, R. L., Thuillier, R. A., Wolf, D. E., and Nitz, K. C., Interregional exchanges of airborn sulfur pollution and deposition in Eastern North America, *in* "Conference Papers, Second Joint Conference on Applications of Air Pollution Meteorology." American Meteorological Society, Boston, 1980, pp. 225–231.
39. Bencala, K. E., and Seinfeld, J. H., *Atmos. Environ.* **13,** 1181–1185 (1979).
40. Fox, D. G., *Bull. Am. Meteorol. Soc.* **62,** 599–609 (1981).
41. Gryning, S. E., and Lyck, E., "Comparison between Dispersion Calculation Methods Based on In-Situ Meteorological Measurements and Results from Elevated- Source Tracer Experiments in an Urban Area." National Agency of Environmental Protection, Air Pollution Laboratory, MST Luft-A40. Riso National Laboratory, Denmark, 1980.
42. Cox, W. M., and Tikvart, J. A., *Atmos. Environ.* **24A,** 2387–2395 (1990).
43. Hanna, S. R., *J. Air Pollut. Control Assoc.* **38,** 406–412 (1988).
44. Weil, J. C., Sykes, R. I., and Venkatram, A., *J. Appl. Meteorol.* **31,** 1121–1145 (1992).
45. Hanna, S. R., *Atmos. Environ.* **23,** 1385–1398 (1989).
46. Bowne, N. E., Validation and performance criteria for air quality models, *in* "Conference Papers, Second Joint Conference on Applications of Air Pollution Meteorology." American Meteorological Society, Boston, 1980, pp. 614–626.
47. U.S. Environmental Protection Agency, "User's Manual for the Single Source (CRSTER) Model," EPA-450/2-77-013. Research Triangle Park, NC, 1977.
48. Turner, D. B., and Irwin, J. S., *Atmos. Environ.* **16,** 1907–1914 (1982).
49. Guzewich, D. C., and Pringle, W. J. B., *J. Air Pollut. Control Assoc.* **27,** 540–542 (1977).
50. Turner, D. B., Zimmerman, J. R., and Busse, A. D., An evaluation of some climatological dispersion models, *in* "Proceedings of the Third Meeting of the Expert Panel on Air Pollution Modeling." North Atlantic Treaty Organization Committee on the Challenges of Modern Society Pub. No. 14. Brussels, 1972. (National Technical Information Service PB 240-574.)
51. Busse, A. D., and Zimmerman, J. R., "User's Guide for the Climatological Dispersion Model." U.S. Environmental Protection Agency Pub. No. EPA-RA-73-024. Research Triangle Park, NC, 1973.
52. Turner, D. B., and Irwin, J. S., Comparison of sulfur dioxide estimates from the model RAM with St. Louis RAPS measurements, *in* "Air Pollution Modeling and Its Application II" (C. de Wispelaere, ed.). Plenum, New York, 1982.
53. Shreffler, J. H., and Schere, K. L., "Evaluation of Four Urban-Scale Photochemical Air Quality Simulation Models." U.S. Environmental Protection Agency Pub. EPA-600/3-82-043. Research Triangle Park, NC, 1982.
54. Lurmann, F., Godden, D., Lloyd, A. C., and Nordsieck, R. A., "A Lagrangian Photochemical Air Quality Simulation Model." Vol. I, "Model Formulation"; Vol. II, "User's Manual." U.S. Environmental Protection Agency Pub. EPA-600/8-79-015a,b. Research Triangle Park, NC, 1979.
55. Killus, J. P., Meyer, J. P., Durran, D. R., Anderson, G. E., Jerskey, T. N., and Whitten, G. Z., "Continued Research in Mesoscale Air Pollution Simulation Modeling," Vol. V, "Refinements in Numerical Analysis, Transport, Chemistry, and Pollutant Removal," Report No. ES77-142. Systems Applications, Inc., San Rafael, CA, 1977.
56. U.S. Environmental Protection Agency, Guideline on Air Quality Models (Revised). EPA-450/4-80-023R. Office of Air Quality Planning and Standards. Research Triangle Park, NC, 1986. (NTIS Accession Number PB86-245 248.)

57. U.S. Environmental Protection Agency, Supplement A to the Guideline on Air Quality Models (Revised). EPA- 450/2-78-027R. Office of Air Quality Planning and Standards. Research Triangle Park, NC, 1987.

58. Brode, R. W., Screening Procedures for Estimating the Air Quality Impact of Stationary Sources. EPA- 450/4-88-010. U. S. Environmental Protection Agency, Research Triangle Park, NC. 1988.

SUGGESTED READING

Benarie, M. M., "Urban Air Pollution Modeling." MIT Press, Cambridge, MA, 1980.

De Wispelaere, C., (ed.), "Air Pollution Modeling and Its Application I." Plenum, New York, 1980.

Electric Power Research Institute, "Survey of Plume Models for Atmospheric Application," Report No. EPRI EA-2243. System Application, Inc., Palo Alto, CA, 1982.

Huber, A. H., Incorporating building/terrain wake effects on stack effluents, pp. 353–356 *in* Preprints, Joint Conference on Applications of Air Pollution Meteorology. November 29–December 2, 1977, Salt Lake City, UT. American Meteorological Society, Boston, MA, 1977.

Huber, A. H., and Snyder, W. H., Building wake effects on short stack effluents, pp. 235–242 *in* Preprints, Third Symposium on Atmospheric Turbulence, Diffusion and Air Quality. October 19–22, 1976, Raleigh, NC. American Meteorological Society, Boston, 1976.

Nieuwstadt, F. T. M., and Van Dop, H., "Atmospheric Turbulence and Air Pollution Modeling." Reidel, Dordrecht, 1982.

Schulman, L. L., and Hanna, S. R., Evaluation of downwash modifications to the Industrial Source Complex model. *J. Air Pollut. Control Assoc.* **36**(3), 258–264 (1986).

Szepesi, D. J., "Compendium of Regulatory Air Quality Simulation Models." Akadémiai Kiadó es Nyomda Vállalat, Budapest, 1989.

Venkatram, A., and Wyngaard, J. C. (eds.), "Lectures on Air Pollution Modeling." American Meteorological Society, Boston, 1988.

Watson, J. G. (ed.), "Receptor Models in Air Resources Management." APCA Transaction Series, No. 14. Air and Waste Management Association, Pittsburgh, 1989.

Wayne, R. P., "Principals and Applications of Photochemistry." Oxford University Press, New York, 1988.

Zannetti, P., "Air Pollution Modeling: Theories, Computational Methods and Available Software." Van Nostrand Reinhold, Florence, KY, 1990.

QUESTIONS

1. Assuming that the buoyancy flux parameter F is greater than 55 in both situations, what is the proportional final plume rise for stack A compared to stack B if A has an inside diameter three times that of B?
2. How much greater is the penetration of a plume through an inversion of 1°C per 100 m than through an inversion of 3°C per 100 m? Assume that the wind speed is 3 m s^{-1}, ambient air temperature is 293 K and the stack characteristics are T_s = 415 K, d = 3 m, and v_s = 20 m s^{-1}.

3. What is the steady-state concentration derived from the box model for a 10-km city with average emissions of 2×10^{-5} g m^{-2} s^{-1} when the mixing height is 500 m and the wind speed is 4 m s^{-1}?
4. In formulating and applying a gradient transfer model, what are two of the major difficulties?
5. What is the advantage in using trajectory models for estimating air pollutant concentrations at specific air monitoring stations?
6. What is a major difficulty in estimating the maximum short-term (hours) impact of two point sources 1 km apart?
7. Using simplified techniques for estimating the concentrations from area sources, what is the annual average particulate matter concentration for a city with an average wind speed of 3.6 m s^{-1} and area emission rate of 8×10^{-7} g s^{-1} m^{-2}?
8. What are the major limitations in modeling pollutant transformations in urban areas?
9. From the results of the application of the EURMAP model to Europe, what pollutant and mechanism seem to cause the least pollution by a nation to itself?
10. Which measure of scatter is likely to be larger, the mean absolute error or the root-mean-square error?
11. Contrast the fractional error for a measurement of 20 and an estimate of 4 to the fractional error for a measurement of 4 and an estimate of 20.

21

Air Pollution Climatology

Climatology refers to averaged or analyzed meteorology over a period of record, usually several years. Air pollution climatology involves meteorological variables that are important in air pollution. Alternatively, it is the interpretation of air pollution data from a meteorological perspective.

I. SOURCES OF DATA

There are numerous sources of meteorological data. Hourly observations, primarily to support forecast programs and aviation operations, are made 24 h a day. Observations throughout the world, including those of over 200 stations in the contiguous United States, are also made at other intervals, when the weather is changing significantly. Since January 1, 1966, when archiving of each hourly U.S. observation in a computer-compatible form was discontinued as an economy move, only every third hour (00 GMT plus every 3 h) has been readily accessible. The other observations are available as reproductions of manually recorded observations and may be specially prepared on magnetic tape at cost for computer use. The U.S. archive for such data is the National Oceanic and Atmospheric Administration's (NOAA) National Climate Center in Asheville, North Carolina. The data available from the hourly observations are listed in Table 21-1.

TABLE 21-1

Hourly Surface Observation Variables

Station number, five digits[a]
Date—year, month, day, six digits[a]
Hour, two digits[a]
Ceiling height, hundreds of feet, three alphanumeric characters[a]
Sky condition, up to four layers, four alphanumeric characters
Visibility, miles, three digits (coded)
Weather and/or obstructions to vision, eight alphanumeric characters
Sea-level pressure, millibars, four digits
Dew point, °F, three digits
Wind direction, tens of degrees azimuth, two digits[a]
Wind speed, knots, two digits[a]
Station pressure, inches of mercury, four digits
Dry bulb temperature, °F, three digits[a]
Wet bulb temperature, °F, three digits
Relative humidity, percent, three digits
Clouds and obscuring phenomena
Total amount, tenths, one coded alphanumeric character[a]
Following for up to four layers:
Amount, tenths, one coded alphanumeric character
Type, one coded alphanumeric character
Height, hundreds of feet, three alphanumeric characters
Amount of opaque cloud cover, tenths, one alphanumeric character[a]

[a] Of particular interest in air pollution work.

Other data, gathered primarily once each day by cooperative observers, consist mostly of temperature and precipitation readings. These are of limited usefulness for air pollution analysis because wind data are generally lacking.

Other sources of data, especially wind data, may be routinely measured by industrial or commercial establishments. Availability of these data must be ascertained through contact with each data collector.

Many city and regional agencies responsible for air pollutant measurements also measure wind and temperature at some of their air pollutant sampling stations. Because exposure at air quality stations is generally considerably less ideal than at airport stations, the data may be representative of extremely local conditions.

Radiosonde balloons are released twice daily, near 00 and 12 GMT. Measurements of temperature and humidity, alternated by a pressure switch, are transmitted by radio signals from the instrument package, which is also tracked by ground-based radio direction-finding equipment at the point of release. This allows computation of wind direction and wind speed at numerous heights above ground. Figure 21-1 shows the locations of radiosonde stations throughout the world, including over 60 locations in the contiguous United States.

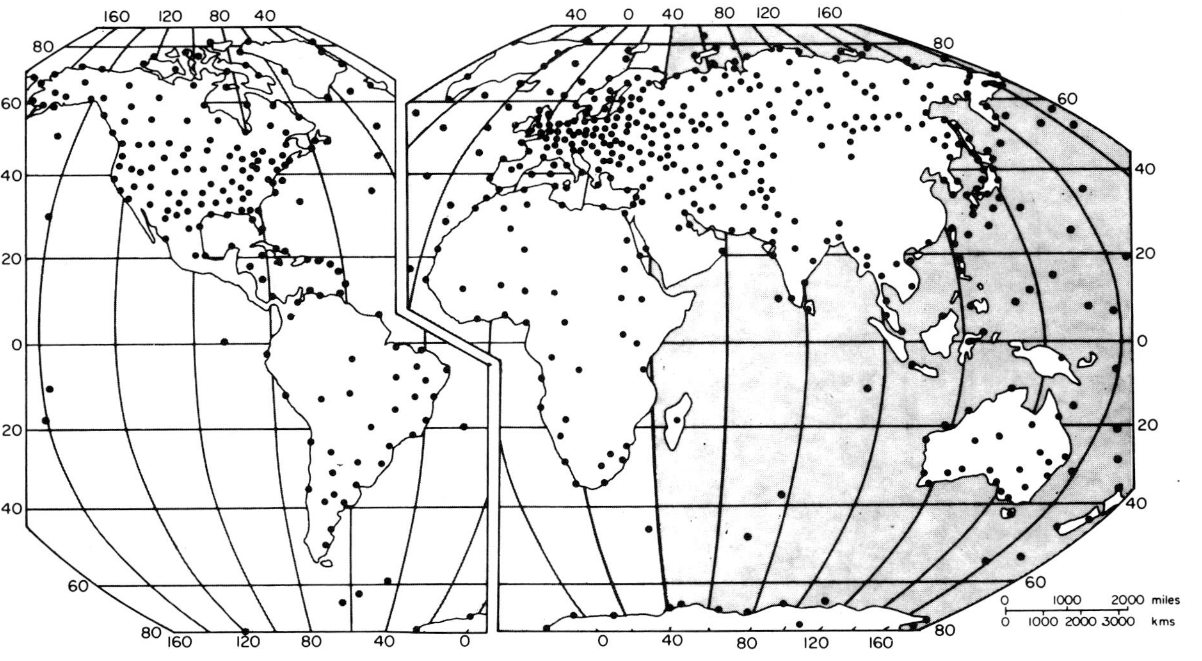

Fig. 21-1. World network of radiosonde stations. Ten stations, one in the Atlantic and nine in the Pacific, are not shown. Each dot represents a station at which an upper-air sounding is made each day at 0000 hr GMT, at 1200 hr GMT, or both. Source: the Secretary-General, World Meteorological Organization, Geneva.

Numerous analyses of data routinely collected in the United States have been performed by the U.S. National Climatic Center, results of these analyses are available at reasonable cost. The joint frequency of Pasquill stability class, wind direction class (primarily to 16 compass points), and wind speed class (in six classes) has been determined for various periods of record for over 200 observation stations in the United States from either hourly or 3-hourly data. A computer program called STAR (STability ARray) estimates the Pasquill class from the elevation of the sun (approximated from the hour and time of year), wind speed, cloud cover, and ceiling height. STAR output for seasons and the entire period of record can be obtained from the Center. Table 21-2 is similar in format to the standard output. This table gives the frequencies for D stability, based on a total of 100 for all stabilities.

Additional tables are furnished for the other stability classes. Note that calms have been distributed among the directions. Such joint frequency data can be used directly in climatological models such as the Climatological Dispersion Model (CDM) (1). The CDM calculates seasonal or annual concentrations at each receptor by considering sources in each wind sector

TABLE 21-2

Relative Frequency of Winds for D Stability (O'Hare Airport, Chicago, 1965–1969)

	Speed (knots)						
Direction	0–3	4–6	7–10	11–16	17–21	>21	Total
N	0.0885	0.7123	1.3492	1.2670	0.1301	0.0411	3.5882
NNE	0.0646	0.5342	1.0547	1.1712	0.2123	0.0959	3.1328
NE	0.0605	0.4589	1.3972	1.0958	0.0959	0.0411	3.1494
ENE	0.0258	0.2123	0.9246	0.7260	0.0616	0.0068	1.9571
E	0.0521	0.3013	0.9109	0.8972	0.0342	0.0068	2.2027
ESE	0.0847	0.5068	0.9109	0.4794	0.0205	0.0068	2.0093
SE	0.0829	0.4726	0.6575	0.3150	0.0137	—	1.5417
SSE	0.0714	0.5274	0.9383	0.5890	0.0616	—	2.1877
S	0.1818	1.1095	2.7190	2.4245	0.2534	0.0479	6.7361
SSW	0.1495	0.7739	1.8423	2.2670	0.2397	0.0548	5.3272
SW	0.0985	0.6301	1.5889	1.4520	0.1781	0.0342	3.9818
WSW	0.1368	0.6712	1.2328	1.1712	0.2603	0.0822	3.5544
W	0.2485	1.0068	1.7191	2.0273	0.3698	0.0753	5.4467
WNW	0.1477	0.7397	1.4109	1.4794	0.2534	0.0274	4.0584
NW	0.1292	0.6643	1.3013	0.9999	0.0753	0.0068	3.1769
NNW	0.0349	0.3835	0.9109	1.0205	0.1781	0.0274	2.5553
Total	1.6574	9.7048	20.8684	19.3822	2.4382	0.5548	
Relative frequency of occurrence of D stability[a]							54.6058
Relative frequency of calms distributed above with D stability = 0.5753							

[a] Total frequency of all stability classes is 100.

(1/16 of the compass), performing a calculation for each wind speed–stability combination occurring for that sector, and weighting the calculation by the frequency for this combination. For annual concentrations, this saves a considerable number of calculations compared with simulating the period hour by hour.

Mixing heights for each day can be calculated from the radiosonce data. Such data for a 5-year period of record, 1960–1964, were calculated and used in a study by Holzworth (2). Figure 21-2 shows the mean annual afternoon mixing height variation across the contiguous United States.

II. REPRESENTATIVENESS

The term *representativeness* in air pollution meteorology usually means the extent to which a particular parameter is measured by instrumentation sited in such a way and with sensitivity and accuracy such that it is useful for the designated purpose. For normal climatological purposes, wind measurements are made at locations relatively free from observation; thus, they are not influenced in different ways by winds coming from different directions and consequently present an unbiased record. A parameter such as wind, which varies with height above ground, must have the height of the measurement reported along with the data.

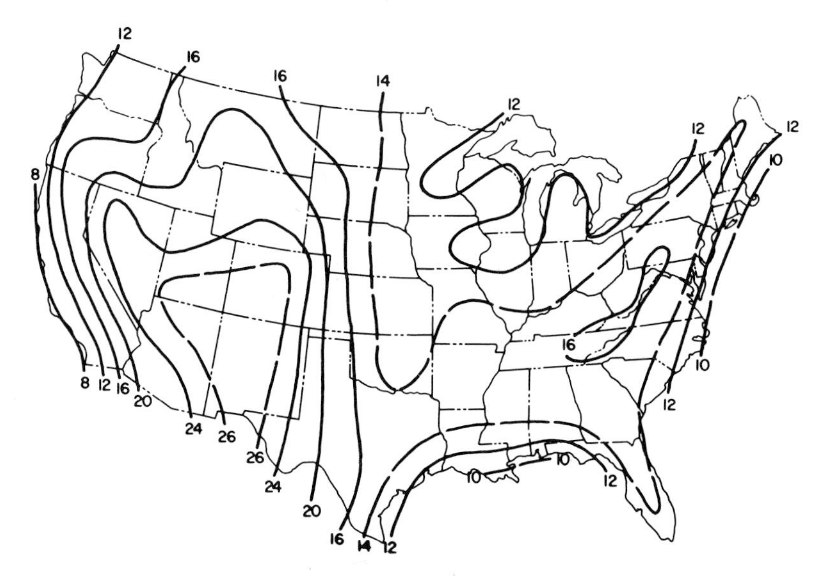

Fig. 21-2. Mean annual afternoon mixing height, in hundreds of meters. Source: Adapted from Holzworth (2).

The use of a measurement generally dictates the circumstances of data collection. For example, to provide a best estimate of plume transport direction, hour by hour, of a release from a 75-m stack, a wind vane at the 100-m level of a tower will probably provide more representative wind direction measurements than a vane at 10 m above ground. If the release has buoyancy so that it rises appreciably before leveling off, even the 100-m measurement may not be totally adequate.

If one is studying the transport of material through the tree canopy of a forest, it is most desirable to disturb the natural environment as little as possible in making a wind measurement in the canopy. An extremely sensitive wind system is necessary because one would expect the winds to be extremely light. Also, it may be necessary to make supporting measurements both above and below the canopy, so that a wind speed profile is obtained.

Frequently, it is necessary to determine whether a record made at one location can be used to infer values of a particular parameter at another location. Many factors help determine whether one location's measurements of a parameter are similar to those made at another location, such as the parameter in question, distance between sites, intervening topography, local topography, surface roughness, local vegetative cover, and the height above ground of the measurement. Generally, no measurements are made at both sites to check for comparability; if there were, there would be little need to use measurements at one site as a substitute for those at the other.

The foregoing factors are not independent. Two sites 50 km apart in Kansas, a state that has quite flat terrain largely planted in field crops, may be much more representative of one another than two sites 5 km apart in more rugged mountainous, wooded terrain, e.g., in the western part of North Carolina.

Data for one full year (1964) for Nashville, Tennessee, and Knoxville, Tennessee, 265 km (165 mi) apart, were compared to determine the extent to which the frequencies of various parameters were similar. Knoxville is located in an area with mountainous ridges oriented southwest–northeast; Nashville is situated in a comparatively flat area. The data available are the number of hours during which each of 36 wind directions (every 10° azimuth) occurred, the average wind speed for each direction, the number of hours of each Pasquill stability class for each direction, and the mean annual wind speed.

Table 21-3 indicates that average wind speeds are lower in Knoxville, where the average wind is 0.86 of that at Nashville. Considering the difference in topography, it is surprising that the difference is not greater. The percentage of calms, slightly greater at Knoxville, is entirely consistent with the wind speeds.

Because of the slight difference in average wind speed, one might expect this to cause a greater frequency of both stable and unstable conditions at

TABLE 21-3

Average Wind Speed and Frequency of Calms

	Nashville	Knoxville
Average wind speed	3.79 m s^{-1}	3.26 m s^{-1}
Percentage of calms	9.21	9.77

Knoxville compared to Nashville. The stability comparisons are given in Table 21-4. The frequencies are very nearly the same, with only A and B stabilities being slightly greater at Knoxville at the expense of the D stability. Stability G is more stable than Pasquill's F.

The maximum number of hours of each stability class in a single wind direction is given in Table 21-5. The total hours for A, C, and D stabilities are nearly the same. The maximum number of hours of B stability, with winds from a single direction, is about 50% higher at Knoxville. For all three stable cases, E, F, and (G), the maximum number of hours at Knoxville is about two- thirds that at Nashville.

The frequency of occurrence of winds from each of the 36 directions for the two sites is given in the second and third columns of Table 21-6. Knoxville has its maximum from 240°, with frequent winds also from adjacent points. There is a secondary maximum across the compass (in the vicinity of 60°). Nashville, on the other hand, seems to have only one principal maximum (from 180°). To explore the magnitude of the differences, the fourth column is the difference for each direction. Summing the absolute values of the differences and expressing them as a percentage of the total number of hours of observations shows that the differences constitute 62% of the total.

TABLE 21-4

Occurrence of Pasquill Stability Classes

Stability class	Nashville		Knoxville	
	Percent	Cumulative	Percent	Cumulative
(G)	7.64	7.64	7.35	7.35
F	14.15	21.79	15.57	22.92
E	14.18	35.97	14.67	37.59
D	45.14	81.11	41.42	79.01
C	12.37	93.48	12.32	91.33
B	5.75	99.23	7.21	98.54
A	0.77	100.00	1.46	100.00

TABLE 21-5

Maximum Hours of Each Stability in One Direction

	Nashville			Knoxville		
Stability	Number of hours	Total for direction	Direction	Number of hours	Total for direction	Direction
A	7	184	22	8	{220 / 364	{12 / 20
B	25	252	16	38	616	6
C	79	736	36	81	616	6
D	324	736	36	332	616	6
E	123	736	36	82	431	22
F	118	736	36	79	342	25
(G)	66	736	36	45	174	31

It is noted that backing the Knoxville frequencies by 60° would result in the maxima occurring together. This is tabulated in the fifth column of Table 21-6. The difference between Nashville and this artificial frequency is then obtained (the sixth column). The sum of the absolute values of the differences expressed as a percentage of the total is 30.2; the differences have been reduced by more than half. This does *not* prove that the frequencies are similar except that the winds at Knoxville are channeled by the major topographical features in the vicinity, but it does imply that this is an explanation.

Certainly the directional frequencies of these sites cannot be considered as representative of each other. However, this comparison would seem to indicate that the percentage of the stability categories may be more conservative over distance than the direction frequencies.

III. FREQUENCY OF ATMOSPHERIC STAGNATIONS

At times when the surface pressure gradient is weak, resulting in light winds in the atmosphere's lowest layers, and there is a closed high-pressure system aloft, there is potential for the buildup of air pollutant concentrations. This is especially true if the system is slow-moving so that light winds remain in the same vicinity for several days. With light winds there will be little dilution of pollutants at the source and not much advection of the polluted air away from source areas.

Korshover (3) studied stagnating anticyclones in the eastern United States over two periods totaling 30 years. He found that for stagnation to occur for 4 days or longer, the high-pressure system had to have a warm core. Korshover's criteria included a wind speed of 15 knots or less, no frontal

TABLE 21-6

Number of Hours of Wind from Each Direction (Calms Are Distributed)

Direction (tens of degrees)	Hours from this direction		Nashville –Knoxville	Knoxville backed 60°	Nashville –Knoxville backed 60°
	Nashville	Knoxville			
1	140	279	−139	342	−202
2	201	364	−163	204	−3
3	158	295	−137	213	−55
4	184	431	−247	119	65
5	138	413	−275	104	34
6	165	426	−261	158	7
7	187	342	−155	174	13
8	134[a]	204	−70	157	−23
9	202	213	−11	81	121
10	146	119	27	86	60
11	167	104	63	67	100
12	170	158	12	137	33
13	155	174	−19	110	45
14	262	157	105	152	110
15	245	81	164	212	33
16	527	86	441	389	138
17	478	67[a]	411	410	68
18	736[b]	137	599	616	120
19	348	110	238	442	−94
20	303	152	151	302	1
21	226	212	14	343	−117
22	283	389	−106	209	74
23	243	410	−167	240	3
24	257	616[b]	−359	220	37
25	195	442	−247	174	21
26	182	302	−120	129	53
27	215	343	−128	129	86
28	174	209	−35	181	−7
29	231	240	−9	187	44
30	231	220	11	289	−58
31	338	174	164	279	59
32	300	129	171	364	−64
33	212	129	83	295	−83
34	252	181	71	431	−179
35	157	187	−30	413	−256
36	242	289	−47	426	−184
	8784	8784	Sum absolute 5450 = 62.0%		Sum absolute 2650 = 30.2%

[a] Minimum.
[b] Maximum.

areas of precipitation, and persistence of these conditions for 4 or more days.

The numbers of occurrences of 4 days or more over the 30-year period (1936–1965) peak in October and September and reach a minimum in February and March. The total number of stagnation days for each part of the study area is shown in Fig. 21-3.

Using criteria of mixing heights of 1500 m or less, with average speed through the mixing height of 4 m s^{-1} or less, no precipitation, and persistence of these conditions for at least 2 days, Holzworth (2) tabulated high air pollution potential for the contiguous United States expressed as the number of days per 5 years (1960–1964) (Fig. 21-4). The pattern over the eastern United States is very similar to that of Korshover. It also shows no occurrences through the central plains. The number of days in the West exceed those in the East, with the maximum in central California.

IV. VENTILATION CLIMATOLOGY

Hosler (4), through study of radiosonde data for the lowest 500 ft at over 70 locations in the United States, determined inversion frequencies. The

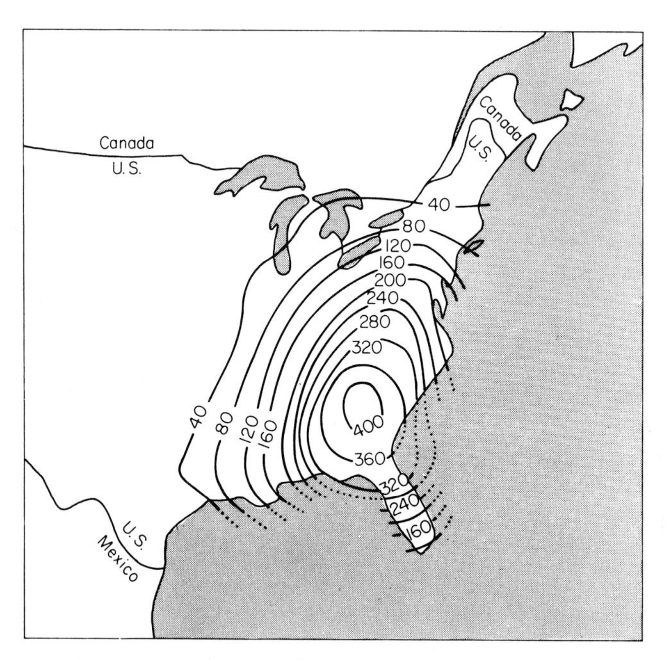

Fig. 21-3. Total number of extreme stagnation days during 1936–1965 east of the Rocky Mountains. Source: Korshover (3).

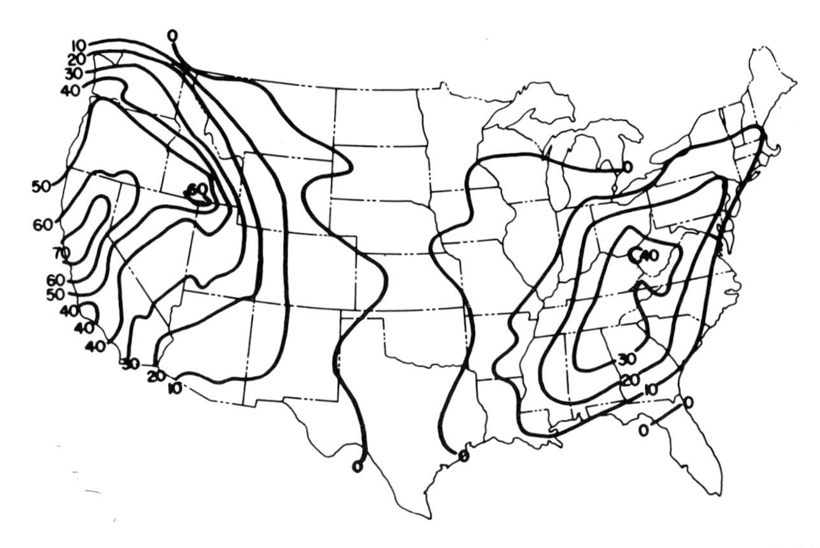

Fig. 21-4. Number of forecast days of high air pollution potential per 5 years (1960–1964). Source: Adapted from Holzworth (2).

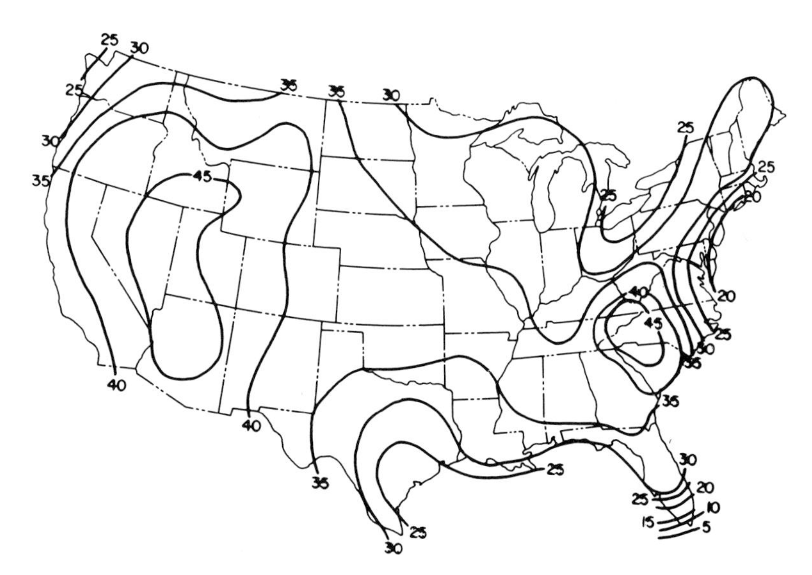

Fig. 21-5. Annual inversion frequency, percent of total hours. Source: Adapted from Hosler (4).

values were used to draw isolines of inversion frequency percentages on U.S. maps for annual values and the four seasons. The percent of total hours of inversions for the annual period is shown in Fig. 21-5. Conditions frequently associated with radiation inversions—light winds and slight cloud cover at night—were also examined in terms of frequency. Both display maxima over the desert Southwest.

The study by Holzworth (2) also examined several other parameters in addition to mixing height. For example, because pollutants are diluted by the wind and mixing height limits the vertical dispersion of pollutants, Holzworth used the radiosonde data to determine the average wind speed through the mixing height for each season and annually. Figure 21-6 shows the distribution of mean annual wind speed averaged through the afternoon mixing layer.

Using the urban model of Miller and Holzworth (5), which requires wind speed and mixing height, Holzworth (2) used the mixing height and wind speed data to calculate concentrations for the median, upper quartile, and upper decile for hypothetical alongwind city lengths of 10 and 100 km. Results for the upper decile for the 10-km city for both the morning and the afternoon are shown in Fig. 21-7.

Another climatological study is of interest. Radiosonde observations for the 5-year period 1960–1964, used previously (2), were analyzed by Holzworth (6) to determine plume rise through the atmosphere's structure for two different stack heights, 50 and 400 m. This encompasses the range of stack heights normally encountered. The annual average effective height

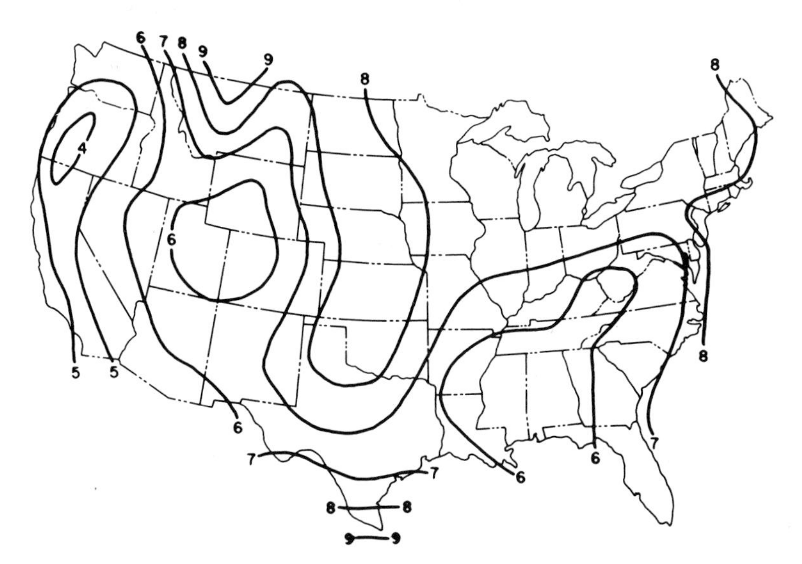

Fig. 21-6. Mean annual wind speed averaged through the afternoon mixing layer. Speeds are in meters per second. Source: Adapted from Holzworth (2).

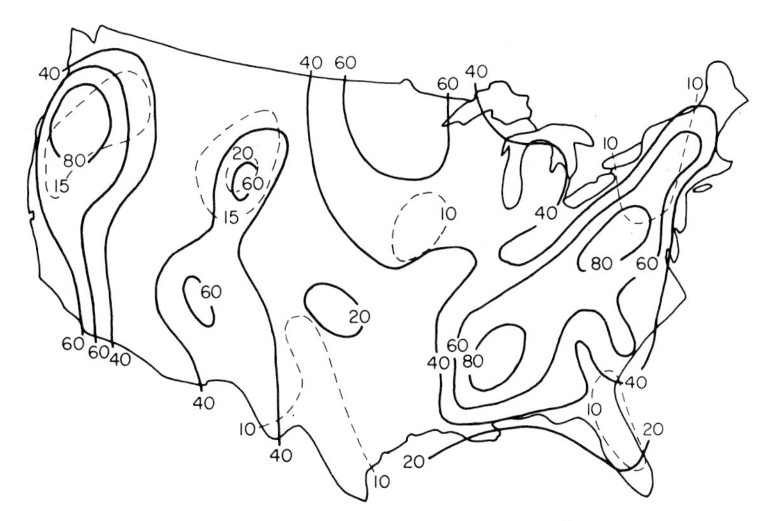

Fig. 21-7. Relative concentration in seconds per meter (s m^{-1}) exceeded in a 10-km city on 10% of all mornings (solid lines) and 10% of all afternoons (dashed lines). Source: After Holzworth (2).

for the morning radiosonde ranged from less than 150 to 200 m for 50-m stacks and from less than 650 to greater than 750 m for 400-m stacks. The frequency of effective heights from 400-m stacks of the morning radiosonde that are exceeded by the afternoon mixing heights ranges from 50 to 60% near coastlines and from 70 to more than 90% throughout the rest of the contiguous United States.

Taylor and Marsh (7) investigated the long-term characteristics of temperature inversions and mixed layers in the lower atmosphere to produce an inversion climatology for the Los Angeles basin. In this area the cooler ocean currents produce an elevated inversion that is nearly always present and traps the pollutants released over the area within a layer seldom deeper than 1200 m and frequently much shallower.

V. WIND AND POLLUTION ROSES

Since wind is circular, it is frequently easier to interpret and visualize the frequency of wind flow subjectively by displaying a wind rose, that is, wind frequencies for each direction oriented according to the azimuth for that direction. Figure 21-8 is a wind rose showing both directional frequencies and wind speed frequencies by six classes from 3-hourly observations for a 5-year period (1965–1969) for O'Hare Airport, Chicago. The highest frequencies are from the south and west, the lowest from the southeast and east.

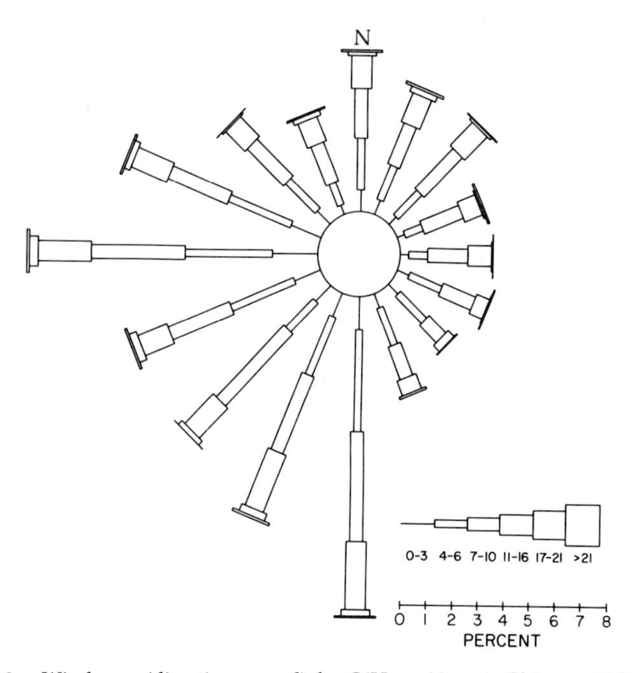

Fig. 21-8. Wind rose (direction–speed) for O'Hare Airport, Chicago, 1965–1969.

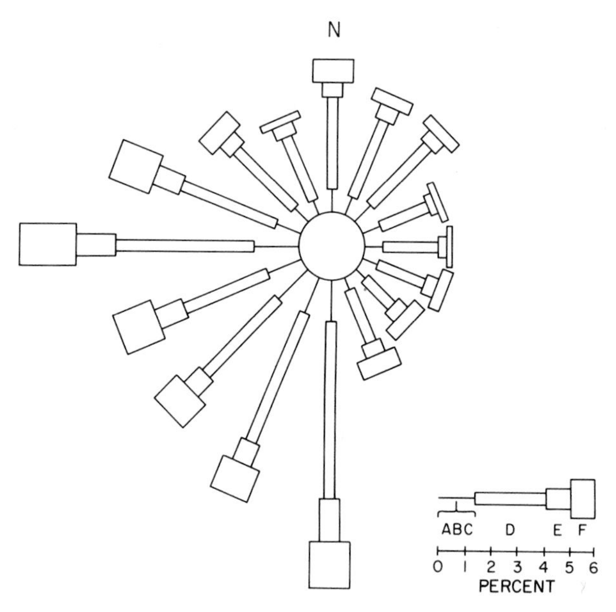

Fig. 21-9. Stability rose (direction–Pasquill stability class) for O'Hare Airport, Chicago, 1965–1969.

Figure 21-9 is a stability wind rose that indicates Pasquill stability class frequencies for each direction. For this location, the various stabilities seem to be nearly a set proportion of the frequency for that direction; the larger the total frequency for that direction, the greater the frequency for each stability. Since the frequencies of A and B stabilities are quite small (0.72% for A and 4.92% for B), all three unstable classes (A, B, and C) are added together and indicated by the single line.

Pollution roses are constructed by plotting either the average concentration for each direction or the frequency of concentrations above some particular concentration. Pollution roses for two pollutants at two times of the year are shown in Fig. 21-10, with wind frequencies by two speed classes

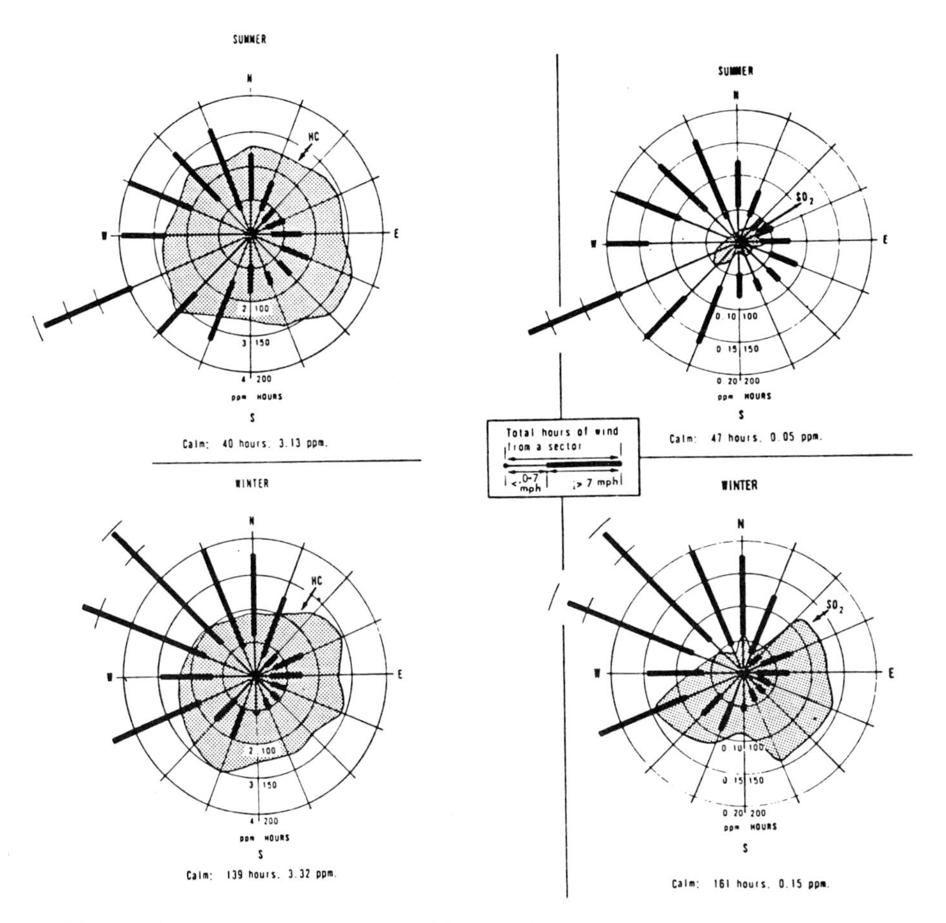

Fig. 21-10. Average concentrations of hydrocarbons and sulfur dioxide for each wind direction and wind direction frequency in two classes (0–7 mi hr^{-1} and greater than 7 mi hr^{-1}), Philadelphia, 1963. Source: U.S. Department of Health, Education and Welfare (8).

(less than 7 mi h^{-1} and greater than 7 mi h^{-1}) superimposed. The average pollutant concentrations are connected so as to be depicted as areas rather than individual lines for each direction. Note that there is little seasonal change for hydrocarbon concentrations and only minor directional variation. SO_2, on the other hand, has very significant seasonal variation, with very low concentrations in the summer. SO_2 also has considerable directional variation.

The behavior of these pollution roses is intuitively plausible, because considerable hydrocarbon emissions come from motor vehicles which are operated in both winter and summer and travel throughout the urban area. On the other hand, sulfur dioxide is released largely from the burning of coal and fuel oil. Space heating emissions are high in winter and low in summer. The SO_2 emissions in summer are probably due to only a few point sources, such as power plants, and result in low average concentrations from each direction as well as large directional variability.

Concentrations resulting from dispersion models can also be depicted using a form of pollution rose. Figure 21-11 is a concentration rose for a

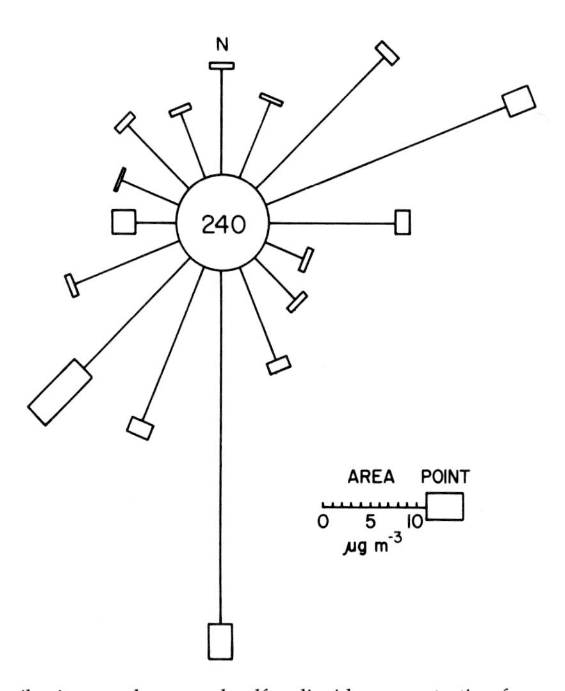

Fig. 21-11. Contributions to the annual sulfur dioxide concentration from each direction at a receptor in New York by area sources (lines) and point sources (rectangles) for 1969 using the Climatological Dispersion Model.

measurement station in New York City for the SO_2 concentration estimates from the Climatological Dispersion Model (1). The number in the circle is the total estimated annual SO_2 concentration from the model. The radial values represent the contribution to the annual concentration from each direction, with the length of the line proportional to the concentration resulting from area sources and the length of the rectangle proportional to the concentration resulting from the point sources. For the monitoring station in Fig. 21-11, the estimated annual concentration is 240 $\mu g\ m^{-3}$. The maximum annual contribution from area sources is from the south (39 $\mu g\ m^{-3}$); the maximum annual contribution from point sources is from the southwest. (7.1 $\mu g\ m^{-3}$). The minimum concentration is from the east–southeast.

An example of frequencies of wind direction when the concentration exceeds a particular value is shown in Fig. 21-12. For this example, the concentration threshold is 0.1 ppm (262 $\mu g\ m^{-3}$). Although the maximum frequency from any one direction is only about 1%, this can be significant. Munn (9) is careful to point out that "The diagram suggests but, of course, does not prove that a major source of SO_2 is situated between the sampling stations" (p. 109).

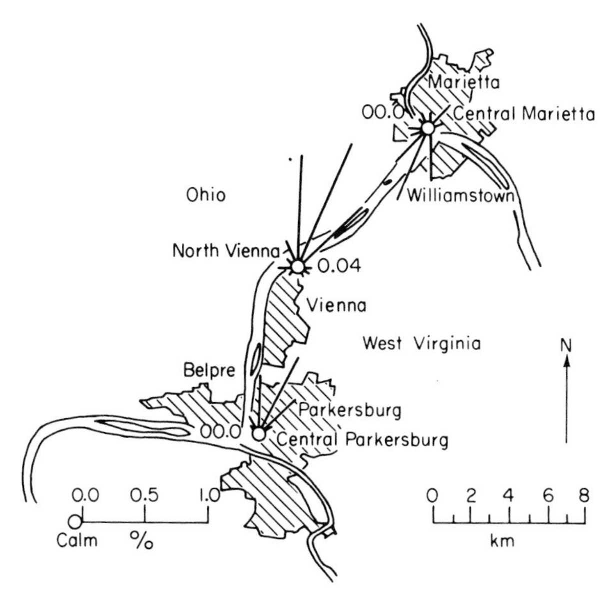

Fig. 21-12. Frequency of wind direction when sulfur dioxide exceeds 0.1 ppm near Parkersburg, West Virginia, Source: Munn (9).

REFERENCES

1. Busse, A. D., and Zimmerman, J. R., "User's Guide for the Climatological Dispersion Model." Environmental Monitoring Series EPA-R4-73-024. U.S. Environmental Protection Agency, Research Triangle Park, NC, 1973.
2. Holzworth, G. C., "Mixing Heights, Wind Speeds and Potential for Urban Air Pollution Throughout the Contiguous United States," Office of Air Programs Pub. No. AP-101. U.S. Environmental Protection Agency, Research Triangle Park, NC, 1972.
3. Korshover, J., "Climatology of Stagnating Anticyclones East of the Rocky Mountains, 1936–1965," Public Health Service Publication No. 999-AP-34. U.S. Department of Health, Education and Welfare, Cincinnati, OH, 1967.
4. Hosler, C. R., *Mon. Weather Rev.* **89,** 319–339 (1961).
5. Miller, M. E., and Holzworth, G. C., *J. Air Pollut. Control Assoc.* **17,** 46–50 (1967).
6. Holzworth, G. C., Climatic data on estimated effective chimney heights in the United States, *in* "Preprints of Joint Conference on Application of Air Pollution Meteorology," Nov. 29–Dec 2, 1977, Salt Lake City, UT. American Meteorological Society, Boston, 1977, pp. 80–87.
7. Taylor, G. H., and Marsh, S. L., Jr., An inversion climatology for the Los Angeles basin, pp. 294–257 *in* "Preprints of Seventh Joint Conference on Application of Air Pollution Meteorology with AWMA," Jan. 14–18, 1991, New Orleans, LA. American Meteorological Society, Boston, 1991.
8. U.S. Department of Health, Education and Welfare, "Continuous Air Monitoring Projects in Philadelphia, 1962–1965," National Air Pollution Control Administration Publication No. APTD 69-14. Cincinnati, OH, 1969.
9. Munn, R. E., "Biometeorological Methods." Academic Press, New York, 1970.

SUGGESTED READING

Doty, S. R., Wallace, B. L., and Holzworth, G. C., "A Climatological Analysis of Pasquill Stability Categories Based on 'STAR' Summaries." National Climatic Center, Environmental Data Service, National Oceanic and Atmospheric Administration, Asheville, NC, 1976.

Holzworth, G. C., and Fisher, R. W., "Climatological Summaries of the Lower Few Kilometers of Rawinsonde Observations," EPA-600/4-79-026. U.S.Environmental Protection Agency, Research Triangle Park, NC, 1979.

Lowry, W. P., "Atmospheric Ecology for Designers and Planners." Peavine Publications. McMinnville, OR, 1988.

Nappo, C. J., Caneill, J. Y., Furman, R. W., Gifford, F. A., Kaimal, J. C., Kramer, M. L., Lockhart, T. J., Pendergast, M. M., Pielke, R. A., Randerson, D., Shreffler, J. H., and Wyngaard, J. C., *Bull. Am. Meteorol. Soc.* **63**(7), 761–764 (1982).

Oke, T. R., "Boundary Layer Climates," 2nd ed. Methuen, New York, 1988.

QUESTIONS

1. What geographical or climatological conditions seem to be associated with the lowest mean annual mixing heights shown in Fig. 21-2? With the highest mixing heights?

2. In attempting to determine the air pollution impact of a small town containing several industrial facilities in a mountainous river valley about 1 km wide with sloping sides extending 400–500 m above the river, what meteorological measurement program would you recommend and what facts would you try to determine before finalizing your recommendation?
3. Over relatively flat terrain, which of the following measurements would be expected to represent conditions most closely at a second site 10 km away: wind velocity or total amount of cloud?
4. What areas of the contiguous United States have the least air pollution potential as defined by mixing heights of less than 1500 m with wind speeds of less than 4 m s^{-1} through the mixing height? Which areas have the greatest air pollution potential?
5. What is the variation in relative concentration exceeded 10% of the time in a 10-km city over the contiguous United States for mornings? For afternoons?
6. From Fig. 21-10, what wind directions are related to highest average winter sulfur dioxide concentrations at this sampling station?
7. If one considers a contribution to the annual concentration from point sources as significant when it exceeds 2 $\mu g\ m^{-3}$, using Fig. 21-11, what wind directions account for significant point source contributions at this station?

Part V

The Regulatory Control of Air Pollution

22

Air Quality Criteria and Standards

I. AIR QUALITY CRITERIA

Air quality criteria are cause–effect relationships, observed experimentally, epidemiologically, or in the field, of exposure to various ambient levels of specific pollutants. The relationships between adverse responses to air pollution and the air quality levels at which they occur have been discussed in Chapter 4 and illustrated in Table 4-5 and Fig. 4-10.

For any pollutant, air quality criteria may refer to different types of effects. For example, Tables 22-1 through 22-6 list effects on humans, animals, vegetation, materials, and the atmosphere caused by various exposures to sulfur dioxide, particulate matter, nitrogen dioxide, carbon monoxide, ozone, and lead. These data are from the Air Quality Criteria for these pollutants published by the U.S. Environmental Protection Agency.

Criteria stipulate conditions of exposure and may refer to sensitive population groups or to the joint effects of several pollutants. Air quality criteria are descriptive. They describe effects that can be expected to occur wherever the ambient air level of a pollutant reaches or exceeds a specific concentration for a particular time period. Criteria will change as new information becomes available.

TABLE 22-1

U.S. Ambient Air Quality Criteria for Carbon Monoxide

Percent of carboxyhemoglobin (CoHb) in blood	Human symptoms associated with this CoHb level
80	Death
60	Loss of consciousness; death if exposure is continued
40	Collapse on exercise; confusion
30	Headache, fatigue; judgment disturbed
20	Cardiovascular damage; electrocardiographic abnormalities
5	Decline (linear with increasing CoHb level) in maximal oxygen uptake of healthy young men undergoing strenuous exercise; decrements in visual perception, manual dexterity, and performance of complex sensorimotor tasks
4	Decrements in vigilance (i.e., ability to detect small changes in one's environment that occur at unpredictable times); decreased exercise performance in both healthy persons and those with chronic obstructive pulmonary disease
3–6	Aggravation of cardiovascular disease (i.e., decreased exercise capacity in patients with angina pectoris, intermittent claudication, or peripheral arteriosclerosis)

Sources: Henderson, Y., and Haggard, H. W., "Noxious Gases." Chemical Catalog Co., New York, 1927; and U.S. Environmental Protection Agency, Research Triangle Park, NC; Air Quality Criteria for Carbon Monoxide EPA/600/8-90/045F, December 1991.

II. CONVERSION OF EFFECTS DATA AND CRITERIA TO STANDARDS

In developing air pollution cause–effect relationships, we must be constantly on guard lest we attribute to air pollution an effect caused by something else. Material damage due to pollution must be differentiated from that due to ultraviolet radiation, frost, moisture, bacteria, fungi, insects, and animals. Air pollution damage to vegetation has to be differentiated from quite similar damage attributable to bacterial and fungal diseases, insects, drought, frost, soil mineral deviations, hail, and cultural practices. In the principal animal disorder associated with air pollution, i.e., fluorosis, the route of animal intake of fluorine is by ingestion, the air being the means for transporting the substance from its source to the forage or hay used for animal feed. However, the water or feed supplements used may also have excess fluorine. Therefore, these sources and disease states, which may have symptoms similar to those of fluorosis, must be ruled out before a cause–effect relationship can be established between ambient air levels of fluorine and fluorosis in animals. Similarly, there are many instances of visibility reduction in the atmosphere by fog or mist for which air pollution is not a causative factor.

TABLE 22-2a

Summary of Lowest Observed Effect Levels for Key Lead-Induced Health Effects in Adults

Lowest observed effect level (PbB)[a] (μg/dl)	Heme synthesis and hematological effects	Neurological effects	Effects on the kidney	Reproductive function effects	Cardiovascular effects
100–120		Encephalopathic signs and symptoms ↑	Chronic nephropathy ↓		
80	Frank anemia				
60				Female reproductive effects	
50	Reduced hemoglobin production	Overt subencephalopathic neurological symptoms ↓		Altered testicular function ↓	
40	Increased urinary ALA and elevated coproporphyrins	Peripheral nerve dysfunction (slowed nerve conduction) ↓			
30					Elevated blood pressure (white males, aged 40–59) ↓
25–30	Erythrocyte protoporphyrin (EP) elevation in males				
15–20	Erythrocyte protoporphyrin (EP) elevation in females				
<10	ALA-D inhibition				?

[a] PbB = blood lead concentrations.

Source: U.S. Environmental Protection Agency, Air Quality Criteria for Lead, EPA-600-8-83/028aF, June 1986.

TABLE 22-2b

Summary of Lowest Observed Effect Levels for Key Lead-Induced Health Effects in Children

Lowest observed effect level (PbB)[a] (μg/dl)	Heme synthesis and hematological effects	Neurological effects	Renal system effects	Gastrointestinal effects
80–100		Encephalopathic signs and symptoms	Chronic nephropathy (aminoaciduria, etc.)	Colic, other overt gastrointestinal symptoms ↓
70	Frank anemia			
60		Peripheral neuropathies ↓ ?		
50				
40	Reduced hemoglobin synthesis	Peripheral nerve dysfunction (slowed NCVs)[d]		
	Elevated coproporphyrin	CNS cognitive effects (IQ deficits, etc.)		
	Increased urinary ALA[c]			
30		?	Vitamin D metabolism interference ↓ ?	
15	Erythrocyte protoporphyrin elevation	Altered CNS electrophysiological response ↓ ?		
10	ALA-D[c] inhibition			
	Py-5-N activity inhibition[b] ↓ ?			

[a] PbB = blood lead concentrations.
[b] Py-5-N = pyrimidine-5′-nucleotidase.
[c] ALA, ALA-D = aminolevulinic acid dehydrase
[d] NCV = nerve conduction velocity
Source: U.S. Environmental Protection Agency, Air Quality Criteria for Lead, EPA-600, EPA-600-8-83/028aF, June 1986.

TABLE 22-3

U.S. Ambient Air Quality Criteria for Sulfur Dioxide

Concentration of sulfur dioxide in air (ppm)	Exposure time	Human symptoms and effects on vegetation
400	—	Lung edema; bronchial inflammation
20	—	Eye irritation; coughing in healthy adults
15	1 hr	Decreased mucociliary activity
10	10 min	Bronchospasm
10	2 hr	Visible foliar injury to vegetation in arid regions
8	—	Throat irritation in healthy adults
5	10 min	Increased airway resistance in healthy adults at rest
1	10 min	Increased airway resistance in asthmatics at rest and in healthy adults at exercise
1	5 min	Visible injury to sensitive vegetation in humid regions
0.5	10 min	Increased airway resistance in asthmatics at exercise
0.5	—	Odor threshold
0.5	1 hr	Visible injury to sensitive vegetation in humid regions
0.5	3 hr	United States National Secondary Ambient Air Quality Standard promulgated in 1973
0.2	3 hr	Visible injury to sensitive vegetation in human regions
0.19	24 hr[a]	Aggravation of chronic respiratory disease in adults
0.14	24 hr	United States National Primary Ambient Air Quality Standard promulgated in 1971[b]
0.07	Annual[a]	Aggravation of chronic respiratory disease in children
0.03	Annual	United States National Primary Ambient Air Quality Standard promulgated in 1971[b]

[a] In the presence of high concentrations of particulate matter.
[b] Sources: Air Quality Criteria for Particulate Matter and Sulfur Oxides, final draft, U.S. Environmental Protection Agency, Research Triangle Park, NC, December 1981; Review of the National Ambient Air Quality Standards for Sulfur Oxides: Assessment of Scientific and Technical Information, Draft OAQPS Staff Paper, U.S. Environmental Protection Agency, Research Triangle Park, NC, April 1982.

To study damage to materials, vegetation, and animals, we can set up laboratory experiments in which most confusing variables are eliminated and a direct cause–effect relationship is established between pollutant dosage and resulting effect. We are limited to low-level exposure experiments under controlled conditions with human beings for ethical reasons. Our cause–effect relationships for humans are based on (1) extrapolation from animal experimentation, (2) clinical observation of individual cases of persons exposed to the pollutant or toxicant (industrially, accidentally, suicidally, or under air pollution episode conditions), and (3) most important, epidemiological data relating population morbidity and mortality to air pollution. There are no human diseases uniquely caused by air pollution. In all air pollution–related diseases in which there is buildup of toxic mate-

TABLE 22-4

U.S. Air Quality Criteria for Nitrogen Dioxide

Concentration of nitrogen dioxide in air (ppm)	Exposure time	Human symptoms and effects on vegetation, materials, and visibility
300	—	Rapid death
150	—	Death after 2 or 3 weeks by bronchiolitis fibrosa obliterans
50	—	Reversible, nonfatal bronchiolitis
10	—	Impairment of ability to detect odor of nitrogen dioxide
5	15 min	Impairment of normal transport of gases between the blood and lungs in healthy adults
2.5	2 hr	Increased airway resistance in healthy adults
2	4 hr	Foliar injury to vegetation
1.0	15 min	Increased airway resistance in bronchitics
1.0	48 hr	Slight leaf spotting of pinto bean, endive, and cotton
0.3	—	Brownish color of target 1 km distant
0.25	Growing season	Decrease of growth and yield of tomatoes and oranges
0.2	8 hr	Yellowing of white fabrics
0.12	—	Odor perception threshold of nitrogen dioxide
0.1	12 weeks	Fading of dyes on nylon
0.1	20 weeks	Reduction in growth of Kentucky bluegrass
0.05	12 weeks	Fading of dyes on cotton and rayon
0.03	—	Brownish color of target 10 km distant
0.003	—	Brownish color of target 100 km distant

Sources: Draft Air Quality Criteria for Oxides of Nitrogen, U.S. Environmental Protection Agency, Research Triangle Park, NC, 1981; Review of the National Ambient Air Quality Standard for Nitrogen Dioxide, Assessment of Scientific and Technical Information, EPA-450/5-82-002. U.S. Environmental Protection Agency, Research Triangle Park, NC, March 1982.

rial in the blood, tissue, bone, or teeth, part or all of the buildup could be from ingestion of food or water containing the material. Diseases which are respiratory can be caused by smoking or occupational exposure. They may be of a bacterial, viral, or fungal origin quite divorced from the inhalation of human-made pollutants in the ambient air. These causes in addition to the variety of congenital, degenerative, nutritional, and psychosomatic causes of disease must all be ruled out before a disease can be attributed to air pollution. However, air pollution commonly exacerbates preexisting disease states. In human health, air pollution can be the "straw that breaks the camel's back."

Air quality standards prescribe pollutant levels that cannot legally be exceeded during a specific time period in a specific geographic area. Air quality standards are based on air quality criteria, with added safety factors as desired.

TABLE 22-5

U.S. Ambient Air Criteria for Ozone

Concentration of ozone in air (ppm)[a]	Human symptoms and vegetation injury threshold
10.0	Severe pulmonary edema; possible acute bronchiolitis; decreased blood pressure; rapid weak pulse
1.0	Coughing; extreme fatigue; lack of coordination; increased airway resistance; decreased forced expiratory volume
0.5	Chest constriction; impaired carbon monoxide diffusion capacity; decrease in lung function without exercise
0.3	Headache; chest discomfort sufficient to prevent completion of exercise; decrease in lung function in exercising subjects
0.25	Increase in incidence and severity of asthma attacks; moderate eye irritation
0.15	For sensitive individuals, reduction in pulmonary lung function; chest discomfort; irritation of the respiratory tract, coughing and wheezing. Threshold for injury to vegetation
0.12	United States National Primary and Secondary Ambient Air Quality Standard, attained when the expected number of days per calendar year with maximum hourly average concentrations above 0.12 ppm is equal to or less than 1, as determined in a specified manner

[a] 1 ppm—1958 $\mu g\ m^{-3}$ ozone.
Sources: Air Quality Criteria for Ozone and Other Photochemical Oxidants, EPA 600/8-78-004. U.S. Environmental Protection Agency, Research Triangle Park, NC, April 1978; Revisions to National Ambient Air Quality Standards for Photochemical Oxidants, *Fed. Reg.* Part V, Feb. 9, 1979, pp. 8202–8237.
40 CFR § 50, July 1992.

The main philosophical question that arises with respect to air quality standards is what to consider an adverse effect or a cost associated with air pollution. Let us examine several categories of receptors to see the judgmental problems that arise.

III. CONVERSION OF PHYSICAL DATA AND CRITERIA TO STANDARDS

Although air quality standards are based predominantly on biological criteria, students should understand the physical criteria that also deserve consideration.

Most materials will deteriorate even when exposed to an unpolluted atmosphere. Iron will rust, metals will corrode, and wood will rot. To prevent deterioration, protective coatings are applied. Their costs are part

TABLE 22-6

U.S. Ambient Air Quality Criteria for Particulate Matter

Concentration of particulate matter in air ($\mu g\ m^{-3}$)				
Total suspended TSP > 25 μm	Thoracic TP > 10 μm	Fine FP > 2.5 μm	Exposure time	Human symptoms and effects on visibility
2000	—	—	2 hr	Personal discomfort
1000	—	—	10 min	Direct respiratory mechanical changes
—	350	—		Aggravation of bronchitis
	150	—	24 hr	United States Primary National Ambient Air Quality Standard as of September, 1987
180	90	—		Increased respiratory disease symptoms
	150	—	24 hr	United States Primary National Ambient Air Quality Standard as of September, 1987
110	55	—	24 hr	Increased respiratory disease risk
	50	—	Annual geometric mean	United States Primary National Air Quality Standard as of September, 1987
	50	—	Annual geometric mean	United States Secondary National Ambient Air Quality Standard as of September, 1987
—	—	22	13 weeks	Usual summer visibility in eastern United States, nonurban sites

Sources: Air Quality Criteria for Particulate Matter and Sulfur Oxides, Draft Final. U.S. Environmental Protection Agency, Research Triangle Park, NC, December 1981; Review of the National Ambient Air Quality Standard for Particulate Matter: Assessment of Scientific and Technical Information, EPA-450/5-82-001. U.S. Environmental Protection Agency, Research Triangle Park, NC, January 1982.
40 CFR § 50, July 1992.

of the economic picture. Some materials, such as railroad rails, are used without protective coatings. There are costs associated with the decrease in their life in a polluted atmosphere as compared to an unpolluted one. One may argue that for materials on which protective coatings are used, only pollution levels that damage such coatings are of concern. One may further argue that some air pollution damage to protection coatings is tolerable, since by their very nature such coatings require periodic replenishment to maintain their protective integrity or appearance; therefore, only coatings that require more frequent replenishment than they would in an unpolluted atmosphere should enter into the establishment of deterioration costs and air quality standards. This argument certainly does hold with respect to the soiling of materials and structures. In fact, it is frequently the protective coatings themselves that require replacement because they become dirty long before their useful life as protectants has terminated. It can readily be shown that there are costs associated with soiling, including the cost of removing soil, the cost of protective coatings to facilitate the removal of soil, the premature disposal of material when it is no longer economical or practicable to remove soil, and the growth inhibition of vegetation due to leaf soiling. However, decision making for air quality standards related to soiling is based less on economic evaluation than on aesthetic considerations, i.e., on subjective evaluation of how much soiling the community will tolerate. This latter determination is judgmental and difficult to make. It may be facilitated by opinion surveys, but even when the limit of public tolerance for soiling is determined, it still has to be restated in terms of the pollution loading of the air that will result in this level of soiling.

An important effect of air pollution on the atmosphere is change in spectral transmission. The spectral regions of greatest concern are the ultraviolet and the visible. Changes in ultraviolet radiation have demonstrable adverse effects; e.g., a decrease in the stratospheric ozone layer permits harmful UV radiation to penetrate to the surface of the earth. Excessive exposure to UV radiation results in increases in skin cancer and cataracts. The worldwide effort to reduce the release of stratospheric ozone–depleting chemicals such as chlorofluorocarbons is directed toward reducing this increased risk of skin cancer and cataracts for future generations.

The fact that after a storm or the passage of a frontal system the air becomes crystal clear and one can see for many kilometers does not give a true measure of year-round visibility under unpolluted conditions. Between storms, even in unpolluted air, natural sources build up enough particulate matter in the air so that on many days of the year there is less than ideal visibility. In many parts of the world, mountains are called "Smoky" or "Blue" or some other name to designate the prevalence of a natural haze, which gives them a smoky or bluish color and impedes visibility. When the Spanish first explored the area that is now Los Angeles, California, they gave it the name "Bay of the Smokes." The Los Angeles definition of air quality before the advent of smog was that "You could see Catalina Island on a clear day." The part of the definition that is lacking is some indication of how many clear days there were before the advent of smog.

There are costs associated with loss of visibility and solar energy. These include increased need for artificial illumination and heating; delays, disruptions, and accidents involving air, water, and land traffic; vegetation growth reduction associated with reduced photosynthesis; and commercial losses associated with the decreased attractiveness of a dingy community or one with restricted scenic views. However, these costs are less likely to be involved in deciding, for air quality standard–setting purposes, how much of the attainable visibility improvement to aim for than are aesthetic considerations. Just as in the previously noted case of soiling, judgment on the limit of public tolerance for visibility reduction still has to be related to the pollutant loading of the atmosphere that will yield the desired visibility. Obviously, the pollutant level chosen for an air quality standards must be the lower of the values required for soiling or visibility, otherwise one will be achieved without the other. Whether the level chosen will not be lower than the atmospheric pollutant level required for prevention of health effects will depend on the aesthetic standards of the jurisdiction.

IV. CONVERSION OF BIOLOGICAL DATA AND CRITERIA TO STANDARDS

There is considerable species variability with respect to damage to vegetation by any specific pollutant. There is also great geographic variability with respect to where these species grow naturally or are cultivated. Because of this, it is possible that in a jurisdiction none of the species particularly susceptible to damage by low levels of pollution may be among those indigenous or normally imported for local cultivation. As an example, the pollution level at which citrus trees are adversely affected, while meaningful in setting air quality standards in California and Florida, is meaningless for this purpose in Minnesota and Wisconsin. In like manner, a jurisdiction may take different viewpoints with respect to indigenous and imported species. It might set its air quality standards low enough to protect its indigenous vegetation even if this level is too high to allow satisfactory growth of imported species. Even if a particularly susceptible species is indigenous, it may be held in such low local esteem commercially or aesthetically that the jurisdiction may be unwilling to let the damage level of that species be the air quality standard discriminator. In other words, the people would rather have that species damaged than assume the cost of cleaning up the air to prevent the damage. This same line of reasoning applies to effects on wild and domestic animals.

A jurisdiction may base part of its decision making regarding vegetation and animal damage on aesthetics. Its citizens may wish to grow certain ornamentals or raise certain species of pet birds or animals and allow these wishes to override the agricultural, forestry, and husbandry economics of the situation. Usually, however, economic considerations predominate in decision making. Costs of air pollution effects on agriculture are the sum of the loss in income from the sale of crops or livestock and the added cost necessary to raise the crops or livestock for sale. To these costs must be added the loss in value of agricultural land as its income potential decreases and the loss suffered by the segments of local industry and commerce that are dependent on farm crops and the farmer for their existence. An interesting sidelight is that when such damage occurs on the periphery of an urban area, it is frequently a precursor to the breakup of such farmland into residential development, with a financial gain rather than a loss to the landowner. When the crop that disappears is an orchard, grove, or vineyard that took years to establish, and when usable farm buildings are torn down, society as a whole suffers a loss to the extent that it will take much time and money to establish a replacement for them at new locations. To some industries, air pollution costs include purchase of farm and ranch land to prevent litigation to recover damages, annual subsidy payments to farmers and ranchers in lieu of such litigation, and maintenance of air quality monitoring systems to protect themselves against unwarranted litigation for this purpose.

There is a range of ambiguity in our human health effects criteria data. In this range there is disagreement among experts as to its validity and interpretation. Thus, from the same body of health effects data, one could adopt an air quality standard on the high side of the range of ambiguity or one on the low side. Much soul searching is required before one accepts the results of questionable human health effects research and is accused of imposing large costs on the public by so doing, or of rejecting these results and being accused of subjecting the public to potential damage of human health.

V. AIR QUALITY STANDARDS

Since air pollution is controlled by air quality and emission standards, the principal philosophical discussions in the field of air pollution control focus on their development and application.

The U.S. Clean Air Amendments of 1977 define two kinds of air quality standards: primary standards, levels that will protect health but not necessarily prevent the other adverse effects of air pollution, and secondary standards, levels that will prevent all the other adverse effects of air pollution (Table 22-7). The amendments also define air quality levels that cannot be exceeded in specified geographic areas for "prevention of significant deterioration" (PSD) of the air of those areas. Although they are called "increments" over "baseline air quality" in the law, they are in effect tertiary standards, which are set at lower ambient levels than either the primary or secondary standards (Table 22-8).

Increments are said to be "consumed" as new sources are given permits that allow pollution to be introduced into these areas. Jurisdictions with authority to issue permits may choose to "allocate" portions of a PSD increment (or of the difference between actual air quality and the primary

TABLE 22-7

U.S. Federal Primary and Secondary Ambient Air Quality Standards

Pollutant	Type of standard	Averaging time	Frequency parameter	Concentration	
				$\mu g/m^3$	ppm
Sulfur oxides (as sulfur dioxide)	Primary	24 hr	Annual maximum[a]	365	0.14
		1 year	Arithmetic mean	80	0.03
	Secondary	3 hr	Annual maximum[a]	1,300	0.5
Particulate matter > 10 μm	Primary	24 hr	Annual maximum[a]	150	—
		24 hr	Annual geometric mean	50	—
	Secondary	24 hr	Annual maximum[a]	150	—
		24 hr	Annual geometric mean	50	—
Carbon monoxide	Primary and secondary	1 hr	Annual maximum[a]	40,000	35.0
		8 hr	Annual maximum[a]	10,000	9.0
Ozone	Primary and secondary	1 hr	Annual maximum[a]	235	0.12
Nitrogen dioxide	Primary and secondary	1 year	Arithmetic mean	100	0.05
Lead	Primary and secondary	3 months	Arithmetic mean	1.5	—

[a] Not to be exceeded more than once per year.
Notes: National primary ambient air quality standards define levels of air quality which the EPA Administrator judges are necessary, with an adequate margin of safety, to protect the public health. National secondary ambient air quality standards define levels of air quality, which the Administrator judges necessary to protect the public welfare from any known or anticipated adverse effects of a pollutant.
Source: 40 CFR §50, July 1992.

or secondary standard) for future consumption, rather than to allow its consumption on a first-come, first-served basis.

The states are required to submit to the federal Environmental Protection Agency (EPA) plans, known as State Implementation Plans (SIP), showing how they will achieve the standards in their jurisdictions within a specified time period. If after that time period there are areas within the states where these standards have not been attained, the states are required to submit and obtain EPA approval of revised plans to achieve the standards in these "nonattainment" areas. EPA also designates certain areas where the standards are being met, but which have the potential for future nonattainment, as Air Quality Maintenance Areas (AQMA). Such regions have stricter requirements than attainment areas for the granting of permits for new sources of the pollutant not in attainment status.

The Canadian Clean Air Act allows the minister to formulate air quality objectives reflecting three ranges of ambient air quality for any contaminant. The *tolerable* range denotes a concentration that requires abatement without

TABLE 22-8

U.S. Federal PSD Concentration Increments

Pollutant	Increment ($\mu g\ m^{-3}$)
Class I areas[a]	
Particulate matter	
TSP, annual geometric mean	5
TSP, 24-hr maximum	10
Sulfur dioxide	
Annual arithmetic mean	2
24-hr maximum	5
3-hr maximum	25
Nitrogen dioxide	
Annual arithmetic mean	2.5
Class II areas[b]	
Particulate matter	
TSP, annual geometric mean	19
TSP, 24-hr maximum	37
Sulfur dioxide	
Annual arithmetic mean	20
24-hr maximum	91
3-hr maximum	512
Nitrogen dioxide	
Annual arithmetic mean	25
Class III areas[c]	
Particulate matter	
TSP, annual geometric mean	37
TSP, 24-hr maximum	75
Sulfur dioxide	
Annual arithmetic mean	40
24-hr maximum	182
3-hr maximum	700
Nitrogen dioxide	
Annual arithmetic mean	50

[a] Class I areas are pristine, e.g., national parks, national seashores, natural wilderness areas.

[b] Class II areas where moderate deterioration is allowed. (Unless otherwise designated, all areas are Class II.)

[c] Class III areas are specifically designated as heavy industrial.

Source: 40 CFR § 51.166, July 1992.

TABLE 22-9

Federal Air Quality Objectives in Canada

Air contaminant ($\mu g\ m^{-3}$)	Desirable range	Acceptable range	Tolerable range
Sulfur dioxide			
1-hr average	0–450	450–900	—
24-hr average	0–150	150–300	300–800
Annual arithmetic mean	0–30	30–60	—
Suspended particulates			
24-hr average	—	0–120	120–400
Annual geometric mean	0–60	60–70	—
Oxidants (ozone)			
1-hr average	0–100	100–160	160–300
24-hr average	0–30	30–50	—
Annual arithmetic mean	—	0–30	—
Nitrogen dioxide			
1-hr average	—	0–400	400–1000
24-hr average	—	0–200	—
Annual arithmetic mean	0–60	60–100	—
Hydrogen sulfide			
1-hr average	0–1	1–15	—
24-hr average	—	0–5	—
Hydrogen fluoride			
24-hr average	0–0.40	0.40–0.85[a]	—
7-day average	0–0.20	0.20–0.55[a]	—
30-day average	—	0–0.35[a]	—
70-day average	—	0–0.20[a]	—
Carbon monoxide ($mg\ m^{-3}$)			
1-hr average	0–15	15–35	—
8-hr average	0–6	6–15	15–20

[a] Proposed.

Source: Air Pollution Control Association, "1981–1982 APCA Directory and Resource Book" (special anniversary edition). Adapted from "Air Quality Objectives, Criteria and Regulations in Canada," Pittsburgh, 1981, pp. 160–161.

TABLE 22-10

Air Quality Standards for Selected Pollutants in Several Countries of the World, $mg\ m^{-3}$

Country	Carbon monoxide	Nitrogen oxides	Suspended particulate	Sulfur oxides
Australia	10/1 hr	0.320/1 hr	0.09/1 yr	0.06/1 yr
Finland	30/1 hr	0.300/1 hr	0.150/24 hr	0.5/1 hr
	10/8 hr	0.150/24 hr	0.060/yr	0.2/24 hr
				0.04/yr
Italy		0.200/1 hr	0.150/24 hr	0.080/yr
Japan	11.1/8 hr	0.100/24 hr	0.20/1 hr	0.3/1 hr
			0.10/24 hr	0.12/24 hr
The Netherlands	40/1 hr	0.135/1 hr	0.150/24 hr	0.83/1 hr
	6/8 hr			0.25/24 hr

Source: Adapted from: IUAPPA, "Clean Air Around the World," Brighton, England, 1988.

delay. The *acceptable* range provides adequate protection against adverse effects. The *desirable* range defines a long-term goal for air quality and provides the basis for a nondegradation policy for unpolluted parts of the country (Table 22-9).

Some examples of air quality standards for other countries are given in Table 22-10.

SUGGESTED READING

Atkisson, A., and Gaines, R. S., eds., "Development of Air Quality Standards." Merrill, Columbus, OH, 1970.

Cochran, L. S., Pielke, R. A., and Kovacs, E., Selected international receptor-based air quality standards. *J. Air Waste Manage. Assoc.* **42,** 1567–1572 (1992).

Kates, R. W., "Risk Assessment of Environmental Hazard." Wiley, Chicester, England, 1978.

Schwing, R. C., and Albers, W. A. (eds.), "Society Risk Assessment." Plenum, New York, 1980.

World Health Organization, "Air Quality Guidelines for Europe." Copenhagen, 1987.

QUESTIONS

1. Why are air quality criteria descriptive?
2. Why are air quality standards prescriptive?
3. Evaluate the use, effectiveness, and equity of local, state, provincial, or national air quality standards in your community.
4. Prepare a table similar in format to Tables 22-1 through 22-6 for another pollutant not yet required by the administrator of the U.S. Environmental Protection Agency to have a criteria document.
5. Discuss the relative merits of stating air quality standards as 1-hr, 3-hr, 8-hr, 24-hr, and annual averages.
6. Discuss the relative merits of national versus local air quality standards.
7. Discuss the differences in approach in using air quality standards (as in the United States), air quality objectives (as in Canada), and air quality goals (as in certain other countries).
8. Discuss the advantages and disadvantages of promulgating only one set of air quality standards (as in most countries) and of employing secondary and tertiary (PSD) standards, as in the United States.
9. Discuss the problem caused by cigarette smoking in the evaluation of epidemiological data on the effect of air pollution on respiratory disease.

23

Indoor Air Quality

I. CHANGING TIMES

Societies' concern with air quality has evolved from medieval times, when breathing smelting fumes was a major hazard, to where we are today (see Chapter 1). In modern society, a parallel effort has been under way to improve air quality in the outside or ambient air, which is the focus of this book, and in the industrial occupational setting in manufacturing and other traditional jobs. A combination of events is moving many countries to consider the quality of air in other locations where we live parts of our lives. Attention is now being refocused on "indoor" air quality.

In developing countries, priorities have often been different. Industrialization, water and food supply and sanitation, infrastructure improvements, and basic health care are often the focus of the leaders of a country. In some areas, the availability of a job is much more problematic than some consideration about the quality of the air in the workplace or the home. Many dwellings in developing countries do not have closable windows and doors, so the outdoor and indoor air quality issues are different. In some houses where cooking is done by firewood or charcoal, the air quality outdoors may be considerably better than that inside the smoky residence.

The evolution of our modern society and the concomitant changes in lifestyle, workplace, and housing improvements place concerns about indoor air quality in a different category than for developing countries and from the times of our ancestors.

For many industrialized countries, efforts to improve the outdoor air quality have been under way for the majority of this century. In many locations around the world, significant improvements have taken place. Air quality in many major cities such as London, New York, and Chicago has improved from the conditions present in the first half of the twentieth century. Mechanisms and control programs are in place in the developed countries to continue the improvement of ambient air quality. Considerable effort and energy have been expended to characterize, evaluate, and control air pollution emissions to the atmosphere.

Buildings and their design have undergone major changes. Fifty years ago, central heating and windows which could be opened and closed depending on the season were the norm for commercial buildings. Now we have multistory buildings with central heating and air conditioning and sealed glass exterior walls. Residential housing has undergone similar design and structural changes, in some cases resulting in dwellings that may have poorer indoor air quality.

New residences and commercial buildings are designed and built with energy conservation as a major design criterion. New materials have been developed and are being used in construction. Although these modifications have helped save energy, a consequence of some of these modifications is slower exchange of air with the outside. This helps considerably with the heating or cooling system because energy must condition this "new" air which is introduced into the structure.

A second consideration is the change in lifestyle for individuals in industrialized societies. We are no longer a society dependent on occupations which require us to be outdoors for a significant part of our day. Over the past two decades, studies of daily activities have consistently shown for urban populations that, on average, we spend about 90% of our time indoors in our homes, cars, offices, factories, public buildings such as restaurants, malls, and others. Any given individual activity profile may differ significantly from this average.

Exposure assessment techniques now attempt to include as many as possible of the locations in which individuals now spend time. The concept involves identification of microenvironments which are important for potential exposure. For example, exposure to CO would include time spent in commuting, parking garages, in residences with gas stoves, as well as time spent outdoors. This approach classifies time spent in these microenvironments and the typical concentrations of CO in these locations.

II. FACTORS INFLUENCING INDOOR AIR QUALITY

Several factors influence the quality of air indoors: the rate of exchange of air with air from outdoors, the concentration of pollutants in outdoor air, the rate of emissions from sources indoors, the rate of infiltration from soil gases, and the rate of removal in the indoor environment (Fig. 23-1).

The source of indoor air pollutants may be inside the building, or they may be transported into the interior space from the outside. Sources located indoors include building materials, combustion sources, furnishings, and pets. Emissions of organic gases are higher with increased temperature and humidity but usually decrease with age of the structure or furnishings. Construction materials and the composition of furnishings inside the building may give off or outgas pollutants into the interior airspace, e.g., glues or adhesives. Natural gas for cooking and kerosene space heaters release NO and CO_2 even when operating properly. Molds may grow in the ventilation ducts and be distributed throughout a building.

Radon from the soil can enter buildings through cracks in the foundation when the pressure inside is lower than in the soil. The rate of infiltration depends on the soil type, the building structure, and the pressure differential between the soil and the building.

Air is exchanged between indoors and outdoors by several ways: natural ventilation, mechanical ventilation, and infiltration or exfiltration. Natural ventilation involves movement of air through building openings like doors, windows, and vents. Mechanical ventilation involves fans and heating and air-conditioning systems. Infiltration and exfiltration represents undesirable movement of air in and out of the structure. Buildings are characterized as "tight" when infiltration rates are low.

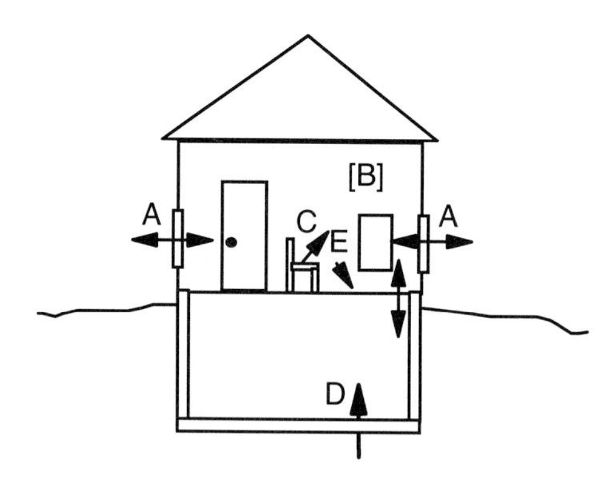

Fig. 23-1. Representation of home with various sources and sinks for indoor air pollutants: (A), exchange; (B), indoor concentration; (C), outgassing of building and furniture materials; (D), infiltration from soils; (E), removal on interior surfaces.

The air exchange rate influences the concentration of indoor pollutants in two ways. At higher air exchange rates, the pollutants inside a structure are removed from the interior. As long as the ambient outside air has lower pollutant concentrations, high exchange rates help lower indoor air pollutant levels. However, if the pollutant concentration outside is elevated, then an increase in the air exchange rate will bring these materials into the building; e.g., an idling vehicle adjacent to an air intake will transfer exhaust fumes into the building. At lower exchange rates, pollutants released from sources inside the building can contribute to higher levels of indoor pollutants.

The concentration of indoor pollutants is a function of removal processes such as dilution, filtration, and destruction. Dilution is a function of the air exchange rate and the ambient air quality. Gases and particulate matter may also be removed from indoor air by deposition on surfaces. Filtration systems are part of many ventilation systems. As air is circulated by the air-conditioning system it passes through a filter which can remove some of the particulate matter. The removal efficiency depends on particle size. In addition, some reactive gases like NO_2 and SO_2 are readily adsorbed on interior surfaces of a building or home.

III. INDOOR AIR POLLUTANTS.

Table 23-1 shows the major categories of indoor air pollutants and sources. Table 23-2 shows a summary of reported indoor air pollutant concentrations compiled by the U.S. Environment Protection Agency. Infor-

TABLE 23-1

Indoor Air Pollutants and Typical Sources

Pollutant	Source
Combustion gases—CO, NO	Combustion—furnace, cooking stove, space heater, etc.
Volatile organic compounds (VOCs)	Outgassing of building materials, coatings, wall and floor coverings, and furnishings
Formaldehyde	Outgassing of pressed wood, insulation foam
Pesticides	Household products
Particulate matter	Combustion
Biological agents—molds, spores, dander	Contaminated ventilation systems, pets
Environmental tobacco smoke	Smoking in building
Radon	Infiltration from soil beneath structure
Asbestos	Construction coatings, tile, insulation

TABLE 23–2

Summary of Reported Indoor Air Pollutant Concentrations

Pollutant	Measured concentration				
	Minimum	Maximum	Mean	Types of building	Reference[d]
Radon			0.8 pCi/l	Residences	EPA (1987d)
	0.5 pCi/l	2000 pCi/l		Residences	EPA (1987b)
	0.14 pCi/l	4.11 pCi/l	—	New pub. bldgs.	Sheldon *et al.* (1988)
	0.3 pCi/l	1.68 pCi/l	—	Old pub. bldgs.	Sheldon *et al.* (1988)
	—	—	1.7–2.4 pCi/l	3 office bldgs	Bayer & Black (1988a)
ETS (as RSP)	—	—	28 $\mu g/m^{3(1)a}$	Residences	NRC (1986b)
	—	—	74 $\mu g/m^{3(2)}$	Residences	NRC (1986b)
	—	—	32 $\mu g/m^{3(3)}$	Residences	DHHS (1986)
	—	—	50 $\mu g/m^{3(4)}$	Residences	DHHS (1986)
(as nicotine)	—	—	0.7–3.2 ppb	3 office bldgs	Bayer & Black (1988a)
Biological contaminants	—	564–5360 CFU/m^{3b}		3 office bldgs	Bayer & Black (1988a)
Formaldehyde	—	131–319 $\mu g/m^3$	78–144 $\mu g/m^3$	Residences	Hawthorne *et al.* (1984)
	ND[c]	192 ppb	—	New pub. bldgs.	Sheldon *et al.* (1987)
	ND	103 ppb	—	Old pub. bldgs.	Sheldon *et al.* (1987)
	—	—	25–39 ppb	3 office bldgs	Bayer & Black (1988a)
Benzene	—	120 $\mu g/m^3$	20 $\mu g/m^3$	Various	Wallace *et al.* (1983)
Carbon tetrachloride	—	14 $\mu g/m^3$	2.5 $\mu g/m^3$	Various	Wallace *et al.* (1983)
Trichloroethylene	—	47 $\mu g/m^3$	3.6 $\mu g/m^3$	Various	Wallace *et al.* (1983)

Tetrachloroethylene	—	250 $\mu g/m^3$	10 $\mu g/m^3$	Various	Wallace *et al.* (1983)
Chloroform	—	200 $\mu g/m^3$	8 $\mu g/m^3$	Various	Wallace *et al.* (1983)
Dichlorobenzenes	—	1200 $\mu g/m^3$	41 $\mu g/m^3$	Various	Wallace *et al.* (1983)
Pesticides					
Diazionon	ND	8.9 $\mu g/m^3$	1.4 $\mu g/m^3$	Residences	Lewis *et al.* (1986)
Chlordane	ND	1.7 $\mu g/m^3$	0.51 $\mu g/m^3$	Residences	Lewis *et al.* (1986)

Source: U.S. Environmental Protection Agency Report to Congress on Indoor Air Quality, EPA/400/1-89/001c, August 1989.

[a] (1) 73 residences without smokers; (2) 73 residences with smokers; (3) nonsmokers not exposed to ETS; (4) nonsmokers exposed to ETS.

[b] Summation of mesophilic bacteria, fungi, and thermophilic bacteria. CFU = colony-forming units.

[c] ND = not detected.

[d] References:

Bayer, C. W., and Black, M. S. (1988a) Indoor Air Quality Evaluations of Three Office Buildings: Two of Conventional Construction Designs and One of a Special Design to Reduce Indoor Air Contaminants. Georgia Institute of Technology, Athens, GA.

Department of Health and Human Services. (1986) The Health Consequences of Involuntary Smoking—A Report to the Surgeon General.

EPA. (1987b) Compendium of Methods for the Determination of Air Pollutants in Indoor Air. Environmental Monitoring Systems Laboratory.

EPA. (1987d) Radon Reference Manual. EPA 520/1-87-20. Office of Radiation Programs.

Hawthorne, A., *et al.* (1987) Models for estimating organic emissions from building materials: formaldehyde example. *Atmos. Environ.* **21,** No. 2.

Lewis, R. G., *et al.* (1986) Monitoring for non-occupational exposure to pesticides in indoor and personal respiratory air. Presented at the 79th Annual Meeting of the Air Pollution Control Association, Minneapolis, MN.

National Research Council. (1986b). "Environmental Tobacco Smoke: Measuring Exposures and Assessing Health Effects." National Academy Press, Washington, DC.

Sheldon, L., Zelon, H., Sickles, J., Eaton, C., and Hartwell, T. (1988) "Indoor Air Quality in Public Buildings," Vol II. Environmental Monitoring Systems Laboratory, Office of Research and Development, U.S. Environmental Protection Agency.

Wallace, L. A., *et al.* (1983) Personal Exposure Assessment Methodology (TEAM) Study: Summary and Analysis: Vol. I.

mation in this table is not meant to be representative of typical indoor concentrations but only examples of measurements obtained by investigators and reported in the literature.

Airborne material affecting the quality of indoor air may be classified as gases or particulate matter. Gases which may be potential problems are radon, CO, NO_2, and hydrocarbons. Particulate matter may come from tobacco smoke, mold spores, animal dander, plant spores, and others as shown in Table 23-1. Other factors interact to influence our perception of indoor air quality, including humidity, temperature, lighting, and sound level.

IV. EFFECTS OF INDOOR AIR POLLUTANTS

Effects of indoor air pollutants on humans are essentially the same as those described in Chapter 7. However, there can be some additional pollutant exposures in the indoor environment that are not common in the ambient setting. From the listing in Table 23-1, radon exposures indoors present a radiation hazard for the development of lung cancer. Environmental tobacco smoke has been found to cause lung cancer and other respiratory diseases. Biological agents such as molds and other toxins may be a more likely exposure hazard indoors than outside.

Radon gas is formed in the process of radioactive decay of uranium. The distribution of naturally occurring radon follows the distribution of uranium in geological formations. Elevated levels have been observed in certain granite-type minerals. Residences built in these areas have the potential for elevated indoor concentrations of radon from radon gas entering through cracks and crevices and from outgassing from well water.

Radon gas is radioactive and emits alpha particles in the decay process. The elements resulting from radon decay are called radon daughters or progeny. These radon daughters can attach to airborne particles, which can deposit in the lung. The evidence supporting the radon risk of lung cancer comes from studies of uranium mine workers, in whom elevated rates of lung cancer have been observed. When an analysis of the potential exposure to radon inside homes was conducted by the U.S. Environmental Protection Agency, an estimate of 5,000 to 20,000 excess lung cancer deaths was projected annually (1). The risk is associated directly with increased lifetime doses; i.e., the longer the time spent living in a residence with elevated levels of radon, the higher the risk. Indoor levels of radon range from less than 1 to 200 picocuries per liter (pCi/l). Levels as high as 12,000 pCi/l have been observed but most levels are much lower. The U.S. Environmental Protection Agency has established an action level of 4 pCi/l for indoor radon, and if a home screening test shows concentrations below 4 pCi/l, no remedial action is suggested.

Environmental tobacco smoke (ETS) represents the exposure to tobacco smoke by individuals other than the smoker. For decades, the U.S. Surgeon General has indicated that smoking is a cause of lung cancer and cardiovascular disease for individuals who smoke. The U.S. Environmental Protection Agency has also concluded that ETS is a lung carcinogen for others breathing it.

The presence of biological contaminants gained widespread recognition with the outbreak of Legionnaires' disease in Philadelphia, Pennsylvania in 1976. In that year 221 persons attending a convention of Legionnaires developed pneumonia symptoms and 34 subsequently died. The agent, a bacterium later named *Legionella pneumophila,* was found in the cooling tower of the hotel's air-conditioning system. This bacterium has subsequently been responsible for other outbreaks of Legionnaires' disease. The bacteria in water supplies may be eliminated by suitable treatment procedures. *Legionnella* represents one of many types of biological agents which can cause allergic reactions and illness in the indoor environment.

One of the more difficult challenges remaining is the characterization of "sick building" syndrome. On numerous occasions, some employees in certain office buildings or other workplaces have developed a combination of symptoms including respiratory problems, dryness of the eyes, nose, and throat, headaches, and other nonspecific complaints. In such situations a substantial portion of the workers may exhibit these symptoms, which decrease in severity or stop when the worker is away from the building over the weekend or for longer periods. Investigations into the cause of these symptoms sometimes provide explanations, uncover ventilation problems, or identify an irritant gas. But many of these problem buildings are very difficult to understand and additional research is necessary to understand the cause-and-effect relationships.

V. CONTROL OF INDOOR AIR POLLUTANTS

The control and regulation of indoor air quality are influenced by individual property rights and a complicated mosaic of federal, state, and local government jurisdiction with conflicts, overlaps, and gaps in addressing these issues. Table 23-3 shows a large number of agencies and departments involved in indoor air quality control efforts at the federal level.

Government can institute certain laws and regulations for the citizens' well-being. Environmental and occupational examples abound, such as clean water and air legislation and workplace safety and health regulations. As the extension of this role into the home occurs, implementation and enforcement become more problematic. Examples of proactive regulatory approaches are building codes, zoning, consumer product standards, and safety requirements. Table 23-4 shows what various parties, from the indi-

TABLE 23–3

Indoor Air Responsibility of Federal Agencies

Point of impact	Agency/Activity	Comments
Direct control of indoor concentrations and/or exposures	OSHA air standards	Limited to industrial environments
	BPA radon action level	Limited to residents in BRA's weatherization program
	NASA air standards	Adopted OSHA standards
Control of emissions by restricting activities or product composition	EPA drinking water MCLs for radon and VOCs	Indoor air exposures considered in determining drinking water levels
	EPA pesticide restrictions	Restricts use and sales of pesticides which may cause indoor air pollution
	CPSC consumer product bans	Bans on use of some potential indoor pollutants in consumer products
	Smoking restrictions imposed by DOD, DOT, and GSA	Restricts smoking in specified indoor environments
	VA restrictions on asbestos use	Restricts use of asbestos in VA buildings
Control through assessment and mitigation procedures	EPA asbestos rules	Provides for the assessment and mitigation of asbestos hazards in schools
	GSA building assessments	Investigates GSA-controlled buildings for indoor air problems
	NIOSH building assessments	Responds to air quality health complaints
	DOD/USAF chlordane assessments	Investigates USAF facilities for chlordane problems
	NASA HVAC system maintenance	Assesses and corrects HVAC operation to optimize indoor air quality
Effort to increase knowledge of indoor air quality problems and controls	Research efforts by EPA, CPSC, DOE, HHS, BPA, DOT, NASA, NIST, NSF, TVA, HUD, and GSA	
	Information dissemination by EPA, CPSC, DOE, HHS, BPA, HUD, TVA, FTC, NASA, NIST, and NIBS	

OSHA, Occupational Safety and Health Administration; BPA, Bonneville Power Administration; NASA, National Aeronautics and Space Administration; MCL, Maximum Contaminant Levels; VOC, Volatile Organic Compounds; CPSC, Consumer Products Safety Commission; DOD, Department of Defense; DOT, Department of Transportation; GSA, General Services Administration; VA, Veterans Administration; NIOSH, National Institute of Occupational Safety and Health; USAF, United States Air Force; HVAC, Heating Ventilation and Air Conditioning; DOE, Department of Energy; HHS, Health and Human Services; NIST, National Institute of Standards and Technology; NSF, National Science Foundation; TVA, Tennessee Valley Authority; HUD, Housing and Urban Development; FTC, Federal Trade Commission; NIBS, National Institute of Building Sciences.

Source: U.S. Environmental Protection Agency Report to Congress on Indoor Air Quality, EPA/400/1-89/001c August 1989.

vidual to the federal government, can do to improve indoor air quality. Many of these efforts focus on education, improved materials, and better design of products and structures.

The technological control strategies are related back to Fig. 23-1. If the hazard is the result of elevated concentrations, then the technological solution is to reduce or remove the sources or dilute or remove the agent.

Control techniques are discussed for agents mentioned earlier—radon, ETS, and biological agents—and also for volatile organic compounds (VOCs). Radon enters the residence by two principal routes: infiltration from soil beneath the structure and outgassing from well water during showers. Elevated levels of radon are generally observed in basements or first-floor rooms. The mitigation techniques available include increased ventilation of the crawl space beneath the first floor, soil gas venting from beneath a basement floor, and sealing of all openings to the subsurface soil. These steps reduce the entry of radon into the home. Elevated levels of radon in well water can be removed by aeration or filtration by absorbent-filled columns.

Control of ETS is more complicated because of the personal behavior of individuals. For public buildings and facilities like offices, restaurants, and malls, many governmental bodies are placing restrictions on smoking in these areas, which can range from complete bans to requiring a restaurant to have a portion of a dining area for smokers and the remainder for nonsmokers. The difficulty for the restaurant owner is ensuring that the nonsmoking section is free of ETS. Education is the primary approach to "control" in the home. Information about the effects of ETS on family members has modified the behavior of some smokers.

Control of biological agents is multifaceted. In the case of *Legionnella*, cleaning and maintenance of heating and air-conditioning systems are generally sufficient to reduce the risk of this disease. In home heating and air-conditioning systems, mold and bacteria may be present and controlled with maintenance procedures. Growth may be inhibited by lower humidity levels. Keeping a house clean lowers the presence of dust mites, pollen, dander, and other allergens.

For VOCs, control options are multiple. Source reduction or removal includes product substitution or reformulation. Particleboard or pressed wood has been developed and used extensively in building materials for cabinet bases and subflooring and in furniture manufacturing for frames. If the product is not properly manufactured and cured prior to use as a building material, VOCs can outgas into the interior of the residence or building. Other sources of VOCs may be paints, cleaning solutions, fabrics, binders, and adhesives. Proper use of household products will lower volatile emissions.

In many of the industries associated with building or household products, efforts are under way to reduce the potential for subsequent VOC release

TABLE 23-4

Stakeholder Interests in Improving Indoor Air Quality

Individuals	Consumer and health professionals	Manufacturers	Building owners and managers	Builders and architects	State and local governments	Federal government
1. Find low-emission products in purchasing decisions.	1. Be knowledgeable of symptoms, effects, and mitigation and advise clients.	1. Adopt test procedures and standards to minimize product and material emissions.	1. Adopt ventilation maintenance procedures to eliminate and prevent contamination and ensure an adequate supply of clean air to building occupants.	1. Adopt indoor air quality as a design objective.	1. Conduct studies of specific problems in state or local area and adopt mitigation strategies.	1. Conduct research and technology transfer programs.
2. Maintain and use products to minimize emissions.	2. Develop information and education programs for constituent publics.	2. Adequately label products as to emission level and proper use and maintenance of products.	2. Use zone ventilation or local exhaust for indoor sources.	2. Ensure compliance with indoor air quality ventilation standards.	2. Establish building codes for design, construction, and ventilation requirements to ensure adequate indoor air quality.	2. Coordinate actions of other sectors.

3. Exercise discretionary control over ventilation to ensure clean air supply.		3. Substitute materials to minimize emissions from products manufactured.	3. Develop specific procedures for use of cleaning solvents, paints, herbicides, insecticides, and other contaminants to protect occupants.	3. Adopt low emission requirements in procurement specifications for building materials from manufacturers.	3. Enforce and monitor code compliance	3. Coordinate actions of other sectors, encourage, or require specific sectors to take actions toward mitigation.
4. Be knowledgeable of indoor air quality problems and take actions to avoid personal exposure.		4. Develop training programs for commercial users to ensure low emissions.	4. Adopt investigatory protocols to respond to occupant complaints.	4. Contain or ventilate known sources.	4. Educate and inform building community, health community, and public about problems and solutions.	
		5. Conduct research to advance mitigation technology.				

Source: U.S. Environmental Protection Agency Report to Congress on Indoor Air Quality, EPA/400/1-89/001c, August 1989.

to the interior of residences or commercial buildings. Modification of the manufacturing process, solvent substitution, product reformulation, and altering installation procedures are a few of the approaches available.

When VOCs are present indoors at elevated concentrations, modification of ventilation rates is a control option for diluting and reducing these concentrations. The American Society of Heating, Refrigerating and Air-Conditioning Engineers (ASHRAE) has established standards for ventilation rates for outside air per individual. The guideline is 15 cfm per person (2). This guideline is designed to bring sufficient fresh air into a building to minimize the buildup of contaminants and odors.

REFERENCES

1. "A Citizen's Guide to Radon: What It Is and What to Do About It." U.S. Environmental Protection Agency and U.S. Department of Health and Human Services, OPA-8-004. Government Printing Office, Washington, DC, 1986.
2. "Ventilation for Acceptable Indoor Air Quality." American Society of Heating, Refrigerating and Air-Conditioning Engineers, Atlanta, 1981.

SUGGESTED READING

Godish, T., "Indoor Air Pollution Control." Lewis Publishers, Chelsea, MI, 1989.

Nazaroff, W. W., and Teichmann, K., Indoor radon. *Environ. Sci. Technol.* **24,** 774–782 (1990).

U.S. Environmental Protection Agency and U.S. Consumer Product Safety Commission, "The Inside Story: A Guide to Indoor Air Quality." EPA/400/1-88/004, September 1988.

QUESTIONS

1. Define "sick building" syndrome.
2. What controls are available to control indoor air quality?
3. Why will the command and control approach not work for residential indoor air quality?
4. Describe an implication of high radon levels during the sale of a home.

24

The U.S. Clean Air Act Amendments of 1990

I. INTRODUCTION

The U.S. Clean Air Act Amendments of 1990 (CAAA90) (1) were a revision of the original U.S. Clean Air Act passed in 1963, amended in 1970, and amended again in 1977 (2). It is one of the most significant pieces of environmental legislation ever enacted. Estimates of the yearly economic impacts of CAAA90 range from U.S. $12 billion to U.S. $53 billion in 1995 and from U.S. $25 billion to U.S. $90 billion by 2005.

CAAA90 is a technology-based program rather than the health-based program used in the original Clean Air Act. The standards and emission limits are based on maximum achievable control technology. The final emission limits will be set forth in permits issued by the individual states.

Stenvaag, in his book, "Clean Air Act 1990 Amendments Law and Practices" (3), has written:

> By far the most important interpretive documents have not yet been written: the countless EPA regulations, decisions, and explanatory preambles that must be published in the Federal Register to carry out the Act. As these rulings are systematically promulgated over the coming decades, flesh and muscle will be added to the massive skeleton of the 1990 amendments and we will all gain a better understanding of what the 101st Congress has wrought.

The Clean Air Act Amendments of 1990 put a heavy burden on the Environmental Protection Agency (EPA) and state agencies to permit and oversee compliance with the Clean Air Act (CAA). Table 24-1 lists the major deadlines affecting industry under the 1990 amendments. Some of the deadlines have already passed without any "action" being promulgated and finalized. The other deadlines may also slip to a later date. Almost no statutory "hammers" have been included in CAAA90.

TABLE 24-1

Deadlines Impacting Industry under the Clean Air Act Amendments of 1990

Date	Action
November 15, 1991	EPA must issue guidelines for cost effectiveness of emission control options.
November 15, 1991	EPA must promulgate a list of categories and subcategories of major sources and area sources of listed hazardous air pollutants.
November 15, 1991	EPA must promulgate permitting regulations.
May 15, 1992	EPA must issue New Source Performance Standards (NSPS) for any solid waste incineration unit not covered by another deadline for issuance of such standards.
November 15, 1993	EPA must issue volatile organic compound (VOC) emission control technique guidelines for categories that make the most significant contributions to the formation of ozone pollution, including Resource Conservation and Recovery Act (RCRA) permitted treatment, storage and disposal (TSD) facilities. Existing control technique guidelines must also be reviewed by this date.
November 15, 1993	EPA is required to identify alternative control technologies for all categories of stationary sources of VOC or NO_x that have the potential to emit 25 tons per year or more of either pollutant.
November 15, 1993	EPA is required to issue guidance on reasonably available control measures (RACM) and best available control measures (BACM) for other sources of particulate matter emissions.
November 15, 1993	States must submit permit programs to EPA for approval.
November 15, 1993	EPA must issue NSPS for solid waste incineration units combusting industrial or commercial waste. (See May 15, 1992 action.)
November 15, 1994	Risk Assessment and Management Commission must report to Congress on the appropriate use and policy implications of risk assessment and risk management in various regulatory programs.
November 15, 1995	If necessary, EPA is required to promulgate regulations to control the atmospheric deposition of hazardous air pollutant to surface and coastal waters.
November 15, 1996	VOC emissions must be reduced by at least 15% from 1990 levels in ozone nonattainment areas.
November 15, 2000	EPA is required to promulgate standards for all listed categories and subcategories of sources of hazardous air pollutants.

II. TITLES

CAAA90 contains 11 "Titles." Some of these are new, and some are greatly expanded from similar titles in the original Clean Air Act.

Title I: Provisions for Attainment and Maintenance of National Ambient Air Quality Standards

Title I changes the existing nonattainment program of the original Clean Air Act Title I, Part D, by establishing new nonattainment requirements for ozone, carbon monoxide (CO), and particulate matter (PM_{10}). Also, the new nonattainment requirements vary from location to location and with the severity of nonattainment. The changes will be accomplished through revisions of the State Implementation Plan (SIP). The SIP must designate "milestones" according to specified timetables. Failure to meet milestones will result in mandatory sanctions.

The most widespread and persistent urban pollution problem is ozone. The causes of this and the lesser problem of CO and PM_{10} pollution in our urban areas are largely due to the diversity and number of urban air pollution sources. One component of urban smog, hydrocarbons, comes from automobile emissions, petroleum refineries, chemical plants, dry cleaners, gasoline stations, house painting, and printing shops. Another key component, nitrogen oxides, comes from the combustion of fuel for transportation, utilities, and industries.

Although there are other reasons for continued high levels of ozone pollution, such as growth in the number of stationary sources of hydrocarbons and continued growth in automobile travel, the remaining sources of hydrocarbons are the most difficult to control. These are the small sources, those that emit less than 100 tons of hydrocarbons per year. These sources, such as auto shops and dry cleaners, may individually emit less than 10 tons per year but collectively emit many hundreds of tons of pollution.

Title I allows the EPA to define the boundaries of "nonattainment" areas for ozone, CO, and PM_{10}. Emission standards for these areas will be based on a new set of "nonattainment categories." EPA has established a classification system for ozone design values (goals) and attainment deadlines. Table 24-2 lists these parameters.

If a nonattainment area is classified as serious, based on ambient ozone measurements, then the state must modify its SIP to bring the area into compliance in 9 years. The CAAA90 also specify the size and, therefore, the number of sources subject to regulatory control as a function of nonattainment classification. Table 24-3 illustrates these requirements for ozone nonattainment classifications of extreme and severe; the state must include

TABLE 24-2

Classification and Attainment Dates for Ozone Nonattainment Areas

Classification	Ozone design values	Attainment deadline (from enactment)
Marginal	0.121 up to 0.138 ppm	1993 (3 years)
Moderate	0.138 up to 0.160 ppm	1996 (6 years)
Serious	0.160 up to 0.180 ppm	1999 (9 years)
Severe	0.180 up to 0.280 ppm	2005 (15 years)
Extreme	0.280 ppm and above	2010 (20 years)

sources with combined NO_x and VOC emissions of 10 tons/year in their control plans

As mentioned, nonattainment areas will have to implement different control measures, depending on their classification. Marginal areas, for example, are the closest to meeting the standard. They will be required to conduct an inventory of their ozone-causing emissions and institute a permit program. Nonattainment areas with more serious air quality problems must implement various control measures. The worse the air quality, the more controls areas will have to implement.

The new law also establishes similar programs for areas that do not meet the federal health standards for the pollutants carbon monoxide and particulate matter. Areas exceeding the standards for these pollutants will be divided into "moderate" and "serious" classifications. Depending on the degree to which they exceed the carbon monoxide standard, areas will be required to implement programs introducing oxygenated fuels and/or enhanced emission inspection programs, among other measures. Depending on their classification, areas exceeding the particulate matter standard will have to implement either Reasonably Available Control Technology (RACT) or Best Available Control Technology (BACT), among other requirements.

TABLE 24-3

Title I Emission Sources Requiring Control

EPA ozone nonattainment classification	Allowable emissions of NO_x and VOC combined (ton/year)
Extreme and severe	10
Serious	50
Moderate and marginal	100

Title II: Provisions Related to Mobile Sources

Title II of the Clean Air Act Amendments of 1990 is related mainly to vehicles that operate on roads and highways. Off-road, or nonroad, engines and vehicles used for site drilling, remediation, or related construction may be regulated if the administrator of EPA determines that some degree of emission reduction is necessary.

The EPA has summarized the provisions related to mobile sources (4) as follows:

> While motor vehicles built today emit fewer pollutants (60% to 80% less, depending on the pollutant) than those built in the 1960s, cars and trucks still account for almost half the emissions of the ozone precursors VOCs and NO_x, and up to 90% of the CO emissions in urban areas. The principal reason for this problem is the rapid growth in the number of vehicles on the roadways and total miles driven. This growth has offset a large portion of the emission reductions gained from motor vehicle controls.
>
> In view of the unforeseen growth in automobile emissions in urban areas combined with the serious air pollution problems in many urban areas, the Congress has made significant changes to the motor vehicle provisions on the 1977 Clean Air Act.
>
> The Clean Air Act of 1990 establishes tighter pollution standards for emissions from automobiles and trucks. These standards will reduce tailpipe emissions of hydrocarbons, carbon monoxide, and nitrogen oxides on a phased-in basis beginning in model year 1994. Automobile manufacturers will also be required to reduce vehicle emissions resulting from the evaporation of gasoline during refueling.
>
> Fuel quality will also be controlled. Scheduled reductions in gasoline volatility and sulfur content of diesel fuel, for example, will be required. New programs requiring cleaner (so-called 'reformulated' gasoline) will be initiated in 1995 for the nine cities with the worst ozone problems. Other cities can 'opt in' to the reformulated gasoline program. Higher levels (2.7%) of alcohol-based oxygenated fuels will be produced and sold in 41 areas during the winter months that exceed the federal standard for carbon monoxide.
>
> The new law also establishes a clean fuel car pilot program in California, requiring the phase-in of tighter emission limits for 150,000 vehicles in model year 1996 and 300,000 by the model year 1999. These standards can be met with any combination of vehicle technology and cleaner fuels. The standards become even stricter in 2001. Other states can 'opt in' to this program, though only through incentives, not sales or production mandates.
>
> Further, twenty-six of the dirtiest areas of the country will have to adopt a program limiting emissions from centrally-fueled fleets of 10 or more vehicles beginning as early as 1998.

Title III: Hazardous Air Pollutants

Toxic air pollutants are pollutants which are hazardous to human health or the environment but which are not specifically regulated by the CAA. These pollutants are typically carcinogens, mutagens, and teratogens. The CAAA of 1977 failed to result in substantial reductions in the emissions of these harmful substances.

The toxic air pollution problem is widespread. Information generated from the Superfund "Right to Know" rule From the Superfund Authorization and Recovery Act (SARA Section 313) indicates that more than 2.7 billion pounds of toxic air pollutants are emitted annually in the United States. EPA studies indicate that exposure to such quantities of air toxics may result in 1000 to 3000 cancer deaths each year.

The CAAA90 offers a comprehensive plan for achieving significant reductions in emissions of hazardous air pollutants from major sources. The new law will improve EPA's ability to address this problem effectively and it will accelerate progress in controlling major toxic air pollutants.

EPA will issue Maximum Achievable Control Technology (MACT) standards for each listed source category according to a prescribed schedule. These standards will be based on the best demonstrated control technology or practices within the regulated industry, and EPA must issue the standards for 40 source categories within 2 years of passage of the new law. The remaining source categories will be controlled according to a schedule which ensures that all controls will be achieved within 10 years of enactment. Companies that voluntarily reduce emissions according to certain conditions can get a 6-year extension to meet the MACT requirements.

Eight years after MACT is installed on a source, EPA must examine the risk levels remaining at the regulated facilities and determine whether additional controls are necessary to reduce unacceptable residual risk.

The EPA developed (4) a one-page summary of the key points of Title III. The following is this summary.

Title III—Air Toxics, Key Points

- List of Pollutants and Source Categories: The law lists 189 hazardous air pollutants. One year after enactment EPA lists source categories (industries) which emit one or more of the 189 pollutants. In two years, EPA must publish a schedule for regulation of the listed source categories.
- Maximum Achievable Control Technology (MACT): MACT regulations are emission standards based on the best demonstrated control technology and practices in the regulated industry. MACT for existing sources must be as stringent as the average control efficiency or the best controlled 12% of similar sources excluding sources which have achieved the Lowest Achievable Emission Rate (LAER) within 18 months prior to proposal or 30 months prior to promulgation. MACT for new sources must be as stringent as the best controlled similar source. For all listed major point sources, EPA must promulgate MACT standards—40 source categories plus coke ovens within two years and 25% of the remainder of the list within four years, an additional 25% in seven years, and the final 50% in ten years.
- Residual Risk: Eight years after MACT standards are established (except for those established two years after enactment), standards to protect against the residual health and environmental risks remaining must be promulgated, if necessary. The standards would be triggered if more than one source in a category exceeds a maximum individual risk of cancer of 1 in 1 million. These residual risk regulations would be based on current CAA language that specifies that standards must achieve an 'ample margin of safety.'

• Accidental Releases: Standards to prevent against accidental release of toxic chemicals are required. EPA must establish a list of at least 100 chemicals and threshold quantities. All facilities with these chemicals on site in excess of the threshold quantities would be subject to the regulations which would include hazard assessments and risk management plans. An independent chemical safety board is established to investigate major accidents, conduct research, and promulgate regulations for accidental release reporting.

Title IV: Acid Deposition Control

The EPA summary (4) of Title IV states the basics of the acid deposition control amendments:

> Acid deposition occurs when sulfur dioxide and nitrogen oxide emissions are transformed in the atmosphere and return to the earth in rain, fog or snow. Approximately 20 million tons of SO_2 are emitted annually in the United States, mostly from the burning of fossil fuels by electric utilities. Acid rain damages lakes, harms forests and buildings, contributes to reduced visibility, and is suspected of damaging health.
>
> The new Clean Air Act will result in a permanent 10 million ton reduction in sulfur dioxide (SO_2) emissions from 1980 levels. To achieve this, EPA will allocate allowances of one ton of sulfur dioxide in two phases. The first phase, effective January 1, 1995, requires 110 powerplants to reduce their emissions to a level equivalent to the product of an emissions rate = (2.5 lbs of SO_2/mm Btu) × (the average mm Btu of their 1985–1987 fuel use). Plants that use certain control technologies to meet their Phase I reduction requirements may receive a two year extension of compliance until 1997. The new law also allows for a special allocation of 200,000 annual allowances per year each of the 5 years of Phase I to powerplants in Illinois, Indiana and Ohio.
>
> The second phase, becoming effective January 1, 2000, will require approximately 2000 utilities to reduce their emissions to a level equivalent to the product of an emissions rate of (1.2 lbs of SO_2/mm Btu) × (the average mm Btu of their 1985–1987 fuel use). In both phases, affected sources will be required to install systems that continuously monitor emissions in order to track progress and assure compliance.
>
> The new law allows utilities to trade allowances within their systems and/or buy or sell allowances to and from other affected sources. Each source must have sufficient allowances to cover its annual emissions. If not, the source is subject to a $2,000/ton excess emissions fee and a requirement to offset the excess emissions in the following year.
>
> Nationwide, plants that emit SO_2 at a rate below 1.2 lbs/mm Btu will be able to increase emissions by 20% between a baseline year and 2000. Bonus allowances will be distributed to accommodate growth by units in states with a statewide average below 0.8 lbs/mm Btu. Plants experiencing increases in their utilization in the last five years also receive bonus allowances. 50,000 bonus allowances per year are allocated to plants in 10 midwestern states that make reductions in Phase I. Plants that repower with a qualifying clean coal technology may receive a 4 year extension of the compliance date for Phase II emission limitations.
>
> The new law also includes specific requirements for reducing emissions of nitrogen oxides, based on EPA regulations to be issued not later than mid-1992 for certain boilers and 1997 for all remaining boilers.

Title IV represents legislation designed to reduce total SO_2 emissions by approximately 50% over a 10-year period. Provisions of the title are designed to introduce economic flexibility for the electric power industry, to recognize controls already implemented by progressive utilities and to reduce the economic impact on high-sulfur coal regions of the United States.

Title V: Permits

The new law introduces an operating permits program modeled after a similar program under the Federal National Pollution Elimination Discharge System (NPDES) law. The purpose of the operating permits program is to ensure compliance with all applicable requirements of the CAAA90 and to enhance EPA's ability to enforce the Act. Air pollution sources subject to the program must obtain an operating permit; states must develop and implement the program; and EPA must issue permit program regulations, review each state's proposed program, and oversee the state's efforts to implement any approved program. EPA must also develop and implement a federal permit program when a state fails to adopt and implement its own program. The final rulemaking for this Title V program was published on July 21, 1992 as Part 70 of Chapter I of Title 40 of the Code of Federal Regulations (57FR32250).

The new program clarifies and makes more enforceable a source's pollution control requirements. Currently, a source's pollution control obligations may be scattered throughout numerous hard-to-find provisions of state and federal regulations, and in many cases, the source is not required, under the applicable State Implementation Plan, to submit periodic compliance reports to EPA or the states. The permit program will ensure that all of a source's obligations with respect to its pollutants will be contained in one permit document and that the source will file periodic reports identifying the extent to which it has complied with those obligations. Both of these requirements will greatly enhance the ability of federal and state agencies to evaluate a source's air quality situation.

In addition, the new program will provide a ready vehicle for states to assume administration, subject to federal oversight, of significant parts of the air toxics program and the acid rain program. Through the permit fee provisions discussed later, the program will greatly augment a state's resources to administer pollution control programs by requiring sources of pollution to pay their fair share of the costs of a state's air pollution program.

Under the new law, EPA was required to issue program regulations by November 15, 1991. By November 15, 1993, each state must submit to EPA a permit program meeting these regulatory requirements. After receiving the state submittal, EPA has 1 year to accept or reject the program. EPA must level sanctions against a state that does not submit or enforce a permit program.

If a state fails to comply, EPA will promulgate and administer the state's program. That could mean lengthy permitting delays and additional paperwork. By supporting the state's permitting authority, a facility can simplify the permitting process and avoid EPA intervention.

Each permit issued to a facility will be for a fixed term of up to 5 years. The new law establishes a permit fee whereby the state collects a fee from the permitted facility to cover reasonable direct and indirect costs of the permitting program.

All sources subject to the permit program must submit a complete permit application within 12 months of the effective date of the program. The state permitting authority must determine whether or not to approve an application within 18 months of the date it receives the application.

To ensure compliance with the standards, the permit also must contain provisions for the inspection, entry, monitoring, certification, and reporting of compliances with the permit conditions.

EPA has 45 days to review each permit and to object to permits that violate the CAAA. If EPA fails to object to a permit that violates the Act or the implementation plan, any person may petition EPA to object within 60 days following EPA's 45-day review period, and EPA must grant or deny the permit within 60 days. Judicial review of EPA's decision on a citizen's petition can occur in the federal court of appeals. The public is guaranteed the right to inspect and review all permit applications and documents. There are provisions for three kinds of permit revisions: administrative amendment, minor permit modification, and significant modification.

These regulations will apply to an estimated 34,000 "major" industrial sources. "Major" sources are defined according to their "potential to emit" and the cutoff levels vary depending on both the pollutant and the local areas' compliance status with the National Ambient Air Quality Standard (NAAQS) for that pollutant. For the present, the EPA has exempted all "nonmajor" sources, of which there are estimated to be about 350,000, from this permitting, until they have studied further the feasibility of permitting them. However, the states can require permitting of some of these sources.

The regulations provide for the collection of fees from permit seekers and the states must require fees sufficient to cover the cost of administering the program. If a state does not submit, properly administer, or enforce a permit program, federal sanctions must be levied including the withholding of highway funds and requiring offset pollution for new sources by reducing emissions by 2 tons from existing sources for every 1 ton of emissions from a proposed new source.

All permits must include a cap on emissions which cannot be exceeded without an approved revision of the permit. Permitted sources must periodically test and monitor their emisisons and report on these activities every 6 months. Civil penalties include fines of not less than $10,000 per day for permit violations and criminal penalties for deliberate false statements or representations, or for rendering inaccurate any monitoring device or method required in the permit.

Title VI: Stratospheric Ozone Protection

The EPA summary (4) for stratospheric ozone and global climate protection lists the basics of the title:

> The new law builds on the market-based structure and requirements currently contained in EPA's regulations to phase out the production of substances that deplete the ozone layer. The law requires a complete phase-out of CFCs and halons with interim reductions and some related changes to the existing Montreal Protocol, revised in June 1990.
>
> Under these provisions, EPA must list all regulated substances along with their ozone-depletion potential, atmospheric lifetimes and global warming potentials within 60 days of enactment.
>
> In addition, EPA must ensure that Class I chemicals be phased out on a schedule similar to that specified in the Montreal Protocol—CFCs, halons, and carbon tetrachloride by 2000; methyl chloroform by 2002—but with more stringent interim reductions. Class II chemicals (HCFCs) will be phased out by 2030. Regulations for Class I chemicals will be required within 10 months, and Class II chemical regulations will be required by December 31, 1999.
>
> The law also requires EPA to publish a list of safe and unsafe substitutes for Class I and II chemicals and to ban the use of unsafe substitutes.
>
> The law requires nonessential products releasing Class I chemicals to be banned within 2 years of enactment. In 1994 a ban will go into effect for aerosols and noninsulating foam using Class II chemicals, with exemptions for flammability and safety. Regulations for this purpose will be required within one year of enactment, to become effective two years afterwards.

Title VII: Provisions Relating to Enforcement

The CAAA90 contains a broad array of authorities to make the law more readily enforceable, thus bringing it up to date with the other major environmental statutes. EPA has new authorities to issue administrative penalty orders up to U.S. $200,000 and field citations up to U.S. $5,000 for lesser infractions. Civil judicial penalties are enhanced. Criminal penalties for knowing violations are upgraded from misdemeanors to felonies, and new criminal authorities for knowing and negligent endangerment will be established.

In addition, sources must certify their compliance, and EPA has authority to issue administrative subpoenas for compliance data. EPA will also be authorized to issue compliance orders with compliance schedules of up to 1 year.

The citizen suit provisions have also been revised to allow citizens to seek penalties against violators, with the penalties going to a U.S. Treasury fund for use by EPA for compliance and enforcement activities. The U.S. government's right to intervene is clarified and citizen plaintiffs will be required to provide the U.S. government with copies of pleadings and draft settlements.

Any atmospheric emissions for which EPA does not develop standards may be regulated by state or regional authorities.

A review of the enforcement and liability provisions of CAAA90 (5) recommends that because of the new enforcement tools available to the federal government, the regulated community should implement effective self-auditing and compliance programs at facilities to reduce the risk of criminal liability. Stenvaag (3) covers the provisions relating to enforcement from a legal standpoint. He states the new language of this title to be "quite confusing, particularly in specifying when criminal sanctions are appropriate."

Title VIII: Miscellaneous Provisions

Section 130, Emission Factors requires revising emission inventory factors every 3 years:

> Within 6 months after enactment of the Clean Air Act Amendments of 1990, and at least every 3 years thereafter, the Administrator shall review and, if necessary, revise, the methods ('emission factors') used for purposes of this Act to estimate the quantity of emissions of carbon monoxide, volatile organic compounds, and oxides of nitrogen from sources of such air pollutants (including area sources and mobile sources). In addition, the Administrator shall permit any person to demonstrate improved emissions estimating techniques, and following approval of such techniques, the Administrator shall authorize the use of such techniques. Any such technique may be approved only after appropriate public participation. Until the Administrator has completed the revision required by this section, nothing in this section shall be construed to affect the validity of emission factors established by the Administrator before the date of the enactment of the Clean Air Act Amendments of 1990.

Title IX: Clean Air Research

Title IX of the Clean Air Act Amendments of 1990 addresses air pollution research areas including monitoring and modeling, health effects, ecological effects, accidental releases, pollution prevention and emissions control, acid rain, and alternative motor vehicle fuels. The provisions require ecosystem studies on the effects of air pollutants on water quality, forests, biological diversity, and other terrestrial and aquatic systems exposed to air pollutants; mandate the development of technologies and strategies for air pollution prevention from stationary and area sources; and call for several major studies. The EPA must improve methods and techniques for measuring individual air pollutants and complex mixtures and conduct research on long- and short-term health effects, including the requirement for a new interagency task force to coordinate these research programs. Finally, the Agency must develop improved monitoring and modeling methods to increase the understanding of tropospheric ozone formation and control.

To implement the research provisions, the EPA plans to conduct research in emissions inventories, atmospheric modeling, source/ambient monitoring, control technologies, health, and ecological monitoring. Both ecological monitoring and ambient monitoring will be done jointly with other agencies who also need these data to meet their mission. Other proposed work includes developing improved risk assessment methods, maintaining existing networks or establishing new ones for aquatic and terrestrial effects monitoring, and continuing work on deposition chemistry. Again, these efforts in particular will be supported, in part, by other agencies.

Title X: Disadvantaged Business Concerns

and

Title XI: Clean Air Employment Transition Assistance

These two final titles were added to cover the subject areas of concern. They relate to procedural matters and direct the appropriate federal agencies to implement and oversee the necessary compliance action.

REFERENCES

1. Public Law 101-549; 101st Congress, November 15, 1990, An Act to amend the Clean Air Act to provide for attainment and maintenance of health protective national ambient air quality standards, and for other purposes.
2. U.S. Public Law No. 88-206, 77 Stat. 392 (1963), U.S. Public Law No. 91-604, 84 Stat. 1676 (1970), U.S. Public Law No. 95-95, 91 Stat. 686 (1977).
3. Stenvaag, J. M., "Clean Air Act 1990 Amendments Law and Practices." Wiley, New York, 1991.
4. Office of Air and Radiation, U.S. Environmental Protection Agency, "The Clean Air Act Amendments of 1990, Summary Materials," November 1990.
5. Elliott, E. D., Schwartz, R. M., Goldman, A. V., Horowitz, A. B., and Laznow, J., The Clean Air Act: new enforcement and liability provisions. *J. Air Waste Manage. Assoc.* **42**(11), 1261–1270 (1992).

SUGGESTED READING

Burke, R. L., "Permitting for Clean Air—A Guide to Permitting under Title V of the Clean Air Act Amendments of 1990." Air and Waste Management Association, Pittsburgh, 1992.

Gas Research Institute briefing on Clean Air Act Amendments of 1990. Presentation by John G. Holmes and Robert W. Crawford, Energy and Environmental Analysis, Inc., December 1990.

"Implementation Strategy for the Clean Air Act Amendments of 1990." EPA Office of Air and Radiation, January 15, 1991.

Lee, B., Highlights of the Clean Air Act Amendments of 1990. *J. Air Waste Manage. Assoc.* **41** (1), 48–55 (1991).

Quarles, J., and Lewis, W. H., Jr., "The New Clean Air Act: A Guide to the Clean Air Program as Amended in 1990," Lewis and Brockius, Washington, DC, 1990.

QUESTIONS

1. Give an example of a health-based air pollution standard with the justification for the standard.
2. Give an example of a technology-based air pollution standard with the justification for the standard.
3. Choose a specific metropolitan area and determine its classification as an ozone nonattainment area. Find the alternative deadline and allowable emissions of NO_x and VOC combined.
4. For the specific metropolitan area in question 3, discuss how the attainable deadline and allowable emissions can be met.
5. List the alternatives that are possible to replace present automobiles with vehicles, or systems, that will reduce emissions of VOCs, NO_x, and CO.
6. List five categories (such as hazardous waste incineration) that may be considered as "major" or "area" sources of hazardous air pollutants.
7. Develop an outline of a permit for a new plastic molding company. Include a schedule of costs and reporting requirements.

25

Emission Standards

I. SUBJECTIVE STANDARDS

Limits on emissions are both subjective and objective. Subjective limits are based on the visual appearance or smell of an emission. Objective limits are based on physical or chemical measurement of the emission. The most common form of subjective limit is that which regulates the optical density of a stack plume, measured by comparison with a Ringelmann chart (Fig. 25-1). This form of chart has been in use for over 90 years and is widely accepted for grading the blackness of black or gray smoke emissions. Within the past four decades, it has been used as the basis for "equivalent opacity" regulations for grading the optical density of emissions of colors other than black or gray.

The original Ringelmann chart was a reflectance chart; the observer viewed light reflected from the chart. More recently, light transmittance charts have been developed for both black (1) and white (2) gradations of optical density which correlate with the Ringelmann chart scale. It is now common practice in the United States to send air pollution inspectors to a "smoke school" where they are trained and certified as being able to read the density of black and white plumes with an accuracy that is acceptable for court testimony.

Before the widespread acceptance of the Ringelmann scale, smoke was regulated by prohibiting the emission of black smoke. Now regulatory

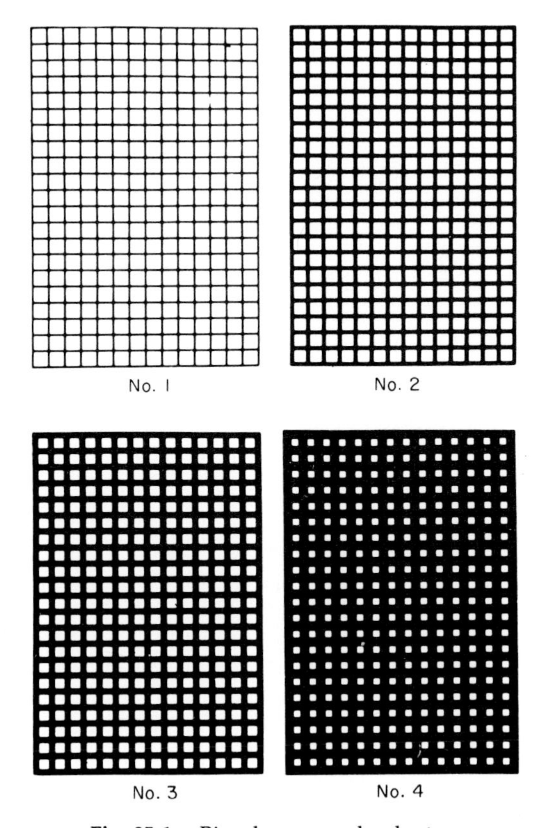

Fig. 25-1. Ringelmann smoke chart.

practice prohibits the emission of smoke whose density is greater than a specified Ringelmann number. Over recent decades, there has been a progressive decrease in the Ringelmann number thus specified, culminating in the specification of number 1 for large new steam power plants in the United States. In addition to subjective observation of smoke density, continuous emission monitoring systems (CEMs) have been developed to measure objectively, by means of a photocell, the decrease in intensity of a beam of light projected through the plume prior to its emission (Fig. 25-2). Such systems are discussed in Chapter 32.

Subjective evaluation of odor emission is made difficult by the phenomenon of odor fatigue, which means that after persons have been initially subjected to an odor, they lose the ability to perceive the continued presence of low concentrations of that odor. Therefore, all systems of subjective odor evaluation rely on preventing olfactory fatigue by letting the observer breathe odor-free air for a sufficient time prior to breathing the odorous air and evaluating its odor content. Usually an activated charcoal bed is

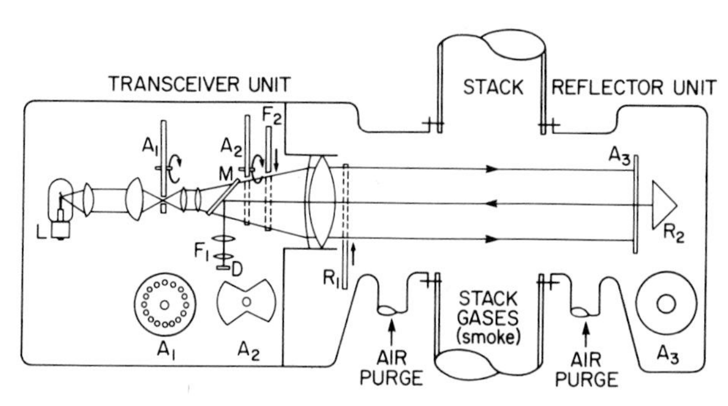

Fig. 25-2. Double-beam, double-pass transmissometer for measuring smoke density in stacks. A_1, chopper wheel; A_2, beam gating wheel; A_3, aperture; D, detector; F_1, spectral filter; F_2, solenoid-activated neutral density filter; L, lamp; M, half-mirror/beam splitter; R_1, solenoid-activated zero calibration reflector; R_2, retroreflector (alignment bullseye not shown). Design patented. Source: Drawing courtesy of Lear Siegler, Inc.

used to clean up the air in order to provide the odor-free air required by the observer.

II. OBJECTIVE STANDARDS

There are two major categories of objective emission limits: those which limit the emission of a specific pollutant, regardless of the process or equipment from which it is emitted, and those which limit the emissions of a specific pollutant from a specific process or type of equipment. Regulations may require the same emission limit for all sources, regardless of size or capacity; or they may vary the allowable emission with the size or capacity of the source. Limits may be stated in absolute terms, i.e., not more than a specified mass of pollutant per unit of time, or in relative terms, i.e., not more than a specified mass of pollutant per unit mass of fuel burned, material being processed, or product produced or per unit of heat released in a furnace. In the case of gaseous pollutants, limits may be stated in volumetric rather than gravimetric terms. Emission limits are sometimes stated as mass of pollutant per unit volume of effluent. Effluent volume varies with gas temperature and pressure, the presence or absence of diluting air, and the amount thereof. Therefore, volumes must be reduced to a specified temperature, pressure, and percent of diluting air. In the case of flue gases from fuel combustion, dilution is usually expressed as percentage of excess air or percentage of carbon dioxide or oxygen in the flue gas.

The pollutants, source categories, and affected facilities for which the United States has established New Source Performance Standards are listed in Table 25-1. Certain categories listed in Table 25-1 are subject to U.S. Prevention of Significant Deterioration (of air quality) (PSD) review if their emission potential of a regulated pollutant exceeds 100 tons per year. In addition, PSD review is required for the following sources, whose emission potential exceeds 100 tons per day: hydrofluoric acid plants, coke oven batteries, furnace process carbon black plants, fuel conversion plants, taconite ore processing plants, gas fiber processing plants, and charcoal production plants.

III. TYPES OF EMISSION STANDARDS

The most common rationale for developing emission limits for stationary sources is the application of the best practicable means for control. Under this rationale, the degree of emission limitation achievable at the best designed and operated installation in a category sets the emission limits for all other installations of that category. As new technology is developed, what was the best practicable means in 1993 may be short of the best attainable in 2003. Because there is thus a moving target, means must be provided administratively to allow the plant which complied with a best practicable means limit to be considered in compliance for a reasonable number of years thereafter. It is important to note that best practicable means limits are set without regard to present or background air quality or the air quality standards for the pollutants involved, the number and location of sources affected by the limit, and the meteorology or topography of the area in which they are located. However, an additional limit on minimum stack height, or buffer zone, based on these noted factors may be coupled with a best practicable means limit on mass or emission.

The other major rationales for developing emission limits have been based on some or all of the noted factors—air quality, air quality standards, number and location of sources, meteorology, and topography. These include the rollback approach, which involves all these factors except source location, meteorology, and topography; and the single-source mathematical modeling approach, which considers only the air quality standard and meteorology, ignoring the other factors listed.

IV. VARIANT FORMS OF EMISSION STANDARDS

Some variants of best practicable means are spelled out in the U.S. Clean Air Act of 1977. One is the requirement that best available control technology (BACT) for a specific pollutant be employed on new "major sources" that are to be located in an area that has attained the National Ambient Air Quality Standard (NAAQS) for that pollutant. BACT is also required for pollutants for which there is no NAAQS [e.g., total reduced sulfur (TRS), for which emission limits are specified by a Federal New Source Performance Standard (NSPS)]. BACT must be at least as stringent as NSPS but is determined on a case-by-case basis.

TABLE 25-1

U.S. National New Source Performance Standards (NSPS)

Source category (in sequence of promulgation)	Affected facility	Pollutants regulated									
		Particulate matter	Opacity	SO_2	NO_x	H_2S	CO	TRS	VOC	Pb	Total F
Fossil fuel and electric utility steam generator[a] [>73 MW (>250 million BTU/hr) input] (264 KJ/hr)	Solid fossil fuel and wood residue-fired	x	x	x	x	—	—	—	—	—	—
	Oil- or gas-fired boilers	x	x	x	x	—	—	—	—	—	—
Incinerators[a]	>45 metric tons/day	x	—	—	—	—	—	—	—	—	—
Portland cement plants[a]	Kiln and clinker cooler	x	x	—	—	—	—	—	—	—	—
	Fugitive emission points	—	x	—	—	—	—	—	—	—	—
Nitric acid plants[a]	Process equipment	—	x	—	x	—	—	—	—	—	—
Sulfuric acid plants	Process equipment	x[b]	x	x	—	—	—	—	—	—	—
Asphalt concrete plants	Dryers, etc.[c]	x	x	—	—	—	—	—	—	—	—
Petroleum refineries[a]	Fluid catalyst regenerator	x	x	—	—	—	x	—	—	—	—
	Fuel gas combustion	—	—	x	—	x	—	—	—	—	—
	Claus sulfur recovery TRS X[d]	—	—	x	—	—	—	x	—	—	—
Storage vessels for petroleum liquids[a]	Storage tanks	—	—	—	—	—	—	—	x	—	—
Secondary lead smelters[a]	Reverberatory and blast furnaces	x	x	—	—	—	—	—	—	—	—
	Pot furnaces	—	x	—	—	—	—	—	—	—	—
Secondary brass and bronze plants[a]	Reverberatory furnaces	x	x	—	—	—	—	—	—	—	—
	Blast and electric furnaces	—	x	—	—	—	—	—	—	—	—
Iron and steel plants[a]	Basic O_2 and electric arc furnaces	x	x	—	—	—	—	—	—	—	—
	Dust-handling equipment	—	x	—	—	—	—	—	—	—	—
Sewage treatment plants	Sludge incineration	x	x	—	—	—	—	—	—	—	—
Primary copper smelters	Dryer	x	x	—	—	—	—	—	—	—	—
	Roaster, furnace, convertor	—	x	x	—	—	—	—	—	—	—

Primary lead smelters[a]	Blast and reverberatory furnaces[e]	x	x	—	—	—	—	—	—	—	—
	Electric furnace converter[f]	—	x	x	—	—	—	—	—	—	—
Primary zinc smelters	Sinter machine	x	x	—	—	—	—	—	—	—	—
	Roaster	—	x	x	—	—	—	—	—	—	—
Primary aluminum reduction plants[a]	Potrooms and anode baking	—	x	—	—	—	—	—	—	—	x
Phosphate fertilizer plants[a]	Wet process phosphoric acid[g]	—	—	—	—	—	—	—	x		
Coal preparation plants[a]	Coal cleaning and dryer	x	x	—	—	—	—	—	—	—	—
	Transfer, loading[h]	—	x	—	—	—	—	—	—	—	—
Ferroalloy production facilities	Electric submerged-arc furnaces	x	x	—	—	—	x	—	—	—	—
	Dust-handling equipment	—	x	—	—	—	—	—	—	—	—
Kraft pulp mills[a]	Smelt tanks and lime kilns	x	—	—	—	—	—	x	—	—	—
	Recovery furnaces	x	x	—	—	—	—	x	—	—	—
	Digester, washer, etc.[i]	—	—	—	—	—	—	x	—	—	—
Glass manufacturing plants	Melting furnaces	x	—	—	—	—	—	—	—	—	—
Grain dryers	Dryers	x	x	—	—	—	—	—	—	—	—
Grain elevators	Truck-loading stations[j]	x	x	—	—	—	—	—	—	—	—
Gas turbines > 1000 hp (10^7 GigaJ/hr)	Simple, regen., and combined	—	—	x	x	—	—	—	—	—	—
	Lime hydrators	x	—	—	—	—	—	—	—	—	—
Automobile and truck surface coating	Prime, guide, and topcoat	—	—	—	—	—	—	—	x	—	—
Ammonium sulfate manufacturing	Dryers	x	x	—	—	—	—	—	—	—	—
Lead acid battery manufacturing	Casting, mixing reclamation and oxide manufacture	—	—	—	—	—	—	—	—	x	—
Phosphate rock plants[a]	Drying, calcining, grinding	x	x	—	—	—	—	—	—	—	—
Rotogravure printing	Each press	—	—	—	—	—	—	—	x	—	—
Large appliance surface coating	Each coating operation	—	—	—	—	—	—	—	x	—	—
Metal coil surface coating	Each coating operation	—	—	—	—	—	—	—	x	—	—
Asphalt processing and roofing manufacture	Saturator	x	x	—	—	—	—	—	—	—	—
	Blowing still	x	—	—	—	—	—	—	—	—	—
	Storage and handling	—	x	—	—	—	—	—	—	—	—

(continued)

TABLE 25-1 (*Continued*)

Source category (in sequence of promulgation)	Affected facility	Pollutants regulated									
		Particulate matter	Opacity	SO_2	NO_x	H_2S	CO	TRS	VOC	Pb	Total F
Beverage can surface coating	Each coating operation	—	—	—	—	—	—	—	x	—	—
Bulk gasoline terminals	All loading racks	—	—	—	—	—	—	—	x	—	—

Note: TRS, total reduced sulfur; VOC, volatile organic carbon; F, fluorides.
[a] Subject to Prevention at Significant Deterioration (PSD) review if emission potential exceeds 90 metric tons/year.
[b] Acid mist.
[c] Screening and weighing systems; storage, transfer, and loading systems; and dust handling equipment.
[d] Reduced sulfur compounds plus H_2S.
[e] Sintering machine discharge.
[f] Sintering machine.
[g] Superphosphoric acid, diammonium phosphate, triple superphosphate and granular triple superphosphate.
[h] Processing and conveying equipment and storage systems.
[i] Evaporator, condensate stripper, and black liquor oxidation systems.
[j] Truck unloading stations, barge, ship, or railcar loading and unloading stations; grain and rack dryers and handling operations.
Source: Pahl, D., *J. Air Pollut. Control Assoc.* **33**, 468–482 (1983).

Another variant is the lowest achievable emission rate (LAER) for a specific pollutant required for a new source of that pollutant to be located in a nonattainment area (i.e., one which has not attained the NAAQS for that pollutant). LAER is the lowest emission rate allowed or achieved anywhere without regard to cost or energy usage. LAER is intended to be more stringent than BACT or NSPS and is also determined on a case-by-case basis.

V. MEANS FOR IMPLEMENTING EMISSION STANDARDS

When owners wish to build a new source which will add a specific amount of a specific pollutant to an area that is innonattainment with respect to that pollutant, they must, under U.S. federal regulations, document a reduction of at least that amount of the pollutant from another source in the area. They can effect this reduction, or "offset," as it is called, in another plant they own in the area or can shut down that plant. However, if they do not own another such plant or do not wish to shut down or effect such reduction in a plant they own, they can seek the required reduction or offset from another owner. Thus, such offsets are marketable credits that can be bought, sold, traded, or stockpiled ("banked") as long as the state or local regulatory agency legitimizes, records, and certificates these transactions. The new source will still have to meet NSPS, BACT or MACT, and/or LAER standards, whichever are applicable.

A. Bubble Concept

When a new source has been added to a group of existing sources under the same ownership in the same industrial complex, the usual practice has been to require the new source independently to meet the offset, NSPS, BACT, and/or LAER, disregarding the other sources in the complex. Under the more recent "bubble concept" (Fig. 25-3) adopted by some states with the approval of the Environmental Protection Agency, the addition of the new source is allowed, whether or not it meets NSPS, BACT, or LAER, provided the total emission of the relevant regulated pollutants from the total complex is decreased. This can be accomplished by obtaining the required offset from one or more of the other sources within the complex, by shutdown, or by improvement in control efficiency. The bubble concept has been subject to litigation and may require a ruling or challenge to make it acceptable.

B. Fugitive and Secondary Sources

In computing the potential for emission of a new source or source complex, it is necessary to consider two other source categories, "fugitive" and

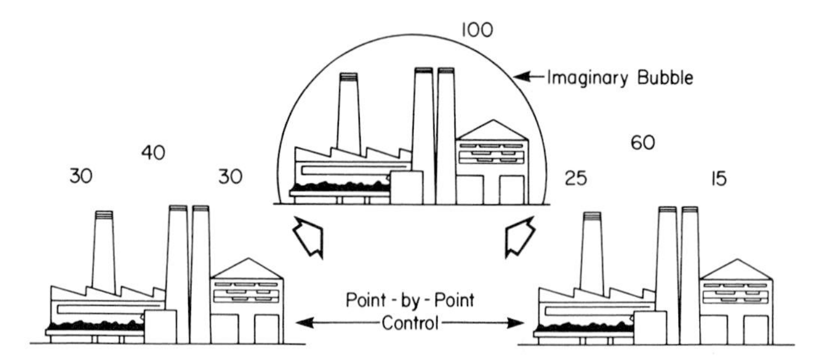

Fig. 25-3. Bubble concept. This pollution control concept places an imaginary bubble over an entire industrial plant, evaluating emissions from the facility as a whole instead of requiring control point-by-point on emission sources. Numbers represent emissions from individual sources, some of which can be fugitive sources, and from the entire industrial plant. Source: Drawing courtesy of the Chemical Manufacturers Association.

"secondary" sources. Fugitive emissions are those from other than point sources, such as unpaved plant roads, outdoor storage piles swept by the wind, and surface mining. Secondary sources are those small sources with emissions of a different character from those of a major source, necessary for the operation of the major source, or source complex.

C. Indirect Sources

The term *indirect sources* is used to describe the type of source created by the parking areas of shopping malls, sports arenas, etc. which attract large numbers of motor vehicles, frequently arriving and leaving over relatively short periods of time. Of somewhat similar character are traffic interchanges, as at an intersection of major highways, each highway being a line source in its own right.

D. Rollback

The rollback approach assumes that emissions and atmospheric concentrations are linearly related, i.e., that a given percentage reduction in emission will result in a similar percentage reduction in atmospheric concentrations. This is most likely a valid assumption for a nonreactive gas such as carbon monoxide, whose principal source is the automobile. The model is

$$R = \frac{g(P) - D}{g(P) - B} \times 100 \tag{25-1}$$

where R is the required percentage reduction in emission, P is present air quality, D is air quality standard, B is background concentration, and g is growth factor in emissions (g is projected to a year in the future when emissions are expected to apply to all vehicles on the road).

An example of a set of emission limits based on the rollback approach is the limits adopted by the United States for carbon monoxide, hydrocarbons, and oxides of nitrogen emissions from new automobiles (Table 25-2).

E. Standard for Hazardous Pollutants

Before the U.S. Clean Air Act Amendments of 1990 (3), hazardous air pollutants were regulated through federal promulgation of the National Emission Standards for Hazardous Air Pollutants (NESHAPS). The EPA listed only eight hazardous air pollutants under NESHAPS.

Title III of the 1990 amendments completely changed the U.S. standards for the hazardous air pollutant control program. Stensvaag (6) has summarized these changes as follows:

- Massively expand the list of hazardous air pollutants. This initial list contained 189 hazardous air pollutants.
- Direct the EPA to list source categories of major sources and area sources.
- Promulgate technology-based emission standards for listed pollutants and sources.
- Set up programs to control accidental releases.
- Establish structures for future risk assessment.
- Require regulation of solid waste combustion facilities.

F. Emission Standards for Existing Installations in the United States

The states and cities of the United States sometimes have emission standards for existing installations, which are usually less restrictive than those

TABLE 25-2

Emission Limits Adopted by the United States for New Light-Duty Automobiles (gm/vehicle mile)

	Vehicle (model year applicable)						
Pollutant emission	1972	1973–4	1975–6	1977–9	1980	1981–1992	1993–1995
Carbon monoxide	39	39	15	15	7	3.4	3.4
Hydrocarbons[a]	3.4	3.4	1.5	1.5	0.41	0.41	0.25
Oxides of nitrogen	—	3	3.1	2.0	2.0	1.0	0.4

[a] Crankcase emissions: no crankcase emissions shall be discharged into the ambient atmosphere. Fuel evaporative emissions: 2 gm per test procedure for vehicles beginning with 1972.

federally required for new installations, i.e., NSPS, BACT, MACT, LAER. The Clean Air Act Amendments of 1990 will change most of these state and local emission standards. The amendments instruct the state air pollution control agencies to submit, by enumerated deadlines, highly detailed revisions to their existing implementation plans. The term "revision" appears almost 250 times in the 1990 amendments (6).

G. Emission Standards for Industrialized Countries of the World

The most recent compilation of emission standards for processes and substances emitted from processes in the industrialized countries of the world was the companion Volume II of the source of Table 22-10 (see Jarrault in Suggested Reading).

H. Fuel Standards

To reduce emissions from fuel-burning sources, one can limit the sulfur, ash, or volatile content of fuels. A listing of such limitations as they existed in 1974 is given in Martin and Stern (see Suggested Reading).

Lead in regular gasoline (usually as tetraethyl lead) is limited in the United States to 1.1 gm of lead per gallon, except that refiners producing less than 10,000 barrels per day may include 1.9 gm per gallon.

The Council of the European Economic Community, established under the 1957 Treaty of Rome, in 1973 issued a declaration on the environment (4), which the European Commission in Brussels has interpreted as giving it authority to issue directives on matters related to the emission of air pollutants, such as one limiting the sulfur content of fuel oil (5).

REFERENCES

1. Rose, A. H., Jr., and Nader, J. S., *J. Air Pollut. Control Assoc.* **8,** 117–119 (1958).
2. Connor, W. D., Smith, C. F., and Nader, J. S., *J. Air Pollut. Control Assoc.* **18,** 748–750 (1968).
3. Public Law No. 101-549, 101st U.S. Congress, "Clean Air Act Amendments of 1990," November 1990.
4. Declaration of Council of the European Communities . . . on the environment. *Official Journal* **16,** C-112 (December 1973).
5. Proposal for a council directive on the use of fuel oils, with the aim of decreasing sulfurous emissions. *Official Journal* **19,** C-54 (March 1976).
6. Stensvaag, J. M., "Clean Air Act 1990 Amendments Law and Practice." Wiley, New York, 1991.

SUGGESTED READING

Jarrault, P. "Limitation des Émission des Pollutants et Qualité de l'Aire-Valeurs Réglemantaires en 1980 dans les principaux Pays Industrielles," Vol. II, "Limitation des Émissions des Pollutants." Institut Francais de l'Energie Publ. I.F.E. No. 66, Paris, 1980.

Marin, W., and Stern, A. C., "The World's Air Quality Management Standards," Vol. I, "Air Quality Management Standards of the World," Vol. II, "Air Quality Management Standards of the United States." Pub. EPA 65019-75-001/002, Miscellaneous Series. U.S. Environmental Protection Agency, Washington, DC, 1974.

Murley, L. (ed.), "Clean Air Around the World. National and International Approaches to Air Pollution Control." International Union of Air Pollution Prevention Associations, Air & Waste Management Association, Pittsburgh, 1991.

"On Prevention of Significant Deterioration of Air Quality." National Academy Press, Washington, DC, 1981.

Stern, A. C., *J. Air Pollut. Control Assoc.* **27,** 440 (1977).

QUESTIONS

1. Discuss remote sensing equipment for measurement of emissions after they have left the stack and have become part of the plume.
2. Discuss the means for determining the strength of an odorous emission from a source and of definitively relating the odor to the presumed source.
3. Discuss the relative merits of the prototype testing of automotive vehicles for certification and of certification based on production line testing of each vehicle produced.
4. Select one source category and affected facility from Table 25-1 and determine the detailed performance standards for the pollutants regulated.
5. Discuss some of the variants of the rollback equation (Eq. 25-1) that have been proposed, and evaluate them.
6. In some countries, the enforcement agency is allowed to exercise judgment on how much emission to allow rather than to adhere to a rigid emission limit. Discuss the advantages and disadvantages of this system.
7. Discuss the availability and reliability of in-stack continuous emission monitors when they are required by U.S. New Source Performance Standards.
8. Discuss the extent to which cost–benefit analysis should be considered in setting emission standards.
9. Trace the history of fuel standards for control of air pollution.

26

The Elements of Regulatory Control

Regulatory control is governmental imposition of limits on emission from sources. In addition to quantitative limits on emissions from chimneys, vents, and stacks, regulations may limit the quantity or quality of fuel or raw material permitted to be used; the design or size of the equipment or process in which it may be used; the height of chimneys, vents, or stacks; the location of sites from which emissions are or are not permitted; or the times when emissions are or are not permitted. Regulations usually also specify acceptable methods of test or measurement.

One instance of an international air pollution control regulation was the cessation of atmospheric testing of nuclear weapons by the United States, the USSR, and other powers signatory to the cessation agreement. Another was the Trail Smelter Arbitration (1) in which Canada agreed to a regulatory protocol affecting the smelter to prevent flow of its stack emissions into the United States. Another example is the Montreal Protocal to protect stratospheric ozone, developed by the United Nations, which will greatly reduce and possibly eliminate the release of chlorofluorocarbons (CFCs) and other ozone-depleting chemicals.

National air pollution control regulations in some instances are preemptive in that they do not allow subsidiary jurisdictions (states, provinces, counties, towns, boroughs, cities, or villages) to adopt different regulations. In other instances, they are not preemptive in that they allow subsidiary jurisdictions to adopt regulations that are not less stringent than the national regulations. In the United States, the regulations for mobile sources are an example of the latter (there is a provision of a waiver to allow more stringent automobile regulations in California); those for stationary sources are an example of the former.

In many countries, provinces or states have enacted air pollution control regulations. Unless or until superseded by national enactment, these regulations are the ones currently in force. In some cases, municipal air pollution control regulatory enactments are the ones currently in force and will remain so until superseded by state, provincial, or national laws or regulations.

A regulation may apply to all installations, to new installations only, or to existing installations only. Frequently, new installations are required to meet more restrictive limits than existing installations. Regulations which exempt from their application installations made before a specified date are called *grandfather clauses,* in that they apply to newer installations but not to older ones. Regulatory enactments sometimes contain time schedules for achieving progressively more restrictive levels of control.

To make regulatory control effective, the regulatory agency must have the right to enter premises for inspection and testing, to require the owner to monitor and report noncompliance, and, where necessary, to do the testing.

I. CONTROL OF NEW STATIONARY SOURCES

In theory, if starting at any date, all new sources of air pollution are adequately limited, all sources installed prior to that date eventually will disappear, leaving only those adequately controlled. The weakness of relying solely on new installation control to achieve community air pollution control is that installations deteriorate in control performance with age and use and that the number of new installations may increase to the extent that what was considered adequate limitation at the time of earlier installations may not prove adequate in the light of these increased numbers. Although in theory old sources will disappear, in practice they take such a long time to disappear that it may be difficult to achieve satisfactory air quality solely by new installation control.

One way to achieve new installation control is to build and test prototype installations and allow the use of only replicates of an approved prototype. This is the method used in the United States for the control of emissions from new automobiles (Table 25-2).

Another path is followed where installations are not replicates of a prototype but rather are each unique, e.g., cement plants. Here two approaches are possible. In one, the owner assumes full responsibility for compliance with regulatory emission limits. If at testing the installation does not comply, it is the owner's responsibility to rebuild or modify it until it does comply. In the alternative approach, the owner makes such changes to the design of the proposed installation as the regulatory agency requires after inspection of the plans and specifications. The installation is then deemed in compliance if, when completed, it conforms to the approved plans and specifications. Most regulatory agencies require both testing and plan approval. In this case, the testing is definitive and the plan filing is intended to prevent less sophisticated owners from investing in installations they would later have to rebuild or modify.

In following the approved replicate route, the regulatory agency has the responsibility to sample randomly and test replicates to ensure conformance with the prototype. In following the owner responsibility route, the agency has the responsibility for locating owners, particularly those who are not aware of the regulatory requirements applicable to them, and then of testing their installations. Locating owners is a formidable task requiring good communication between governmental agencies concerned with air pollution control and those which perform inspections of buildings, factories, and commercial establishments for other purposes. Testing installations for emissions is also a formidable task and is discussed in detail in Chapter 32. Plan examination requires a staff of well-qualified plan examiners.

Because of the foregoing requirements, a sophisticated organization is needed to do an effective job of new installation control. Prototype testing is best handled at the national level; installation testing and plan examinations at the state, provincial, or regional level. Location of owners is a local operation with respect to residential and commercial structures but may be a state or provincial operation for industrial sources.

II. CONTROL OF EXISTING STATIONARY SOURCES

Existing installations may be controlled on a retrospective, a present, or a prospective basis. The retrospective basis relies on the receipt of complaints from the public and their subsequent investigation and control. It follows the theory that the squeaking wheel gets the grease. Complaint investigation and control can become the sole function of a small air pollution control organization to the detriment of its overall air pollution control effectiveness. Activities of this type may improve an agency's short-term image by appeasing the more vocal elements of the public, but this may be accomplished at the expense of the agency's ability to achieve long-range goals.

On a present basis, an agency's staff can be used to investigate and bring under control selected source categories, selected geographic areas, or both. The selection is done, not on the basis of the number or intensity of complaints, but rather on an analysis of air quality data, emission inventory data, or both. An agency's field activity may be directed to the enforcement of existing regulatory limits or to the development of data from which new regulatory limits may be set and later enforced.

On a prospective basis, an agency can project its source composition and location and their emissions into the future and by the use of mathematical models and statistical techniques determine what control steps have to be taken now to establish future air quality levels. Since the future involves a mix of existing and new sources, decisions must be made about the control levels required for both categories and whether these levels should be the same or different.

Regulatory control on a complaint basis requires the least sophisticated staffing and is well within the capability of a local agency. Operation on present basis requires more planning expertise and a larger organization and therefore is better adapted to a regional agency. To operate on a prospective basis, an agency needs a still higher level of planning expertise, such as may be available only at the state or provincial level. To be most effective, an air pollution agency must operate on all three bases, retrospective, present, and prospective, because no agency can afford to ignore complaints, to neglect special source or area situations, or to fail to do long-range planning.

III. CONTROL OF MOBILE SOURCES

Mobile sources include railroad locomotives, marine vessels, aircraft, and automotive vehicles. Over the past 100 years, we have gained much experience in regulating smoke and odor emission from locomotives and marine craft. Methods of combustion equipment improvement, firefighter training, and smoke inspection for these purposes are well documented. This type of control is best at the local level.

Regulation of aircraft engine emissions has been made a national responsibility by law in the United States. The Administrator of the Environmental Protection Agency is responsible for establishing emission limits of aircraft engines, and the Secretary of Transportation is required to prescribe regulations to ensure compliance with these limits.

In the United States, regulation of emissions from new automotive vehicles has followed the prototype-replicate route. The argument for routine annual automobile inspection is that cars should be regularly inspected for safety (brakes, lights, steering, and tires) and that the additional time and cost required to check the car's emission control system during the same inspection will be minimal. Such an inspection certainly pinpoints cars whose emission control system has been removed, altered, damaged, or deteriorated and force such defects to be remedied. The question is whether the improvement in air quality that results from correcting these defects is worth the cost to the public of maintaining the inspection system. Another way of putting the question is whether the same money would be better invested in making the prototype test requirements more rigid with respect to the durability of the emission control system (with the extra cost added to the cost of the new car) than in setting up and operating an inspection system for automotive emissions from used cars. A final question in this regard is whether or not the factor of safety included in the new car emission standards is sufficient to allow a percentage of all cars on the road to exceed the emission standards without jeopardizing the attainment of the air quality standard.

IV. AIR QUALITY CONTROL REGIONS

Workers in the field of water resources are accustomed to thinking in terms of watersheds and watershed management. It was these people who introduced the term *airshed* to describe the geographic area requiring unified management for achieving air pollution control. The term airshed was not well received because its physical connotation is wrong. It was followed by the term *air basin,* which was closer to the mark but still had the wrong physical connotation, since unified air pollution control management is needed in flat land devoid of valleys and basins. The term that finally evolved was *air quality control region,* meaning the geographic area including the sources significant to production of air pollution in an urbanized area and the receptors significantly affected thereby. If long averaging time isopleths (i.e., lines of equal pollution concentration) of a pollutant such as suspended particulate matter are drawn on the map of an area, there will be an isopleth that is at essentially the same concentration as the background concentration. The area within this isopleth meets the description of an air quality control region.

For administrative purposes, it is desirable that the boundaries of an air quality control region be the same as those of major political jurisdictions. Therefore, when the first air quality control regions were officially designated in the United States by publication of their boundaries in the *Federal Register,* the boundaries given were those of the counties all or part of which were within the background concentration isopleth.

When about 100 such regions were designated in the United States, it was apparent that only a small portion of the land area of the country was in officially designated regions. For uniformity of administration of national air pollution legislation, it became desirable to include all the land area of the nation in designated air quality control regions. The Clean Air Amendments of 1970 therefore gave the states the option of having each state considered an air quality control region or of breaking the state into smaller air quality control regions mutually agreeable to the state and the U.S. Environmental Protection Agency. The regions thus created need bear no relation to concentration isopleths, but rather represent contiguous counties which form convenient administrative units. Therefore, for purposes of air pollution control, the United States is now a mosaic of multicounty units, all called air quality control regions, some of which were formed by drawing background concentration isopleths and others of which were formed for administrative convenience. Some of the former group are interstate, in that they include counties in more than one state. All of the latter are intrastate.

None of the interstate air quality control regions operates as a unified air pollution control agency. Their control functions are all exercised by their separate intrastate components.

V. TALL STACKS AND INTERMITTENT AND SUPPLEMENTARY CONTROL SYSTEMS

For years, it was an item of faith that the higher the stack from which pollution was emitted, the better the pollution would be dispersed and the lower the ground-level concentrations resulting form it. As a result, many tall stacks were built. Recently, this practice has come under attack on the basis that discharge from tall stacks is more likely to result in long-range transport with the associated problems of interregional and international transport, fumigation, and acidic deposition. It is argued that, to prevent these problems, pollutants must be removed from the effluent gases so that, even if plumes are transported for long distances, they will not cause harm when their constituents eventually reach ground level, and that, if such cleanup of the emission takes place, the need for such tall stacks disappears. Despite these arguments, the taller the stack, the better the closer-in receptors surrounding the source are protected from its effluents.

The ability of a tall stack to inject its plume into the upper air and disperse its pollutants widely depends on prevalent meteorological conditions. Some conditions aid dispersion; others retard it. Since these conditions are both measurable and predictable, intermittent and supplementary control systems (ICS/SCS) have been developed to utilize these measurements and predictions to determine how much pollution can be released from a stack before ground-level limits are exceeded. If such systems are used primarily to protect close-in receptors by decreasing emissions when local meteorological dispersion conditions are poor, they tend to allow relatively unrestricted release when the upper-air meteorology transports the plume for a long distance from the source. This leads to the same objections to ICS/SCS as were noted to the use of tall stacks without a high level of pollutant removal before emission. Therefore, here again it is argued that if a high level of cleanup is used, the need to project the effluent into the upper air is reduced and the benefits of operating an ICS/SCS (which is costly) in lieu of installing more efficient pollutant removal equipment disappear. The U.S. 1990 Clean Air Act Amendment (2) does not allow ICS/SCS to be substituted for pollutant removal from the effluent from fossil fuel–fired steam-powered plants. However, ICS/SCS may still be useful for other applications.

REFERENCES

1. Dean, R. S., and Swain, R. E., Report submitted to the Trail Smelter Arbitral Tribunal, U.S. Department of the Interior, Bureau of Mines Bulletin 453, U.S. Government Printing Office, Washington, DC, 1944.
2. Public Law No. 101-549, U.S. 101st Congress, "Clean Air Act Amendments of 1990," November 1990.

SUGGESTED READING

"Air Toxics Issues in the 1990s. Policies, Strategies and Compliance." Air & Waste Management Association, Pittsburgh, 1991.

Eisenbud, M., "Environment, Technology and Health—Human Ecology in Historical Perspective." New York University Press, New York, 1978.

Friedlaender, A. F. (ed.), "Approaches to Controlling Air Pollution." MIT Press, Cambridge, MA, 1978.

Powell, R. J., and Wharton, L. M., *J. Air Pollut. Control. Assoc.* **32,** 62 (1982).

Stern, A. C., *J. Air Pollut. Control. Assoc.* **32,** 44 (1982).

QUESTIONS

1. What are the geographic boundaries of the air quality region (or its equivalent in countries other than the United States) in which you reside?
2. Are there regulatory limits in the jurisdiction where you reside which are different from your state, provincial, or national air quality or emission limits? If so, what are they?
3. Discuss the application of the prototype testing–replicate approval approach to stationary air pollution sources.
4. One form of air pollution control regulation limits the pollution concentration at the owner's "fence line." Find an example of this type of regulation and discuss its pros and cons.
5. How are pollutant emissions to the air from used automobiles classified where you reside? Discuss the merits and the extent of such regulation.
6. Limitation of visible emission was the original form of control of air pollution a century ago. Has this concept outlived its usefulness? Discuss this question.

7. Discuss the use of data telemetered to the office of the air pollution control agency from automatic instruments measuring ambient air quality and automatic instruments measuring pollutant emissions to the atmosphere as air pollution control regulatory means.
8. The ICS/SCS control system has been used for control of emissions from nonferrous smelters. Discuss at least one such active system in terms of its success or failure.
9. Discuss the reasons why the interstate air pollution control region concept has failed in the United States.

27

Organization for Air Pollution Control

The best organizational pattern for an air pollution control agency is that which most effectively and efficiently performs all its functions. There are many functions a control agency or industrial organization could conceivably perform. The desired budget and staff for the agency or organization are determined by listing the costs of performing all desired functions. The actual functions performed by the agency or organization are determined by limitations on staff, facilities, and services imposed by its budget.

I. FUNCTIONS

The most elementary function of an air pollution control agency is its *control* function, which breaks down into two subsidiary functions: *enforcement* of the jurisdiction's air pollution control laws, ordinances, and regulations and *evaluation* of the effectiveness of existing regulations and regulatory practices and the need for new ones.

The enforcement function may be subdivided in several ways, one of which is control of *new sources* and *existing sources*. New-source control can involve all or some of the following functions:

1. *Registration* of new sources.
2. *Filing* of applicants' plans, specifications, air quality monitoring data, and mathematical model predictions.
3. *Review* of applicant's plans, specifications, air quality monitoring data, and mathematical model predictions for compliance with emission and air quality limitations.
4. *Issuance* of certificates of approval for construction.
5. *Inspection* of construction.
6. *Testing* of installation.
7. *Issuance* of certificates of approval for operation.
8. *Receipt* of required fees for the foregoing services.
9. *Appeal* and *variance* hearings and actions.
10. *Prosecution* of violations.

Control of existing sources can involve all or some of the following functions:

1. *Visible emission* inspection.
2. *Complaint* investigation.
3. *Periodic* and *special* industrial category or geographic area inspections.
4. *Bookkeeping* of offsets and offset trades.
5. *Fuel* and fuel dealer inspection and testing.
6. *Testing* of installations.
7. *Renewal* of certificates of operation.
8. *Receipt* of required fees for the foregoing services.
9. *Appeal* and *variance* hearings and actions.
10. *Prosecution* of violations.

Note that items 6–10 of the proceeding two lists are the same.

The evaluation function may be subdivided into *retrospective evaluation* of existing regulations and practices and *prospective planning* for new ones. Retrospective evaluation involves the following functions:

1. *Air quality monitoring* and surveillance.
2. *Emission inventory* and monitoring.
3. *Statistical analysis* of air quality and emission data and of agency activities.
4. *Analytic evaluation.*
5. *Recommendation* of required regulations and regulatory practices.

Prospective planning involves the following activities:

1. *Prediction* of future trends.
2. *Mathematical modeling.*

3. *Analytical evaluation.*
4. *Recommendation* of required regulations and regulatory practices.

Meteorological services are closely related to both retrospective evaluation and prospective planning. The last two items on the proceeding two lists are the same.

Functions such as those already noted require extensive technical and administrative support. The *technical* support functions required include the following:

1. *Technical* information services—libraries, technical publications, etc.
2. *Training* services—technical.
3. *Laboratory* services—analytical instrumentation, etc.
4. *Computer* services.
5. *Shop* services.

An agency requires, either within its own organization or readily available from other organizations, provision of the following *administrative* support functions:

1. *Personnel.*
2. *Procurement.*
3. *Budget*—finance and accounting.
4. *General* services—secretarial, clerical, reproduction, mail, phone, building maintenance, etc.

Since an air pollution control agency must maintain its *extramural* relationships, the following functions must be provided:

1. *Public relations* and information.
2. *Public education.*
3. *Liaison* with other agencies.
4. *Publication* distribution.

An agency needs *legal* services for prosecution of violations, appeal and variance hearings, and the drafting of regulations. In most public and private organizations, these services are provided by lawyers based in organizational entities outside the agency. However, organizationally, it is preferable that the legal function be provided within the agency.

One category that has been excluded from all of the proceding lists is research and development (R & D). R & D is not a necessary function for an air pollution control agency at the state, provincial, regional, or municipal level. It is sufficient if the national agency, e.g., the U.S. Environmental Protection Agency, maintains an R & D program sufficient for the nation's needs and encourages each industry to undertake the R & D required to solve its particular problems. National agencies tend to give highest priority to problems common to a number of areas in the country. However, where major problems of a state, province, region, or municipality are unique to its locality and not likely to have high national priority, the area may have to undertake the required R & D.

It is apparent that some of the functions listed overlap. An example is the overlap of the laboratory function in technical support and the testing and air quality monitoring functions in control. It is because of such overlaps that different organizational structures arise. Many of the functions listed for a control agency are not required in an industrial air pollution control organization. A control agency must emphasize its enforcement function, sometimes at the expense of its evaluation function. The converse is the case for an industrial organization. Also, in an industrial organization, the technical and administrative support, legal, and R & D functions are likely to be supplied by the parent organization.

II. ORGANIZATION

In the foregoing section functions were grouped into categories. The most logical way to organize an air pollution control or industrial organization is along these categorical lines (Figs. 27-1 and 27-2), deleting from the organizational structure those functions and categories with which the agency or organization is presently not concerned. When budget and staff are small, one person is required to cover all the agency's or organization's activities in more than one function and, in very small agencies or organizations, in more than one category.

Agencies and organizations which cover a large geographic area must decide to what extent they will either centralize or decentralize functions and categories. Centralization consolidates the agency's or organization's technical expertise, facilitating the resolution of technical matters particularly in relation to large or complex sources. Decentralization facilitates the agency's ability to deal with a larger number of smaller sources. The ultimate decentralization is delegation of certain of an agency's functions to lesser jurisdictions. This can lead to a three-tiered structure, with certain functions fully centralized, certain ones delegated, and certain ones decentralized to regional offices or laboratories.

In an industrial organization the choice is between centralization in corporate or company headquarters and decentralization to the operating organizations or individual plans. The usual organization is a combination of headquarters centralization and company decentralization.

A major organizational consideration is where to place an air pollution control agency in the hierarchy of government. As state, provincial, and municipal governmental structure evolved during the nineteenth century, smoke abatement became a function of the departments concerned with buildings and with boilers. In this century, until the 1960s, air pollution control shifted strongly to national, state, provincial, and municipal health

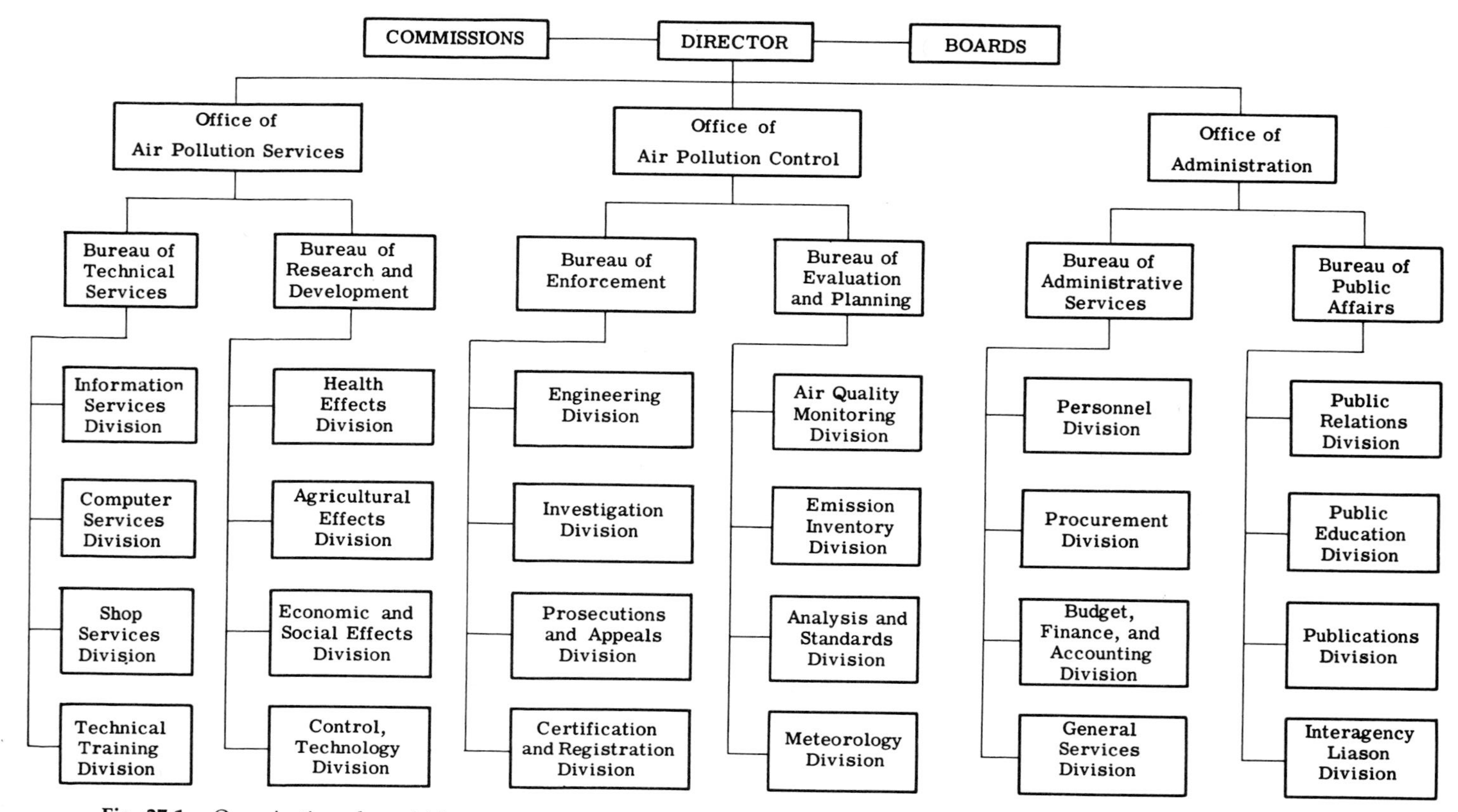

Fig. 27-1. Organization plan which encompasses all functions likely to be required of a governmental air pollution control agency.

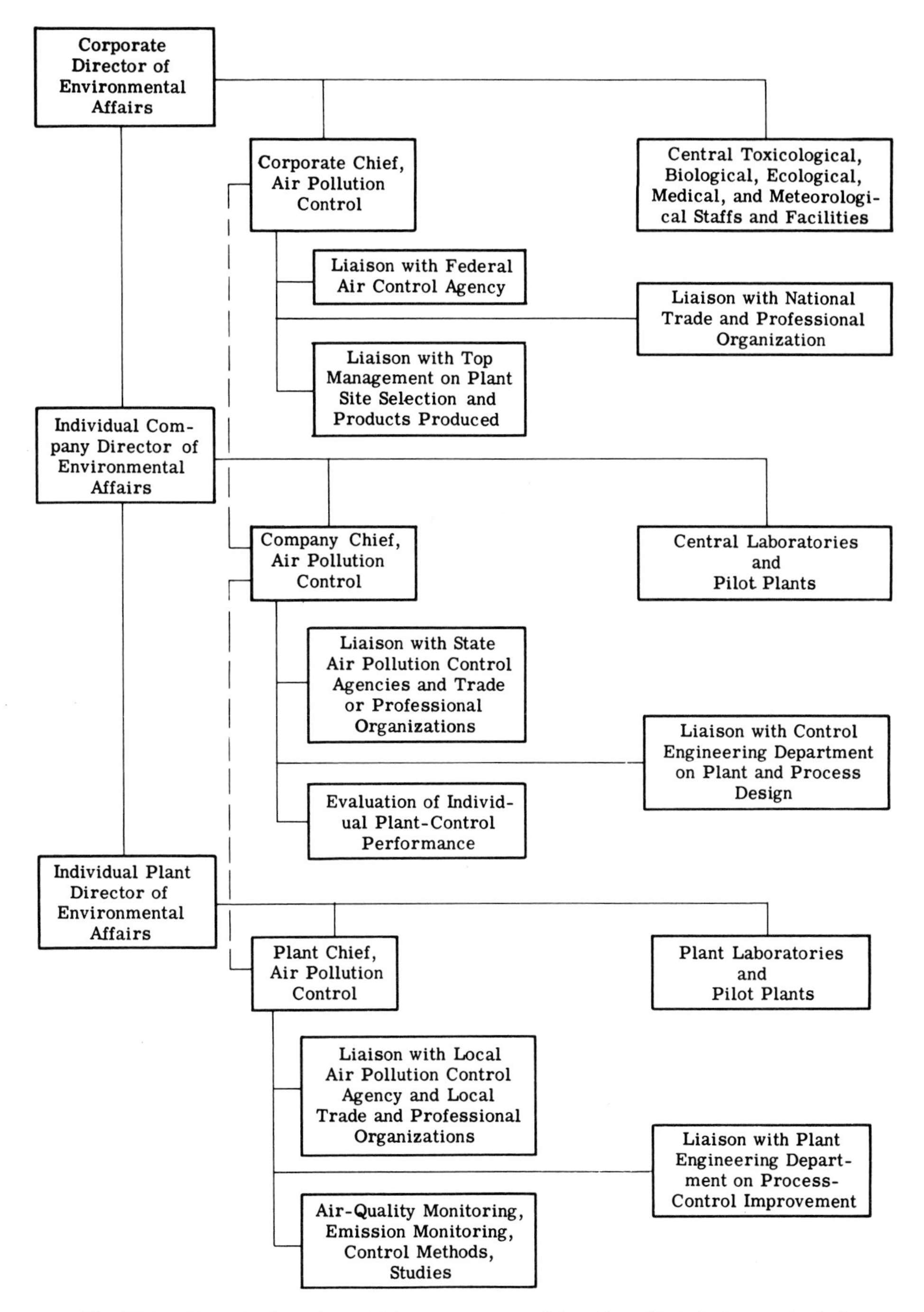

Fig. 27-2. Organization plan which encompasses all functions likely to be required of an industrial air pollution organization.

agencies. However, since the mid-1960s there has been an increasing tendency to locate air pollution control in agencies concerned with natural resources and the environment.

III. FINANCE

A. Fines and Fees

Governmental air pollution control agencies are primarily tax supported. Most agencies charge fees for certain services, such as plan filing, certificate issuance, inspection, and tests. In some agencies, these fees provide a substantial fraction of the agency's budget. In general, this is not a desirable situation because when the continuity of employment of an agency's staff depends on the amount of fees collected, there is an understandable tendency to concentrate on fee-producing activities and resist their deemphasis. This makes it difficult for an agency to change direction with changes in program priorities. Even when these fees go into the general treasury, rather than being retained by the agency, such collection gives some leverage to the agency in its budget request and has the same effect as it would if the fees were retained by the agency.

In most jurisdictions, violators of air pollution control laws and regulations are subject to fines. In the past, it has been considered good administrative practice for these fines to go into the general treasury in order to discourage imposition of fines as a means of supporting the agency's program, which might result if the agency retained the fines. Collection of fines could become an end in itself if the continued employment of the agency staff were dependent on it. Moreover, as in the case of fees, an agency can use its fine collection to give it leverage in budget allocation.

There has been extensive recent rethinking of the role of fees and fines as means of influencing industrial decision making with regard to investment in pollution control equipment and pollution-free processes. In their new roles, fees and fines take the form of tax write-offs and credits for pollution control investment; taxes on the sulfur and lead content of fuels; continuing fines based on the pollution emission rate; and effluent fees on the same basis. Tax write-offs and credits tend to be resisted by treasury officials because they diminish tax income. Air pollution control agencies tend to look with favor on such write-offs and credits because they result in air pollution control with minimal effort on the part of their staffs and with minimal effect on their budget.

One problem with fuel taxes, continuing fines, and effluent fees is how to use the funds collected most effectively. Among the ideas which have been developed are to use the funds as the basis for loans or subsidies for air pollution control installations by industrial or domestic sources and for financing air pollution control agency needs of jurisdictions subsidiary to those which collect and administer the fines or fees. By ensuring that the continuity of employment of agency personnel is divorced from the fine and fee setting, collection, and administrative process, it is argued that the use of fines and fees can be a constructive rather than a stultifying process.

There is a discussion in Chapter 5, Section IV, of financial incentives to supplement or replace regulations.

B. Budget

An agency's budget consists of the following categories:

- Personnel
  - Salaries
  - Fringe benefits
  - Consultant fees
- Space, furniture, and office equipment
  - Acquisition or rental
  - Operation
  - Maintenance
- Technical and laboratory equipment
  - Acquisition or rental
  - Operation
  - Maintenance
- Transportation equipment
  - Acquisition
  - Operation
  - Maintenance
- Travel and sustenance
- Supplies
- Reproduction services
- Communication services
- Contractual services
- Other—miscellaneous

The principal precaution is to avoid allocating too high a percentage of an agency's budget to the personnel category. When this happens, the ability of the agency staff to perform its functions is unduly restricted.

An industrial organization must decide whether to charge off air pollution control costs as a corporate charge, so that the plant manager does not include them in the accounts; or the reverse, so that the plant manager must show a profit for the plant as a cost center, after including costs of air pollution control.

IV. PERSONNEL

The personnel needs of an agency or industrial organization can be assessed qualitatively and quantitatively (see the "1993 Government Agencies Directory" in the Suggested Reading list). Qualitatively, the professional categories required by an agency or organization staff are determined by the functions included in their scope of work. Table 27-1 lists the types of professional training usually specified for exercise of the principal functions. For smaller agencies or organizations in which one person handles several functions or categories, the person trained by institutions with curricula in air pollution or by internship in a large air pollution control agency or organization is especially valuable.

Quantitative personnel needs are determined by the answer to the following questions: How many people should there be ideally to perform all functions? Within specific budget restrictions, how should personnel be allocated among the several functions? Personnel needs in air pollution control have been the subject of several studies, from which Tables 27-2 and 27-3 have been developed to help answer these two questions.

Persons are trained for two purposes: To perform their assigned job and to be promotable to a higher job classification. The means of accomplishing such training also follow two paths—on-the-job (in-service or intramural) training and extramural training, which ranges from intensive certificate-granting short courses to extensive training in degree-granting graduate programs. In the United States, the principal organization offering short courses is the Air Pollution Training Institute of the U.S. Environmental Protection Agency.

Tables 27-1 to 27-3 have concentrated on the personnel makeup of control agencies. For a broader look at places of employment, Table 27-4 shows where 8037 members of the Air Pollution Control Association (APCA) of the United States and Canada worked in 1982. (This list includes foreign as well as domestic members of APCA but does not include the membership of the air pollution control associations of other countries.) This table shows that only 10.7% of the members work in control agencies. This table gives a somewhat distorted picture because in many air pollution organizations only the senior executive, professional, and scientific personnel belong to APCA, whereas the total North American workforce in air pollution includes several times the 8037 membership total who are in junior, technical, service, or manual sectors and are not association members. These numbers could be still greater if those engaged in this work outside North America were included. The Air Pollution Control Association changed its name to the Air and Waste Management Association in 1988. The Air and Waste Management Association had a membership of over 14,000 in 1993, but only a portion of the members were active in the air pollution profession.

TABLE 27-1

Professional Categories Required in Air Pollution Control Agencies

Control agency function	Professional category of persons usually responsible for functions
Administrative services	Business managers
Air quality monitoring	Chemists
Analytic evaluation	Administrators
Budget, finance, accounting	Accountants
Certificate issuance and renewal	Business managers
Computer services	Computer scientists
Construction inspection	Engineers
Data processing and analysis	Data processing specialists
Education	Educators
Emission inventory	Engineers
Equipment maintenance and service	Technicians
Fee receipt	Accountants
Hearings and prosecution	Attorneys
Inspection (supervision)	Engineers
Inspection (field force)	Inspectors, sanitarians
Installation testing	Chemists, engineers
Investigation	Chemists, engineers
Laboratory services	Chemists
Legal	Attorneys
Liaison	Administrators
Mathematical modeling	Meteorologists
Meteorological services	Meteorologists
New source registration and filing	Business managers
Office staff	Clerical personnel
Personnel	Business managers
Plan review	Engineers
Planning	Planners
Procurement	Business managers
Public relations	Public information specialists
Regulation recommendation	Administrators
Research and development	
Agricultural effects	Biological scientists
Atmospheric effects	Physical scientists
Control technology	Engineers
Economic and social effects	Social scientists
Health effects	Medical scientists
Shop services	Technicians
Statistical analysis	Statisticians
Technical information	Librarians
Training services	Educators
Trend prediction	Statisticians

TABLE 27-2

Personnel Levels in United States and Canadian Air Pollution Control Agencies, 1981

	Number of persons employed[a]					Percent of agencies employing persons in category				Median number of persons per agency using category[b]			
Personnel category[c]	State, provincial, territorial	County, multicounty, county–city, regional	City, multicity	All agencies	Percent of part-time employees, all agencies[d]	State, provincial, territorial	County, multicounty, county–city, regional	City, multicity	All agencies	State, provincial, territorial	County, multicounty, county–city, regional	City, multicity	All agencies
Administrators[e]	317	421	124	862	14	100	99	100	99	3	2	2	2
Attorneys	93	62	22	157	68	87	53	57	61	1	1	1	1
Chemists	191	143	49	383	14	83	44	43	52	3	1	1	1
Data processing specialists[f]	136	50	6	192	25	75	20	16	31	2	1	1	1
Engineers	533	460	145	1138	8	94	66	55	70	11	2	2	4
Inspectors[g]	141	388	192	721	17	51	52	71	56	5	2	2	3
Meteorologists	58	23	6	87	10	54	10	9	20	1	1	1	1
Office staff[h]	741	464	142	1347	14	93	89	84	89	6	1	2	2
Planners[i]	45	20	0	65	2	25	5	0	9	2	1	0	1
Public information specialists[j]	43	32	5	80	53	70	25	11	32	1	1	1	1
Sanitarians	98	185	47	330	26	11	35	38	30	10	2	1	3
Scientists[k]	72	7	19	98	6	24	2	4	7	1	1	3	1
Specialists[l]	532	60	10	602	5	41	12	9	17	8	3	2	4
Technicians[m]	727	364	176	1267	9	87	55	46	60	8	2	2	3
Total	3707	2679	943	7329									
Number of agencies	63	167	56	286									
Average number of persons per agency	59	16	17	26									

Median number of persons per agency	40	9	9	14

[a] Equivalent full-time employees; a part-time employee is counted as half a person.
[b] This is not the median number of persons in all agencies, since some agencies do not utilize categories.
[c] Actual agency employment titles may differ from these categories (see footnotes [e] through [m]).
[d] Most of these part-time employees are full-time employees of the parent organization of the air pollution control agency, but devote as much of their time to the work of the air pollution control agency as required.
[e] Accountants, administrators, agricultural commissioners, air monitoring, supervisors, coordinators (environmental health project, grants, quality assurance, special project and state implementation plan), grants analysts, officers (enforcement, hearings, staff services, technical services and training), supervisors, and technical advisors.
[f] Air quality and data analysts and data processing specialists.
[g] Environmental investigators, environmental, field, and unspecified inspectors.
[h] Staff (office and unspecified).
[i] Plan analysts and planners (air pollution control, environmental, urban, and unspecified).
[j] Librarians, public information specialists.
[k] Air pollution control statistical planners, agricultural biologists, biologists, computer specialists, economists, management analysts, mathematicians, microbiologists, physicists, phytotoxicologists, researchers, research analysts, research scientists, research specialists, scientists (environmental and unspecified), statisticians, and statistical analysts.
[l] Air pollution and environmental officers, environmental protection associates, environmental quality managers and environmentalists, and specialists (air pollution, air pollution control, air quality, air quality control, air resource, environmental control, environmental health, environmental management, environmental quality, highway transportation, monitoring, pollution control, program, public participation, quality assurance, and unspecified).
[m] Aides (administrative, data reduction, engineering, environmental health, laboratory, sanitarian, and unspecified), assistants (administrative, fiscal, laboratory, and legal), draftspersons, laboratory workers, mechanics, project illustrators, samplers, and technicians (air pollution control, air quality monitoring station, electronic, engineering, instrument, and unspecified).

Source: "1982 Directory—Governmental Air Pollution Control Agencies." Air Pollution Control Association, Pittsburgh, and U.S. Environmental Protection Agency, Washington, DC.

TABLE 27-3

Personnel Levels in Ideal Governmental Air Pollution Control Agencies[a]

Category	Positions per million population
Administrators	3
Attorneys	1
Chemists	2
Data processing specialists	1
Engineers	15
Inspectors (including sanitarians)	5
Meteorologists	1
Office staff	5
Planners	1
Public information specialists	1
Scientists	1
Specialists (other categories)	4
Technicians	10
Total	50

[a] State, provincial, territorial, county, multicounty, county–city, regional, city, and multicity.

Table 27-5 gives the results of a survey sponsored by the Industrial Gas Cleaning Institute of peak employment in the air pollution control industry, showing that the primarily American segment of the industry represented by Institute members created about 111,383 person-years of employment in 1982. Obviously, this number is greater worldwide. Although this total includes many with only a part-time commitment to the air pollution industry, e.g., shop and field workers who build pumps, fans, and motors, prepare foundations, erect structures, and provide clerical and accounting services, it shows that a good proportion of the total are dedicated, full-time air pollution professionals, whose education is the purpose of this text.

V. ADVISORY GROUPS

It is common practice, at least in the United States, for air pollution control agencies to be associated with nonsalaried groups who meet from time to time in capacities ranging from official to advisory. The members of such groups may be paid fees or expenses for their days of service, or they may contribute their time and expenses without cost to the agency. The members may be appointed by elected officials, legislative bodies, department heads, program directors, or directors of program categories. They may have to be sworn into office, and the records of their meetings

TABLE 27-4

Occupational Distribution of 8037 Members of the Air Pollution Control Association, 1982

Sector	Percent by	
	Major categories	Minor categories
Consulting and control agency	34.1	
Consulting		23.4
Control agency		10.7
Industrial	34.1	
Chemical		8.9
Power generation		7.0
Petroleum		4.2
Smelting and refining		2.6
Pulp and paper		2.1
Steel		2.1
Mining and gas production		1.6
Motor vehicle manufacture		1.6
Cement/glass		1.2
Food processing		0.7
Foundries		0.6
Pharmaceutical		0.4
Other		1.1
Research, education, publishing, and printing	16.8	
Research (institutional)		6.7
Education		5.0
Research (governmental)		4.8
Publishing and printing		0.3
Manufacturing, fabrication, and distribution	15.0	
Air pollution control equipment		8.2
General		2.9
Instrumentation		2.6
Contracting		1.3
Total	100%	100%

Source: "Survey of the Membership of the Air Pollution Control Association." McManis Associates, Washington, DC, 1982.

may be official documents; or their appointments and meetings may be informal.

In general, when such bodies are official, laws or ordinances set a statutory base for their creation. This base may be as specific as the statutory requirement that a specified group be appointed in a particular manner with specified duties and authorities, or as broad as a general statutory statement that advisory groups may be created, without enumerating any details of their creation or duties. Such groups may provide their own secretariat and have their own budget authority, or be completely dependent on the agency for services and costs.

TABLE 27-5

Peak Employment in the Air Pollution Control Industry, 1982

Labor associated with:	Persons employed	Percent share	Person-years
Direct and auxiliary materials and equipment	155,000	37.2	14,988
Fabrication, design, and installation of APC equipment	121,635	29.2	40,787
Services	34,000	8.2	11,825
Operation, maintenance, and repair	31,000	7.5	10,500
Architects and engineers	30,840	7.4	8,269
Government	15,500	3.7	13,780
Monitoring and testing equipment	12,300	3.0	6,150
Independent attorneys	8,000	1.9	1,600
Manufacturer's representatives	5,100	1.2	1,684
Additional support staff	2,500	0.6	1,500
Other R&D	600	0.1	300
Total peak people	416,475	100.0	111,383
Correction factor[a]	103,370		
Total adjusted peak	313,105		

[a] Adjustment required to eliminate double counting of people.
Source: Orem, S. J., *J. Air. Pollut. Control Assoc.* **32**(6), 673–674 (1982).

The highest levels of such groups are commissions or boards, which can promulgate standards and regulations, establish policy, issue variances, and award funds. The next highest level consists of the hearings or appeals boards, which can issue variances but which neither promulgate standards or regulations nor set policy. Next in rank are the boards, panels, or committees which formally recommend standards or regulations but do not have authority to promulgate them. Next in line are formally organized groups which review requests for funds and make recommendations but lack final award authority. These groups are particularly prevalent in federal programs, such as that of the United States, where large amounts of money are awarded as grants, fellowships, or contracts. These review groups tend to be specialized, e.g., a group of cardiologists to review requests for funds to do cardiovascular research. Although such groups do not have final award authority, the award process tends to follow their recommendations, thereby making their real authority greater than their statutory authority.

At the lowest level are groups that review program content and make recommendations. The impact of groups in this category depends on whether their reports and recommendations are or are not published and the ground rules for determining whether or not they shall be published. If the groups are creations of the agency whose programs they review and the decision on whether or not to publish rests with the agency, minimal impact can be expected. Where review groups are relatively independent of the agency whose programs they review and have their own authority and means for publication, maximum impact can be expected. An example of the latter is the statutory requirement in Section 202 of the U.S. Clean Air Act Amendments of 1977 that automobile exhaust standards be reviewed by the National Academy of Sciences.

SUGGESTED READINGS

Jones, C. O., "Clean Air." University of Pittsburgh Press, Pittsburgh, 1975.

Kokoszka, L. C., and Flood, J. W., "Environmental Management Handbook," Marcel Decker, New York, 1989.

Mallette, F., Minutes of the Proceedings, Institution of Mechanical Engineers (GB), **N68**(2), 595–625 (1954).

Marsh, A., "Smoke: The Problem of Coal and the Atmosphere." Faber and Faber, London, 1947.

"1993 Government Agencies Directory," Section 4, Directory and Resource Book. Air and Waste Management Association, Pittsburgh, 1993.

Portney, P. R. (ed.), "Public Policies for Environmental Protection." Resources for the Future, Washington, DC, 1990.

QUESTIONS

1. How is the air pollution control agency in your state or province organized? Where is its principal office? Who is its head?
2. Is there a local air pollution control agency in your city, county, or region? How is it organized? Where is its principal office? Who is its head?
3. Draw an organization chart for a governmental air pollution control organization which is limited by budget to 10 professional persons.
4. Write a job specification for the chief of the Bureau of Enforcement in Fig. 23-1.
5. Discuss the relative roles of the staff of an air pollution control agency, its advisory board, and its chief executives in the development and promulgation of air quality and emission standards.
6. Discuss the alternatives mentioned in last paragraph of Section III,B regarding allocation of costs of air pollution control in an industrial organization.
7. How does a governmental air pollution control agency or an industrial air pollution control organization organize to ensure that its registration of new sources does not miss significant new sources?
8. Draw an organizational chart for an air pollution organization with a number of plants limited by budget to three professional persons.
9. How does the job market for air pollution control personnel respond to changes in regulatory requirements and to the state of the economy?

Part VI

The Engineering Control of Air Pollution

28

Engineering Control Concepts

I. INTRODUCTION

The application of control technology to air pollution problems assumes that a source can be reduced to a predetermined level to meet a regulation or some unknown minimum value. Control technology cannot be applied to an uncontrollable source, such as a volcano, nor can it be expected to control a source completely to reduce emissions to zero. The cost of controlling any given air pollution source is usually an exponential function of the percentage of control and therefore becomes an important consideration in the level of control required (1). Figure 28-1 shows a typical cost curve for control equipment.

If the material recovered has some economic value, the picture is different. Figure 28-2 shows the previous cost of control with the value recovered curve superimposed on it. The plant manager looking at such a curve would want to be operating in the area to the left of the intersection of the two curves, whereas the local air pollution forces would insist on operation as far to the right of the graph as the best available control technology would allow.

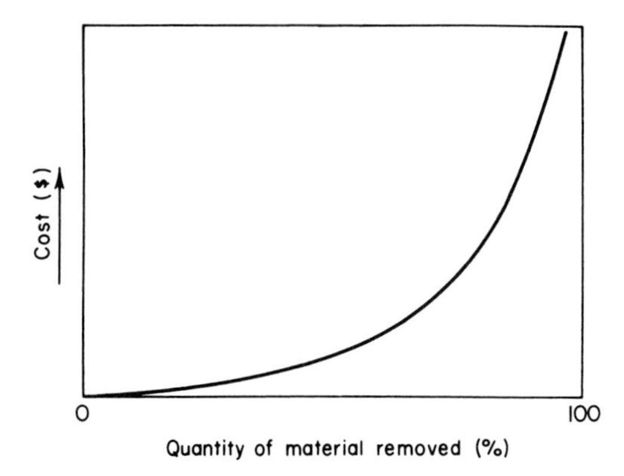

Fig. 28-1. Air pollution control equipment cost.

Control of any air pollution source requires a complete knowledge of the contaminant and the source. The engineers controlling the source must be thoroughly familiar with all available physical and chemical data on the effluent from the source. They must know the rules and regulations of the control agencies involved, including not only the air pollution control agency but also any agencies which may have jurisdiction over the construction, operation, and final disposal of the waste from the source (2).

In many cases, heating or cooling of the gaseous effluent will be required before it enters the control device. The engineer must be thoroughly aware of the gas laws, thermodynamic properties, and reactions involved to secure a satisfactory design. For example, if a gas is cooled there may be condensation if the temperature drops below the dewpoint. If water is sprayed into

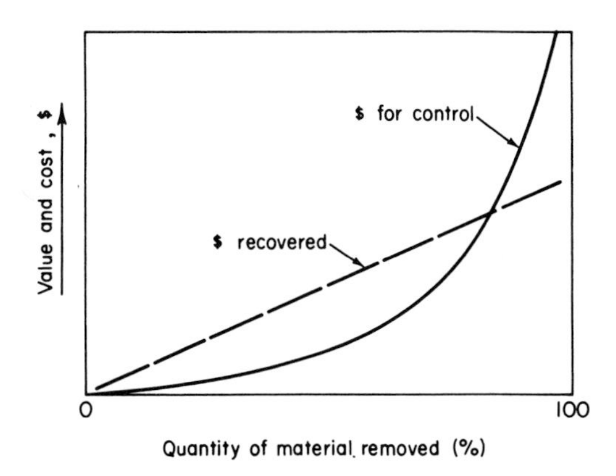

Fig. 28-2. Control equipment cost with value recovered.

the hot gas for cooling, it adds greatly to the specific volume of the mixture. As the gases pass through hoods, ducts, fans, and control equipment, temperatures and pressures change and hence, also, specific volumes and velocities (3).

The control of atmospheric emissions from a process will generally take one of three forms depending on the process, fuel, types, availability of control equipment, etc. The three general methods are (1) process change to a less polluting process or to lowered emission from the existing process through a modification or change in operation, (2) fuel change to a fuel which will give the desired level of emissions, and (3) installation of control equipment between the point of pollutant generation and its release to the atmosphere. Control may consist of either removal of the pollutant or conversion to a less polluting form (3).

II. PROCESS CHANGE

A process change can be either a change in operating procedures for an existing process or the substitution of a completely different process. In recent years this has been labeled "pollution prevention." Consider a plant manager who for years has been using solvent A for a degreasing operation. By past experimentation, it has been found that with the conveyor speed at 100 units per hour, with a solvent temperature of 80°C, one gets maximum cleaning with a solvent loss that results in the lowest overall operating cost for the process.

A new regulation is passed requiring greatly reduced atmospheric emissions of organic solvents, including solvent A. The manager has several alternatives:

1. Change to another more expensive solvent, which by virtue of its lower vapor pressure would emit less organic matter.
2. Reduce the temperature of the solvent and slow down the conveyor to get the same amount of cleaning. This may require the addition of another line or another 8-hr shift.
3. Put in the necessary hooding, ducting, and equipment for a solvent recovery system which will decrease the atmospheric pollution and also result in some economic solvent recovery.
4. Put in the necessary hooding, ducting, and equipment for an afterburner system which will burn the organic solvent vapors to a less polluting emission, but with no solvent recovery.

In some cases, the least expensive control is achieved by abandoning the old process and replacing it with a new, less polluting one. Any increased production and/or recovery of material may help offset a portion of the cost. It has proved to be cheaper to abandon old steel mills and to replace them with completely new furnaces of a different type than to modify the old systems to meet pollution regulations. Kraft pulp mills found that the least costly method of meeting stringent regulations was to replace the old, high-emission recovery furnaces with a new furnace of completely different design. The kraft mills have generally asked for, and received, additional plant capacity to offset partially the cost of the new furnace type.

III. FUEL CHANGE

In the past, for many air pollution control situations, a change to a less polluting fuel offered the ideal solution to the problem. If a power plant was emitting large quantities of SO_2 and fly ash, conversion to natural gas was cheaper than installing the necessary control equipment to reduce the pollutant emissions to the permitted values. If the drier at an asphalt plant was emitting 350 mg of particulate matter per standard cubic meter of effluent when fired with heavy oil of 4% ash, it was probable that a switch to either oil of a lower ash content or natural gas would allow the operation to meet an emission standard of 250 mg per standard cubic meter.

Fuel switching based on meteorological or air pollution forecasts was, in the past, a common practice to reduce the air pollution burden at critical times. Some control agencies allowed power plants to operate on residual oil during certain periods of the year when pollution potential was low. Some large utilities for years have followed a policy of switching from their regular coal to a more expensive but lower-sulfur coal when stagnation conditions were forecast.

Caution should be exercised when considering any change in fuels to reduce emissions. This is particularly true considering today's fuel costs. Specific considerations might be the following:

1. What are current and potential fuel supplies? In many areas natural gas is already in short supply. It may not be possible to convert a large plant with current allocations or pipeline capacity.
2. Most large boilers use a separate fuel for auxiliary or standby purposes. One actual example was a boiler fired with wood residue as the primary fuel and residual oil as the standby. A change was made to natural gas as the primary fuel, with residual oil kept for standby. This change was made to lower particulate emissions and to achieve a predicted slightly lower cost. Because of gas shortages, the plant now operates on residual oil during most of the cold season, and the resulting particulate emission greatly exceeds that of the previously burned wood fuel. In addition, an SO_2 emission problem exists with the oil fuel that never occurred with the wood residue. Overall costs have not been lowered because natural gas rates have increased since the conversion.

3. Charts or tables listing supplies or reserves of low-sulfur fuel may not tell the entire story. For example, a large percentage of low-sulfur coal is owned by steel companies and is therefore not generally available for use in power generating stations even though it is listed in tables published by various agencies.

4. Strong competition exists for low-pollution fuels. While one area may be drawing up regulations to require use of natural gas or low-sulfur fuels, it is probable that other neighboring areas are doing the same. Although there may have been sufficient premium fuel for one or two areas, if the entire region changes, not enough exists. Such a situation has resulted in extreme fuel shortages during cold spells in some large cities. The supply of low-sulfur fuels was exhausted during period of extensive use.

The increasing number of atomic reactors used for power generation has been questioned from several environmental points of view. A modern atomic plant, as shown in Fig. 28-3, appears to be relatively pollution free compared to the more familiar fossil fuel–fired plant, which emits carbon monoxide and carbon dioxide, oxides of nitrogen and sulfur, hydrocarbons, and fly ash. However, waste and spent-fuel disposal problems may offset the apparent advantages. These problems (along with steam generator leaks) caused the plant shown in Fig. 28-3 to close permanently in 1993.

IV. POLLUTION REMOVAL

In many situations, sufficient control over emissions cannot be obtained by fuel or process change. In cases such as these, the levels of the pollutants of concern in the exhaust gases or process stream must be reduced to allowable values before they are released to the atmosphere.

The equipment for the pollutant removal system includes all hoods, ducting, controls, fans, and disposal or recovery systems that might be necessary. The entire system should be engineered as a unit for maximum efficiency and economy. Many systems operate at less than maximum efficiency because a portion of the system was designed or adapted without consideration of the other portions (4).

Efficiency of the control equipment is normally specified before the equipment is purchased. If a plant is emitting a pollutant at 500 kg/hr and the regulations allow an emission of only 25 kg/hr, it is obvious that at least 95% efficiency is required of the pollution control system. This situation requires the regulation to state "at least 95% removal on a weight basis." The regulation should further specify how the test will be made to determine the efficiency. Figure 28-4 shows the situation as it exists.

Fig. 28-3. Trojan nuclear power plant. Source: Portland General Electric Company.

The efficiency for the device shown in Fig. 28-4 may be calculated in several ways:

$$\text{Efficiency, \%} = 100\left(\frac{C}{A}\right), \quad \text{but since } A = B + C \tag{28-1}$$

$$\text{Efficiency, \%} = 100\left(\frac{C}{B + C}\right) \quad \text{or} \quad 100\left(\frac{A - B}{A}\right) \quad \text{or} \quad 100\left(\frac{A - B}{B + C}\right) \tag{28-2}$$

The final acceptance test would probably be made by measuring two of the three quantities and using the appropriate equation. For a completely valid efficiency test the effect of hold-up (D) and loss (E) must also be taken into account.

To remove a pollutant from the carrying stream, some property of the pollutant that is different from the carrier must be exploited. The pollutant may have different size, inertia, electrical, or absorption properties. Removal requires that the equipment be designed to apply the scientific principles necessary to perform the separation.

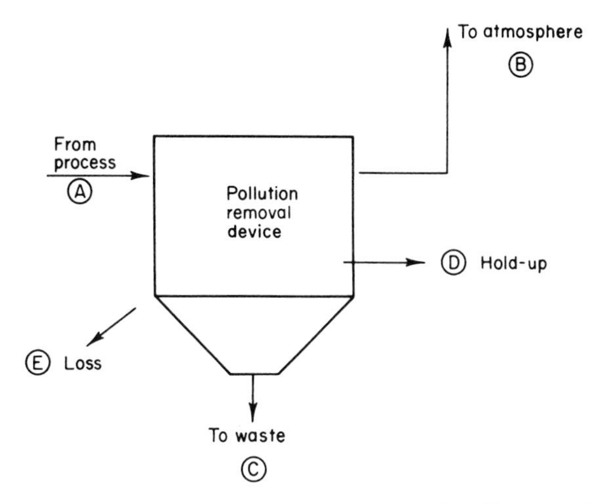

Fig. 28-4. Typical pollution control device as shown for efficiency calculations.

V. DISPOSAL OF POLLUTANTS

If a pollutant is removed from the carrying gas stream, disposal of the collected material becomes of vital concern. If the collected material is truly inert, it may be disposed of in a sanitary landfill. If it is at the other end of the scale, it is probably considered as a toxic waste and strict laws governing its disposal apply. Disposal of hazardous wastes is regulated by governmental agencies.

In the United States, the Resource Conservation and Recovery Act of 1976 (RCRA) is the major legislation covering the disposal of solid and hazardous wastes (2). This act provides a multifaceted approach to solving the problems associated with the generation of approximately 5 billion metric tons of solid waste each year in the United States. It places particular emphasis on the regulation of hazardous wastes. This law established the Office of Solid Waste within the Environmental Protection Agency and directed the agency to publish hazardous waste characteristics and criteria.

If a waste is designated as hazardous under the RCRA, regulations are applied to generators, transporters, and those who treat, store, or dispose of that waste. Regulations regarding hazardous wastes are enforced by the federal government, while the individual states are responsible for enforcing the provisions of the RCRA which apply to nonhazardous wastes. The act also provides for research, development, and demonstration grants for waste disposal.

The U.S. Environmental Protection Agency, Office of Solid Waste Management Programs, defines hazardous waste as "wastes or combinations of wastes which pose a substantial present or potential hazard to human health or living organisms because they cause, or tend to cause, detrimental cumulative effects."

Hazardous wastes can be categorized in a way which shows their potential or immediate effect. A common system is: (1) toxic substances (acute or chronic damage to living systems), (2) flammable, (3) explosive or highly reactive, (4) irritating and/or sensitizing, (5) corrosive (strong oxidizing agents), (6) radioactive, (7) bioaccumulative and/or biomagnified substances (with toxic effects), and (8) genetically interactive substances (mutagenic, teratogenic, and carcinogenic). It is possible for a substance to be placed in any number of these categories, but placement in only one category is sufficient for it to be considered hazardous.

Table 28-1 indicates the four main types of hazardous material, with examples of substances of each type. Not presented in Table 28-1 are radioactive materials, which are considered as a separate type of hazardous waste (5).

TABLE 28-1

Hazardous Material Types

Miscellaneous inorganics	Halogens and interhalogens	Miscellaneous organics	Organic halogen compounds
Metals	Bromine pentafluoride	Acrolein	Aldrin
Antimony	Chlorine	Dinitrophenol	Chlordane
Bismuth	Chlorine pentafluoride	Tetrazene	DDD, DDT
Cadmium	Chlorine trifluoride	Nitroglycerine	Dieldrin
Chromium	Fluorine	Nitroaniline	Endrin
Cobalt	Perchloryl fluoride	Chloroacetophenone	Potassium cyanide
Copper		(CN tear gas)	Heptachlor
Lead			Lindane
Mercury			Parathion
Nickel			Methyl bromide
Selenium			Polychlorinated
Silver			biphenyls
Tellurium			(PCBs)
Thallium			
Tin			
Zinc			
Nonmetallics			
Cyanide			
(ion)			
Hydrazine			
Fluorides			
Phosgene			

Table 28-2 lists some of the currently used pretreatments and ultimate disposal methods for hazardous wastes (6). *Pretreatment* refers almost entirely to thickening or dewatering processes for liquids or sludges. This process not only reduces the volume of the waste but also allows easier handling and transport.

The general purpose of ultimate disposal of hazardous wastes is to prevent the contamination of susceptible environments. Surface water runoff, ground water leaching, atmospheric volatilization, and biological accumulation are processes that should be avoided during the active life of the hazardous waste. As a rule, the more persistent a hazardous waste is (i.e., the greater its resistance to breakdown), the greater the need to isolate it from the environment. If the substance cannot be neutralized by chemical treatment or incineration and still maintains its hazardous qualities, the only alternative is usually to immobilize and bury it in a secure chemical burial site.

VI. POLLUTION PREVENTION

"Pollution prevention" is currently the most popular term used in the United States to describe strategies and technologies that reduce the generation of pollutants at their source (7). Other terms such as "waste minimization" and "waste reduction" are also used, and more positive terms such as "sustainable growth" and "total resource management" may find increasing use. The U.S. EPA defines "pollution prevention" as "the use of materials, processes or practices that reduce or eliminate the creation of pollutants or wastes at the source. It includes practices that reduce the use of hazardous materials, energy, and water and promotes resources and practices that protect natural resources through conservation or more efficient use" (8).

The idea underlying the promotion of pollution prevention is that it makes far more sense for a generator not to produce waste than to develop extensive treatment schemes to ensure that the waste poses no threat to the quality of the environment (9).

In the fall of 1990, the U.S. Congress passed the Pollution Prevention Act of 1990 (10). This act established a national policy that (11):

- Pollution should be prevented or reduced at the source whenever feasible.
- Pollution that cannot be prevented should be recycled in an environmentally safe manner whenever feasible.
- Pollution that cannot be prevented or recycled should be treated in an environmentally safe manner whenever feasible.

TABLE 28-2

Ultimate Waste Disposal Methods

Process	Purpose	Wastes	Problems (remarks)
Cementation and vitrification	Fixation Immobilization Solidification	Sludges Liquids	Expensive
Centrifugation	Dewatering Consolidation	Sludges Liquids	
Filtration	Dewatering Volume reduction	Sludges Liquids	Expensive
Thickening (various methods)	Dewatering Volume reduction	Sludges Liquids	
Chemical addition (polyelectrolytes)	Precipitation Fixation Coagulation	Sludges Liquids	Can be used in conjunction with other processes
Submerged combustion	Dewatering	Liquids	Acceptable for aqueous organics
Major Ultimate Disposal Methods			
Deep well injection	Partial removal from biosphere Storage	Oil field brines; low toxicity, low-persistence wastes; refinery wastes	Monitoring difficulty Need for special geological formations Ground water contamination
Incineration	Volume reduction	Most organics	If poor process control,

	Toxicity destruction		unwanted emissions produced
			Can produce NO_x, SO_x, halo acids
Recovery	Reuse	Metals Solvents	Sometimes energy prohibitive
Landfill	Storage	Inert to radioactive	Volatilization
	Major Waste Disposal Methods		
Land application	Isolation		Leaching to ground water
Land burial	Dispersal		Access to biota
Ocean disposal	Dispersal Dilution Neutralization Isolation (?)	Acids, bases Explosives Chemical war agents Radioactive wastes	Contact with ocean ecosystem Containers unstable
	Minor Disposal Methods		
Biological degradation	Reduction of concentration Oxidation	Biodegradable organics	Most hazardous wastes do not now qualify
Chemical degradation (chlorination)	Conversion Oxidation	Some persistent pesticides	
Electrolytic processes	Oxidation	Organics	
Long-term sealed storage	Isolation Storage	Radioactive	How good are containers?
Salt deposit disposal	Isolation Storage	Radioactive	Are salt deposits stable in terms of waste lifetimes?

- Disposal or other release into the environment should be employed only as a last resort and should be conducted in an environmentally safe manner.

The U.S. government is promoting pollution prevention at federal facilities. Cooperating with the EPA are the U.S. Departments of Agriculture, Defense, Energy, Interior, Transportation, Treasury, and Veterans Affairs. The U.S. Postal Service has also committed to an extensive pollution prevention program.

The U.S. EPA reports in its "Pollution Prevention 1991: Progress on Reducing Industrial Pollutants" (12) that 24 major companies have made extensive progress in pollution prevention. The 3M Corporation, which instituted its "Pollution Prevention Pays" (3P) Program in 1975, has saved over $500 million that can be directly traced to the 3P Program (13).

REFERENCES

1. Stern, A. C. (ed.), "Air Pollution," 3rd ed., Vol. IV. Academic Press, New York, 1977.
2. Arbuckle, J. G., Frick, G. W., Miller, M. L., Sullivan, T. F. P., and Vanderver, T. A., "Environmental Law Handbook," 6th ed. Governmental Institutes, Inc., Washington, DC, 1979.
3. Strauss, W., "Industrial Gas Cleaning," 2nd ed. Pergamon, Oxford, 1975.
4. Bvonicore, A. J., and Davis, W. T. (eds.), "Air Pollution Engineering Manual." Van Nostrand Reinhold, New York, 1992.
5. Schieler, P., "Hazardous Materials." Reinhold, New York, 1976.
6. Powers, P. W., "How to Dispose of Toxic Substances and Industrial Wastes." Noyes Data Corp., Park Ridge, NJ, 1976.
7. Freeman, H., Harten, T., Springer, J., Randall, P., Curran, M. A., and Stone, K., Industrial pollution prevention: a critical review." *J. Air Waste Manage. Assoc.* **42,** 618 (1992).
8. Environmental Protection Agency, Pollution Prevention Directive, U.S. EPA, May 13, 1990.
9. Freeman, H., Hazardous waste minimization: a strategy for environmental improvement. *J. Air Waste Manage. Assoc.* **38,** 59 (1988).
10. Pollution Prevention Act of 1990, 42nd U.S. Congress, 13101.
11. Pollution Prevention News, U.S. EPA Office of Pollution Prevention and Toxics, June 1992.
12. "Pollution Prevention 1991: Progress on Reducing Industrial Pollutants." U.S. EPA, 21P-3003, October 1991.
13. Bringer, R. P., and Benforado, D. M., "3P Plus: Total Quality Environmental Management," Proceedings of the 85th Annual Meeting, Air & Waste Management Association, Kansas City, June 1992.

SUGGESTED READING

"Estimation of Permissible Concentrations of Pollutants for Continuous Exposure," Report 600/2-76-155. U.S. Environmental Protection Agency, Research Triangle Park, NC, 1976.

Fischhoff, B., Slovic, P., and Lichtenstein, S., *Environment* **21**(4) (1979).

Henstock, M. E., and Biddulph, M. W. (eds.), "Solid Waste as a Resource." Pergamon, Oxford, 1978.

Lowrance, W. W., "Of Acceptable Risk." Kaufmann, Los Altos, CA, 1976.

QUESTIONS

1. For a given process at a plant, the cost of control can be related to the equation: dollars for control = 10,000 + $10e^x$, where x = percent of control ÷ 10. The material collected can be recovered and sold and the income determined from the equation: dollars recovered = (1000) (percent of control).
   a. At what level of control will the control equipment just pay for itself?
   b. At what level of control will the dollars recovered per dollars of control equipment be the maximum?
   c. What would be the net cost to the process for increased control from 97.0 to 99.5%?
2. Give three examples of conversion of a pollutant to a less polluting form or substance.
3. List the advantages and disadvantages of a municipal sanitary landfill and a municipal incinerator.
4. List the advantages and disadvantages of recovering energy, in the form of steam, from a municipal incinerator.
5. Show by means of a flow diagram or sketch how you would treat and dispose of the fly ash collected from a municipal incinerator. The fly ash contains toxic and nontoxic metals, nonmetallic inorganics, and organic halogen compounds.

29

Control Devices and Systems

I. INTRODUCTION

One of the methods of controlling air pollution mentioned in the previous chapter was pollution removal. For pollution removal to be accomplished, the polluted carrier gas must pass through a control device or system, which collects or destroys the pollutant and releases the cleaned carrier gas to the atmosphere. The control device or system selected must be specific for the pollutant of concern. If the pollutant is an aerosol, the device used will, in most cases, be different from the one used for a gaseous pollutant. If the aerosol is a dry solid, a different device must be used than for liquid droplets.

Not only the pollutant itself but also the carrier gas, the emitting process, and the operational variables of the process affect the selection of the control system. Table 29-1 illustrates the large number of variables which must be considered in controlling pollution from a source (1–4).

Once the control system is in place, its operation and maintenance become a major concern. Important reasons for an operation and maintenance (O&M) program (2) are (1) the necessity of continuously meeting emission regulations, (2) prolonging control equipment life, (3) maintaining produc-

TABLE 29-1

Key Characteristics of Pollution Control Devices and/or Systems

Factor considered	Characteristic of concern
General	Collection efficiency
	Legal limitations such as best available technology
	Initial cost
	Lifetime and salvage value
	Operation and maintenance costs
	Power requirement
	Space requirements and weight
	Materials of construction
	Reliability
	Reputation of manufacturer and guarantees
	Ultimate disposal/use of pollutants
Carrier gas	Temperature
	Pressure
	Humidity
	Density
	Viscosity
	Dewpoint of all condensibles
	Corrosiveness
	Inflammability
	Toxicity
Process	Gas flow rate and velocity
	Pollutant concentration
	Variability of gas and pollutant flow rates, temperature, etc.
	Allowable pressure drop
Pollutant (if gaseous)	Corrosiveness
	Inflammability
	Toxicity
	Reactivity
Pollutant (if particulate)	Size range and distribution
	Particle shape
	Agglomeration tendencies
	Corrosiveness
	Abrasiveness
	Hygroscopic tendencies
	Stickiness
	Inflammability
	Toxicity
	Electrical resistivity
	Reactivity

tivity of the process served by the control device, (4) reducing operation costs, (5) promoting better public relations and avoiding community alienation, and (6) promoting better relations with regulatory officials.

The O&M program has the following minimum requirements: (1) an equipment and record system with equipment information, warranties, instruction manuals, etc.; (2) lubrication and cleaning schedules; (3) planning and scheduling of preventive maintenance; (4) a storeroom and inventory system for spare parts and supplies; (5) listing of maintenance personnel; (6) costs and budgets for O&M; and (7) storage of special tools and equipment.

II. REMOVAL OF DRY PARTICULATE MATTER

Dry aerosols, or particulate matter, differ so much from the carrying gas stream that their removal should present no major difficulties. The aerosol is different physically, chemically, and electrically. It has vastly different inertial properties than the carrying gas stream and can be subjected to an electric charge. It may be soluble in a specific liquid. With such a variety of removal mechanisms that can be applied, it is not surprising that particulate matter, such as mineral dust, can be removed by a filter, wet scrubber, or electrostatic precipitator with equally satisfactory results.

A. Filters

A filter removes particulate matter from the carrying gas stream because the particulate impinges on and then adheres to the filter material. As time passes, the deposit of particulate matter becomes greater and the deposit itself then acts as a filtering medium. When the deposit becomes so heavy that the pressure necessary to force the gas through the filter becomes excessive, or the flow reduction severely impairs the process, the filter must either be replaced or cleaned.

The filter medium can be fibrous, such as cloth; granular, such as sand; a rigid solid, such as a screen; or a mat, such as a felt pad. It can be in the shape of a tube, sheet, bed, fluidized bed, or any other desired form. The material can be natural or man-made fibers, granules, cloth, felt, paper, metal, ceramic, glass, or plastic. It is not surprising that filters are manufactured in an infinite variety of types, sizes, shapes, and materials.

The theory of filtration of aerosols from a gas stream is much more involved than the sieving action which removes particles in a liquid medium. Figure 29-1 shows three of the mechanisms of aerosol removal by a filter. In practice, the particles and filter elements are seldom spheres or cylinders.

Direct interception occurs when the fluid streamline carrying the particle passes within one-half of a particle diameter of the filter element. Regardless of the particle's size, mass, or inertia, it will be collected if the streamline passes sufficiently close. Inertial impaction occurs when the particle would miss the filter element if it followed the streamline, but its inertia resists the change in direction taken by the gas molecules and it continues in a

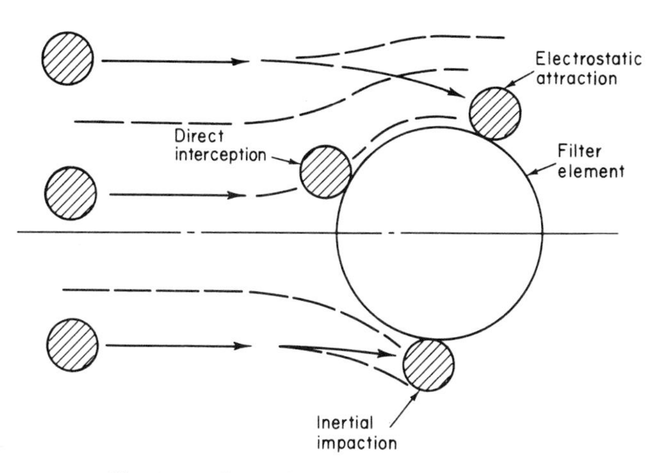

Fig. 29-1. Particulate removal by filter element.

direct enough course to be collected by the filter element. Electrostatic attraction occurs because the particle, the filter, or both possess sufficient electrical charge to overcome the inertial forces; the particle is then collected instead of passing the filter element.

Other lesser mechanisms that result in aerosol removal by filters are (1) gravitational settling due to the difference in mass of the aerosol and the carrying gas, (2) thermal precipitation due to the temperature gradient between a hot gas stream and the cooler filter medium which causes the particles to be bombarded more vigorously by the gas molecules on the side away from the filter element, and (3) Brownian deposition as the particles are bombarded with gas molecules that may cause enough movement to permit the aerosol to come in contact with the filter element. Brownian motion may also cause some of the particles to miss the filter element because they are moved away from it as they pass by. For practical purposes, only the three mechanisms shown in Fig. 29-1 are normally considered for removal of aerosols from a gas stream.

Regardless of the mechanism which causes the aerosol to come in contact with the filter element, it will be removed from the gas stream only if it adheres to the surface. Aerosols arriving later at the filter element may then, in turn, adhere to the collected aerosol instead of the filter element. The result is that actual aerosol removal seldom agrees with theoretical calculations. One should also consider that certain particles do not adhere to the filter element even though they touch it. As time passes, the heavier deposits on the filter surface will be dislodged more easily than the light deposits, resulting in increased reentrainment. Because of plugging of the filter with time, the apparent size of the filter element increases, causing more interception and impaction. The general effect of all of these variables on the particle buildup and reentrainment is shown in Fig. 29-2.

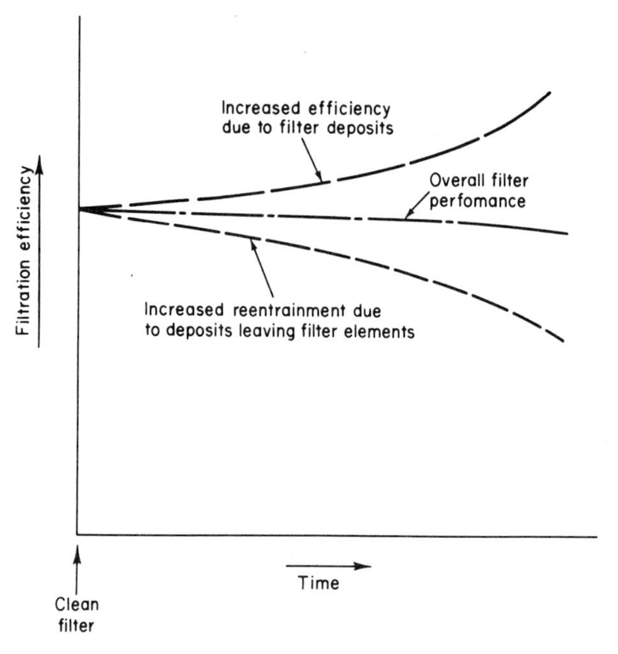

Fig. 29-2. Filtration efficiency change with time.

The pressure drop through the filter is a function of two separate effects. The clean filter has some initial pressure drop. This is a function of filter material, depth of the filter, the superficial gas velocity, which is the gas velocity perpendicular to the filter face, and the viscosity of the gas. Added to the clean filter resistance is the resistance that occurs when the adhering particles form a cake on the filter surface. This cake increases in thickness as approximately a linear function of time, and the pressure difference necessary to cause the same gas flow also becomes a linear function with time. Usually, the pressure available at the filter is limited so that as the cake builds up the flow decreases. Filter cleaning can be based, therefore, on (1) increased pressure drop across the filter, (2) decreased volume of gas flow, or (3) time elapsed since the last cleaning.

Industrial filtration systems may be of many types. The most common type is the baghouse shown in Fig. 29-3. The filter bags are fabricated from woven material, with the material and weave selected to fit the specific application. Cotton and synthetic fabrics are used for relatively low temperatures, and glass cloth fabrics can be used for elevated temperatures, up to 290°C.

The filter ratio for baghouses, also called the *gas to cloth ratio,* varies from 0.6 to 1.5 m^3 of gas per minute per square meter of fabric. The pressure drop across the fabric is a function of the filter ratio; it ranges from about

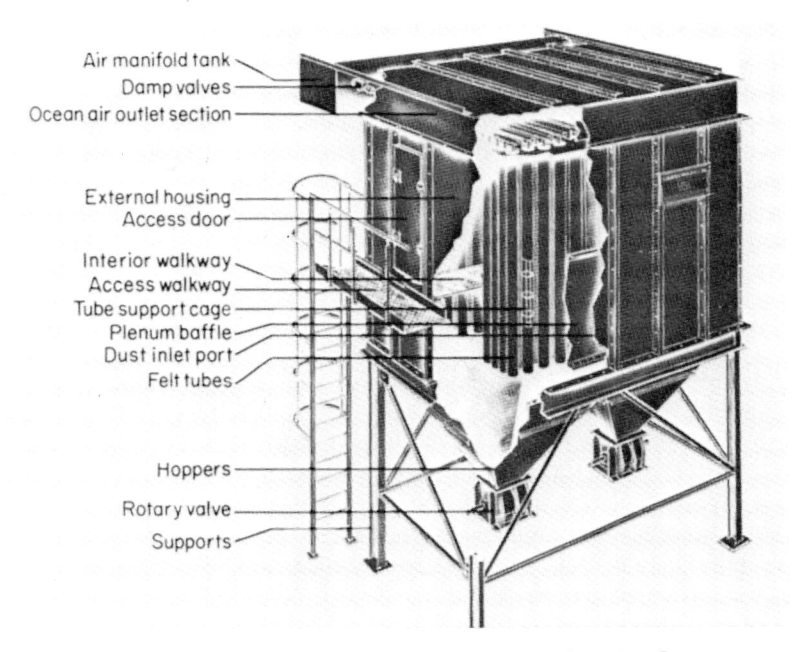

Fig. 29-3. Industrial baghouse. Source: Joy Manufacturing Co.

80 mm of water for the lower filter ratios up to about 200 mm of water for the higher ratios. Before selecting any bag filter system, a thorough engineering study should be made, followed by a consultation with different bag and baghouse manufacturers.

The bags must be periodically cleaned to remove the accumulated particulate matter. Bag cleaning methods vary widely with the manufacturer and with baghouse style and use. Methods for cleaning include (1) mechanical shaking by agitation of the top hanger, (2) reverse flow of gas through a few of the bags at a time, (3) continuous cleaning with a reverse jet of air passing through a series of orifices on a ring as it moves up and down the clean side of the bag, and (4) collapse and pulsation cleaning methods.

The cleaning cycles are usually controlled by a timing device which deactivates the section being cleaned. The dusts removed during cleaning are collected in a hopper at the bottom of the baghouse and then removed, through an air lock or star valve, to a bin for ultimate disposal.

Other types of industrial filtration systems include (1) fixed beds or layers of granular material such as coke or sand; some of the original designs for cleaning large quantities of gases from smelters and acid plants involved passing the gases through such beds; (2) plain, treated, or charged mats or pads (common throw-away air filters used for hot air furnaces and for air conditioners are of this type); (3) paper filters of multiple plies and folds to increase filter efficiency and area (the throwaway dry air filters used on automotive engines are of this type); (4) rigid porous beds which can be made of metal, plastic, or porous ceramic (these materials are most efficient for removal of large particles such as the 30-μm particles from a wood sanding operation); and (5) fluidized beds in which the granular material of the bed is made to act as a fluid by the gas passing through it. Most fluidized beds are used for heat or mass transfer. Their use for filtration has not been extensive.

B. Electrostatic Precipitators

High-voltage electrostatic precipitators (ESPs) have been widely used throughout the world for particulate removal since they were perfected by Fredrick Cottrell early in the twentieth century (5). Most of the original units were used for recovery of process materials, but today gas cleaning for air pollution control is often the main reason for their installation. The ESP has distinct advantages over other aerosol collection devices: (1) it can easily handle high-temperature gases, which makes it a likely choice for boilers, steel furnaces, etc.; (2) it has an extremely small pressure drop, so that fan costs are minimized; (3) it has an extremely high collection efficiency if operated properly on selected aerosols (many cases are on record, however, in which relatively low efficiencies were obtained because of unique or unknown dust properties); (4) it can handle a wide range of particulate sizes and dust concentrations (most precipitators work best on particles smaller than 10 μm, so that an inertial precleaner is often used to remove the large particles); and (5) if it is properly designed and constructed, its operating and maintenance costs are lower than those of any other type of particulate collection system. Figure 29-4 shows a large ESP installed on a cement kiln.

Three of the disadvantages of ESPs are as follows: (1) the initial cost is the highest of any particulate collection system, (2) a large amount of space is required for the installation, and (3) ESPs are not suitable for combustible particles such as grain or wood dust.

The ESP works by charging dust with ions and then collecting the ionized particles on a surface. The collection surfaces may consist of either tubular or flat plates. For cleaning and disposal, the particles are then removed from the collection surface, usually by rapping the surface.

A high-voltage (30 kV or more) DC field is established between the central wire electrode and the grounded collecting surface. The voltage is high enough that a visible corona can be seen at the surface of the wire. The result is a cascade of negative ions in the gap between the central wire and the grounded outer surface. Any aerosol entering this gap is both bombarded and charged by these ions. The aerosols then migrate to the collecting surface because of the combined effect of this bombardment and

Fig. 29-4. Large ESP for recovery of cement dust.

the charge attraction. When the particle reaches the collecting surface, it loses its charge and adheres because of the attractive forces existing. It should remain there until the power is shut off and it is physically dislodged by rapping, washing, or sonic means.

In a tube type ESP, the tubes are 8–25 cm in diameter and 1–4 m long. They are arranged vertically in banks with the central wires, about 2 mm in diameter, suspended in the center with tension weights at the bottom. Many innovations, including square, triangular, and barbed wires, are used by different manufacturers.

A plate-type ESP is similar in principle to the tubular type except that the air flows across the wires horizontally, at right angles to them. The particles are collected on vertical plates, which usually have fins or baffles to strengthen them and prevent dust reentrainment. Figure 29-5 illustrates a large plate-type precipitator. These precipitators are usually used to control and collect dry dusts.

Problems with ESPs develop because the final unit does not operate at ideal conditions. Gas channeling through the unit can result in high dust loadings in one area and light loads in another. The end result is less than optimum efficiency because of much reentrainment. The resistivity of the dust greatly affects its reentrainment in the unit. If a high-resistance dust collects on the plate surface, the effective voltage across the gap is decreased. Some power plants burning high-ash, low-sulfur coal have re-

Fig. 29-5. Commercial plate-type ESP. Source: Joy Manufacturing Co.

ported very low efficiency from the precipitator because the ash needed more SO_2 to decrease its resistivity. The suggestion that precipitator efficiency could be greatly improved by *adding* SO_2 or SO_3 to the stack gases has not yet been accepted.

C. Inertial Collectors

Inertial collectors, whether cyclones, baffles, louvers, or rotating impellers, operate on the principle that the aerosol material in the carrying gas stream has a greater inertia than the gas. Since the drag forces on the particle are a function of the diameter squared and the inertial forces are a function of the diameter cubed, it follows that as the particle diameter increases, the inertial (removal) force becomes relatively greater. Inertial collectors, therefore, are most efficient for larger particles. The inertia is also a function of the mass of the particle, so that heavier particles are more efficiently removed by inertial collectors. These facts explain why an inertial collector will be highly efficient for removal of 10-μm rock dust and very inefficient for 5-μm wood particles. It would be very efficient, though, for 75-μm wood particles.

The most common inertial collector is the cyclone, which is used in two basic forms, the tangential inlet and the axial inlet. Figure 29-6 shows the two types.

In actual industrial practice, the tangential inlet type is usually a large (1–5 m in diameter) single cyclone, while the axial inlet cyclone is relatively

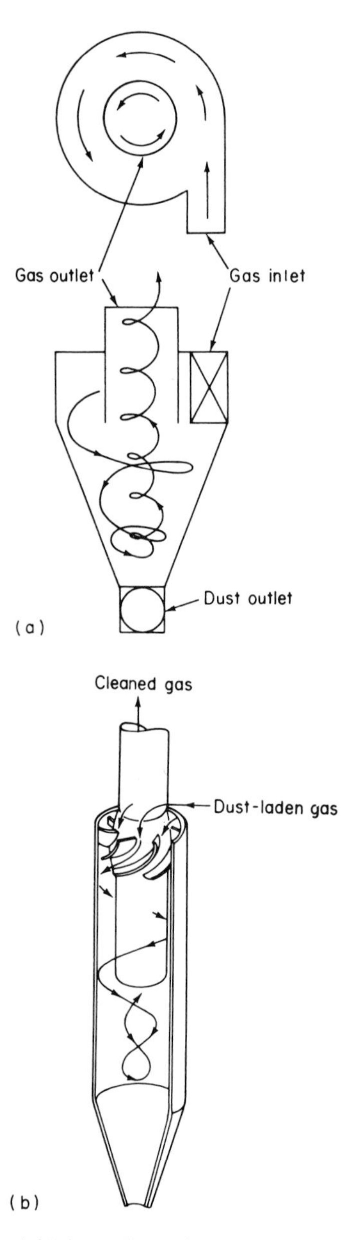

Fig. 29-6. (a) Tangential inlet cyclone. (b) Axial inlet cyclone.

small (about 20 cm in diameter and arranged in parallel units for the desired capacity).

For any cyclone, regardless of type, the radius of motion (curvature), the particle mass, and the particle velocity are the three factors which determine the centrifugal force exerted on the particle. This centrifugal force may be expressed as

$$F = MA \tag{29-1}$$

where F = force (centrifugal), M = mass of the particle, and A = acceleration (centrifugal) and

$$A = \frac{V^2}{R} \tag{29-2}$$

where V = velocity of particle and R = radius of curvature. Therefore:

$$F = \frac{MV^2}{R} \tag{29-3}$$

Other types of inertial collectors which might be used for particulate separation from a carrying gas stream depend on the same theoretical principles developed for cyclones. Table 29-2 summarizes the effect of the common variables on inertial collector performance.

It should further be pointed out that while decreasing the radius of curvature and increasing the gas velocity both result in increased efficiency, the same changes cause increased pressure drop through the collector. Design of inertial collectors for maximum efficiency at minimum cost and minimum pressure drop is a problem which lends itself to computer optimization. Unfortunately, many inertial collectors, including the majority of the large single cyclones, have been designed to fit a standard-sized sheet of metal rather than a specific application and gas velocity. As tighter emission standards are adopted, inertial collectors will probably become precleaners for the more sophisticated gas-cleaning devices.

TABLE 29-2

Effect of Independent Variables on Inertial Collector Efficiency

Independent variable of concern	Increase or decrease to improve efficiency
Radius of curvature	Decrease
Mass of particle	Increase
Particle diameter	Increase
Particle survace/volume ratio	Decrease
Gas velocity	Increase
Gas viscosity	Decrease

D. Scrubbers

Scrubbers, or wet collectors, have been used as gas-cleaning devices for many years. The process has two distinct mechanisms which result in the removal of the aerosol from the gas stream. The first mechanism involves wetting the particle by the scrubbing liquid. As shown in Fig. 29-7, this process is essentially the same whether the system uses a spray to atomize the scrubbing liquid or a diffuser to break the gas into small bubbles. In either case, it is assumed that the particle is trapped when it travels from the supporting gaseous medium across the interface to the liquid scrubbing medium. Some relative motion is necessary for the particle and liquid–gas interface to come in contact. In the spray chamber, this motion is provided by spraying the droplets through the gas so that they impinge on and make contact with the particles. In the bubbler, inertial forces and severe turbulence achieve this contact. In either case, the smaller the droplet or bubble, the greater the collection efficiency. In the scrubber, the smaller the droplet, the greater the surface area for a given weight of liquid and the greater the chance for wetting the particles. In a bubbler, smaller bubbles mean not only that more interface area is available but also that the particles have a shorter distance to travel before reaching an interface where they can be wetted.

The second mechanism important in wet collectors is removal of the wetted particles on a collecting surface, followed by their eventual removal from the device. The collecting surface can be in the form of a bed or simply a wetted surface. One common combination follows the wetting section with an inertial collector which then separates the wetted particles from the carrying gas stream.

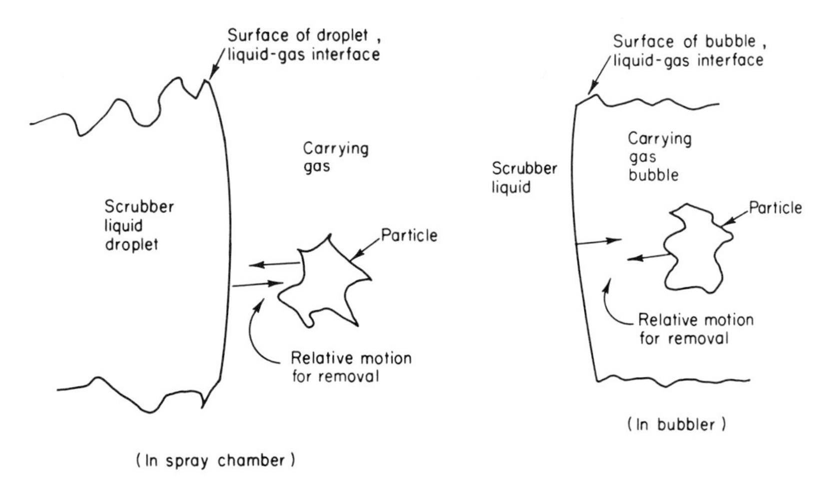

Fig. 29-7. Wetting of aerosols in a spray chamber or bubbler.

Increasing either the gas velocity or the liquid droplet velocity in a scrubber will increase the efficiency because of the greater number of collisions per unit time. The ultimate scrubber in this respect is the venturi scrubber, which operates at extremely high gas and liquid velocities with a very high pressure drop across the venturi throat. Figure 29-8 illustrates a commercial venturi scrubber unit.

E. Dry Scrubbers

Dry scrubber is a term that has been applied to gravel bed filters that recirculate the gravel filter medium using some type of external cleaning or washing system. Some units also use an electrostatic field across the gravel bed to enhance removal of the particulate material. The dry scrubber may have to be followed by a baghouse to clean the effluent to acceptable standards. The advantage of dry scrubbers is their ability to remove large quantities of particulate pollutants, such as fly ash, from hot gas streams.

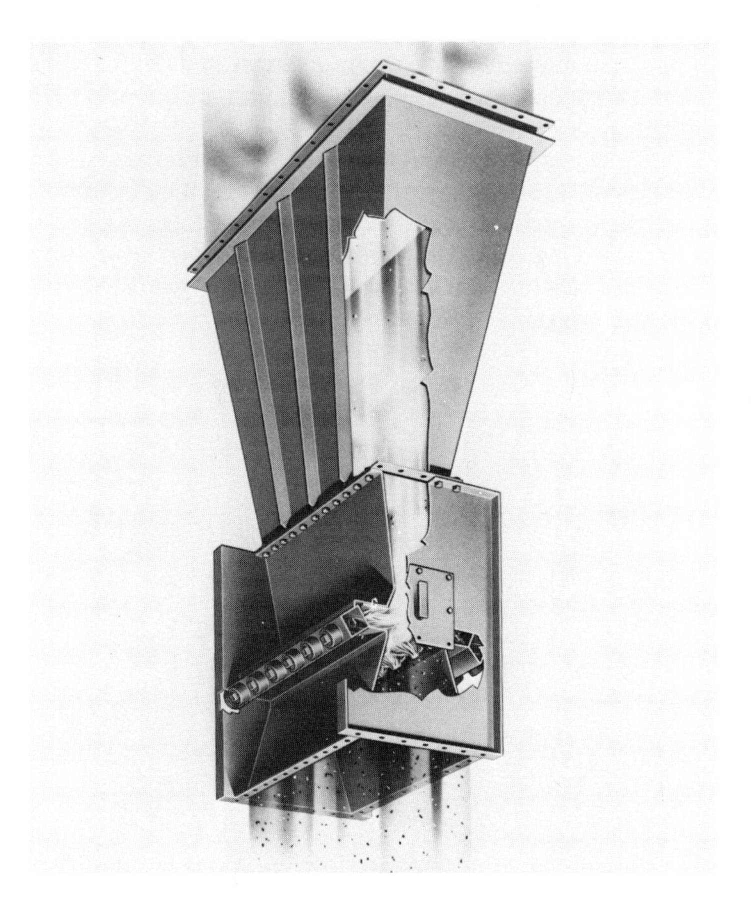

Fig. 29-8. Venturi scrubber. Source: American Air Filter Co.

F. Comparison of Particulate Removal Systems

When selecting a system to remove particulate from a gas stream, many choices concerning equipment can be made. The selection could be made on the basis of cost, gas pressure drop, efficiency, temperature, resistance, etc. Table 29-3 summarizes these factors for comparative purposes. The tabular values must not be considered absolute because great variations occur between types and manufacturers. No table is a substitute for a qualified consulting engineer or a reputable manufacturer's catalog.

TABLE 29-3

Comparison of Particulate Removal Systems

Type of collector	Particle size range (μm)	Removal efficiency	Space required	Max. temp. (°C)	Pressure drop (cm H^2O)	Annual cost (U.S. $ per year/m^3)[a]
Baghouse (cotton bags)	0.1–0.1	Fair	Large	80	10	28.00
	1.0–10.0	Good	Large	80	10	28.00
	10.0–50.0	Excellent	Large	80	10	28.00
Baghouse (Dacron, nylon, Orlon)	0.1–1.0	Fair	Large	120	12	34.00
	1.0–10.0	Good	Large	120	12	34.00
	10.0–50.0	Excellent	Large	120	12	34.00
Baghouse (glass fiber)	0.1–1.0	Fair	Large	290	10	42.00
	1.0–10.0	Good	Large	290	10	42.00
	10.0–50.0	Good	Large	290	10	42.00
Baghouse (Teflon)	0.1–1.0	Fair	Large	260	20	46.00
	1.0–10.0	Good	Large	260	20	46.00
	10.0–50.0	Excellent	Large	260	20	46.00
Electrostatic precipitator	0.1–1.0	Excellent	Large	400	1	42.00
	1.0–10.0	Excellent	Large	400	1	42.00
	10.0–50.0	Good	Large	400	1	42.00
Standard cyclone	0.1–1.0	Poor	Large	400	5	14.00
	1.0–10.0	Poor	Large	400	5	14.00
	10.0–50.0	Good	Large	400	5	14.00
High-efficiency cyclone	0.1–1.0	Poor	Moderate	400	12	22.00
	1.0–10.0	Fair	Moderate	400	12	22.00
	10.0–50.0	Good	Moderate	400	12	22.00
Spray tower	0.1–1.0	Fair	Large	540	5	50.00
	1.0–10.0	Good	Large	540	5	50.00
	10.0–50.0	Good	Large	540	5	50.00
Impingement scrubber	0.1–1.0	Fair	Moderate	540	10	46.00
	1.0–10.0	Good	Moderate	540	10	46.00
	10.0–50.0	Good	Moderate	540	10	46.00
Venturi scrubber	0.1–1.0	Good	Small	540	88	112.00
	1.0–10.0	Excellent	Small	540	88	112.00
	10.0–50.0	Excellent	Small	540	88	112.00
Dry scrubber	0.1–1.0	Fair	Large	500	10	42.00
	1.0–10.0	Good	Large	500	10	42.00
	10.0–50.0	Good	Large	500	10	42.00

[a] Includes water and power cost, maintenance cost, operating cost, capital and insurance costs.

III. REMOVAL OF LIQUID DROPLETS AND MISTS

The term *mist* generally refers to liquid droplets from submicron size to about 10 μm. If the diameter exceeds 10 μm, the aerosol is usually referred to as a *spray* or simply as *droplets.* Mists tend to be spherical because of their surface tension and are usually formed by nucleation and the condensation of vapors (6). Larger droplets are formed by bursting of bubbles, by entrainment from surfaces, by spray nozzles, or by splash-type liquid distributors. The large droplets tend to be elongated relative to their direction of motion because of the action of drag forces on the drops.

Mist eliminators are widely used in air pollution control systems to prevent free moisture from entering the atmosphere. Usually, such mist eliminators are found downstream from wet scrubbers. The recovered mist is returned to the liquid system, resulting in lowered liquid makeup requirements.

Since mist and droplets differ significantly from the carrying gas stream, just as dry particulates do, the removal mechanisms are similar to those employed for the removal of dry particulates. Control devices developed particularly for condensing mists will be discussed separately. Mist collection is further simplified because the particles are spherical and tend to resist reentrainment, and they agglomerate after coming in contact with the surface of the collecting device.

A. Filters

Filters for mists and droplets have more open area than those used for dry particles. If a filter is made of many fine, closely spaced fibers, it will become wet due to the collected liquid. Such wetting will lead to matting of the fibers, retention of more liquid, and eventual blocking of the filter. Therefore, instead of fine, closely spaced fibers, the usual wet filtration system is composed of either knitted wire or wire mesh packed into a pad. A looser filtration medium results in a filter with a lower pressure drop than that of the filters used for dry particulates. The reported pressure drop across wire mesh mist eliminators is 1–2 cm of water at face velocities of 5 m sec^{-1}. The essential collection mechanisms employed for filtration of droplets and mists are inertial impaction and, to a lesser extent, direct interception.

B. Electrostatic Precipitators

ESPs for liquid droplets and mists are essentially of the wetted wall type. Figure 29-9 shows a wet wall precipitator with tubular collection electrodes (1). The upper ends of the tubes form weirs, and water flows over the tube ends to irrigate the collection surface.

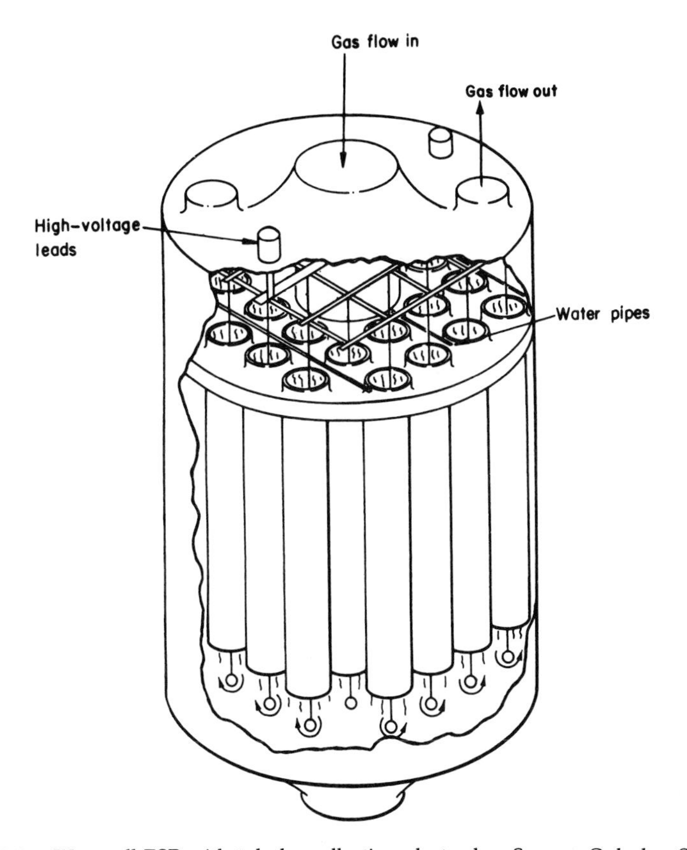

Fig. 29-9. Wet-wall ESP with tubular collection electrodes. Source: Oglesby, S., Jr., and Nichols, G. B., Electrostatic precipitators, in "Air Pollution," 3rd ed., Vol. IV (A. C. Stern, ed.), p. 238, Academic Press, New York, 1977.

Figure 29-10 shows an alternative type of wet precipitator with plate-type collection electrodes. In this design, sprays located in the ducts formed by adjacent collecting electrodes serve to irrigate the plates (1). These are often supplemented by overhead sprays to ensure that the entire plate surface is irrigated. The design of such precipitators is similar to that of conventional systems except for the means of keeping insulators dry, measures to minimize corrosion, and provisions for removing the slurry.

C. Inertial Collectors

Inertial collectors for mists and droplets are widely used. They include cyclone collectors, baffle systems, and skimmers in ductwork. Inertial devices can be used as primary collection systems, precleaners for other

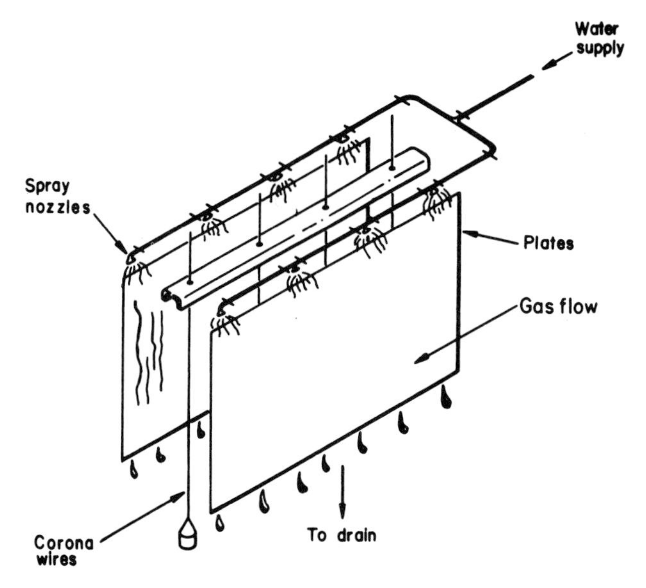

Fig. 29-10. Wet ESP with plate collection electrodes. Source: Oglesby, S., Jr., and Nichols, G. B., Electrostatic precipitators in "Air Pollution," 3rd ed., Vol. IV (A. C. Stern, ed.), p. 239, Academic Press, New York, 1977.

devices, or mist eliminators. The systems are relatively inexpensive and reliable and have low pressure drops.

Cyclone mist eliminators and collectors have virtually the same efficiency for both liquid aerosols and solid particles. To avoid reentrainment of the collected liquid from the walls of the cyclone, an upper limit is set to the tangential velocity that can be used. The maximum tangential velocity should be limited to the inlet velocity. Even at this speed, the liquid film may creep to the edge of the exit pipe, from which the liquid is then reentrained.

Baffle separators of the venetian blind, V, W, and wave types are widely used for spray removal. They have small space requirements and low pressure drops. They operate by diverting the gas stream and ejecting the droplets onto the collector baffles. Efficiencies of single stages may be only 40–60%, but by adding multiple stages, efficiencies approaching 100% may be obtained.

D. Scrubbers

A widely used type of scrubber for mists and droplets is the venturi scrubber. It has been used for the collection of sulfuric acid and phosphoric acid mists with very high efficiency. The scrubbing contact is made at the venturi throat, where very small droplets of the scrubbing liquid (usually water) are injected. At the throat, gas velocities as high as 130 m sec^{-1} are used to increase collision efficiencies. Water, injected for acid mist control, ranges from 0.8 to 2.0 liters m^{-3} of gas. Collection efficiencies approaching 100% are possible, but high efficiencies require a gas pressure drop of 60–90 cm of water across the scrubber. Normal operation, with a submicron mist, is reported to be in the 90–95% efficiency range (1).

One problem in using scrubbers to control mists and droplets is that the scrubber also acts as a condenser for volatile gases. For example, a hot plume containing volatile hydrocarbon gases, such as the exhaust from a gas turbine, may be cooled several hundred degrees by passing through a scrubber. This cooling can cause extensive condensation of the hydrocarbons, resulting in a plume with a high opacity. Teller (7) reports that cooling of exhaust gases from a jet engine in a test cell by the use of water sprays can result in droplet loadings 10–100% greater than those measured at the engine exhaust plane because of the condensation of hydrocarbons which were normally exhausted as gases.

E. Other Systems

Many unique systems have been proposed, and some used, to control the release of mists and droplets. Included are the following:

1. Ceramic candles, which are thimble-shaped, porous, acid-resistant ceramic tubes. Although efficiencies exceeding 98% have been reported, the candles have high maintenance requirements because they are very fragile.
2. Electric cyclones, which utilize an electrode in the center of the cyclone to establish an electric field within the cyclone body. This device is more efficient than the standard cyclone. It is probably more applicable to mists and droplets than to dry particulates, due to possible fire or explosion hazards with combustible dusts.
3. Sonic agglomerators, which have been used experimentally for sulfuric acid mists and as mist eliminators. Commercial development is not projected at this time because the energy requirements are considerably greater than those for venturi scrubbers of similar capacity.

IV. REMOVAL OF GASEOUS POLLUTANTS

Gaseous pollutants may be easier or harder to remove from the carrying gas stream than aerosols, depending on the individual situation. The gases may be reactive to other chemicals, and this property can be used to collect them. Of course, any separation system relying on differences in inertial properties must be ruled out. Four general methods of separating gaseous pollutants are currently in use. These are (1) absorption in a liquid, (2) adsorption on a solid surface, (3) condensation to a liquid, and (4) conversion into a less polluting or nonpolluting gas.

A. Absorption Devices

Absorption of pollutant gases is accomplished by using a selective liquid in a wet scrubber, packed tower, or bubble tower. Pollutant gases commonly controlled by absorption include sulfur dioxide, hydrogen sulfide, hydrogen chloride, chlorine, ammonia, oxides of nitrogen, and low-boiling hydrocarbons.

The scrubbing liquid must be chosen with specific reference to the gas being removed. The gas solubility in the liquid solvent should be high so that reasonable quantities of solvent are required. The solvent should have a low vapor pressure to reduce losses, be noncorrosive, inexpensive, nontoxic, nonflammable, chemically stable, and have a low freezing point. It is no wonder that water is the most popular solvent used in absorption devices. The water may be treated with an acid or a base to enhance removal of a specific gas. If carbon dioxide is present in the gaseous effluent and water is used as the scrubbing liquid, a solution of carbonic acid will gradually replace the water in the system.

In many cases, water is a poor scrubbing solvent. Sulfur dioxide, for example, is only slightly soluble in water, so a scrubber of very large liquid capacity would be required. SO_2 is readily soluble in an alkaline solution, so scrubbing solutions containing ammonia or amines are used in commercial applications.

Chlorine, hydrogen chloride, and hydrogen fluoride are examples of gases that are readily soluble in water, so water scrubbing is very effective for their control. For years hydrogen sulfide has been removed from refinery gases by scrubbing with diethanolamine. More recently, the light hydrocarbon vapors at petroleum refineries and loading facilities have been absorbed, under pressure, in liquid gasoline and returned to storage. All of the gases mentioned have economic importance when recovered and can be valuable raw materials or products when removed from the scrubbing solvent.

B. Adsorption Devices

Adsorption of pollutant gases occurs when certain gases are selectively retained on the surface or in the pores or interstices of prepared solids. The process may be strictly a surface phenomenon with only molecular forces involved, or it may be combined with a chemical reaction occurring at the surface once the gas and adsorber are in intimate contact. The latter type of adsorption is known as *chemisorption*.

The solid materials used as adsorbents are usually very porous, with extremely large surface-to-volume ratios. Activated carbon, alumina, and silica gel are widely used as adsorbents depending on the gases to be removed. Activated carbon, for example, is excellent for removing light hydrocarbon molecules, which may be odorous. Silica gel, being a polar material, does an excellent job of adsorbing polar gases. Its characteristics for removal of water vapor are well known.

Solid adsorbents must also be structurally capable of being packed into a tower, resistant to fracturing, and capable of being regenerated and reused after saturation with gas molecules. Although some small units use throw-away canisters or charges, the majority of industrial adsorbers regenerate the adsorbent to recover not only the adsorbent but also the adsorbate, which usually has some economic value.

The efficiency of most adsorbers is very near 100% at the beginning of operation and remains extremely high until a breakpoint occurs when the adsorbent becomes saturated with adsorbate. It is at the breakpoint that the adsorber should be renewed or regenerated. This is shown graphically in Fig. 29-11.

Industrial adsorption systems are engineered so that they operate in the region before the breakpoint and are continually regenerated by units.

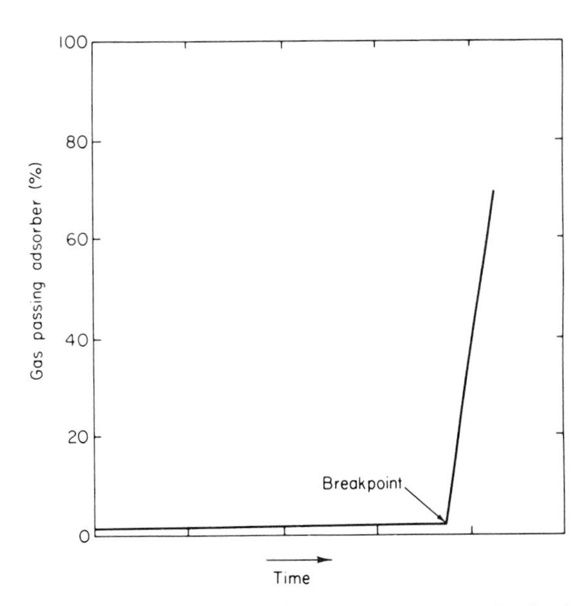

Fig. 29-11. Adsorbent breakpoint at saturation with adsorbate.

Figure 29-12 shows a schematic diagram of such a system, with steam being used to regenerate the saturated adsorbent. Figure 29-13 illustrates the actual system shown schematically in Fig. 29-12.

C. Condensers

In many situations, the most desirable control of vapor-type discharges can be accomplished by condensation. Condensers may also be used ahead of other air pollution control equipment to remove condensable components. The reasons for using condensers include (1) recovery of economically valuable products, (2) removal of components that might be corrosive or damaging to other portions of the system, and (3) reduction of the volume of the effluent gases.

Although condensation can be accomplished either by reducing the temperature or by increasing the pressure, in practice it is usually done by temperature reduction only.

Condensers may be of one or two general types depending on the specific application. Contact condensers operate with the coolant, vapors, and condensate intimately mixed. In surface condensers, the coolant does not come in contact with either the vapors or the condensate. The usual shell-and-tube condenser is of the surface type. Figure 29-14 illustrates a contact condenser which might be used to clean or preclean a hot corrosive gas.

Table 29-4 lists several applications of condensers currently in use. For most operations listed, air and noncondensable gases should be kept to a minimum, as they tend to reduce condenser capacity.

D. Conversion to Nonpollutant Material

A widely used system for the control of organic gaseous emissions is oxidation of the combustible components to water and carbon dioxide.

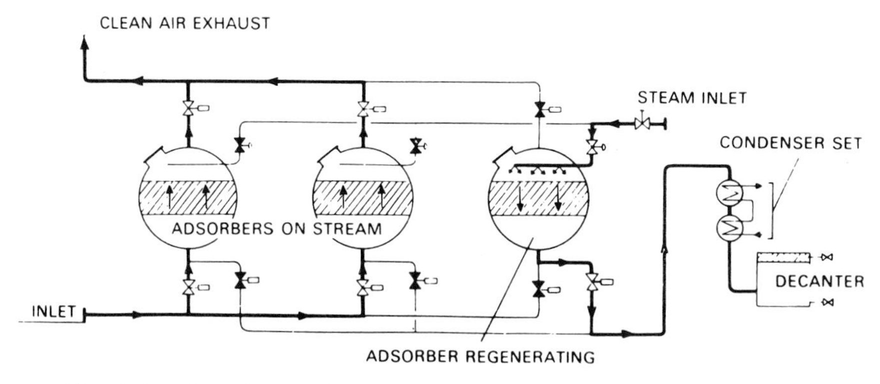

Fig. 29-12. Flow diagram for adsorber. Source: The British Ceca Company, Ltd.

Fig. 29-13. Adsorption system shown schematically in Fig. 29-13. Source: The British Ceca Company, Ltd.

Other systems such as the oxidation of H_2S to SO_2 and H_2O are also used even though the SO_2 produced is still considered a pollutant. The trade-off occurs because the SO_2 is much less toxic and undesirable than the H_2S. The odor threshold for H_2S is about three orders of magnitude less than that for SO_2. For oxidation of H_2S to SO_2, the usual device is simply an open flare with a fuel gas pilot or auxiliary burner if the H_2S is below the stoichiometric concentration.

Afterburners are widely used as control devices for oxidation of undesirable combustible gases. The two general types are (1) direct-flame afterburners, in which the gases are oxidized in a combustion chamber at or above the temperature of autogenous ignition, and (2) catalytic combustion systems, in which the gases are oxidized at temperatures considerably below the autogenous ignition point.

Direct-flame afterburners are the most commonly used air pollution control device in which combustible aerosols, vapors, gases, and odors are to be controlled. The components of the afterburner are shown in Fig. 29-15.

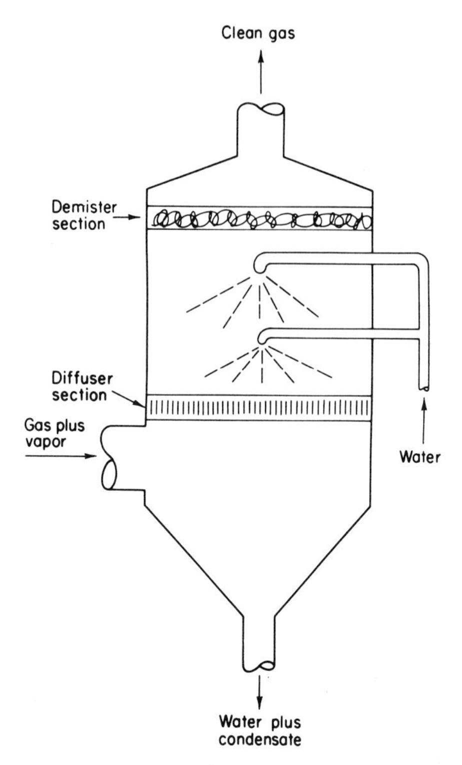

Fig. 29-14. Contact condenser.

TABLE 29-4

Representative Applications of Condensers in Air Pollution Control

Petroleum refining	Petrochemical manufacturing	Basic chemical manufacture	Miscellaneous industries
Gasoline accumulator	Polyethylene gas vents	Ammonia	Dry cleaning
Solvents	Styrene	Chlorine solutions	Degreasers
Storage vessels	Copper naphthenates		Tar dipping
Lube oil refining	Insecticides		Kraft paper
	Phthalic anhydride		
	Resin reactors		
	Solvent recover		

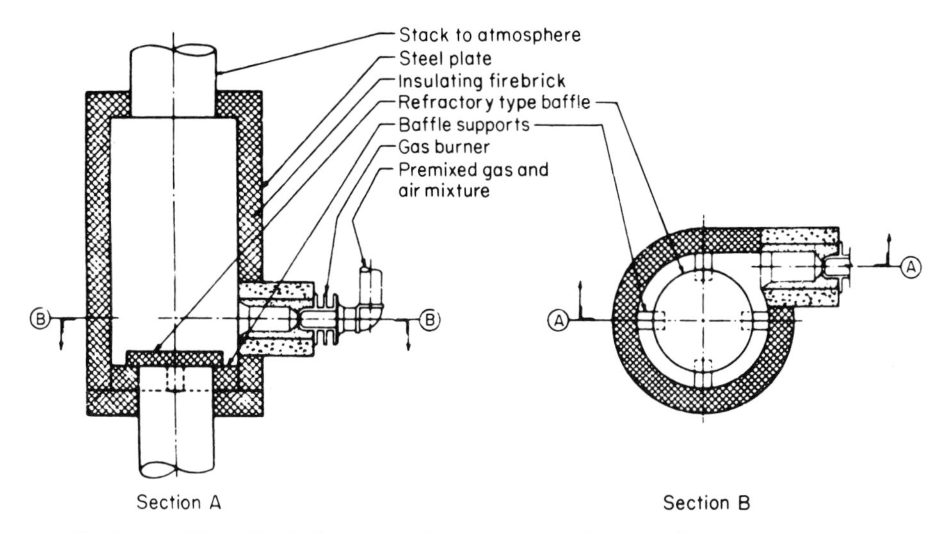

Fig. 29-15. Direct-fired afterburner. Source: Los Angeles Air Pollution Control District.

They include the combustion chamber, gas burners, burner controls, and exit temperature indicator. Usual exit temperatures for the destruction of most organic materials are in the range of 650°–825°C, with retention times at the elevated temperature of 0.3–0.5 sec.

Direct-flame afterburners are nearly 100% efficient when properly operated. They can be installed for approximately $350–700 per cubic meter of gas flow. Operating and maintenance costs are essentially those of the auxiliary gas fuel. On larger installations, the overall cost of the afterburner operation may be considerably reduced by using heat recovery equipment as shown in Fig. 29-16. In many industrial situations, boilers or kilns are used as entirely satisfactory afterburners for gases generated in other areas or processes.

Catalytic afterburners are currently used primarily in industry for the control of solvents and organic vapor emissions from industrial ovens. They are used as emission control devices for gasoline-powered automobiles (see Chapter 31).

The main advantage of the catalytic afterburner is that the destruction of the pollutant gases can be accomplished at a temperature range of about 315°–485°C, which results in considerable savings in fuel costs. However, the installed costs of the catalytic systems are higher than those of the direct-flame afterburners because of the expense of the catalyst and associated systems, so the overall annual costs tend to balance out.

In most catalytic systems there is a gradual loss of activity due to contamination or attrition of the catalyst, so the catalyst must be replaced at regular

Fig. 29-16. Afterburner with heat recovery. A, Fume inlet to insulated forced draft fan (310 m^3/min at 230°C). B, Regenerative shell-and-tube heat exchanger (55% effective recovery). C, Automatic bypass around heat exchanger for temperature control (required for excess hydrocarbons in fume steam under certain process conditions). D, Fume inlet and burner chamber internally insulated (fume steam raised to 425°C by heat exchanger). E, Combustion chamber, refractory lined for 815°C duty (operating at 760°C for required fume oxidation to meet local regulations). F, Discharge stream leaving regenerative heat exchanger at 520°C enters ventilating air heat exchanger for further waste heat recovery. G, Ventilating air fan and filter (310 m^3/min of outside air). H, Automatic bypass with dampers for control of ventilating air temperature. I, Heated air for winter comfort heating requirements leaves at controlled temperature. J, Discharge stack (470°C). K, Combustion safeguard system with dual burner manifold and controls for high turndown. L, Remote control panel with electronic temperature controls. Source: Hirt Combustion Engineers.

intervals. Other variables that affect the proper design and operation of catalytic systems include gas velocities through the system, amount of active catalyst surface, residence time, and preheat temperature necessary for complete oxidation of the emitted gases.

E. Comparison of Gaseous Removal Systems

As with particulate removal systems, it is apparent that many choices are available for removal of gases from effluent streams. Table 29-5 presents some of the factors that should be considered in selecting equipment.

For the control of SO_2, several systems are currently in development and use. Table 29-6 briefly explains these systems, which are presented in greater detail in Chapter 30.

TABLE 29-5

Comparison of Gaseous Pollutant Removal Systems

Type of equipment	Pressure drop (cm H_2O)	Installed cost (U.S. $ per m^3)	Annual operating cost (U.S. $ per m^3)
Scrubber	10	9.80	14.00
Absorber	10	10.40	28.00
Condenser	2.5	28.00	7.00
Direct flame afterburner	1.2	8.20	8.40 + gas
Catalytic afterburner	2.5	11.60	28.00 + gas

V. REMOVAL OF ODORS

An odor can be described as a physiological response to activation of the sense of smell (1). It can be caused by a chemical compound (e.g., H_2S) or a mixture of compounds (e.g., coffee roasting). Generally, if an odor is objectionable, any perceived quantity greater than the odor threshold will be cause for complaint. The control of odors, therefore, becomes a matter of reducing them to less than their odor thresholds, preventing them from entering the atmosphere, or converting them to a substance that is not odorous or has a much higher odor threshold. Odor masking is not recommended for a practical, long-term odor control system.

TABLE 29-6

Possible Sulfur Dioxide Control Systems

Method	Remarks
Limestone or lime injection (dry)	Calcined limestone or lime reacts with sulfur oxides. They are then removed with a dry particulate control system.
Limestone or lime injection (wet)	Calcined limestone or lime reacts with sulfur oxides, which are then removed by wet scrubbers.
Sodium carbonate	Sodium carbonate reacts with sulfur oxides in a dry scrubber to form sodium sulfite and CO_2. Sodium sulfite is then removed with a baghouse.
Citrate process	Citrate is added to scrubbing water to enhance SO_2 solution into water. Sulfur is then removed from the citrate solution.
Copper oxide adsorption	Oxides of sulfur react with copper oxide to form copper sulfate. Removal with a dry particulate control system follows.
Caustic scrubbing	Caustic neutralizes sulfur oxides. This method is used on small processes.

Source: Teller (8).

A. Odor Reduction by Dilution

If the odor is not a toxic substance and has no harmful effects at concentrations below its threshold, dilution may be the least expensive control technique. Dilution can be accomplished either by using tall stacks or by adding dilution air to the effluent. Tall stacks may be more costly if only capital costs are considered, but they do not require the expenditure for energy that is necessary for dilution systems.

The odor threshold for most atmospheric pollutants may be found in the literature (1). By properly applying the diffusion equations, one can calculate the height of a stack necessary to reduce the odor to less than its threshold at the ground or at a nearby structure. A safety factor of two orders of magnitude is suggested if the odorant is particularly objectionable.

Odor control by the addition of dilution air involves a problem associated with the breakdown of the dilution system. If a dilution fan, motor, or control system fails, the odorous material will be released to the atmosphere. If the odor is objectionable, complaints will be noted immediately. Good operation and maintenance of the dilution system becomes an absolute requirement, and redundant systems should be considered.

B. Odor Removal

It is sometimes possible to close an odorous system in order to prevent the release of the odor to the atmosphere. For example, a multiple-effect evaporator can be substituted for an open contact condenser on a process emitting odorous, noncondensable gases.

Another possible solution to an odor problem is to substitute a less noxious or more acceptable odor within a process. An example of this type of control is the substitution of a different resin in place of a formaldehyde-based resin in a molding or forming process.

Many gas streams can be deodorized by using solid adsorption systems to remove the odor before the stream is released to the atmosphere. Such procedures are often both effective and economical.

C. Odor Conversion

Many odorous compounds may be converted to compounds with higher odor thresholds or to nonodorous substances. An example of conversion to another compound is the oxidation of H_2S, odor threshold 0.5 ppb, to SO_2, odor threshold 0.5 ppm. The conversion results in another compound with an odor threshold three orders of magnitude greater than that of the original compound.

An example of conversion to a nonodorous substance would be the passage of a gas stream containing butyraldehyde, $CH_3CH_2CH_2CHO$, with an odor threshold of 40 ppb, through a direct-fired afterburner which converts it to CO_2 and H_2O, both nonodorous compounds. It should be noted that using a direct-fired afterburner, particularly one without heat recovery, to destroy 40 ppb is not an economical use of energy, and some other odor control system may be more desirable.

REFERENCES

1. Stern, A. C. (ed.), "Air Pollution," 3rd ed., Vol. IV. Academic Press, New York, 1977.
2. Theodore, L., and Buonicore, A. J., "Air Pollution Control Equipment: Selection, Design, Operation and Maintenance." Prentice-Hall, Englewood Cliffs, NJ, 1982.
3. Strauss, W., "Industrial Gas Cleaning," 2nd ed. Pergamon, Oxford, 1975.
4. Buonicore, A. J., and Davis, W. T. (eds.), "Air Pollution Engineering Manual." Van Nostrand Reinhold, New York, 1992.
5. Katz, J., "Electrostatic Precipitation." Precipitator Technology, Munhall, PA, 1979.
6. Bell, C. G., and Strauss, W., *J. Air Pollut. Control Assoc.* **23**(11), 967–969 (1973).
7. Teller, A. J., *J. Air Pollut. Control Assoc.* **27**(2), 148–149 (1977).
8. Teller, A. J., "Controlling Sulfur Oxides," EPA Research Summary EPA-600/8-80-029. U.S. Environmental Protection Agency, Research Triangle Park, NC, 1980.

SUGGESTED READING

"A Competitive Assessment of U.S. Industrial Air Pollution Control Equipment Industry." U.S. Department of Commerce, Washington, DC, 1991.

"Field Operations and Enforcement Manual for Air Pollution Control," Vol. II, "Control Technology and General Source Inspection," APTD-1101. U.S. Environmental Protection Agency, Research Triangle Park, NC, 1972.

"Operation and Maintenance of Electrostatic Precipitators." Specialty Conference Proceedings, Air Pollution Control Association, Pittsburgh, 1978.

Englund, H. M., and Beery, W. T. (eds.), "Electrostatic Precipitation of Fly Ash," by Harry J. White. APCA Reprint Series, Air Pollution Control Association, Pittsburgh, 1977.

Kester, B. E. (ed.), "Design, Operation, and Maintenance of High Efficiency Particulate Control Equipment." Specialty Conference Proceedings, Air Pollution Control Association, Pittsburgh, 1973.

Szabo, M. F., and Gerstle, R. W., "Operation and Maintenance of Particulate Control Devices on Coal-Fired Utility Boilers," EPA-600/2-77-129. U.S. Environmental Protection Agency, Research Triangle Park, NC, 1977.

QUESTIONS

1. List the similarities and differences of pollution control systems for solid particulate matter and liquid droplets.
2. You wish to design a baghouse to clean 3000 m^3 at a filter ratio of 3 $m^3\ m^{-2}$ of cloth. The filter bags are 15 cm in diameter by 3 m long. If you design a "square" baghouse with

the bags on 30-cm centers, what would be the exterior dimensions, neglecting ductwork? An alternative system uses 15-mm-diameter porous plastic tubes 1 m long on 25-mm centers. For the same filter ratio and flow, what would be the exterior dimensions for a "square" enclosure?

3. For a given cyclone collector, plot centrifugal force as a function of particle specific gravity (0.50–3.00), gas velocity (175–1750 m min^{-1}), and radius of curvature (30–250 cm).
4. List the advantages and disadvantages of using a baghouse, wet scrubber, or ESP for particulate collection from an asphalt plant drying kiln. The gases are at 250°C and contain 450 mg m^{-3} of rock dust in the 0.1–10 μm size range. Gas flow is 2000 m^3 min^{-1}. Consider initial and operation cost, space requirement, ultimate disposal, etc.
5. Suppose a gaseous process effluent of 30 m^3 min^{-1} is at 200°C and 50% relative humidity. It is cooled to 65°C by spraying with water that was initially at 20°C. What volume of saturated gas would you have to design for at 65°C? How much water per cubic meter would the system require? How much water per cubic meter would you have to remove from the system?
6. If an ESP is 90% efficient for particulate removal, what overall efficiency would you expect for two of the ESPs in series? Would the cost of the two in series be double the cost of the single ESP? List two specific cases in which you might use two ESPs in series.
7. The gaseous effluent from a process is 30 m^3 min^{-1} at 65°C. How much natural gas at 8900 kg cal m^{-3} would have to be burned per hour to raise the effluent temperature to 820°C? Natural gas requires 10 m^3 of air for every cubic meter of gas at a theoretical air/fuel ratio. Assume the air temperature is 20°C and the radiation and convection losses are 10%.
8. For question 7, if heat recovery equipment were installed to raise the incoming effluent to 425°C, how much natural gas would have to be burned per hour?
9. Choose a representative area (a city, county, region, etc.) and prepare a table showing the change in air pollution emission if natural gas were used as a fuel instead of oil and coal.
10. Why are "oxides of nitrogen" and "oxides of sulfur" usually reported in emission inventory tables rather than the actual oxidation states?
11. For a given area estimate the yearly pollutants emitted by automobiles using the figures for gallons of gasoline sold supplied by (a) the gasoline dealers association and (b) the local taxation authorities.

30

Control of Stationary Sources

I. INTRODUCTION

Control of stationary sources of air pollution requires the application of either the control concepts mentioned in Chapter 28 of the control devices mentioned in Chapter 29. In some cases, more than one system or device must be used to achieve satisfactory control. The three general methods of control are (1) process change to a less polluting process or to a lowered emission from the existing process through modification of the operation, (2) fuel change to a fuel which will give the desired level of emissions, and (3) installation of control equipment. It is more efficient to engineer air pollution control into the source when it is first considered than to leave it until the process is operational and found to be in violation of emission standards. Most new, large stationary sources of air pollution in the United States are regulated under the Clean Air Act Amendments of 1990 (1) and are legally required to comply with the Amendments.

Existing stationary sources may require modification of existing systems or installation of newer, more efficient control devices to meet more restrictive emission standards. Such changes are often required by control agencies when it can be shown that a new control technology is superior to older systems or devices being used. This is usually referred to as application of the best available control technology (BACT).

Installation of control systems may have a positive economic benefit which will offset a portion of their cost (2). Such benefits include (1) tax deduction provisions, (2) recovery of materials previously emitted, (3) depreciation schedules favoring the owner of the source, and (4) banking or sale of the emission offset credits if the source is in a nonattainment area.

II. ENERGY, POWER, AND INCINERATION

Thermal energy, power generation, and incineration have several factors in common. All rely on combustion, which causes the release of air pollutants; all exhaust their emissions at elevated temperatures; and all produce large quantities of ash when they consume solid or residual fuels. The ratio of the energy used to control pollution to the gross energy produced can be a deciding factor in the selection of the control system. These processes have important differences which influence the selection of specific systems and devices for individual facilities.

A. Energy-Producing Industries

Stationary energy-producing systems are of two general types, residential and commercial space heating and industrial steam generation. The smaller systems (residential and commercial heating) are usually regulated only with respect to their smoke emission, even though they may produce appreciable amounts of other air pollutants (3). The large industrial systems which generate steam for process use and space heating (where superheated or high-temperature saturated steam is used for a process, e.g., cogeneration, exhausted from the process at a lower energy level, and then introduced into a space heating system, where it gives up a large amount of latent energy, condensing to hot water) are required to comply with rigid standards in most countries.

Control of air pollution from energy-producing industries is a function of the fuel used and the other variables of the combustion process. The system must be thoroughly analyzed before a control system is chosen. The important variables are listed in Table 29-1. For particulate matter control, the variables of process control such as improved combustion, fuel cleaning, fuel switching, and load reduction through conservation should be considered before choosing an add-on control system. If the particulate matter emission is still found to be in excess of standards, then control devices must be used. These include inertial devices (such as multiple cyclones), baghouses, wet and dry scrubbers, electrostatic precipitators

(ESPs), and some of the previously discussed novel devices. Series combinations of control devices may be necessary to achieve the required level of particulate matter emission. A commonly used system is a multiple cyclone followed by a fine-particle control system, such as a baghouse, scrubber, or ESP.

Opacity reduction is the control of fine particulate matter (less than 1 μm). It can be accomplished through the application of the systems and devices discussed for control of particulate matter and by use of combustion control systems to reduce smoke and aerosol emission. In addition, operational practices such as continuous soot blowing and computerized fuel and air systems should be considered.

Sulfur dioxide reduction to achieve required emission levels may be accomplished by switching to lower-sulfur fuels. Use of low-sulfur coal or oil, or even biomass such as wood residue as a fuel, may be less expensive than installing an SO_2 control system after the process. This is particularly true in the wood products industry, where wood residue is often available at a relatively low cost.

If an SO_2 control device is necessary, the first decision is whether to use a wet or dry system. Many times this decision is based on the local situation regarding the disposal of the collected residue (sludge or dry material). Wet scrubbing systems, using chemical additions to the scrubbing liquid, are widely used. Various commercial systems have used lime or limestone, magnesium oxide, or sodium hydroxide slurries to remove the SO_2. Dry removal of SO_2 can be accomplished by adding the same chemicals used in wet scrubbers, but adding them in a spray drier and then removing the spent sulfates with a baghouse or ESP. This results in the collection of a dry material which may either be disposed of by landfilling or used as a raw material for other processes. The electric power generating industry has had many years of experience with SO_2 control systems (4). These methods are discussed in more detail in the next section.

Control of oxides of nitrogen can be accomplished by catalysts or absorbants, but most control systems have concentrated on changing the combustion process to reduce the formation of NO_x. Improved burners, change in burner location, staged combustion, and low-temperature combustion utilizing fluidized-bed systems are all currently in use. These combustion improvement systems do not generate waste products, so no disposal problems exist.

B. Power Generation

In general, plants producing electric power are much larger than those producing steam for space heating or process use. Therefore, the mass of emissions is much greater and the physical size of the control equipment larger.

The extensive control of particulate matter, opacity, SO_2, and NO_x required on new power plants is very expensive. The high projected cost of environmental control for a new coal-fired electric generating plant has made utility companies reluctant to risk the billions of dollars necessary to use coal as a major fuel, particularly when the standards are being constantly redefined and changed (5). It does appear that the 1990 Amendments will require more complex emission controls as an integrated part of the plant rather than the previous approach of adding control devices independently. Figure 30-1 illustrates the complexity of the current technology for the various alternative systems.

Figure 30-1A presents the integrated environmental control potential for maximum control of particulate matter and SO_2. Cooling tower water blowdown and treatment by-products may be used to satisfy scrubber makeup requirements. Fly ash and scrubber sludge will be produced separately. If the catalytic NO_x process is required, the integration issues will be increased significantly.

Figure 30-1B is similar to Fig. 30-1A except that an ESP is used for particulate control. This represents the most common approach for compliance when configured without a catalytic NO_x unit.

Figure 30-1C is distinctly different from the first two in the type of SO_2 control processes used and the sequence of the particulate matter and SO_2 controls. It is a promising approach for up to 90% SO_2 control of western United States coal, and there is a single waste product. Other features include the collection of particulate matter at temperatures below 90°C and the possibility for spray dryer cooling tower water integration. This system may or may not include a catalytic NO_x unit.

Figure 30-1D represents the simplest, least expensive, and lowest water consumer of all the alternatives. There is a single solid waste product. The key element is the integrated SO_2/particulate control process. Using sodium-based sorbants, compliance may be achievable for western United States coals.

Figure 30-1E includes a hot ESP for fly ash collection prior to a catalytic NO_x unit. Having a hot ESP dictates the use of a conventional wet scrubber and perhaps the need for a second particulate matter control device at the end of the system. Fly ash and scrubber sludge would be separate by-products, but sludge could be contaminated with NH_4 from the catalytic NO process.

Figure 30-2 illustrates a wet SO_2 desulfurization system using a spray tower absorber. Figure 30-3 illustrates a rotary atomizer injecting an alkaline slurry into a spray dryer for SO_2 control.

Selection and installation of an integrated air pollution control system do not end the concern of the utility industry. Maintenance and operational problems of the system are considered by many engineers to be the weak link in the chain of power generation equipment (6). The reliability of the

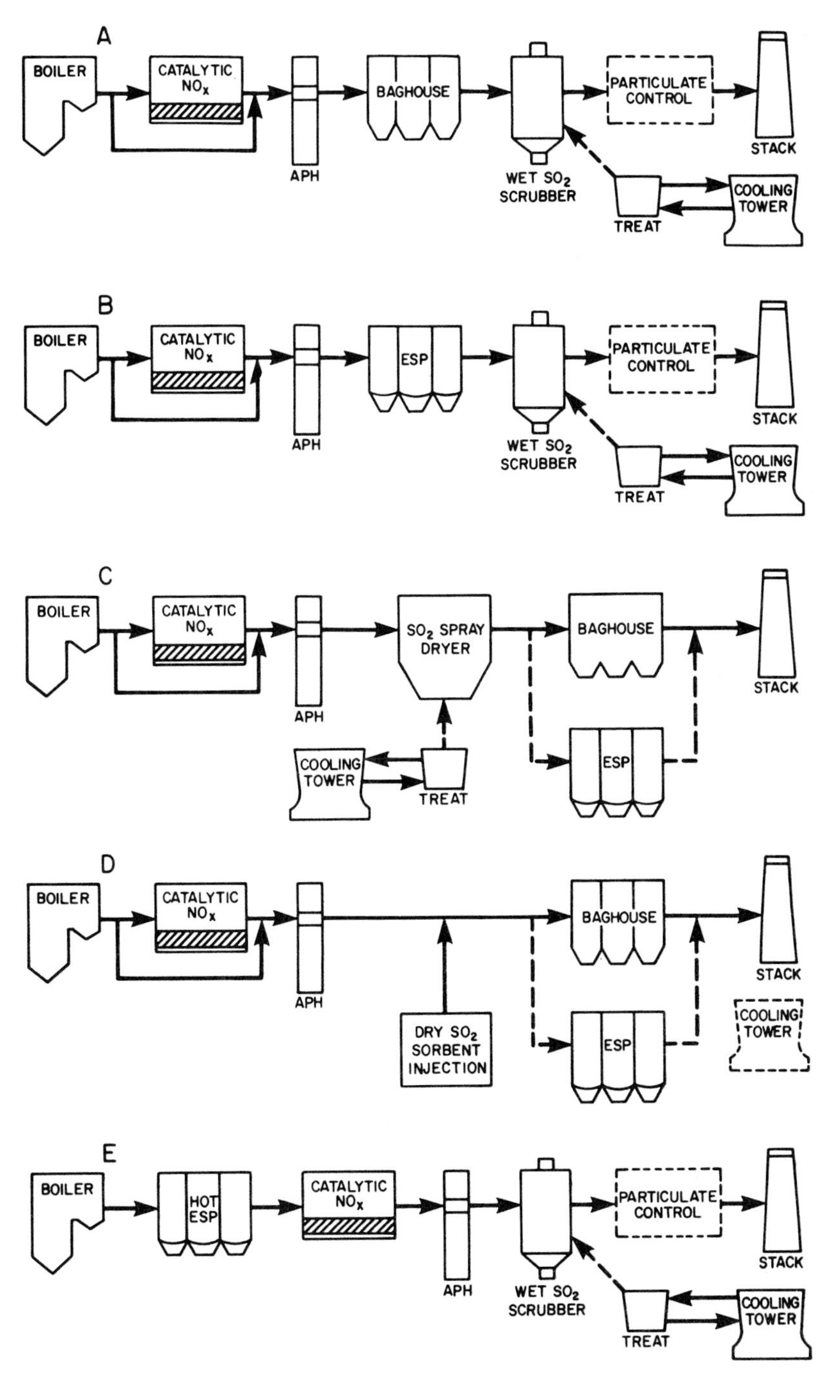

Fig. 30-1. Integrated environmental control for electric generating plants (5). APH, air preheater; ESP, electrostatic precipitator; TREAT, water treatment system.

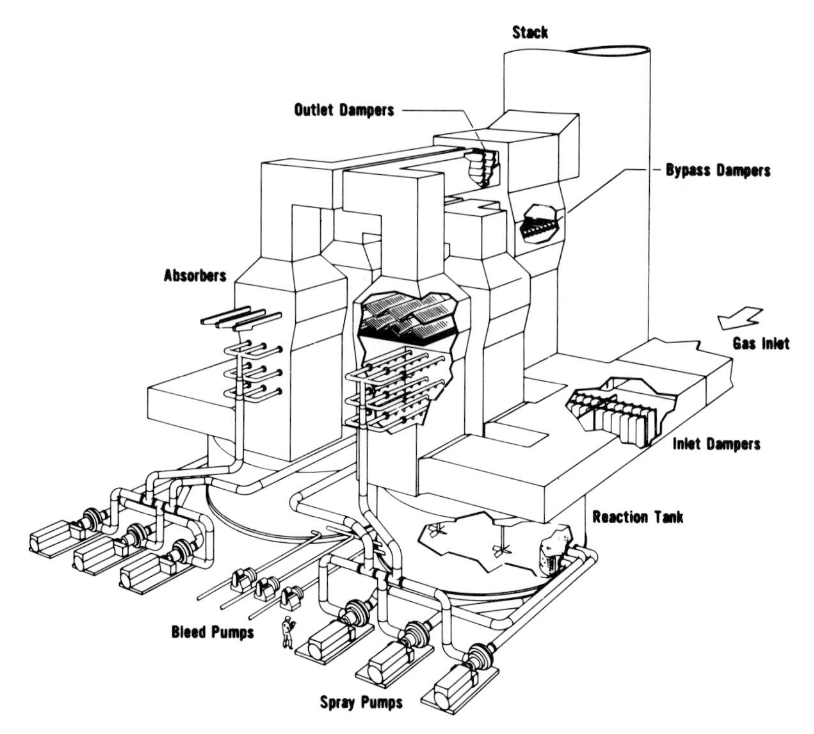

Fig. 30-2. Cutaway drawing of a flue gas desulfurization spray tower absorber. Source: CE Power Systems, Combustion Engineering, Inc.

flue gas desulfurization system (FGDS) is defined as the time the system operates properly divided by the time it should have operated. For large U.S. power plants, this has been determined by the U.S. Environmental Protection Agency to be 83.3% in 1980 and 82.0% in 1981. These figures are cause for concern even though the consequences to the utility may not have been extreme because the FGDS usually could be bypassed. The primary factors cited for low FGDS reliability are (1) plugging, scaling, and corrosion of scrubber internals, mist eliminators, and reheaters; (2) need for open-loop operation and blowdown of scrubber liquor to reduce the corrosive substance concentration and dissolved solids content; (3) corrosion and failure of stack liners; and (4) plugging and failure of piping, pumps, and valves. Probably more critical are failures of dampers, ducts, and baffles, because this may require plant shutdown to perform maintenance.

Other methods which should be mentioned because they show potential benefits for pollution reduction from utility stacks include (1) coal cleaning

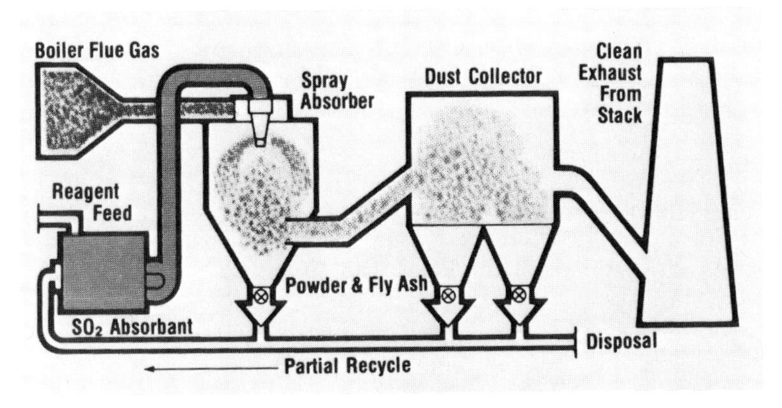

Fig. 30-3. Alkaline slurry SO_2 spray dryer. Source: Niro Atomizer, Inc.

and treatment, (2) atmospheric pressure and pressurized fluidized-bed combustors, (3) conversion of solid fuels to liquid or gaseous fuels, and (4) combustion modification through staged combustors or other systems.

Tall stacks are no longer considered to be an acceptable alternative for controlling emissions from electric power generating plants. (See further discussion in Chapter 26, Section V.)

C. Incineration

Incineration is similar to combustion-generated energy and power processes in that fuel combines with oxygen. The incineration process, however, is designed as a waste disposal process, and if any energy is recovered it is considered as a secondary system. Ideally, incineration will reduce combustors, (3) conversion of solid fuels to liquid or gaseous fuels, and (4) combustion modification through staged combustors or other systems.

An incinerator will usually have a fuel of varying chemical composition and physical properties, as well as varying moisture content and heating value. Also, the fuel fired in one locality may be vastly different from that fired by an incinerator of similar size and design in another locality. Refuse production in the United States has been estimated to average 2.5 kg per person per day in 1970, increasing to 10 kg per person per day by the year 2000.

The air pollutants from incinerators consist of particulate matter (fly ash, carbon, metals and metal oxides, and visible smoke), combustible gases such as CO, organics, polynuclear organic material (POM), and noncombustible gases such as oxides of nitrogen, oxides of sulfur, and hydrogen chloride. The oxides of nitrogen are formed by two mechanisms: thermal NO_x, in which atmospheric nitrogen and oxygen combine at high furnace temperatures, and fuel NO_x, when nitrogen-bearing compounds are incinerated. Hydrogen chloride (HCl) emissions are causing concern because of the increased amounts of halogenated polymers, notably polyvinyl chloride (PVC), in the refuse; 1 kg of pure PVC yields about 0.6 kg of HCl.

POM emissions appear to be a function of the degree of combustion control, decreasing with increasing incinerator size (larger incinerators are more thoroughly instrumented and controlled). Table 30-1 shows the measured emission rates for POM and CO from various-sized incinerators.

Air pollution control systems using wet scrubbers will remove some water-soluble gases, but the removal of particulate matter is the primary concern for a control system. The air pollution control system, therefore, is usually a single device such as a wet scrubber, small-diameter multiple cyclones, fabric filters, or ESPs. The multicyclones are the least expensive system and the ESPs the most expensive.

Some novel methods of incineration offer the possibility of reduced emissions. These include "slagging" (operating at such a high temperature that incombustible materials are melted and removed as a fluid slag); fluidized beds (which are useful only on homogeneous or well-classified refuse); suspension burning in cylindrical combustion chambers, which may or may not result in slagging; and pyrolysis, which is destructive distillation in the absence of oxygen.

The emission control requirements set for municipal incinerators by the U.S. Clean Air Act Amendments of 1990 (1) are extensive and complex. Many of the final standards have not been established as of the date of publication of this book. A thorough study of the regulations is necessary for any person dealing with incinerator technology and control.

TABLE 30-1

Generation and Emission Rates for Polynuclear Organic Material (POM) and Carbon Monoxide (CO) from Incinerators

Incinerator type, size, and control	POM (gm/metric ton)	CO (gm/kg)
Municipal, 227 metric tons per day, before settling chamber	0.032	0.35
Municipal, 45 metric tons per day, before scrubber	0.258	2.00
Municipal, 45 metric tons per day, after scrubber	0.014	1.00
Commercial multiple chamber, 3 metric tons per day, no control system	1.726	12.50

Source: Ref. (7).

III. CHEMICAL AND METALLURGICAL INDUSTRIES

The chemical and metallurgical industries of the world are so varied and extensive that it is impossible to cover all of the processes, emissions, and controls in a single chapter.

A. Chemical Industries

The term *chemical industry* applies to a group of industries which range from small, single owner–employee operations to huge complexes employing thousands of people. The number of environmental regulations that the chemical industry must comply with is so extensive that specialized consulting firms have been formed to aid the industry in handling them (8).

1. *Inorganic Chemical Processes*

Production of major inorganic chemicals in the United State exceeds 200 million metric tons per year produced in over 1300 plants (9). These inorganic chemicals may be categorized as acids, bases, fertilizer, chlorine, and bromine.

a. Acids The major acids produced are hydrochloric, hydrofluoric, nitric, phosphoric, and sulfuric. The emissions and usual control methods for the various acid and manufacturing processes are shown in Table 30-2.

b. Bases Major bases and caustics produced by the chemical industry are calcium oxide (lime), sodium carbonate (soda ash), and sodium hydroxide (caustic soda). The emissions and usual control methods for the various bases and their manufacturing processes are shown in Table 30-3.

c. Fertilizer Fertilizer production is dependent on the production of phosphates and nitrates. Phosphate rock preparation generates some dry particulate matter during drying, grinding, and transferring of the rock. These emissions are controlled by wet scrubbers and baghouses. The atmospheric emissions and control methods for the production processes are shown in Table 30-4.

d. Ammonium Nitrate Fertilizer Ammonium nitrate fertilizer is produced by the neutralization of nitric acid with ammonia. The primary emission is the dust or fume of ammonium nitrate from the prill tower. The material is of submicron size and, therefore, highly visible. Control is usually performed by a wet scrubber followed by a mist eliminator.

TABLE 30-2

Air Pollution Emissions and Controls: Inorganic Acid Manufacture

Acid	Manufacturing process	Air pollutant emissions	Control methods in use
Hydrochloric	By-product of organic chlorination, salt process, and synthetic HCl	HCl	Absorption
Hydrofluoric	Fluorspar–sulfuric acid	SiF_4, HF	Scrubber (some with caustic)
Nitric	Pressure process and direct strong acid	NO, NO_2, N_2O_4	Catalytic reduction, adsorption, absorption
Phosphoric	Elemental phosphorus	Particulate matter, fluorides	Baghouse
	Thermal process	H_3PO_4, H_2S	Mist eliminators, alkaline scrubbers
	Wet process	SiF_4, HF	Scrubber
	Superphosphoric	Fluorides	Scrubber
Sulfuric	Contact	SO_2, acid mist	Scrubbers with mist eliminators, ESPs

e. Chlorine Most of the chlorine manufactured is produced by two electrolytic methods, the diaphragm cell and the mercury cell processes. Both processes emit chlorine to the atmosphere from various streams and from handling and loading facilities. If the gas streams contain over 10% chlorine, the chlorine is recovered by absorption. If the chlorine concentration is less, the usual practice is to scrub the vent gases with an alkaline solution.

TABLE 30-3

Air Pollution Emissions and Controls: Inorganic Base Manufacture

Base	Manufacturing process	Air pollutant emissions	Control methods in use
Calcium oxide (lime)	Rotary kilns, vertical and shaft kilns, fluidized bed furnaces	Particulate matter	Cyclones plus secondary collectors (baghouse, ESP, wet scrubbers, granular bed filters, wet cyclones)
Sodium carbonate (soda ash)	Solvay (ammonia–soda)	Particulate matter	Wet scrubbers
Sodium hydroxide, caustic soda	Electrolytic	Chlorine Mercury	Alkaline scrubbers Chemical scrubbing and adsorbers

TABLE 30-4

Air Pollution Emissions and Controls: Phosphate Fertilizer Plants

Process	Air pollutant emissions	Control methods in use
Normal superphosphate	SiF_4, HF	Venturi or cyclonic scrubber
	Particulate matter	Wet scrubber of baghouse
Diammonium phosphate	Gaseous F, NH_3	Venturi or cyclonic scrubber with 30% phosphoric acid
	Particulate matter	Cyclone followed by scrubber
Triple superphosphate, run of pile	SiF_4, HF	Venturi or cyclonic scrubber
Triple superphosphate, granular	SiF_4, HF, particulate matter	Venturi or packed scrubber

Mercury is emitted from the mercury cell process from ventilation systems and by-product streams. Control techniques include (1) condensation, (2) mist elimination, (3) chemical scrubbing, (4) activated carbon adsorption, and (5) molecular sieve absorption. Several mercury cell (chloralkali) plants in Japan have been converted to diaphragm cells to eliminate the poisonous levels of methyl mercury found in fish (9).

f. Bromine All of the bromine produced in the United States is extracted from naturally occurring brines by steam extraction. The major air pollution concern is H_2S from the stripper if H_2S is present in the brine. The H_2S can either be oxidized to SO_2 in a flare or sent to a sulfur recovery plant.

2. *Organic (Petrochemical) Processes*

Most petrochemical processes are essentially enclosed and normally vent only a small amount of fugitive emissions. However, the petrochemical processes that use air-oxidation-type reactions normally vent large, continuous amounts of gaseous emissions to the atmosphere (10). Six major petrochemical processes employ reactions using air oxidation. Table 30-5 lists the atmospheric emissions from these processes along with applicable control measures.

B. Metallurgical Industries

The metallurgical industry offers some of the most challenging air pollution control problems encountered. The gas volumes are huge, and the gas may be at a high temperature. These large, hot gas volumes may convey large quantities of dust or metal oxide fumes, some of which may be highly toxic. Also, gaseous pollutants such as SO_2 or CO may be very highly concentrated in the carrying stream. The process emissions to the atmo-

TABLE 30-5

Air Pollution Emissions and Controls: Petrochemical Processes

Petrochemical process	Air pollutant emissions	Control methods in use
Ethylene oxide (most emissions from purge vents)	Ethane, ethylene, ethylene oxide	Catalytic afterburner
Formaldehyde (most emissions from exit gas stream of scrubber)	Formaldehyde, methanol, carbon monoxide, dimethyl ether	Wet scrubber for formaldehyde and methanol only; afterburner for organic vent gases
Phthalic anhydride (most emissions from off-gas from switch condensers)	Organic acids and anhydrides, sulfur dioxide, carbon monoxide, particulate matter	Venturi scrubber followed by cyclone separator and packed countercurrent scrubber
Acrylonitrile (most emissions from exit gas stream from product absorber)	Carbon monoxide, propylene, propane, hydrogen cyanide, acrylonitrile, acetonitrile	Thermal incinerators (gas-fired afterburners or catalytic afterburners)
	NO_x from by-product incinerator	None
Carbon black (most emissions from exit gas stream from baghouse, some fugitive particulate)	Hydrogen, carbon monoxide, hydrogen sulfide, sulfur dioxide, methane, acetylene	Waste heat boiler or flare (no control for SO_2)
	Particulate matter (carbon black)	Baghouse
Ethylene dichloride (most emissions from exit gas stream of solvent scrubber)	Carbon monoxide, methane, ethylene, ethane, ethylene dichloride, aromatic solvent	None at present, but could use a waste heat boiler or afterburner, followed by a caustic scrubber for hydrochloric acid generated by combustion

sphere may have harmful effects on visibility, vegetation, animals, and inert materials, as well as being detrimental to human health. It is no wonder that the metallurgical industries have spent huge sums to control emissions.

1. Nonferrous Metallurgical Operations

Nonferrous metallurgy is as varied as the ores and finished products. Almost every thermal, chemical, and physical process known to engineers is in use. The general classification scheme that follows gives an understanding of the emissions and control systems: aluminum (primary and secondary), beryllium, copper (primary and secondary), lead (primary and secondary), mercury, zinc, alloys of nonferrous metals (primary and secondary), and other nonferrous metals.

a. Aluminum (Primary) The emissions from primary aluminum reduction plants may come from the primary control system, which vents the electrolytic cells through control devices, or from the secondary system, which controls the emissions from the buildings housing the cells.

Hydrogen fluoride accounts for about 90% of the gaseous fluoride emitted from the electrolytic cell. Other gaseous emissions are SO_2, CO_2, CO, NO_2, H_2S, COS, CS_2, SF_6, and various gaseous fluorocarbons. Particulate fluoride is emitted directly from the process and is also formed from condensation and solidification of the gaseous fluorides.

The fluoride removal efficiency of the control equipment at primary aluminum reduction plants is shown in Table 30-6. The removal efficiency for total fluorides is a matter of great concern.

Emission rates using BACT on the three electrolytic cell types are shown in Table 30-7.

b. Beryllium Beryllium is extracted from the ore in the form of beryllium hydroxide, which is then converted to the desired product, metal, oxide, or alloy (12). Some of these products are extremely toxic. Table 30-8 lists the emissions from the various beryllium production steps, along with the control measures commonly used.

c. Copper Most copper is removed from low-grade sulfide ores using pryrometallurgical processes. The copper is first concentrated and then dewatered and filtered. The copper smelting process consists of roasters,

TABLE 30-6

Fluoride Removal Efficiencies: Selected Aluminum Industry Primary and Secondary Control Systems

Control system	Total fluoride removal efficiency (%)
Coated filter baghouse	94
Fluid bed dry scrubber	99
Injected alumina baghouse	98
Wet scrubber + wet ESP	99+
Dry ESP + wet scrubber	95
Floating bed	95
Spray screen (secondary)	62–77
Venturi scrubber	98
Bubbler scrubber + wet ESP	99

Source: Iverson (11).

TABLE 30-7

Emissions for Three Electrolytic Aluminum Reduction Cell Types Using Best Available Control Technology (BACT)

Cell type	Emissions with primary and secondary control	
	Fluorides (gm/kg Al)	Particulate (gm/kg Al)
Prebaked	0.8	3.0
Vertical stud soderberg	1.4	4.6
Horizontal stud soderberg	1.8	6.2

reverberatory furnaces, and converters. Some copper is further refined electrolytically to eliminate impurities.

Some fugitive particulate emissions occur around copper mines, concentrating, and smelting facilities, but the greatest concern is with emissions from the ore preparation, smelting, and refining processes. Table 30-9 gives the emissions of SO_2 from the smelters.

Tall stacks for SO_2 dispersion have been used in the past but are no longer acceptable as the sole means of SO_2 control. Acid plants have been installed at many smelters to convert the SO_2 to sulfuric acid, even though it may not be desirable from an economic standpoint.

The emission of volatile trace elements from roasting, smelting, and converting processes is undesirable from both an air pollution and an economic standpoint. Gravity collectors, cyclones, and ESPs are used to attain collection efficiencies of up to 99.7% for dust and fumes.

Treatment of slimes for economic recovery of silver, gold, selenium, tellurium, and other trace elements requires fusion and oxidation in a furnace. The furnace gases are exhausted through a wet scrubber followed by an ESP to recover the metals.

d. Lead Smelting and Lead Storage Battery Manufacture Lead ores are crushed, ground, and concentrated in a manner similar to the processing of copper

TABLE 30-8

Air Pollution Emissions and Controls: Beryllium Processing

Process	Air pollutant emissions	Control methods in use
Extraction	Beryllium salts	Baghouses
	Acids	Wet collectors
	Beryllium oxides	Baghouses or high-efficiency particulate air (HEPA) filters
	Beryllium dust, fume, mist	Cyclones and baghouses
Machining	Beryllium dust	Baghouses and HEPA filters
	Beryllium oxide dust	Baghouses and HEPA filters

TABLE 30-9

Sulfur Dioxide Emission Rates from Primary Copper Smelters

Process	Emission (gm SO_2 per kg of copper)
Roasting	325–675
Reverberatory furnaces	150–475
Converters	975–1075
Reverberatory furnaces[a]	275–800
Converters[a]	850–1800

[a] Without roasting
Source: Nelson *et al.* (13).

ores. Fugitive emissions from these processes include dusts, fumes, and trace metals. Smelting is usually accomplished in a blast furnace after the concentrated ore is sintered. Sintering removes up to 85% of the sulfur.

Gases from the sintering process contain SO_2, dust, and metal oxide fumes. The blast furnace gases contain similar particulates plus SO_2 and CO. Table 30-10 indicates the expected SO_2 emissions.

e. Mercury Mercury is produced commercially by processing mercury sulfide (cinnabar). The mercury sulfide is thermally decomposed in a retort or roaster to produce elemental mercury and sulfur dioxide. The off-gases are cleaned by being passed through cyclonic separators, and the mercury is then condensed. The SO_2 is removed by scrubbers before the exhaust gases are released to the atmosphere. Any mercury vapors that escape are collected in refrigerated units and, usually, recovered with a baghouse or ESP. Other systems use absorption with sodium hypochlorite and sodium chloride or adsorption on activated carbon or proprietary adsorbents. The U.S. Environmental Protection Agency has placed a limit of 2300 gm of mercury emission per 24 hr on any mercury smelter or process.

TABLE 30-10

Sulfur Dioxide Emission Rates from Primary Lead Smelters

Process	Emission (gm SO_2 per kg of lead)
Sintering	575–1075
Blast furnace	2.5–5
Dross reverberatory furnace	2.5–5

Source: Nelson *et al.* (13).

f. Zinc Zinc is processed very similarly to copper and lead. The zinc is bound in the ore as ZnS, sphalerite. Zinc is also obtained as an impurity from lead smelting, in which it is recovered from the blast furnace slag.

Dusts, fumes, and SO_2 are evolved during sintering, retorting, and roasting, as shown in Table 30-11.

Particulate emissions from zinc processing are collected in baghouses or ESPs. SO_2 in high concentrations is passed directly to an acid plant for production of sulfuric acid by the contact process. Low-concentration SO_2 streams are scrubbed with an aqueous ammonia solution. The resulting ammonium sulfate is processed to the crystalline form and marketed as fertilizer.

g. Other Nonferrous Metals and Alloys Nonferrous metals of lesser significance include arsenic, cadmium, and refractory metals such as zirconium and titanium. Air pollution emissions from the manufacture of these metals do not constitute a major problem, although severe local problems may exist near the facility. Control of emissions is usually accomplished by a single device at the exit of the process. In many cases, the material removed by the control device has some value, either as the primary product or as a by-product. Table 30-12 shows some of the atmospheric emissions and control systems used on these metallurgical processes.

Alloys of nonferrous metals, primarily the brasses (copper and zinc) and the bronzes (copper and tin), can cause an air pollution problem during melting and casting. The type and degree of emissions depend on the furnace and the alloy. Control systems consist of hoods over the furnaces and pouring stations to collect the hot gases, ducts and fans, and baghouses or ESPs.

h. Secondary Metals Secondary metals are those recovered from scrap. Copper (including brass and bronze), lead, zinc, and aluminum are the

TABLE 30-11

Air Pollution Emissions from Primary Zinc Processing

	Emissions to the atmosphere			
Process	Dust (gm/dscm)	Percent of particles less than 10 μm	SO_2 (%)	SO_2 (gm/kg zinc)
Sintering	10	100	4.5–7.0	—
Horizontal retort	0.1–0.3	100	—	—
Roasting	—	—	—	825–1200

Source: Nelson *et al.* (13).

TABLE 30-12

Air Pollution Emissions from Miscellaneous Nonferrous Metallurgical Processes

Metal	Type of process	Air pollutant emissions	Control methods in use
Arsenic	By-product of copper and lead smelters	Arsenic trioxide	Baghouses or ESPs
Cadmium	By-product of zinc and lead smelters	Cadmium, cadmium oxide	Baghouses
Refractory metals Zirconium Hafnium Titanium	Kroll process, chlorination, and magnesium reduction	Chlorine, chlorides, $SiCl_4$	Wet scrubbers
Columbium Tantalum Vanadium Tungsten Molybdenum	Separation process	Ammonia	Conversion to ammonium sulfate fertilizer

principal nonferrous secondary metals. Emissions from the recovery processes are similar to those from the primary metallurgical operations except that little or no SO_2 or fluorides are evolved. Baghouses and ESPs are the commonly used control devices.

2. *Ferrous Metallurgical Operations*

Iron and steel industries are generally grouped as steel mills, which produce steel sheets or shapes, and foundries, which produce iron or steel castings. Some steel mills use electric arc furnaces with scrap steel as the raw material, but the majority are large, integrated mills with facilities for coke making, sintering, iron making, steelmaking, rolling, processing, and finishing (14).

a. Coke Making Coke is produced from blended coals by either the nonrecovery beehive process or the by-product process. The by-product process produces the majority of the coke. Air pollutants from the coke-making process vary according to the point of release from the process and the time the process has been in operation. Table 30-13 shows the emissions, and their control, from the different stages of the by-product process.

b. Sintering Sintering consists of mixing moist iron ore fines with a solid fuel, usually coke, and then firing the mixture to eliminate undesirable elements and produce a product of relatively uniform size, physically and chemically stable, for charging the blast furnace. Air pollutants are emitted at different points in the process, as indicated in Table 30-14.

TABLE 30-13

Air Pollution Emissions and Controls: By-Product Coke Making

Process	Air pollutant emissions	Control methods in use
Coal and coke handling	Fugitive particulate matter	Enclose transfer points and duct to baghouses; pave and water roadways
Coke oven charging	Hydrocarbons, carbon, coal dust	Aspiration systems to draw pollutants into oven, venturi scrubbers
Coke oven discharging (pushing)	Hydrocarbons, coke dust	Hoods to fans and venturi scrubbers, low-energy scrubbers followed by ESPs (may use water spray at oven outlet)
Coke quenching	Particulate matter	Baffles and water sprays
Leaking oven doors	Hydrocarbons, carbon	Door seals with proper operation and maintenance
By-product processing	Hydrogen sulfide	Conversion to elemental sulfur or sulfuric acid by liquid absorption, wet oxidation to elemental sulfur, combustion to SO_2

c. Iron Making Iron making is the term used to describe how iron is produced in large, refractory-lined structures called *blast furnaces*. The iron ore, limestone, and coke are charged, heated, and then reacted to form a reducing gas, which reduces the iron oxide to metallic iron. The iron is tapped from the furnace along with the slag which contains the impurities. A modern alternative to the blast furnace is continuous casting of iron instead of intermittent tapping. The blast furnace gas is exhausted from

TABLE 30-14

Air Pollution Emissions and Controls: Sintering

Process	Air pollutant emissions	Control methods in use
Waste gases (main stack gases)	Particulates, CO, SO_2, chlorides, fluorides, ammonia, hydrocarbons, arsenic	Gravity separators to cyclones; then ESPs or wet scrubbers with pH adjustment
Sinter machine discharge	Particulate matter	Multiple cyclones, baghouse, or low-energy wet scrubber
Materials handling	Fugitive particulate matter	Hoods over release points to baghouse or multiple cyclones
Sinter cooler	Particulate matter (trace)	Baghouse (if required)

Source: Steiner (14).

the top of the furnace, cooled, and cleaned of dust before it is used to fire the regenerative stoves for heating the blast furnace. The atmospheric emissions from the iron making process are listed in Table 30-15.

d. Steelmaking The open hearth steelmaking process produced 80–90% of the steel in the United States until the 1960s, when the basic oxygen process came into wide use. By 1990 less than 5% of U.S. steel was produced by the open hearth process. Particulate emissions are highest during oxygen lancing with a hot metal charge in the furnace. Particulate matter loadings are reported to be in the range of 6–11 kg per metric ton of steel. Most of the particulate matter is iron or iron oxide. Control of the open hearth particulate matter emissions is accomplished by ESPs or high-energy scrubbers. Only small quantities of SO_2 are emitted, but if venturi scrubbers are used for particulate matter control, they will also reduce the SO_2 emissions. However, severe corrosion problems have been reported for wet scrubbers on open hearth furnaces (14).

Basic oxygen furnaces (BOFs) have largely replaced open hearth furnaces for steelmaking. A water-cooled oxygen lance is used to blow high-purity oxygen into the molten metal bath. This causes violent agitation and rapid oxidation of the carbon, impurities, and some of the iron. The reaction is exothermic, and an entire heat cycle requires only 30–50 min. The atmospheric emissions from the BOF process are listed in Table 30-16.

The electric arc furnace process accounted for about 25% of the 1982 U.S. steelmaking capacity (14). Most of the raw material used for the process is steel scrap. Pollutants generated by the electric furnace process are primarily particulate matter and CO. The furnaces are hooded, and the gas stream containing the particulate matter is collected, cooled, and passed to a baghouse for cleaning. Venturi scrubbers and ESPs are used as control devices at some mills. Charging and tapping emissions are also collected by hoods and ducted to the particulate matter control device.

After the steel is tapped from the furnace, it is poured into ingots or continuously cast into slabs or billets. Many metallurgical processes are

TABLE 30-15

Air Pollution Emissions and Controls: Iron Making

Process	Air pollutant emissions	Control methods in use
Blast furnace exhaust gases	Particulate matter	Multiple cyclone plus wet scrubber or wet ESP, two-stage wet scrubber
Slag handling	H_2S, SO_2 (trace)	None
Casting	Particulate matter	Baghouse

Source: Steiner (14).

TABLE 30-16

Air Pollution Emissions and Controls: Basic Oxygen Furnace

Process	Air pollutant emissions	Control methods in use
Hot metal transfer	Graphite and iron oxide particulate matter	Multiple cyclones plus baghouses
Charging and tapping	Particulate matter	Baghouse or venturi scrubber
Furnace waste gases	Particulate matter (7–30 kg per metric ton of steel)	ESP or venturi scrubber
	Carbon monoxide	Flare

required between the furnace and the finished product. Reheat furnaces cause no air pollution problems. Scarfing processes create a fine iron oxide fume on the steel surface and also release the same fume, which must be controlled by wet scrubbers or ESPs before it reaches the atmosphere. Pickling may result in the release of acid mists. Scrubbing with special solutions may be required for control. Galvanizing is the process of applying a zinc coating and can result in release of zinc oxide emissions to the atmosphere. Control is accomplished through local collection hoods followed by ESPs, wet scrubbers, or baghouses.

e. Ferrous foundry operations Ferrous foundry operations produce castings of iron or steel. Many foundry air pollution problems are similar to those of steel mills but on a smaller scale. Potential emissions from foundry operations, along with the usual control methods, are shown in Table 30-17.

TABLE 30-17

Air Pollution Emissions and Controls: Iron and Steel Foundries

Process	Air pollutant emissions	Control methods in use
Cupolas	Particulate matter (5–22 kg/ton)	Baghouses, wet scrubbers, and ESPs
	Carbon monoxide	Afterburner
	SO_2 (25–250 ppm)	If necessary, wet scrubber
Sand conditioning, shakeout, molding	Particulate matter	Medium-energy wet scrubbers, baghouses
Core making	Hydrocarbons	Afterburners (thermal or catalytic)

Source: Steiner (14).

f. Ferroalloy production Ferroalloys are used to add various elements to iron or steel for specific purposes. Examples are chromium (in the form of ferrochrome) and manganese (in the form of ferromanganese) added to steel to improve its strength or hardness, or nickel and chromium added to steel to increase its corrosion resistance. In the electrolytic production of nickel, iron is not removed from nickel because nickel will be used for the production of stainless steel. The product is marketed as "ferronickel" rather than nickel. Ferrochrome, ferromanganese, and ferrosilicon are produced in high-temperature furnaces which emit copious quantities of metallic fume and particulate matter. Roasting and concentrating the ore prior to ferroalloy production produce particulate matter and oxide emissions; SO_2 and CO are released during reduction; and casting produces metal oxides and fumes.

IV. AGRICULTURE AND FOREST PRODUCTS INDUSTRIES

The agricultural and forest products industries are dependent on renewable resources for their existence. They are also acutely aware that air pollution can damage vegetation and, therefore, threaten their existence. Both industries have been exempt from many air pollution regulations in the past, but now they are finding these exemptions questioned and in some cases withdrawn (15).

A. Agriculture

The term *agriculture* refers to the operations involved in growing crops or raising animals. Dusts, smoke, gases, and odors are all emissions form various agricultural operations.

1. *Agronomy*

The preparation of soils for crops, planting, and tilling raises dust as a fugitive emission. Such operations are still exempt from air pollution regulations in most parts of the world. The application of fertilizers, pesticides, and herbicides is also exempt from air pollution regulations, but other regulations may cover the drift of these materials or runoff into surface waters. This is particularly true of the materials are hazardous or toxic.

2. *Open Burning*

A major source of particulate matter, carbon monoxide, and hydrocarbons is open burning of agricultural residue. Over 2.5 million metric tons of particulate matter per year are added to the atmosphere over the United States from burning rice, grass straw and stubble, wheat straw and stubble, weeds, prunings, and range bush. Figure 30-4 illustrates an open burn of grass straw and stubble following the harvest of the seed crop.

The major effect of such open burning is the nuisance caused by the smoke, but health effects are noticed by sensitive individuals downwind from the burn. Table 30-18 lists the pollutant emissions from grass field burning (15).

If the open burning of agricultural residue is permitted, it should be scheduled to minimize the effect on populated areas. This requires burning when the wind is blowing away from the population centers, not burning during inversion periods, burning dry residue to establish a strong convection column rather than a smoldering fire, and burning only a certain number of acres at a time, so that the atmosphere does not become overloaded.

3. *Orchard Heating*

The practice of smudging is still carried out in many areas to protect orchards from frost. Petroleum products are burned in pots, producing both heat and smoke. Since the heat is the desirable product, smokeless heaters with return ducts to reburn the smoke are required by most air pollution control agencies. Some control agencies have passed regulations limiting the smoke to 0.5 or 1.0 gm per minute per burner.

Fig. 30-4. Open burning of a field after a grass seed harvest.

TABLE 30-18

Air Pollution Emission Factors for Agricultural Field Burning

Pollutant	Emission (kg/metric ton)
Particulate matter	8.5
Carbon monoxide	50
Hydrocarbons (as CH_4)	10
Nitrogen oxides (as NO_2)	1

Source: Faith (15).

Replacement of orchard heaters by wind machines is the most desirable control measure. These large propellers force the warmer air aloft to the ground, where it mixes with the cold air, minimizing frost formation.

4. *Alfalfa Dehydration and Pelletizing*

Alfalfa dehydration is carried out in a direct-fired rotary dryer. The dried product is transported pneumatically to an air cooler and then to a collecting cyclone. The collected particles are ground or pelletized and then packaged for shipment. The major atmospheric emission from the process is particulate matter, which is controlled by baghouses. Odors may also be a problem, but they disperse rapidly and are no longer a problem at distances of over 1 km.

5. *Animal Production*

Feeding of domestic animals on a commercial basis results in large quantities of excreta, both liquid and solid. This produces obnoxious odors, which, in turn, produce complaints from citizens of the area. If the animals are concentrated in a feedlot, the odors may become so extreme that odor counteractants are necessary. However, if the feedlots are paved and regularly washed down, the odors may be kept to a satisfactory minimum with much less expense.

Manure is often recycled as a solid organic fertilizer or mixed with water and sprayed as a liquid fertilizer. If the manure is repeatedly used upwind of populated areas, complaints are sure to be filed with the air pollution control agency.

6. *Feed and Grain Milling and Handling*

All grain milling involves grinding and handling of dried grain. Air streams are used for transport of the raw material and the finished products. The result is atmospheric emissions of grain dust. Originally, the control method used was cyclones; today most systems use baghouses following

TABLE 30-19

Particulate Matter Emissions from Cotton Ginning

Process	Emissions (kg bale)[a]
Unloading fan	2.20
Cleaner	.45
Stick and burr machine	1.35
Miscellaneous	1.35
Total	5.35

[a] One bale weighs 277 kg.
Source: Ref. (16).

the cyclones. Caution must be exercised in all phases of baghouse construction, operation, cleaning, etc., as the grain dust is explosive and can cause fires accompanied by loss of property and even of lives. Particulate emissions from uncontrolled grain-processing plants range from 0.1 to 2.0 kg of particulate matter per metric ton of grain (16).

7. Cotton Processing

The processing of cotton, from the field to the cloth, releases both inorganic and organic particulate matter to the atmosphere. Also, adhering pesticide residues may be emitted at the cotton gin exhaust. Table 30-19 lists the emission factors for particulate matter from cotton ginning operations.

8. Meat and Meat Products

The control of odors from holding pens and yards is similar to that discussed in Section IV,A,5. Odors can arise during rendering, cooking, smoking, and processing. Since most of the emissions from the meat products industry are odorous organics, afterburners are used successfully as control devices. Some processors have tried to use wet scrubbers or ESPs, but the emissions are often sticky and can cause severe cleaning problems. Fish processing has similar problems and solutions.

9. Fruit and Vegetable Processing

The most severe environmental problem of fruit and vegetable processors is the potential for water pollution if the liquid wastes are not handled properly. Cooking can cause odors, which are usually controlled by using furnaces as afterburners.

One processing problem is presented by the roasting of coffee. This releases smoke, odor, and particulate matter. The particulate matter, primarily dusty and chaff, can be removed with a cyclone. The smoke and odors are usually consumed by passing the roaster exhaust gases through an afterburner. Heat recovery may be desirable if the afterburner is large enough to make it economical.

10. Miscellaneous Agricultural Industries

Other industries of interest are (1) the manufacturing of spices and flavorings, which may use activated carbon filters to remove odors from their exhaust stream; (2) the tanning industry, which uses afterburners or activated carbon for odor removal and wet scrubbers for dust removal; and (3) glue and rendering plants, which utilize sodium hypochlorite scrubbers or afterburners to control odorous emissions.

B. Forest Products

The forest products industry encompasses a broad spectrum of operations which range from the raising of trees, through cutting and removing the timber, to complete utilization of the wood residue (17).

1. Open Burning

The forest products industry (as well as governmental agencies such as the U.S. Forest Service) practices open, prescribed, burning of logging residue (slash) as a forest management tool and as an economical means of residue disposal (18). This burning is usually done when meteorological conditions and fuel variables, such as moisture content, can give as clean a burn as possible with a minimum effect on populated areas. On a worldwide basis, it has been estimated that approximately 90 million metric tons of particulate matter from wild and controlled forest and range fires enter the atmosphere each year. Table 30-20 lists the pollutant emissions from forest burning (16).

TABLE 30-20

Air Pollution Emission Factors for Forest Burning

Pollutant	Emission (kg/metric ton)
Particulate matter	8.5
Carbon monoxide	22
Hydrocarbons (as CH_4)	2
Nitrogen oxides (as NO_2)	1

Source: Ref. (16).

2. *Wood-Fired Power Boilers*

Wood-fired power boilers are generally found at the mills where wood products are manufactured. They are fired with waste materials from the process, such as "hogged wood," sander dust, sawdust, bark, or process trim. Little information is available on gaseous emissions from wood-fired boilers, but extensive tests of particulate matter emissions are reported (19). These emissions range from 0.057 to 1.626 gm per dry standard cubic meter, with an average of 0.343 reported for 135 tests. Collection devices for particulate matter from wood-fired boilers are shown in Table 30-21.

3. *Driers*

Driers are used in the forest products industry to lower the moisture content of the wood product being processed. Drying of dimension lumber gives it dimensional stability. This type of drying is done in steam kilns and is a batch process. No appreciable pollutants are released.

Veneer for the manufacture of plywood is dried on a continuous line in a veneer drier to assure that only dry veneer goes to the layup and gluing process. Glue will not bond if the veneer contains too much moisture. Emissions from the veneer driers are fine particulate and condensed organic material. The condensed organic material is of submicron size and appears as a blue haze coming from the stack. Control is accomplished by means of (1) a wet scrubber or (2) ducting the emissions to a wood-fired boiler, where they are burned. All of these systems must be carefully sized and operated in order to meet a 20% opacity regulation.

Wood particle and fiber driers are used to dry the raw material for particleboard and similar products (20). Just as with the veneer for plywood, the particles must be dried before being mixed with the resins and formed into board. Drying is accomplished in a gas-fired drier, a direct wood-fired drier, or steam coil driers. Many different types of driers are used in the industry. Emissions are fine particles and condensible hydrocarbons, which produce

TABLE 30-21

Particulate Matter Collection Devices for Wood-Fired Boilers

Pollutant control device	Efficiency (%)
Multiple cyclone	51
Wet scrubber	67
Dry scrubber	85–97 (depending on fuel)
Baghouse	99+

Source: Boubel (19).

a highly visible plume. Control is accomplished with multiple cyclones or wet scrubbers. Fires in the control equipment and ductwork are quite common and must be expected periodically.

4. *Kraft Process Pulp and Paper Plants*

The kraft process has become the dominant process for pulp production throughout the world, primarily because of the recovery of the pulping chemicals. A schematic diagram of the kraft pulping process, with the location of atmospheric emission sources, is shown in Fig. 6-11.

Control of air pollutant emissions in modern kraft mills is accomplished by (1) proper operation of the entire mill, (2) high-efficiency ESPs on the recovery furnace (up to 99.7% efficiency for particulate matter, which is recovered and returned to the process), (3) collection of noncondensible gases from several vent points (digesters, blow tanks, washers) and ducting them to the lime kiln, where they are completely burned, and (4) high-energy wet scrubbers on the lime kiln exhaust to remove the particulate matter and sulfurous gases. A more complete analysis of the kraft process, including the emissions and their control, may be found in Reference 17.

V. OTHER INDUSTRIAL PROCESSES

Many industries operated throughout the world do not fall into the previous categories. Some of these are universal, such as asphalt batching plants, whereas others are regional, such as bagasse-fired boilers. Each has its own emission and control problems and requires knowledgeable analysis and engineering. Some of the more widely used processes are examined in this section.

A. Mineral Products

Conversion of minerals to useful products is a major worldwide industry. Mining or quarrying of minerals can produce fugitive emissions, which may be controlled by paving work and traffic areas, wetting the materials being removed or handled, or using collection and exhaust systems at the site where the particulate matter is being generated. The usual air pollution control device is a multiple cyclone or a baghouse at the system exit. The same control techniques can be applied at other points in the process where the minerals are transported, stored, crushed and ground, concentrated, dried, and mixed (21).

1. *Asphaltic Concrete Plants*

Two types of asphaltic concrete plants are in common use, batch mix plants and continuous mix plants. Figure 30-5 shows a batch mix asphalt

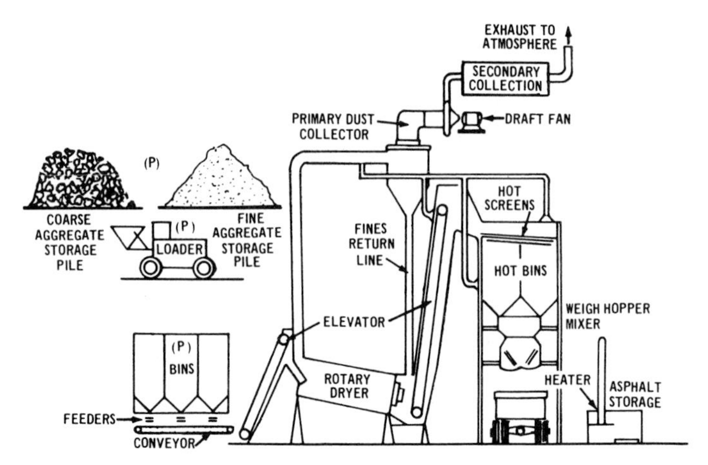

Fig. 30-5. Batch mix asphalt plant; P, denotes fugitive particulate matter emissions. Source: Ref. (16).

plant. Fugitive emissions occur at the handling areas and at the bin loading facility. The emissions of greatest concern, however, occur at the rotary drier, hot aggregate elevator, and hot aggregate handling systems. Each has the potential for releasing large quantities of uncontrolled particulate matter. Table 30-22 illustrates the large range of emissions from uncontrolled and controlled asphaltic concrete plants.

TABLE 30-22

Particulate Matter Emission Factors for Asphaltic Concrete Plants

Type of control	Emissions (kg/metric ton)
Uncontrolled[a]	22.5
Cyclone precleaner	7.5
High-efficiency cyclone	0.85
Spray tower scrubber	0.20
Multiple centrifugal scrubber	0.15
Baffle spray tower scrubber	0.15
Orifice-type scrubber	0.02
Baghouse[b]	0.05

[a] Almost all plants have at least a precleaner following the rotary drier. The fines collected are returned and are an important part of the mix.
[b] Emissions from a properly designated, installed, operated, and maintained baghouse collector can be as low as 0.0025–0.010 kg per metric ton.
Source: Ref. (16).

2. *Cement Plants*

Portland cement manufacture accounts for about 98% of the cement production in the United States. The raw materials are crushed, processed, proportioned, ground, and blended before going to the final process, which may be either wet or dry. In the dry process, the moisture content of the raw material is reduced to less than 1% before the blending process occurs. The dry material is pulverized and fed to the rotary kiln. Further drying, decarbonating, and calcining take place as the material passes through the rotary kiln. The material leaves the kiln as clinker, which is cooled, ground, packaged, and shipped.

For the wet process, a slurry is made by adding water during the initial grinding. The homogeneous wet mixture is fed to the kiln as a wet slurry (30–40% water) or as a wet filtrate (20% water). The burning, cooling, grinding, packaging, and shipping are the same as for the dry process.

Particulate matter emissions are the primary concern with cement manufacture. Fugitive emissions and uncontrolled kiln emissions are shown in Table 30-23.

Control of particulate matter emissions from the kilns, dryers, grinders, etc. is by means of standard devices and systems: (1) multiple cyclones (80% efficiency), (2) ESPs (95% + efficiency), (3) multiple cyclones followed by ESPs (97.5% efficiency), and (4) baghouses (99.8% efficiency).

3. *Glass Manufacturing Plants*

Soda-lime glass accounts for about 90% of the U.S. production. It is produced in large, direct-fired, continuous-melting furnaces in which the

TABLE 30-23

Air Pollution Emission Factors for Portland Cement Manufacturing without Controls

	Emissions (kg/metric ton)			
	Dry process		Wet process	
Pollutant	Kilns	Dryers, grinders, etc.	Kilns	Dryers, grinders, etc.
Particulate matter	122.0	48.0	114.0	16.0
Sulfur dioxide[a]	5.1			
Mineral source	Neg	—	5.1	—
Gas combustion	$2.1 \times S^b$	—	Neg	—
Oil combustion	$3.4 \times S$	—	$2.1 \times S^b$	—
Coal combustion		—	$3.4 \times S$	—
Nitrogen oxides	1.3	—	1.3	—

[a] If a baghouse is used as control device, reduce SO_2 by 50% because of reactions with an alkaline filter cake.

[b] S is the percent of sulfur in the fuel.

Source: Ref. (16).

blended raw materials are melted at 1480°C to form glass. The emissions from soda-lime glass melting are 1.0 kg of particulate matter per metric ton of glass and 2 × (fluoride percentage) for the fluoride emissions in kilograms per metric ton. For effective control of the emissions, baghouses are used.

Fiberglass is manufactured primarily from borosilicate glass by drawing the molten glass into fibers. Two fiberglass products are produced, textile and glass wool. The emissions from the two processes are shown in Table 30-24 (16). Control is achieved through proper design and operation of the manufacturing operations rather than by add-on devices.

B. Petroleum Refining and Storage

A modern petroleum refinery is a complex system of chemical and physical operations. The crude oil is first separated by distillation into fractions such as gasoline, kerosene, and fuel oil. Some of the distillate fractions are converted to more valuable products by cracking, polymerization, or reforming. The products are treated to remove undesirable components, such as sulfur, and then blended to meet the final product specifications. A detailed analysis of the entire petroleum production process, including emissions and controls, is obviously well beyond the scope of this text.

TABLE 30-24

Air Pollution Emission Factors for Fiber Glass Manufacturing without Controls

	Emissions (kg/metric ton)				
Type of process	Particulate matter	Sulfur oxides as SO_2	Carbon monoxide	Nitrogen oxides as NO_2	Fluorides
Textile products					
Glass furnace					
Regenerative	8.2	14.8	0.6	4.6	1.9
Recuperative	13.9	1.4	0.5	14.6	6.3
Forming	0.8	—	—	—	—
Curing ovens	0.6	—	0.8	1.3	—
Wool products					
Glass furnace					
Regenerative	10.8	5.0	0.13	2.5	0.06
Recuperative	14.2	4.8	0.13	0.9	0.06
Electric	0.3	0.02	0.03	0.14	0.01
Forming	28.8	—	—	—	—
Curing ovens[a]	1.8	—	0.9	0.6	—
Cooling[a]	0.7	—	0.1	0.1	—

[a] In addition, 0.05 kg per metric ton for phenol and 1.7 kg per metric ton for aldehyde during curing and cooling.

Source: Ref. (16).

The reader is referred to References 16, 22, and 23. Reference 16 presents an extensive tabulation of the emission sources for all processes involved in petroleum refining and production, some of which are summarized in Table 30-25.

Control of atmospheric emissions from petroleum refining can be accomplished by process change, installation of control equipment, and improved housekeeping and maintenance. In many cases, recovery of the pollutants will result in economic benefits. Table 30-26 lists some of the control measures that can be used at petroleum refineries.

C. Sewage Treatment Plants

The concern with atmospheric emissions from sewage treatment plants involves gases and odors from the plant itself, particulate matter and gaseous emissions from the sludge incinerator if one is used, and all three pollutants (gases, odors, and particulate matter) if sludge disposal is conducted at the site. The gases and odors are combustible, so afterburners or flares are used. Some plants use the sewage gas to fire small stationary boilers or fuel gas-diesel engines for plant energy. Particulate matter from sludge incinerators is usually scrubbed with treated water from the plant, and the effluent is returned to the incoming plant stream. If the odors are too persistent, masking agents are sometimes specified to lessen the objections of the public.

TABLE 30-25

Sources of Emissions from Oil Refining

Type of emission	Source
Hydrocarbons	Air blowing, barometric condensers, blind changing, blowdown systems, boilers, catalyst regenerators, compressors, cooling towers, decoking operations, flares, heaters, incinerators, loading facilities, processing vessels, pumps, sampling operations, tanks, turnaround operations, vacuum jets, waste effluent handling equipment
Sulfur oxides	Boilers, catalyst regenerators, decoking operations, flares, heaters, incinerators, treaters, acid sludge disposal
Carbon monoxide	Catalyst regenerators, compressor engines, coking operations, incinerators
Nitrogen oxides	Boilers, catalyst regenerators, compressor engines, flares
Particulate matter	Boilers, catalyst regenerators, coking operations, heaters, incinerators
Odors	Air blowing, barometric condensers, drains, process vessels, steam blowing, tanks, treaters, waste effluent handling systems
Aldehydes	Catalyst regenerators, compressor engines
Ammonia	Catalyst regenerators

Source: Ref. (23).

D. Coal Preparation Plants

Coal preparation plants are used to reduce noncombustibles and other undesirable materials in coal before it is burned.

E. Gas Turbines

Gas turbines are used as prime movers for pumps, electric generators, and large rotating machinery. Their main economic advantage is driving

TABLE 30-26

Control Measures for Air Pollutants from Petroleum Refining

Source	Control method
Storage vessels	Vapor recovery systems; floating roof tanks; pressure tanks; vapor balance; painting tanks white
Catalyst regenerators	Cyclones–precipitator–CO boiler; cyclones–water scrubber; multiple cyclones
Accumulator vents	Vapor recovery; vapor incineration
Blowdown systems	Smokeless flares–gas recovery
Pumps and compressors	Mechanical seals; vapor recovery; sealing glands by oil pressure; maintenance
Vacuum jets	Vapor incineration
Equipment valves	Inspection and maintenance
Pressure relief valves	Vapor recovery; vapor incinceration; rupture disks; inspection and maintenance
Effluent waste disposal	Enclosing separators; covering sewer boxes and using liquid seals; liquid seals on drains
Bulk loading facilities	Vapor collection with recovery or incineration; submerged or bottom loading
Acid treating	Continuous-type agitators with mechanical mixing; replace with catalytic hydrogenation units; incinerate all vented cases; stop sludge burning
Acid sludge storage and shipping	Caustic scrubbing; incineration, vapor return system
Spent caustic handling	Incineration; scrubbing
Doctor treating	Steam strip spent doctor solution to hydrocarbon recovery before air regeneration; replace treating unit with other, less objectionable units (Merox)
Sour water treating	Use sour water oxidizers and gas incineration; conversion to ammonium sulfate
Mercaptan disposal	Conversion to disulfides; adding to catalytic cracking charge stock; incineration; using material in organic synthesis
Asphalt blowing	Incineration; water scrubbing (nonrecirculating type)
Shutdowns, turnarounds	Depressure and purge to vapor recovery

Source: Ref. (22).

high-horsepower, consistent loads. Many stationary gas turbines use the same core engine as their jet engine counterpart.

REFERENCES

1. Public Law 101-549, 101st Congress—"Clean Air Act Amendments of 1990," November 1990.
2. Kurtzweg, J. A., and Griffin, C. N., *J. Air Pollut. Control Assoc.* **31,** 1155–1162 (1981).
3. Cooper, J. A. (ed.), "Residential Solid Fuels: Environmental Impacts and Solutions." Oregon Graduate Center, Beaverton, OR, 1981.
4. Preston, G. T., and Miller, M. J., *Pollut. Eng.* **14**(4), 25–29 (1982).
5. Moskowitz, J., *Mech. Eng.* **104**(4), 68–71 (1981).
6. Ellison, W., and Lefton, S. A., *Power* **126**(4), 71–76 (1982).
7. Brunner, C. R., "Handbook of Incineration Systems." McGraw-Hill, New York, 1991.
8. Arbuckle, J. G., Frick, G. W., Miller, M. L., Sullivan, T. F. P., and Vanderver, T. A., "Environmental Law Handbook." Government Institutes, Inc., Washington, DC, 1979.
9. *Chem. Week* **113,** 31–32 (1973).
10. Cuffe, S. T., Walsh, R. T., and Evans, L. B., Chemical industries, *in* "Air Pollution" (A. C. Stern, ed.), 3rd ed., Vol. IV. Academic Press, New York, 1977.
11. Iverson, R. E., *J. Metals* **25**, 19–23 (1973).
12. "Control Techniques for Beryllium Air Pollutants," EPA Pub. No. AP-116. U.S. Environmental Protection Agency, Research Triangle Park, NC, 1973.
13. Nelson, K. W., Varner, M. O., and Smith, T. J., Nonferrous metallurgical operations *in* "Air Pollution" (A. C. Stern, ed.), 3rd ed., Vol. IV. Academic Press, New York, 1977.
14. Steiner, B. A., Ferrous metallurgical operations, *in* "Air Pollution" (A. C. Stern, ed.), 3rd ed., Vol. IV. Academic Press, New York, 1977.
15. Faith, W. L., Agriculture and agricultural-products processing, *in* "Air Pollution" (A. C. Stern, ed.), 3rd ed., Vol. IV. Academic Press, New York, 1977.
16. "Compilation of Air Pollutant Emission Factors," 2nd ed. (with supplements), EPA Pub. No. AP-42. U.S. Environmental Protection Agency, Research Triangle Park, NC, 1980.
17. Hendrickson, E. R., The forest products industry, *in* "Air Pollution" (A. C. Stern, ed.), 3rd ed., Vol. IV. Academic Press, New York, 1977.
18. Ward, D. E., Nelson, R. M., and Adams, D. F., Proceedings of the 70th Annual Meeting of the Air Pollution Control Association, Pittsburgh, 1977.
19. Boubel, R. W., "Control of Particulate Emissions from Wood-Fired Boilers," Stationary Source Enforcement Series, EPA 340/1-77-026. U.S. Environmental Protection Agency, Washington, DC, 1977.
20. Junge, D. C., and Boubel, R. W., "Analysis of Control Strategies and Compliance Schedules for Wood Particle and Fiber Dryers," EPA Contract Report No. 68-01-3150. PEDCO Environmental Specialists, Cincinnati, 1976.
21. Sussman, V. H., Mineral products industries, *in* "Air Pollution" (A. C. Stern, ed.), 3rd ed., Vol. IV. Academic Press, New York, 1977.
22. Buonicore, A. J., and Davis, W. T. (eds.), "Air Pollution Engineering Manual." Van Nostrand Reinhold, New York, 1992.

23. Elkins, H. F., Petroleum refining, *in* "Air Pollution" (A. C. Stern, ed.), 3rd ed., Vol. IV. Academic Press, New York, 1977.

SUGGESTED READING

Power **126**(4), 375–410 (1982).

Elliot, T. C., *Power* **118**, 5.1–5.24, 1974.

Gage, S. J., "Controlling Sulfur Oxides," Research Summary, EPA-600/8-80-029. U.S. Environmental Protection Agency, Washington, DC, 1980.

Hesketh, H. E., "Air Pollution Control, Traditional and Hazardous Pollutants." Technomic, Lancaster, PA, 1991.

Kokkinos, A., Cicnanowicz, J. E., Hall, R.E., and Jedman, C. B., Stationary combustion NO_x control: a summary of the 1991 symposium. *J. Air Waste Manage. Assoc.* **41**(9) (September 1991).

Lutz, S. J., McCoy, B. C., Mullingan, S. W., Christman, R. C., and Slimak, K. M., "Evaluation or Dry Sorbants and Fabric Filtration for FGD," EPA-600/7-79-005. U.S. Environmental Protection Agency, Research Triangle Park, NC, 1979.

Theodore, L., and Bunicore, A. J., "Air Pollution Control Equipment." Prentice-Hall, Englewood Cliffs, NJ, 1982

Vatauk, W. M., "Estimating Costs of Air Pollution Control." Lewis, Chelsea, MI, 1990.

QUESTIONS

1. Rank the control systems shown in Fig. 30-1 according to their relative capital construction and operation costs.
2. Justify the statement, "Tall stacks are no longer considered as an acceptable alternative for controlling emissions from electric power generating plants."
3. For the NSPS for incinerators, only particulate matter emissions are covered. Devise a standard which would also include POM, CO, and NO_x for large municipal incinerators.
4. What advantage is gained by oxidizing H_2S to SO_2? (Consider toxicity and odors.)
5. Why are total fluoride emissions from an aluminum smelter of more concern than gaseous or solid fluoride emissions?
6. What systems can be used to detect and prevent grain dust explosions?
7. Would you expect wood-fired boilers to emit more or less CO per metric ton of fuel than coal-fired boilers? More or less NO_x? More or less SO_2?
8. What would be the ultimate disposal of dry material collected by an ESP at a cement plant kiln outlet? What would be the ultimate disposal of wet sludge from a scrubber on a cement plant kiln outlet?
9. Petroleum plants may have a disposal problem with sulfur removed from their products. What are some potential uses for this sulfur?

31

Control of Mobile Sources

I. INTRODUCTION

Because mobile sources of air pollution are capable of moving from one local jurisdiction to another, they are usually regulated by the national government. In the United States, state or local agencies can have more restrictive standards, if they choose. Through 1990, only the state of California had established standards more restrictive than the U.S. federal standards, and these only for gasoline-powered automobiles.

II. GASOLINE-POWERED VEHICLES

Gasoline-powered motor vehicles outnumber all other mobile sources combined in the number of vehicles, the amount of energy consumed, and the mass of air pollutants emitted. It is not surprising that they have received the greatest share of attention regarding emission standards and air pollution control systems. Table 25-2 shows the U.S. federal emission control requirements for gasoline-powered passenger vehicles.

Crankcase emissions in the United States have been effectively controlled since 1963 by positive crankcase ventilation systems which take the gases from the crankcase, through a flow control valve, and into the intake manifold. The gases then enter the combustion chamber with the fuel–air mixture, where they are burned.

Figure 31-1 shows a cross section of a gasoline engine with the positive crankcase ventilation (PCV) system.

Evaporative emissions from the fuel tank and carburetor have been controlled on all 1971 and later model automobiles sold in the United States. This has been accomplished by either a vapor recovery system which uses the crankcase of the engine for the storage of the hydrocarbon vapors or an adsorption and regeneration system using a canister of activated carbon to trap the vapors and hold them until such time as a fresh air purge through the canister carries the vapors to the induction system for burning in the combustion chamber.

The exhaust emissions from gasoline-powered vehicles are the most difficult to control. These emissions are influenced by such factors as gasoline formulation, air–fuel ratio, ignition timing, compression ratio, engine speed and load, engine deposits, engine condition, coolant temperature, and combustion chamber configuration. Consideration of control methods must be based on elimination or destruction of unburned hydrocarbons, carbon monoxide, and oxides of nitrogen. Methods used to control one pollutant may actually increase the emission of another requiring even more extensive controls.

Control of exhaust emissions for unburned hydrocarbons and carbon monoxide has followed three routes.

1. Fuel modification in terms of volatility, hydrocarbon types, or additive content. Some of the fuels currently being used are liquefied petroleum gas (LPG), liquefied natural gas (LNG), compressed natural gas (CNG), fuels with alcohol additives, and unleaded gasoline. The supply of some of these fuels is very limited. Other fuel problems involving storage, distribution, and power requirements have to be considered.

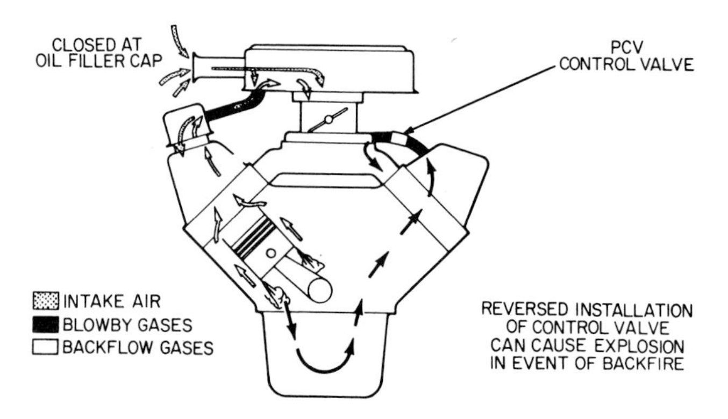

Fig. 31-1. Positive crankcase ventilation (PCV) system.

2. Minimization of pollutants from the combustion chamber. This approach consists of designing the engine with improved fuel–air distribution systems, ignition timing, fuel–air ratios, coolant and mixture temperatures, and engine speeds for minimum emissions. The majority of automobiles sold in the United States now use an electronic sensor/control system to adjust these variables for maximum engine performance with minimum pollutant emissions.

3. Further oxidation of the pollutants outside the combustion chamber. This oxidation may be either by normal combustion or by catalytic oxidation. These systems require the addition of air into the exhaust manifold at a point downstream from the exhaust valve. An air pump is employed to provide this air. Figure 31-2 illustrates an engine with an air pump and distribution manifold for the oxidation of CO and hydrocarbons (HC) outside the engine.

Beginning with the 1975 U.S. automobiles, catalytic converters were added to nearly all models to meet the more restrictive emission standards. Since the lead used in gasoline is a poison to the catalyst used in the converter, a scheduled introduction of unleaded gasoline was also required. The U.S. petroleum industry simultaneously introduced unleaded gasoline into the marketplace.

In order to lower emissions of oxides of nitrogen from gasoline engines, two general systems were developed. The first is exhaust gas recirculation (EGR), which mixes a portion of the exhaust gas with the incoming fuel–air charge, thus reducing temperatures within the combustion chamber. This recirculation is controlled by valving and associated plumbing and electronics, so that it occurs during periods of highest NO_x production, when some power reduction can be tolerated: a cruising condition at highway speed. Other alternatives are to use another catalytic converter, in series with the

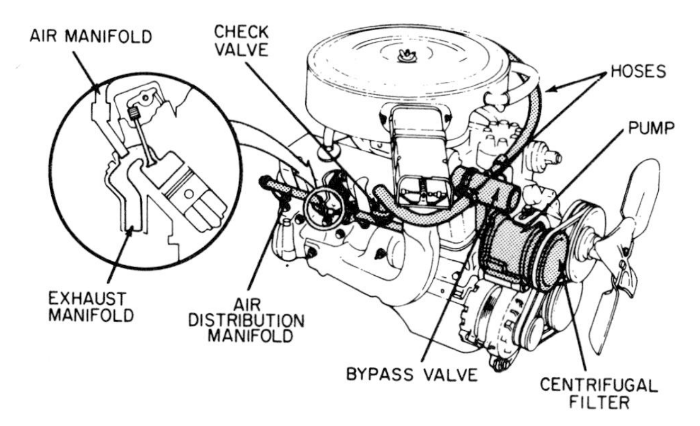

Fig. 31-2. Manifold air oxidation system.

HC/CO converter, which decomposes the oxides of nitrogen to oxygen and nitrogen before the gases are exhausted from the tailpipe.

III. DIESEL-POWERED VEHICLES

The diesel (compression ignition) cycle is regulated by fuel flow only, air flow remaining constant with engine speed. Because the diesel engine is normally operated well on the lean side of the stoichiometric mixture (40:1 or more), emission of unburned hydrocarbons and carbon monoxide is minimized. The actual emissions from a diesel engine are (1) oxides of nitrogen, as for spark ignition engines; (2) particulate matter, mainly unburned carbon, which at times can be excessive; (3) partially combusted organic compounds, many of which cause irritation to the eyes and upper respiratory system; and (4) oxides of sulfur from the use of sulfur-containing fuels. A smoking diesel engine indicates that more fuel is being injected into the cylinder than is being burned and that some of the fuel is being only partially burned, resulting in the emission of unburned carbon.

Control of diesel-powered vehicles is partially accomplished by fuel modification to obtain reduced sulfur content and cleaner burning and by proper tuning of the engine using restricted fuel settings to prevent overfueling.

Effective with the 1982 model year, particulate matter from diesel vehicles was regulated by the U.S. Environmental Protection Agency for the first time, at a level of 0.37 gm km^{-1}. Diesel vehicles were allowed to meet an NO_x level of 0.93 gm km^{-1} under an Environmental Protection Agency waiver. These standards were met by a combination of control systems, primarily exhaust gas recirculation and improvements in the combustion process. For the 1985 model year, the standards decreased to 0.12 gm of particulate matter per kilometer and 0.62 gm of NO_x per kilometer. This required the use of much more extensive control systems (1). The Clean Air Act Amendments of 1990 (2) have kept the emission standards at the 1985 model level with one exception: diesel-fueled heavy trucks shall be required to meet an NO_x standard of 4.0 gm per brake horsepower hour.

IV. GAS TURBINES AND JET ENGINES

The modified Brayton cycle is used for both gas turbines and jet engines. The turbine is designed to produce a usable torque at the output shaft, while the jet engine allows most of the hot gases to expand into the atmosphere, producing usable thrust. Emissions from both turbines and jets are similar, as are their control methods. The emissions are primarily unburned hydrocarbons, unburned carbon which results in the visible exhaust, and oxides of nitrogen. Control of the unburned hydrocarbons and the unburned carbon may be accomplished by redesigning the fuel spray nozzles and reducing cooling air to the combustion chambers to permit more complete combustion. U.S. airlines have converted their jet fleets to lower-emission engines using these control methods. NO_x emissions may be minimized by reduction of the maximum temperature in the primary zone of the combustors.

U.S. Environmental Protection Agency regulations for commercial, jet, and turbine-powered aircraft (3) are based on engine size (thrust) and pressure ratio (compressor outlet/compressor inlet) for the time in each mode of a standardized takeoff and landing cycle. Once the aircraft exceeds an altitude of 914 m, no regulations apply.

The gas turbine engine for automotive or truck use could be either a simple turbine, a regenerative turbine, a free turbine, or any combination. Figure 31-3 shows the basic types which have been successfully tried in automotive and truck use.

V. ALTERNATIVES TO EXISTING MOBILE SOURCES

The atmosphere of the world cannot continue to accept greater and greater amounts of emissions from mobile sources as our transportation systems expand. The present emissions from all transportation sources in the United States exceed 50 billion kg of carbon monoxide per year, 20 billion kg per year of unburned hydrocarbons, and 20 billion kg of oxides of nitrogen. If presently used power sources cannot be modified to bring their emissions to acceptable levels, we must develop alternative power sources or alternative transportation systems. All alternatives should be considered simultaneously to achieve the desired result, an acceptable transportation system with a minimum of air pollution.

One modified internal combustion engine which shows promise is the stratified-charge engine. This is a spark ignition engine using fuel injection in such a manner as to achieve selective stratification of the air/fuel ratio in the combustion chamber. The air/fuel ratio is correct for ignition at the spark plug, and the mixture is fuel lean in other portions of the combustion chamber. Only air enters the engine on the intake stroke, and the power output is controlled by the amount of fuel injected into the cylinder. Stratified-charge engines have been operated experimentally and used in some production vehicles (4). They show promise as relatively low-emission engines. The hydrocarbon emission levels from this engine are quire variable, the CO levels low, and the NO_x levels variable but generally high.

An external combustion engine that has been widely supported as a low-emission power source is the Rankine cycle steam engine. Many different types of expanders can be used to convert the energy in the working fluid

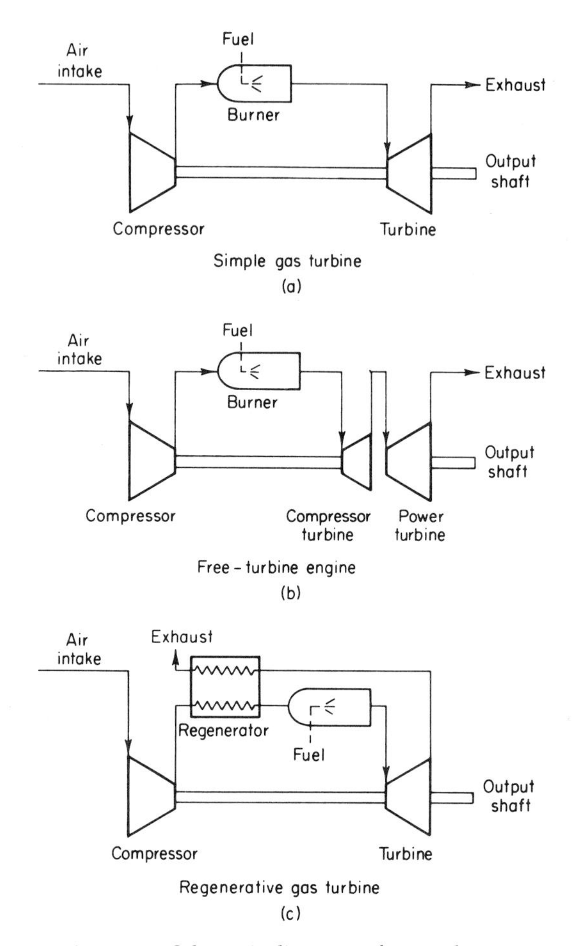

Fig. 31-3. Schematic diagrams of gas turbines.

into rotary motion at a drive shaft. Expanders that have been tried or proposed are reciprocating piston engines, turbines, helical expanders, and all possible combinations of these. The advantage of the steam engine is that the combustion is continuous and takes place in a combustor with no moving parts. The result is a much lower release of air pollutants, but emissions are still not completely zero. Present technology is capable of producing a satisfactory steam-driven car, truck, or bus, but costs, operating problems, warmup time, and weight and size must be considered in the total evaluation of the system. A simple Rankine cycle steam system is shown diagrammatically in Fig. 31-4.

Electric drive systems have been tried as a means of achieving propulsion without harmful emissions. Currently, most battery-operated vehicles use

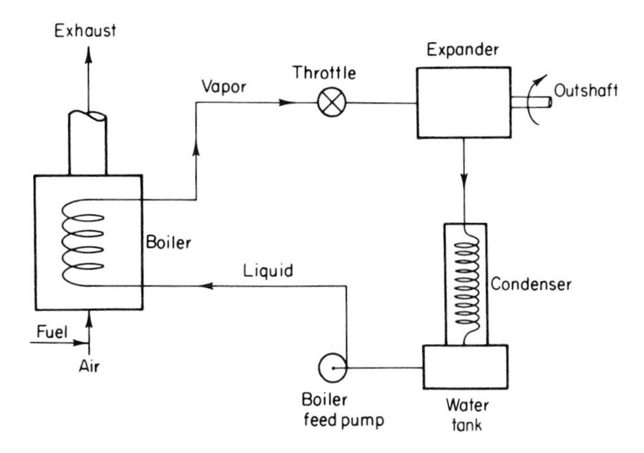

Fig. 31-4. Rankine cycle system.

lead–acid batteries, which give low power, limited range, and require frequent recharging. In power shortage areas this could be a severe additional load on the electrical system. Sulfuric acid, hydrogen, and oxygen emissions from millions of electric vehicles using lead–acid storage batteries for an energy source would be appreciable. Other types of batteries offer some promise, but their present cost makes them prohibitive for automotive use.

Hybrid systems consisting of two or more energy-conversion processes may offer the greatest promise for lower-emission automobiles. A constant speed and load internal combustion engine driving a generator with a small battery for load surges could be made to emit less hydrocarbons, CO, and NO_x than a standard automobile engine, but the cost would be much higher. Other hybrid systems which have been proposed are steam-electric and turbine-electric. The problem associated with hybrid systems is the cost of the two engines plus the cost of the added controls and system integration (5).

Fuel cells, which rely on electrochemical generation of electric power, could be used for nonpolluting sources of power for motor vehicles. Since fuel cells are not heat engines, they offer the potential for extremely low emissions with a higher thermal efficiency than internal combustion engines. Their lack of adoption by mobile systems has been due to their cost, large size, weight, lack of operational flexibility, and poor transient response. It has been stated that these problems could keep fuel cells from the mass-produced automobile market until after the year 2010 (5).

Probably the ultimate answer to the problem of emissions from millions of private automobiles is an alternative transportation system. It must be remembered, however, that even rail systems and bus systems do emit some air pollution. Rail systems are expensive and lack flexibility. A quick calculation of the number of passengers carried per minute past a single point on a freeway in private automobiles will illustrate the difficulties of a rail system in replacing the automobile. Buses offer much greater flexibility at lower cost than rail systems, but in order to operate efficiently and effectively, they would require separate roadway systems and loading stations apart from automobile traffic.

REFERENCES

1. "1982 General Motors Public Interest Report." General Motors Corp., Detroit, 1982.
2. Public Law No. 101-549, U.S. 101st Congress, "Clean Air Act Amendments of 1990," November 1990.
3. EPA Proposed Revisions to Gaseous Emissions Rules for Aircraft and Aircraft Engines. *Fed. Regist.* **43**, 12615 (1978).
4. Olson, D. R., The control of motor vehicle emissions, *in* "Air Pollution" (A. C. Stern, ed.), 3rd ed., Vol. IV. Academic Press, New York, 1977.
5. Harmon, R., Alternative vehicle-propulsion systems. *Mech. Eng.* **105**(4), 67–74. (March 1992).

SUGGESTED READING

Chang, T. Y., Alternative transportation fuels and air quality. *Environ. Sci. Technol.* **25,** 1190 (1991).

Dasch, J. M., Nitrous oxide emissions from vehicles. *J. Air Waste Manage. Assoc.* **42**(1) 32–38 (January 1992).

"Diesel Combustion and Emissions," Proceedings P-86. Society of Automotive Engineers, Warrendale, PA, 1980.

"Diesel Combustion and Emissions, Part III," SP-495. Society of Automotive Engineers, Warrendale, PA, 1981.

Faith, W. L., and Atkisson, A. A., Jr., "Air Pollution," 2nd ed. Wiley (Interscience), New York, 1972.

"Fuel and Combustion Effects on Particulate Emissions," SP-502. Society of Automotive Engineers, Warrendale, PA, 1981.

Wolf, G. T., and Frosch, R. A., Impact of alternative fuels on vehicle emissions of greenhouse gases. *J. Air Waste Manage. Assoc.* **41**(12) 1172–1176 (December 1991).

QUESTIONS

1. Would you expect to find the same chemical composition of the hydrocarbons from the exhaust of a gasoline-powered automobile as that of gasoline in the vehicle's tank? Why?
2. What would be the effect on emissions from a gasoline-powered vehicle if it was designed to be operated on leaded fuel and an unleaded fuel was used?
3. What would be the effect on emissions from a gasoline-powered vehicle if it was designed to be operated on unleaded fuel and a leaded fuel was used?

4. Why might you expect exhaust gas recirculation on a diesel engine to increase the particulate matter emissions?
5. Considering the wide range of sizes of automotive engines, do you feel that an emission standard in gm km^{-1} rather than $\mu g\ m^{-3}$ or ppm is equitable to all automobile manufacturers?
6. Discuss the emissions you would expect from a jet aircraft compared to those from a piston engine aircraft.
7. If a major freeway with four lanes of traffic in one direction passes four cars per second at 100 km hr^{-1} during the rush period and each car carries two people, how often would a commuter train of five cars carrying 100 passengers per car have to be operated to handle the same load? Assume the train would also operate at 100 km hr^{-1}.
8. An automobile traveling 50 km hr^{-1} emits 1% CO from the exhaust. If the exhaust rate is 80 $m^3\ min^{-1}$, what is the CO emission in grams per kilometer?
9. List the following in increasing amounts from the exhaust of an idling automobile: O_2, NO_x, SO_x, N_2, unburned hydrocarbons, CO_2, and CO.

32

Source Sampling and Monitoring

I. INTRODUCTION

Air pollutants released to the atmosphere may be characterized by qualitative descriptions or quantitative analysis. A plume may be characterized as brown, dense smoke, or 60% opacity. These are qualitative descriptions made by observing the effluent as it entered the atmosphere. Probably of more concern are the quantitative data regarding the effluent. How many parts per million? How many kilograms per hour? How many kilograms per year? To obtain these numbers, it becomes necessary to sample or monitor the effluent. Sampling and monitoring, therefore, are necessary for air pollution evaluation and control. In any situation concerning atmospheric emission of pollutants, source sampling or monitoring is necessary to obtain accurate data. Figure 32-1 shows a simple source test being conducted.

II. SOURCE SAMPLING

The purpose of source sampling is to obtain as accurate a sample as possible of the material entering the atmosphere at a minimum cost. This

Fig. 32-1. Source test.

statement needs to be examined in light of each source test conducted. The following issues should continually be considered: (1) Is the sampling and collecting of the material representative? Is this the material entering the atmosphere? Sampling at the base of a tall stack may be much easier than sampling at the top, but the fact that a pollutant exists in the breeching does not mean that it will eventually be emitted to the atmosphere. Molecules can also undergo both physical and chemical changes before leaving the stack. (2) Maximum accuracy in sampling is desirable. Is maximum accuracy attainable? Decisions regarding the total effluent will be based on what was found from a relatively small sample. Only if the sample accurately represents the total will the extrapolation to the entire effluent be valid. (3) Collecting a sample is a costly and time-consuming process. The economics of the situation must be considered and the costs minimized consistent with other objectives. It makes little sense to spend $5000 on an extensive stack testing analysis to decide whether to purchase a $10,000 scrubber of 95% efficiency or to try to get by with a $7000 scrubber of 90% efficiency.

The reasons for performing a source test differ. The test might be necessary for one or more of the following reasons: (1) To obtain data concerning the emissions for an emission inventory or to identify a predominant source in the area. An example of this would be determination of the hydrocarbon release from a new type of organic solvent used in a degreasing tank.

(2) To determine compliance with regulations. If authorization is obtained to construct an incinerator and the permit states that the maximum allowable particulate emission is 230 mg per standard cubic meter corrected to 12% CO_2, a source test must be made to determine compliance with the permit. (3) To gather information which will enable selection of appropriate control equipment. If a source test determines that the emission is 3000 mg of particulate per cubic meter and that it has a weight mean size of 5 μm, a control device must be chosen which will collect enough particulate to meet some required standard, such as 200 mg per cubic meter. (4) To determine the efficiency of control equipment installed to reduce emissions. If a manufacturer supplies a device guaranteed to be 95% efficient for removal of particulate with a weight mean size of 5 μm, the effluent stream must be sampled at the inlet and outlet of the device to determine if the guarantee has been met.

III. STATISTICS OF SAMPLING

When one takes a sample at the rate of 0.3 liter min^{-1} from a stack discharging 2000 $\text{m}^3\,\text{min}^{-1}$ to the atmosphere, the chances for error become quite large. If the sample is truly representative, it is said to be both accurate and unbiased. If it is not representative, it may be biased because of some consistent phenomenon (some of the hydrocarbons condense in the tubing ahead of the trap) or in error because of some uncontrolled variation (only 1.23 gm of sample was collected, and the analytical technique is accurate to ±0.5 gm) (1).

For all practical purposes, source testing can be considered as simple random sampling (2). The source may be considered to be composed of such a large population of samples that the population N is infinite. From this population, n units are selected in such a manner that each unit of the population has an equal chance of being chosen. For the sample, determine the sample mean, $\bar{y}$:

$$\bar{y} = \frac{y_1 + y_2 + \cdots + y_n}{n} \tag{32-1}$$

If the sample is unbiased, estimate the source mean, so that

$$\bar{Y} = \bar{y} \tag{32-2}$$

For example, take six samples of carbon monoxide from the exhaust of an idling automobile and obtain the CO percentages as shown in Table 32-1. The sample mean is

$$\bar{y} = \frac{1.8 + 1.6 + 1.8 + 1.9 + 1.7 + 1.8}{6} = 1.767$$

TABLE 32-1

Idling Internal Combustion Engine, CO Percentages

Test no.	CO, %
1	1.8
2	1.6
3	1.8
4	1.9
5	1.7
6	1.8

The source mean is assumed to be the same if the sample is unbiased, as seen by

$$\bar{Y} = \bar{y} = 1.767 \quad (32\text{-}3)$$

The variance of the sample and the population (source) may also be assumed equal if the sample is unbiased. The variance is S^2, defined as

$$S^2 = \frac{\sum_1^n (y_i - \bar{y})^2}{n - 1} \quad (32\text{-}4)$$

The variance of the source is usually calculated by the formula

$$s^2 = \frac{1}{n - 1}\left[\sum y_i^2 - \frac{(\sum y)^2}{n}\right] \quad (32\text{-}5)$$

For the preceding example,

$$\sum y_i^2 = 18.78, \qquad \sum y_i = 10.6, \qquad n = 6$$

$$s^2 = \frac{1}{6 - 1}\left[18.78 - \frac{(10.6)^2}{6}\right] = 0.01067$$

The standard deviation of the sample is defined as the square root of the variance. For the example,

$$s = (s^2)^{1/2} = (0.01067)^{1/2} = 0.103$$

The sample represents a population (source) which, if normally distributed, has a mean of 1.767% and a standard deviation of 0.103%. This can be illustrated as shown in Fig. 32-2.

The inference from the statistical calculations is that the true mean value of the carbon monoxide from the idling automobile has a 66.7% chance of being between 1.664% and 1.870%. The best single number for the carbon monoxide emission would be 1.767% (the mean value).

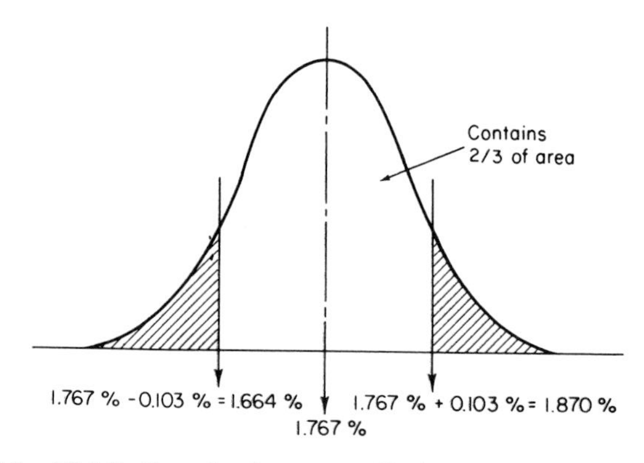

Fig. 32-2. Distribution of carbon monoxide from an automotive source.

Further statistical procedures can be applied to determine the confidence limits of the results. Generally, only the values for the mean and standard deviation would be reported. The reader is referred to any good statistical text to expand on the brief analysis presented here.

IV. THE SOURCE TEST

A. Test Preliminaries

The first thing that must be done for a successful source test is a complete review of all background material. The test request may come in either verbal or written form. If it is verbal, it should be put into writing for the permanent record. The request may contain much or little information, but it is important to verify that it is complete and understood. Questions to ask are (1) Why should the test be made? Is it to measure a specific pollutant such as SO_2, or is it to determine what is causing the odor problem in the new residential area? (2) What will the test results be used for? Will it be necessary to go to court, or are the results for general information only? This may make a difference regarding the test method selected. (3) What equipment or process is to be tested? What are its operational requirements? (4) What methods would be preferred by the analytical group? Are the analytical methods standard or unique? A literature search regarding the process and test should be conducted unless the test crew is thoroughly familiar with the source and all possible test methods. It is important to check the regulations regarding the process and specific test procedures as a part of the search (3).

When all the background material has been reviewed, it is time to inspect the source to be tested. The inspector should be accompanied by the plant manager or someone who knows the process in detail. It is also important that any technicians or mechanics be contacted at this time regarding necessary test holes, platforms, scaffolding, power requirements, etc. During this inspection, checks should be made for environmental conditions and space requirements at the sampling site. Testing in a noisy or dusty place at elevated temperatures is certainly uncomfortable and possibly hazardous. Rough estimates of several important factors should be made at this time. These estimates can be noted in writing during the inspection. A simple check sheet, such as the one shown in Fig. 32-3, should be a great aid.

The information obtained during the background search and from the source inspection will enable selection of the test procedure to be used. The choice will be based on the answers to several questions: (1) What are the legal requirements? For specific sources there may be only one acceptable method. (2) What range of accuracy is desirable? Should the sample be collected by a procedure that is ±5% accurate, or should a statistical technique be used on data from eight tests at ±10% accuracy? Costs of different test methods will certainly be a consideration here. (3) Which sampling and analytical methods are available that will give the required accuracy for the estimated concentration? An Orsat gas analyzer with a sensitivity limit of ±0.02% would not be chosen to sample carbon monoxide

SOURCE TEST PRELIMINARY VISIT CHECK LIST

Plant ____________

Location ____________

By ____________ Date ____________

1. Gas flow at test point, m/min ________ , m^3/min ________
2. Gas temperature, °C ____________
3. Gas pressure, mm of water (±) ____________
4. Gas humidity, R. H., % ____________
5. Pollutants of concern ____________
6. Estimate of concentration ____________
7. Any toxic materials? ____________
8. Test crew needed ____________
9. Site check:

   Electric power ________ Test holes ________

   Ambient temperature ________ Illumination ________

   Platform ________ Scaffolding ________

   Hoist ________ Ladders ________

   Test date ____________
10. Environmental or safety gear ____________
11. Personnel involved (names)

    Plant manager or foreman ____________

    Mechanic or electrician ____________

Fig. 32-3. Source test checklist.

at 50–100 ppm. Conversely, an infrared gas analyzer with a full-scale deflection of 1000 ppm would not be chosen to sample CO_2 from a power boiler. (4) Is a continuous record required over many cycles of source operation, or will one or more grab samples suffice? If a source emits for only a short period of time, a method would not be selected which requires hours to gather the required sample.

The test must be scheduled well in advance for the benefit of all concerned. The plant personnel, as well as the test crew, should be given the intended date and time of the test. It is also a good idea to let the chemist or analytical service know when the testing will be conducted so that they can be ready to do their portion of the work. It may be necessary to schedule or rent equipment in advance, such as boom trucks or scaffolding. When scheduling the test, make sure that the source will be operating in its normal manner. A boiler may be operating at only one-third load on weekends because the plant steam load is off the line and only a small heating load is being carried.

B. Gas Flow Measurement

Gas flow measurement is a very important part of source testing. The volume of gaseous effluent from a source must be determined to obtain the mass loading to the atmosphere. Flow measurement through the sampling train is necessary to determine the volume of gas containing the pollutant of interest. Many of the sampling devices used for source testing have associated gas flow indicators which must be continually checked and calibrated.

Gas flows are often determined by measuring the associated pressures. Figure 32-4 illustrates several different pressure measurements commonly made on systems carrying gases. Static pressure measurements are made to adjust the absolute pressure to standard conditions specified in the test procedure.

The quantity of gaseous effluent leaving a process is usually calculated from the continuity equation, which for this use is written as

$$Q = AV \tag{32-6}$$

where Q = flow at the specified conditions of temperature, pressure, and humidity; A = area through which the gas flows; and V = velocity of the effluent gas averaged over the area.

A is commonly measured, and V determined, to calculate Q. The velocity V is determined at several points, in the center of equal duct areas, and averaged. Table 32-2 shows one commonly accepted method of dividing stacks or ducts into equal areas for velocity determinations.

For rectangular ducts, the area is evenly divided into the necessary number of measurement points. For circular ducts, Table 32-3 can be used to

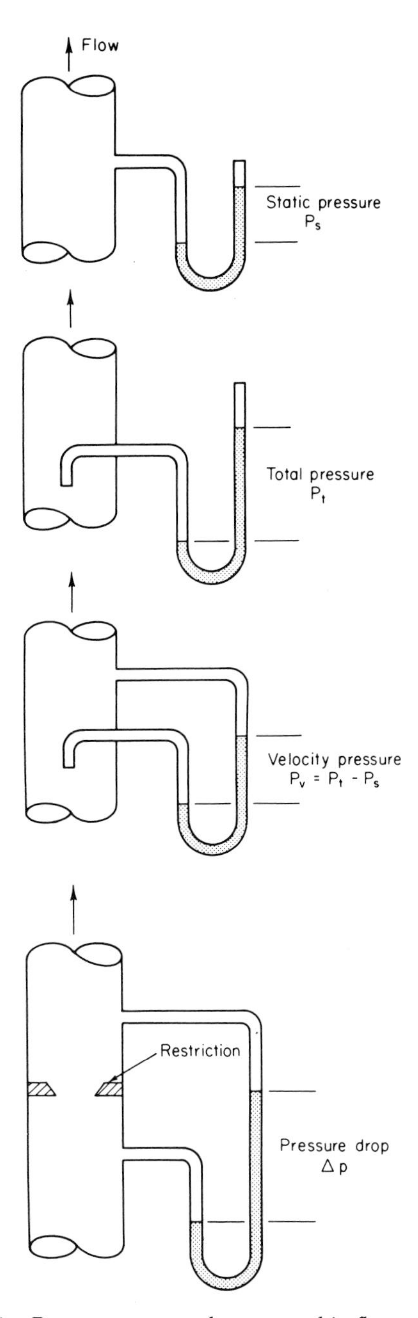

Fig. 32-4. Pressures commonly measured in flow systems.

TABLE 32-2

Number of Velocity Measurement Points

Stack diameter or (length + width)/2 (m)	Number of velocity measurement points
0.0–0.3	8
0.3–0.6	12
0.6–1.3	16
1.3–2.0	20
2.0→	24

determine the location of the traverse points. In using this table, realize that traverses are made along two diameters at right angles to each other, as shown in Fig. 32-5.

In most source tests, the measurement of velocity is made with a pitot-static tube, usually referred to simply as a *pitot tube.* Figure 32-6 illustrates the two types of pitot tubes in common use.

The standard type of pitot tube shown in this figure does not need to be calibrated, but it may be easily plugged in some high-effluent loading streams. The type S pitot tube shown in Fig. 32-6 does not plug as easily, but it does need calibration to assure its accuracy. The type S pitot tube is also more sensitive to alignment with the gas flow to obtain the correct

TABLE 32-3

Velocity Sampling Locations, Diameters from Inside Wall to Traverse Point

Point number	Number of equal areas to be sampled				
	2	3	4	5	6
1	0.067	0.044	0.033	0.025	0.021
2	0.250	0.147	0.105	0.082	0.067
3	0.750	0.295	0.194	0.146	0.118
4	0.933	0.705	0.323	0.226	0.177
5		0.853	0.677	0.342	0.250
6		0.956	0.806	0.658	0.355
7			0.895	0.774	0.645
8			0.967	0.854	0.750
9				0.918	0.823
10				0.975	0.882
11					0.933
12					0.979

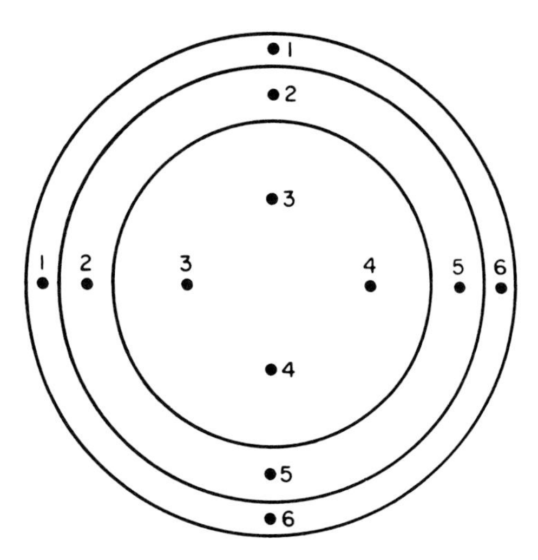

Fig. 32-5. Circular duct divided into three equal areas, as described in Table 29-3. Numbers refer to sampling points.

reading. The velocity pressure of the flowing gas is read at each point of the traverse, and the associated gas velocity is calculated from the formula

$$V = 420.5[(P_v/\rho)^{1/2}] \tag{32-7}$$

where V = velocity in meters per minute, P_v = velocity pressure in millimeters of water, and ρ = gas density in kilograms per cubic meter. The velocities are averaged for all points of the traverse to determine the gas velocity in the duct. Velocity pressures should not be averaged, as a serious error results.

Gas velocities can also be measured with anemometers (rotating vane, hot wire, etc.), from visual observations such as the velocity of smoke puffs, or from mass balance data (knowing the fuel consumption rate, air/fuel ratio, and stack diameter).

In the sampling train itself, the gas flow must be measured to determine the sample volume. Particulates and gases are measured as micrograms per cubic meter. In either case, determination of the fraction requires that the gas volume be measured for the term in the denominator. Some sample trains contain built-in flow-indicating devices such as orifice meters, rotometers, or gas meters. These devices require calibration to assure that they read accurately at the time of the test and under test conditions.

To determine the volume through the sampling train, a positive displacement system can be used. A known volume of water is displaced by gas

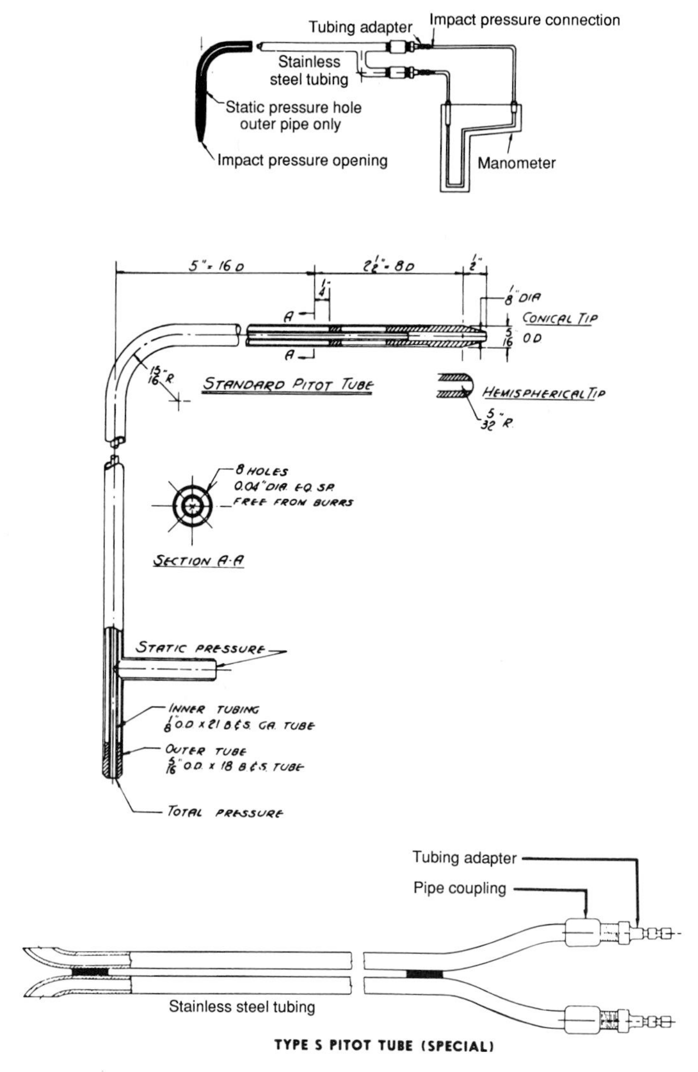

Fig. 32-6. Pitot tubes for velocity determination. Source: "Annual Book of Standards" (3). Note: English units were used by the American Society for Testing and Materials.

containing the sample. Another inexpensive procedure that works well consists of measuring the time needed for the gas to fill a plastic bag to a certain static pressure. The volume of the bag can be accurately measured under the same conditions and hence the flow determined by dividing the bag volume by the time required to fill it.

C. Collection of the Source Sample

A typical sample train is shown in Fig. 32-7. This shows the minimum number of components, but in some systems the components may be combined. Extreme care must be exercised to assure that no leaks occur in the train and that the components of the train are identical for both calibration and sampling. The pump shown in Fig. 32-7 must be both oil-less and leakproof. If the pump and volume measurement devices are interchanged, the pump no longer needs to be oil-less and leakproof, but the volume measurement will be in error unless it is adjusted for the change in static pressure. Some sampling trains become very complex as additional stages with controls and instruments are added. Many times the addition of components to a sampling train makes it so bulky and complicated that it becomes nearly impossible to use. A sampling train developed in an air-conditioned laboratory can be useless on a shaky platform in a snowstorm.

Standard sampling trains are specified for some tests. One of these standards is the system specified for large, stationary combustion sources (4). This train was designed for sampling combustion sources and should not be selected over a simpler sampling train when sampling noncombustion sources such as low-temperature effluents from cyclones, baghouses, filters, etc. (5).

Before taking the sample train to the test site, it is wise to prepare the operating curves for the particular job. With most factory-assembled trains, these curves are a part of the package. If a sampling train is assembled from components, the curves must be developed. The type of curves will vary from source to source and from train to train. Examples of useful operating curves include (1) velocity versus velocity pressure at various temperatures (6), (2) probe tip velocity versus flowmeter readings at various temperatures, and (3) flowmeter calibration curves of flow versus pressure drop. It is much easier to take an operating point from a previously prepared curve than to take out a calculator and pad to make the calculations at the

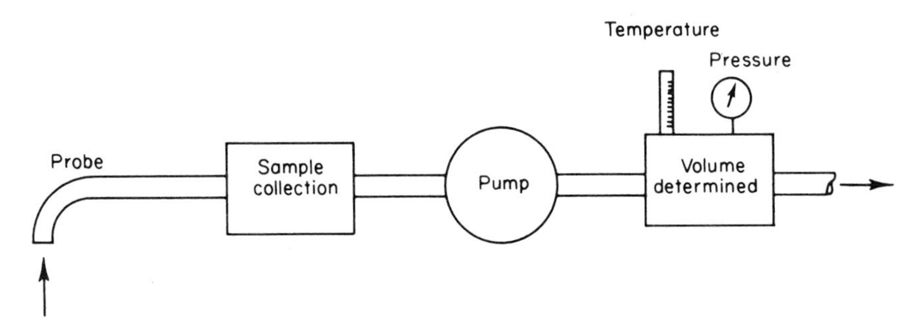

Fig. 32-7. Sampling train.

moment of the test. Remember, too, that time may be a factor and that settings must be made as rapidly as possible to obtain the necessary samples.

For sampling particulate matter, one is dealing with pollutants that have very different inertial and other characteristics from the carrying gas stream. It becomes important, therefore, to sample so that the same velocity is maintained in the probe tip as exists in the adjacent gas stream. Such sampling is called *isokinetic.* Isokinetic sampling, as well as anisokinetic sampling, is illustrated in Fig. 13-3.

If the probe velocity is less than the stack velocity, particles will be picked up by the probe, which should have been carried past it by the gas streamlines. The inertia of the particles allows them to continue on their path and be intercepted. If the probe velocity exceeds the stack velocity, the inertia of the particles carries them around the probe tip even though the carrying gases are collected. Adjustment of particulate samples taken anisokinetically to the correct stack values is possible if all of the variables of the stack gas and particulate can be accounted for in the appropriate mathematical equations.

Modern transducers and microprocessors have been used successfully to automate particulate sampling trains in order to eliminate the operating curves and manual adjustments (7). The automated samplers adjust continuously to maintain isokinetic conditions. In addition, the microprocessor continuously calculates and displays both instantaneous sampling conditions and the total sample volume collected at any given moment. The use of the automated system with the microprocessor, therefore, eliminates both operator and calculation errors.

Several separating systems are used for particulate sampling. All rely on some principle of separating the aerosol from the gas stream. Many of the actual systems use more than one type of particulate collection device in series. If a size analysis is to be made on the collected material, it must be remembered that multiple collection devices in series will collect different size fractions. Therefore, size analyses must be made at each device and mathematically combined to obtain the size of the actual particulate in the effluent stream. In any system the probe itself removes some particulate before the carrying gas reaches the first separating device, so the probe must be cleaned and the weight of material added to that collected in the remainder of the train.

Care should be exercised when sampling for aerosols that are condensable. Some separating systems, such as wet impingers, may remove the condensables from the gas stream, whereas others, such as electrostatic precipitators, will not. Of equal concern should be possible reactions in the sampling system to form precipitates or aerosols which are not normally found when the stack gases are exhausted directly to the atmosphere. SO_3

plus other gaseous products may react in a water-filled impinger to form particulate matter not truly representative of normal SO_3 release.

When sampling particulate matter from combustion processes, it is necessary to take corresponding CO_2 readings of the effluent. Emission standards usually require combustion stack gases to be reported relative to either 12% CO_2 or 50% excess air. Adjusting to a standard CO_2 or excess air value normalizes the emission base. Also, emission standards require that the loadings be based on weight per standard cubic volume of air (usually at 20°C and 760 mm Hg). In most regulations, the agency requires that the standard volume be dry, but this is not always specified.

For sampling of gases, the sample can be collected by any of several devices. Some commonly used manual methods include Orsat analyzers, absorption systems, adsorption systems, bubblers, reagent tubes, condensers, and traps. Continuous analyzers are now more widely used than manual methods. Some types of continuous analyzers include infrared and ultraviolet instruments; flame ionization detectors; mass spectrometers; calorimetric systems; gas, liquid, and solid chromatography; coulometric and potentiometric systems; chemiluminescense; and solid-state electronic systems. Since gases undergoing analysis do not need to be sampled isokinetically, it is only necessary to insert a probe and withdraw the sample. Usually, the gas sample should be filtered to remove any accompanying particulate matter which could damage the analytical instrumentation.

For the detection and intensity of odorous substances, the nose is still the instrument usually relied upon. Since odors are gaseous, they may be sampled by simply collecting a known volume of effluent and performing some manipulation to dilute the odorous gas with known volumes of "pure" air. The odor is detected by an observer or a panel of observers. The odor-free air for dilution can be obtained by passing air through activated carbon or any other substance that removes all odors while not affecting the other gases that constitute normal air. The odor-free air is then treated by adding more and more of the odorous gas until the observer just detects the odor. The concentration is then recorded as the odor threshold as noted by that observer. The test is not truly quantitative, as much variation between observers and samples is common.

If the compound causing the odor is known and can be chemically analyzed, it may be possible to get valid quantitative data by direct gas sampling. An example would be a plant producing formaldehyde. If the effluent were sampled for formaldehyde vapor, this could be related, through proper dispersion formulas, to indicate whether the odor would cause any problems in residential neighborhoods adjacent to the plant.

Extreme care should be taken in transporting and storing the samples between the time of collection and the time of analysis. Some condensable hydrocarbon samples have been lost because the collection device was subjected to elevated temperatures during shipment. Equally disastrous is placing the sample in an oven at 105°C to drive off the moisture, only to discover that the particles of interest had a very low vapor pressure and also departed the sample. At such times, source sampling can be very frustrating.

A very important analytical tool that is overlooked by many source-testing personnel is the microscope. Microscopic analysis of a particulate sample can tell a great deal about the type of material collected as well as its size distribution. This analysis is necessary if the sample was collected to aid in the selection of a piece of control equipment. All of the efficiency curves for particulate control devices are based on fractional sizes. One would not try to remove a submicron-size aerosol with a cyclone collector, but unless a size analysis is made on the sampled material, one is merely guessing at the actual size range. Figure 32-8 is a photomicrograph of material collected during a source test.

D. Calculations and Report

Calculations that are repeatedly made can be made more accurately, and at lower cost, by using a computer. If, for example, automotive emissions are continually tested over a standardized driving cycle, a computer program to analyze the data is a necessity. Otherwise, days would be spent calculating the data obtained in hours.

Fig. 32-8. Photomicrograph of particulate from a source test of a wood-fired boiler.

For sampling a relatively small number of sources, a simplified calculation form may be used. Such forms enable the office personnel to perform the arithmetic necessary to arrive at the answers, freeing the technical staff for proposals, tests, and reports. Many of the manufacturers of source-testing equipment include example calculation forms as part of their operating manuals. Some standard sampling methods include calculation forms as a part of the method (8). Many control agencies have developed standard forms for their own use and will supply copies on request.

The source test report is the end result of a large amount of work. It should be thorough, accurate, and written in a manner understandable to the person who intends to use it. It should state the purpose of the test, what was tested, how it was tested, the results obtained, and the conclusions reached. The actual data and calculations should be included in the appendix of the source test report so that they are available to substantiate the report if questioned.

V. SOURCE MONITORING

The monitoring of pollutant concentration or mass flow of pollutants is of interest to both plant owners and control agencies. Industry uses such measurements to keep a record of process operations and emissions for its own use and to meet regulatory requirements. Control officials use the

TABLE 32-4

System Concept of Stationary Source Measurements

Operation	Objective
Sampling site selection	Representative sampling consistent with intended interpretation of measurement
Sample transport (when applicable)	Spatial and temporal transfer of sample extract with minimum and/or known effects on sample integrity
Sample treatment (when applicable)	Physical and/or chemical conditioning of sample consistent with analytical operation, with controlled and/or known effects on sample integrity
Sample analysis	Generation of qualitative and quantitative data on pollutant or parameter of interest
Data reduction and display	Calibration and processing of analog data and display of final data in a format consistent with measurement objectives
Data interpretation	Validly relating the measurement data to the source environment within the limitations of the sampling and analytical operations

Source: Nader (9).

TABLE 32-5

Source Emissions Requiring Continuous Monitoring by United States New Source Performance Standards

	Pollutant						Scrubber pressure loss and water pressure	Flow rate
Source	SO_2	NO_x	CO	Opacity	H_2S	Total reduced sulfur		
Electric power plants	x	x		x				
Sulfuric acid plants	x							
Onshore natural gas processing: SO_2 emissions	x							
Nitric acid plants		x						
Petroleum refineries	x		x	x	x	x		
Iron and steel mills (BOF)[a]							x	
Steel mills (electric arc)				x				x
Ferroalloy production				x				
Glass manufacturing plants				x				
Portland cement plants				x				x
Primary copper smelters	x			x				
Primary zinc smelters	x			x				
Primary lead smelters	x			x				
Coal preparation plants							x	
Wet process phosphoric acid							x	
plants							x	
Superphosphoric acid plants							x	
Diammonium phosphate plants							x	
Triple superphosphate plants							x	
Granular triple superphosphate plants							x	
Phosphate rock plants				x			x	
Metallic mineral processing plants							x	
Nonmetallic mineral processing plants							x	
Kraft pulp mills				x		x	x	x
Gas turbines[b]								
Lime kilns[c]				x		x	x	
Ammonium sulfate plants							x	
Lead–acid battery manufacture							x	

[a] BOF, Basic oxygen furnace.
[b] Monitor sulfur and nitrogen content of fuel and water/fuel ratio.
[c] Also monitor scrubber liquid flow rate.
Source: Code of Federal Regulations (8).

information for compiling emission inventories, modeling of air sheds, and in some cases for enforcement.

A monitoring system is selected to meet specific needs and is tailored to the unique properties of the emissions from a particular process. It is necessary to take into account the specific process, the nature of the control devices, the peculiarities of the source, and the use of the data obtained (8).

Source monitoring can best be treated as a system concept ideally consisting of six unit operations, as shown in Table 32-4. In the United States, installation and operation of monitoring systems have been prescribed for a number of industries, as shown in Table 32-5.

A. Types of Monitors

Continuous emission monitors (CEMs) for plume opacity have been required on all utility, fossil fuel–fired, steam generators (over 264 megajoules) constructed in the United States since December 1971. These monitors are *in situ* opacity meters which measure the attenuation of a light beam projected across the stack (see Fig. 25-2). Remote-sensing monitors have been developed, but these have not yet been approved as equivalent to the *in situ* opacity monitors. CEMs for gaseous emissions are also available and required for certain facilities. Figure 32-9 illustrates the various approaches to monitoring particulate opacity and gaseous emissions.

B. Quality Assurance in Monitoring

In order to assure that the source is being accurately monitored, several requirements must be met (1). Some of these requirements, which assure representative, noncontaminated samples, are shown in Table 32-6.

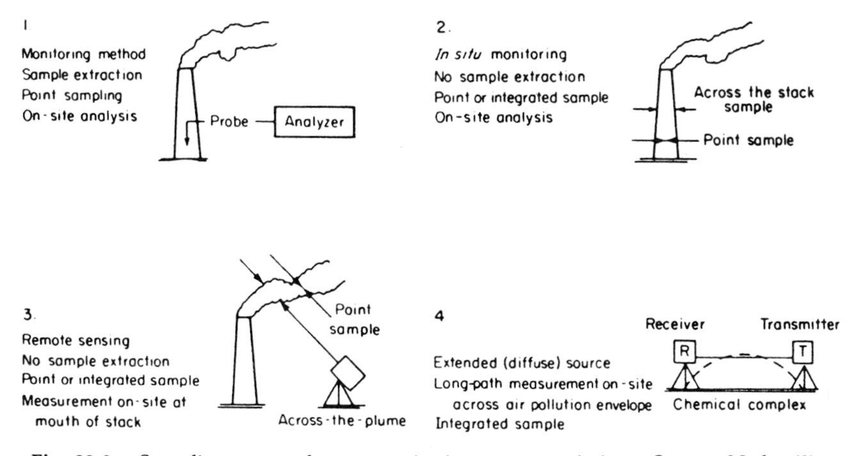

Fig. 32-9. Sampling approaches to monitoring source emissions. Source: Nader (9).

TABLE 32-6

Stationary Source Monitoring Requirements

Requirement	Method of attainment
Maintain gas temperature above the water or acid dewpoint	Heat lines or dilute with dry air
Remove water before sample enters instrument	Refrigerate or desiccate sample
Remove particulate matter before sample enters instrument	Use cyclone or filter in sample line
Dilute sample to lower the temperature to an acceptable level for the instrument	Air-dilute with necessary blowers, flow measurement, and control systems
Maintain integrity of particulate sample (mass, size, chemical composition)	Use isokinetic sampling, refrigerated sample transport, and careful handling to minimize physical or chemical changes

C. Monitoring of Particulate Emissions

The most common monitoring of particulate matter is for light attenuation (opacity). Less frequently used methods exist for monitoring mass concentration, size distribution, and chemical composition.

Opacity is a function of light transmission through the plume. Opacity is defined as follows:

$$\text{Opacity} = (1 - I/I_0 \times 100 \tag{32-8}$$

where I_0 is the incident light flux and I is the light flux leaving the plume. Techniques for monitoring visible emissions (opacity) are listed in Table 32-7.

D. Monitoring of Gaseous Emissions

Gas-monitoring systems are more widely used than particulate monitoring systems. They can also be used for both emission compliance monitors and process control systems. Gas monitors may be of either the *in situ* or

TABLE 32-7

Opacity Monitoring Techniques

Method of analysis	Measurement system
In-stock opacity	Optical transmissometer
Plume opacity	Lidar (light detection and ranging)
Selective opacity (for fine particles)	Extractive with light-scattering determination

the extractive type and use the approaches illustrated in Fig. 32-9. Table 32-8 lists the various types of gas-monitoring systems.

E. Data Reduction and Presentation

Continuous monitors usually indicate the pollutant concentration on both an indicator and a chart recording. This provides a visual indication of the instantaneous emissions, along with a permanent record of the quantitative emissions over a period of time. The monitoring system may also be equipped with an alarm device to signal the operator if the allowable emission level is being exceeded. Data-logging systems coupled with micropro-

TABLE 32-8

Gas Emission Monitoring Systems

Analytical scheme	Sampling approach[a]	Pollutant capability[b]
Chemielectromagnetic		
Colorimetry	E	SO_2, NO_x, H_2S, TS
Chemiluminescent	E	NO_x
Electromagnetic/electrooptical		
Flame photometry	E	SO_2, H_2S, TRS, TS
Nondispersive infrared	E	SO_2, NO, CO, HC, CO_2
Nondisperisive, ultraviolet and visible	E	SO_2, NO_x, NH_3, H_2S
Dispersive, infrared and ultraviolet	I	SO_2, NO, CO, CO_2, HC, H_2S
Dispersive, infrared and ultraviolet	E	SO_2, CO, CO_2, HC, NO
Correlation, ultraviolet	I	SO_2
Correlation, ultraviolet and visible	R	SO_2, NO_x
Derivative, ultraviolet	I	SO_2, NO_2, NO, O_2, NH_3, CO
Fluorescence, ultraviolet	E	SO_2
Electrical		
Conductivity	E	SO_2, NH_3, HCl
Coulometry	E	SO_2, H_2S, TRS
Electrochemical	E,I	SO_2, NO_x, CO, H_2S, O_2
Flame ionization	E	HC
Thermal		
Oxidation	E	CO
Conductivity	E	SO_2, NH_3, CO_2
Hybrid		
Gas chromatography, flame ionization	E	CO, HC
Gas chromatography, flame photometry	E	Sulfur compounds

[a] E, extractive; I, *in situ;* R, remote.
[b] HC, hydrocarbons; TRS, total reduced sulfur; TS, total sulfur; NO_x, total oxides of nitrogen, but system may be specific for NO, NO_2, or both.
Source: Nader (9).

cessors are popular. These systems can give instantaneous values of the variables and pollutants of interest, along with the averages or totals for the period of concern.

REFERENCES

1. "Quality Assurance Handbook for Air Pollutant Measurement Systems," Vol. III, "Stationary Source Specific Methods," EPA-600/4-77-027b. U.S. Environmental Protection Agency, Research Triangle Park, NC, 1977.
2. Stern, A. C. (ed.), "Air Pollution," 3rd ed., Vol. III. Academic Press, New York, 1977.
3. "1981 Annual Book of ASTM Standards, Part 26, Gaseous Fuels; Coal and Coke; Atmospheric Analysis." American Society for Testing and Materials, Philadelphia, 1981.
4. Environmental Protection Agency—Standards of Performance for New Stationary Sources. *Fed. Regist.* **36**(247), 2488 (1971).
5. Boubel, R. W., *J. Air. Pollut. Control Assoc.* **21**, 783–787 (1971).
6. Baumeister, T. (ed.), "Marks' Standard Handbook for Mechanical Engineers," 8th ed. McGraw-Hill, New York, 1978.
7. Boubel, R. W., Hirsch, J. W., and Sadri, B., Particulate sampling has gone automatic, Proceedings of the 68th Annual Meeting of the Air Pollution Control Association, Pittsburgh, 1975.
8. Code of Federal Regulations, Chapter I, Part 60, Standards of performance for new stationary sources, 1992.
9. Nader, J. S., Source monitoring, *in* "Air Pollution" (A. C. Stern, ed.), 3rd ed., Vol III. Academic Press, New York, 1976.

SUGGESTED READING

Brenchley, D. L., Turley, C. D., and Yarmac, R. F. "Industrial Source Sampling." Ann Arbor Science Publishers, Ann Arbor, MI, 1973.

Cooper, H. B. H., and Rossano, A. T. "Source Testing for Air Pollution Control." Environmental Sciences Services Division, Wilton, CT, 1970.

Frederick, E. R. (ed.), "Proceedings, Continuous Emission Monitoring: Design, Operation, and Experience, Specialty Conference." Air Pollution Control Association, Pittsburgh, 1981.

Morrow, N. L., Brief, R. S., and Bertrand, R. R., *Chem. Eng.* **79,** 85–98 (1972).

QUESTIONS

1. During a pitot traverse of a duct, the following velocity pressures, in millimeters of water, were measured at the center of equal areas: 13.2, 29.1, 29.7, 20.6, 17.8, 30.4, 28.4, 15.2. If the flowing fluid was air at 760 mm Hg absolute and 85^0C, what was the average gas velocity?
2. Would you expect errors of the same magnitude when sampling anisokinetically at 80% of stack velocity as when sampling anisokinetically at 120% of stack velocity? Explain.

3. Suppose a particulate sample from a stack is separated into two fractions by the sampling device. Both are sized microscopically and found to be lognormally distributed. One has a count mean size of 5.0 μm and a geometric deviation of 2.0. The other has a count mean size of 10.0 μm and a geometric deviation of 2.2. Two grams of the smaller-sized material were collected for each 10 gm of the larger. What would be reported for the weight mean size and geometric deviation of the stack effluent?
4. A particulate sample was found to weigh 0.0216 gm. The sample volume from which it was collected was 0.60 m^3 at 60°C, 760 mm Hg absolute, and 90% relative humidity. What was the stack loading in milligrams per standard cubic meter?
5. A particulate sample was found to contain 350 mg m^{-3}. The CO_2 during the sampling period averaged 7.2%. If the exhaust gas flow was 2000 m^3 min^{-1}, what would be the particulate loading in both milligrams per cubic meter and kilograms per hour, corrected to 12% CO_2?
6. Give an example of how opacity monitoring of a coal-fired boiler could be used to improve combustion efficiency.
7. An opacity monitor is set so that the incident light is 100 units. Prepare a graph of the percentage of opacity versus the light flux leaving the plume (opacity, 0–100%; exiting light flux, 0–100 units).
8. List the advantages and disadvantages of both *in situ* and extractive gas monitors.
9. Discuss the advantages and disadvantages of one-time source testing for a specific emission versus continuous monitoring of the same emission.

Index

B

D

E

G

H

I

J

K

L

M

N

O

P

Q

R